Advanced Materials for Clothing and Textile Engineering

Advanced Materials for Clothing and Textile Engineering

Editors

Dubravko Rogale
Snježana First Rogale

MDPI • Basel • Beijing • Wuhan • Barcelona • Belgrade • Manchester • Tokyo • Cluj • Tianjin

Editors
Dubravko Rogale
University of Zagreb/Faculty
of Textile Technology
Croatia

Snježana Firšt Rogale
University of Zagreb/Faculty
of Textile Technology
Croatia

Editorial Office
MDPI
St. Alban-Anlage 66
4052 Basel, Switzerland

This is a reprint of articles from the Special Issue published online in the open access journal *Materials* (ISSN 1996-1944) (available at: https://www.mdpi.com/journal/materials/special_issues/materials_clothing).

For citation purposes, cite each article independently as indicated on the article page online and as indicated below:

LastName, A.A.; LastName, B.B.; LastName, C.C. Article Title. *Journal Name* **Year**, *Volume Number*, Page Range.

ISBN 978-3-0365-7828-6 (Hbk)
ISBN 978-3-0365-7829-3 (PDF)

Cover image courtesy of Dubravko Rogale

© 2023 by the authors. Articles in this book are Open Access and distributed under the Creative Commons Attribution (CC BY) license, which allows users to download, copy and build upon published articles, as long as the author and publisher are properly credited, which ensures maximum dissemination and a wider impact of our publications.

The book as a whole is distributed by MDPI under the terms and conditions of the Creative Commons license CC BY-NC-ND.

Contents

About the Editors vii

Snježana Firšt Rogale and Dubravko Rogale
Advanced Materials for Clothing and Textile Engineering
Reprinted from: *Materials* **2023**, *16*, 3407, doi:10.3390/ma16093407 1

Ana Šaravanja, Tanja Pušić and Tihana Dekanić
Microplastics in Wastewater by Washing Polyester Fabrics
Reprinted from: *Materials* **2022**, *15*, 2683, doi:10.3390/ma15072683 5

Tanja Pušić, Bosiljka Šaravanja and Krešimir Malarić
Electromagnetic Shielding Properties of Knitted Fabric Made from Polyamide Threads Coated with Silver
Reprinted from: *Materials* **2021**, *14*, 1281, doi:10.3390/ma14051281 21

Hyun-Ah Kim
Moisture Vapor Permeability and Thermal Wear Comfort of Ecofriendly Fiber-Embedded Woven Fabrics for High-Performance Clothing
Reprinted from: *Materials* **2021**, *14*, 6205, doi:10.3390/ma14206205 33

Vânia Pais, Carlos Mota, João Bessa, José Guilherme Dias, Fernando Cunha and Raul Fangueiro
Study of the Filtration Performance of Multilayer and Multiscale Fibrous Structures
Reprinted from: *Materials* **2021**, *14*, 7147, doi:10.3390/ma14237147 57

Goran Čubrić, Ivana Salopek Čubrić, Dubravko Rogale and Snježana Firšt Rogale
Mechanical and Thermal Properties of Polyurethane Materials and Inflated Insulation Chambers
Reprinted from: *Materials* **2021**, *14*, 1541, doi:10.3390/ma14061541 73

Marianna Halász, Jelka Geršak, Péter Bakonyi, Gabriella Oroszlány, András Koleszár and Orsolya Nagyné Szabó
Study on the Compression Effect of Clothing on the Physiological Response of the Athlete
Reprinted from: *Materials* **2022**, *15*, 169, doi:10.3390/ma15010169 87

Anica Hursa Šajatović, Sandra Flinčec Grgac and Daniela Zavec
Investigation of Flammability of Protective Clothing System for Firefighters
Reprinted from: *Materials* **2022**, *15*, 2384, doi:10.3390/ma15072384 101

Tea Kaurin, Tanja Pušić, Tihana Dekanić and Sandra Flinčec Grgac
Impact of Washing Parameters on Thermal Characteristics and Appearance of Proban®—Flame Retardant Material
Reprinted from: *Materials* **2022**, *15*, 5373, doi:10.3390/ma15155373 117

Željko Knezić, Željko Penava, Diana Šimić Penava and Dubravko Rogale
The Impact of Elongation on Change in Electrical Resistance of Electrically Conductive Yarns Woven into Fabric
Reprinted from: *Materials* **2021**, *14*, 3390, doi:10.3390/ma14123390 135

Tetiana Ielina, Liudmyla Halavska, Daiva Mikucioniene, Rimvydas Milasius, Svitlana Bobrova and Oksana Dmytryk
Development of 3D Models of Knits from Multi-Filament Ultra-Strong Yarns for Theoretical Modelling of Air Permeability
Reprinted from: *Materials* **2021**, *14*, 3489, doi:10.3390/ma14133489 153

Tea Jovanović, Željko Penava and Zlatko Vrljičak
Impact of the Elastane Percentage on the Elastic Properties of Knitted Fabrics under Cyclic Loading
Reprinted from: *Materials* **2022**, *15*, 6512, doi:10.3390/ma15196512 . . . . **167**

Antoneta Tomljenović, Juro Živičnjak and Ivan Mihaljević
Usage Durability and Comfort Properties of Socks Made from Differently Spun Modal and Micro Modal Yarns
Reprinted from: *Materials* **2023**, *16*, 1684, doi:10.3390/ma16041684 . . . . **183**

Catarina Pimenta, Carla Costa Pereira and Raul Fangueiro
Textile Pattern Design in Thermal Vision—A Study on Human Body Camouflage
Reprinted from: *Materials* **2021**, *14*, 4364, doi:10.3390/ma14164364 . . . . **211**

Hafeezullah Memon, Eldana Bizuneh Chaklie, Hanur Meku Yesuf and Chengyan Zhu
Study on Effect of Leather Rigidity and Thickness on Drapability of Sheep Garment Leather
Reprinted from: *Materials* **2021**, *14*, 4553, doi:10.3390/ma14164553 . . . . **227**

Slavenka Petrak, Maja Mahnić Naglić, Dubravko Rogale and Jelka Geršak
Analysis of Polygonal Computer Model Parameters and Influence on Fabric Drape Simulation
Reprinted from: *Materials* **2021**, *14*, 6259, doi:10.3390/ma14216259 . . . . **239**

Andreja Rudolf, Zoran Stjepanovič and Andrej Cupar
Study Regarding the Kinematic 3D Human-Body Model Intended for Simulation of Personalized Clothes for a Sitting Posture
Reprinted from: *Materials* **2021**, *14*, 5124, doi:10.3390/ma14185124 . . . . **255**

Slavica Bogović and Ana Čorak
A New Method for Testing the Breaking Force of a Polylactic Acid-Fabric Joint for the Purpose of Making a Protective Garment
Reprinted from: *Materials* **2022**, *15*, 3549, doi:10.3390/ma15103549 . . . . **279**

Martinia Glogar, Tanja Pušić, Veronika Lovreškov and Tea Kaurin
Reactive Printing and Wash Fastness of Inherent Flame Retardant Fabrics for Dual Use
Reprinted from: *Materials* **2022**, *15*, 4791, doi:10.3390/ma15144791 . . . . **291**

Eglė Kumpikaitė, Eglė Lapelytė and Stasė Petraitienė
Method of Predicting the Crimp of Jacquard-Woven Fabrics
Reprinted from: *Materials* **2021**, *14*, 5157, doi:10.3390/ma14185157 . . . . **309**

Eglė Kumpikaitė, Indrė Tautkutė-Stankuvienė, Lukas Simanavičius and Stasė Petraitienė
The Influence of Finishing on the Pilling Resistance of Linen/Silk Woven Fabrics
Reprinted from: *Materials* **2021**, *14*, 6787, doi:10.3390/ma14226787 . . . . **319**

Habiba Halepoto, Tao Gong, Saleha Noor and Hafeezullah Memon
Bibliometric Analysis of Artificial Intelligence in Textiles
Reprinted from: *Materials* **2022**, *15*, 2910, doi:10.3390/ma15082910 . . . . **331**

About the Editors

Dubravko Rogale

Prof. Dubravko Rogale, Ph.D., is a full professor and scientific advisor at the University of Zagreb Faculty of Textile Technology, where he graduated in 1982, obtained a master's degree in 1987 and a doctorate in 1994 in the field of technical sciences and branches of textile technology and field clothing technology. His scientific and research activities are related to the development of intelligent clothing, garment manufacturing processes, process parameters, new methods of computer design and garment modeling, improvement of methodological methods in garment development, modern joining techniques, etc. He was previously the vice president of the Croatian Academy of Engineering Sciences. He is a member of the Scientific Council for Technological Development of the Croatian Academy of Sciences and Arts. He was the Head of the Department of Clothing Technology for three terms, and the Vice Dean for Teaching of the Faculty of Textile Technology and Dean of the Faculty of Textile Technology for two terms.

Snježana Firšt Rogale

Prof. Snježana Firšt Rogale, Ph.D., worked in a clothing manufacturer factory as a technologist, where she gained experience and basic practical knowledge in the field of technological processes of clothing production. She is a professor at the University of Zagreb Faculty of Textile Technology. She conducts research in the field of clothing technology, which is related to the field of technological processes of clothing production and development of conventional and smart clothing, and the study of the construction and thermal properties of clothing. She was the Head of the Department for Clothing Technology for two terms and is now the Head of Doctoral Study. She is a member of the Academy of Technical Sciences of Croatia.

Editorial

Advanced Materials for Clothing and Textile Engineering

Snježana Firšt Rogale * and Dubravko Rogale

University of Zagreb Faculty of Textile Technology, 10000 Zagreb, Croatia; dubravko.rogale@ttf.unizg.hr
* Correspondence: sfrogale@ttf.unizg.hr

Citation: Firšt Rogale, S.; Rogale, D. Advanced Materials for Clothing and Textile Engineering. *Materials* **2023**, *16*, 3407. https://doi.org/10.3390/ma16093407

Received: 21 April 2023
Accepted: 25 April 2023
Published: 27 April 2023

Copyright: © 2023 by the authors. Licensee MDPI, Basel, Switzerland. This article is an open access article distributed under the terms and conditions of the Creative Commons Attribution (CC BY) license (https://creativecommons.org/licenses/by/4.0/).

The main objective of this Special Issue is to showcase outstanding papers presenting advanced materials for clothing and textile engineering. Advanced materials have improved the performance of conventional and protective clothing and accelerated the development of e-clothing and smart clothing. The main directions of progress have become visible in recent years and are reflected in the research on and development of new materials and the improvement of their properties. Special treatment methods for textile materials are being developed to improve their use properties as well as textile care. This is of great importance for textile and clothing engineering.

This Special Issue brings together several outstanding articles on a wide range of topics, including the impact of textile production and waste on the environment, the properties of yarns and textile materials, and the wearing comfort of textiles and clothing.

Environmental pollution is a major problem and a very important research topic. Microplastics have become one of the major environmental hazards. Thus, it is imperative to investigate the main potential sources of microplastic pollution on the environment. Šaravanja et al. gave an overview of washable polyester materials. They stated that the main goals are to produce polyester with optimal structural parameters and to switch to recycled production as much as possible, including the use of the most appropriate agents that protect the structure of polymer while preventing the release of microplastics during washing. Another goal is to optimize the washing and drying processes of synthetic materials [1].

The importance of cleaning is evident in the use of textile materials for protection against electromagnetic radiation, which is suitable for composite structures of garments and for technical and interior applications. The authors evaluated the stability of the materials' shielding properties against EM radiation after the application of apolar and polar solvents, in synergy with a cyclic process involving the parameters of wet and dry cleaning [2].

The consumption of eco-friendly fibers with a decrease in synthetic fibers, which can reduce pollution generated by the textile industry, was studied in [3].

Nanotechnology, more specifically electrospun nanofibers, has been identified as a potential solution for developing efficient filtration systems. In one study, the electrospinning technique for producing polyamide nanofibers was optimized by varying several parameters, such as polymer concentration, flow rate, and needle diameter. The optimized polyamide nanofibers were combined with polypropylene and polyester microfibers to construct a multilayer and multiscale system with increased filtration efficiency [4].

The comfort of wearing textiles and clothing is the focus of a few papers. Čubrić et al. examined the mechanical properties of polyurethane materials used to construct inflated thermal-expanding insulation chambers that serve as adaptive thermal layers in intelligent clothing, as well as their efficiency in providing thermal protection [5].

An increasing number of companies are developing advanced high-tech sportswear, often with high compression, for professional athletes. Analysis of the effects of sportswear compression on the physiological comfort of athletes is a very important topic [6].

The moisture vapor permeability and thermal wear comfort of bamboo and Tencel as environmentally friendly fibers, together with fiber core and polypropylene, nylon, and Coolmax® as core filaments, were investigated in [3].

The flammability of protective clothing systems for firefighters is increasingly being investigated. Hursa Šajatović et al. presented studies on a clothing system for protection against heat and flames using a fire manikin and systematically analyzed the damage caused after testing [7].

The effects of washing parameters on the thermal properties and appearance of Proban® flame-retardant material were investigated in one study. This research focused on the effects of washing conditions on the effluent composition and durability of the flame-retardant material's properties and the appearance of Proban® cotton fabrics [8].

Knezić et al. analyzed the effect of the number of electrically conductive yarns woven into a fabric on the values of electrical resistance. They observed how the direction of action of the elongation force affects the change in electrical resistance of the electrically conductive fabric, taking into account the position of the interwoven electrically conductive yarns [9].

Ielina et al. studied the geometric parameters of a knitted loop. In this paper, a mathematical description of the coordinates of the characteristic points of the loop and an algorithm for calculating the coordinates of the control vertices of the second-order spline, which determine the configuration of the yarn axes in the loop, are presented [10].

Knitted fabrics are subjected to dynamic loading due to body movements during their use. The influence of elastane content on the elastic properties of knitted fabrics under dynamic loading was investigated in [11].

Stockings must be made of high-quality material and provide comfort to wearers. Stockings that were knitted in a plain single jersey pattern and made with the highest percentage of ring, rotor, and air-spun modal or micromodal yarns of the same linear density in full plating with various textured polyamide 6.6 yarns were investigated in [12].

A very interesting area of research is the new approach to the creation process in fashion design that results from the use of thermal camouflage in the design of clothing. In one study, the main variation factors of thermal images were determined by analyzing their color behavior in a daytime and nighttime outdoor environment in the presence and absence of a dressed human body through the use of a thermal imaging camera [13].

Drape is one of the most important characteristics associated with the quality and attractiveness of a garment. Memon et al. investigated how bending stiffness and thickness affect the drapeability of garment leathers. This study also has practical implications as it can help practitioners better understand these elements and select appropriate materials for garment companies and customers [14].

Petrak et al. presented an analysis of the parameters of a polygonal computer model that affect fabric drape simulations. The fabric drape simulations were performed using the 2D/3D CAD system for computer clothing design on a disk model, which corresponded with the real tests using a drape tester; the aim was to perform a correlation analysis between the values of the drape parameters of the simulated fabrics and the realistically measured values for fabrics [15].

Virtual prototyping is a technique in the apparel development process that involves the use of computer-aided design to develop apparel and their virtual prototypes. Virtual simulations of trousers and real-trouser prototypes were compared to investigate their fit and comfort on scanned and kinematic 3D body models, as well as on a real body. By changing the body posture, the real and virtual body circumferences were changed, which affected the fit and comfort of the virtual and real trousers [16].

Bogović et al. described a study on the use of 3D printed knee protectors intended for wheelchair users. The construction of clothing and the 3D modeling of the elements integrated into the garment were interdependent, and the design solutions were found to provide adequate and reusable garments, especially for sensitive target groups such as people with disabilities [17].

The properties of flame-retardant fabrics and the possibility of their finishing in the processes of dyeing and printing were studied. The possibility of reactive printability on protective flame-resistant fabrics varying in the composition of weft threads and weave was studied, and the washfastness of the printed samples was analyzed [18].

Kumpikaitė et al. investigated the distribution of crimp in new jacquard fabric structures combining single- and double-ply weaves and fabric width to provide a method for predicting crimp [19].

The pilling resistance of fashion fabrics is a fundamentally important and common problem when wearing clothes. The aim of one study published in this Special Issue was to evaluate the pilling behavior of linen/silk fabrics with different mechanical and chemical finishes and to determine the influences of the raw materials and the specifics of dyeing and digital printing with different dyes [20].

In another study, the first descriptive bibliometric analysis to study the most influential journals, institutions, and countries in the field of artificial intelligence in the textile industry was conducted. The analysis covered all major areas of artificial intelligence, including data mining and machine learning [21].

The articles published in this Special Issue show that the field of textile and clothing engineering is experiencing extraordinary development dynamics. We are pleased that the Editorial Office of *Materials* has recognized this trend and made possible the publication of this Special Issue. The published articles highlight new trends and address many important issues in the development of advanced materials for textile and clothing engineering.

Acknowledgments: The guest editors of this Special Issue would like to thank all authors whose valuable work has contributed to the creation of this Special Issue. A special thanks goes to the reviewers for their constructive comments and thoughtful suggestions. Finally, the editors would like to thank the Editorial Office of *Materials* for the kind support.

Conflicts of Interest: The authors declare no conflict of interest.

References

1. Šaravanja, A.; Pušić, T.; Dekanić, T. Microplastics in Wastewater by Washing Polyester Fabrics. *Materials* **2022**, *15*, 2683. [CrossRef] [PubMed]
2. Pušić, T.; Šaravanja, B.; Malarić, K. Electromagnetic Shielding Properties of Knitted Fabric Made from Polyamide Threads Coated with Silver. *Materials* **2021**, *14*, 1281. [CrossRef] [PubMed]
3. Kim, H.-A. Moisture Vapor Permeability and Thermal Wear Comfort of Ecofriendly Fiber-Embedded Woven Fabrics for High-Performance Clothing. *Materials* **2021**, *14*, 6205. [CrossRef] [PubMed]
4. Pais, V.; Mota, C.; Bessa, J.; Dias, J.G.; Cunha, F.; Fangueiro, R. Study of the Filtration Performance of Multilayer and Multiscale Fibrous Structures. *Materials* **2021**, *14*, 7147. [CrossRef] [PubMed]
5. Čubrić, G.; Salopek Čubrić, I.; Rogale, D.; Firšt Rogale, S. Mechanical and Thermal Properties of Polyurethane Materials and Inflated Insulation Chambers. *Materials* **2021**, *14*, 1541. [CrossRef]
6. Halász, M.; Geršak, J.; Bakonyi, P.; Oroszlány, G.; Koleszár, A.; Nagyné Szabó, O. Study on the Compression Effect of Clothing on the Physiological Response of the Athlete. *Materials* **2022**, *15*, 169. [CrossRef]
7. Hursa Šajatović, A.; Flinčec Grgac, S.; Zavec, D. Investigation of Flammability of Protective Clothing System for Firefighters. *Materials* **2022**, *15*, 2384. [CrossRef]
8. Kaurin, T.; Pušić, T.; Dekanić, T.; Flinčec Grgac, S. Impact of Washing Parameters on Thermal Characteristics and Appearance of Proban®—Flame Retardant Material. *Materials* **2022**, *15*, 5373. [CrossRef]
9. Knezić, Ž.; Penava, Ž.; Penava, D.Š.; Rogale, D. The Impact of Elongation on Change in Electrical Resistance of Electrically Conductive Yarns Woven into Fabric. *Materials* **2021**, *14*, 3390. [CrossRef]
10. Ielina, T.; Halavska, L.; Mikucioniene, D.; Milasius, R.; Bobrova, S.; Dmytryk, O. Development of 3D Models of Knits from Multi-Filament Ultra-Strong Yarns for Theoretical Modelling of Air Permeability. *Materials* **2021**, *14*, 3489. [CrossRef]
11. Jovanović, T.; Penava, Ž.; Vrljičak, Z. Impact of the Elastane Percentage on the Elastic Properties of Knitted Fabrics under Cyclic Loading. *Materials* **2022**, *15*, 6512. [CrossRef] [PubMed]
12. Tomljenović, A.; Živičnjak, J.; Mihaljević, I. Usage Durability and Comfort Properties of Socks Made from Differently Spun Modal and Micro Modal Yarns. *Materials* **2023**, *16*, 1684. [CrossRef] [PubMed]
13. Pimenta, C.; Pereira, C.C.; Fangueiro, R. Textile Pattern Design in Thermal Vision—A Study on Human Body Camouflage. *Materials* **2021**, *14*, 4364. [CrossRef]

14. Memon, H.; Chaklie, E.B.; Yesuf, H.M.; Zhu, C. Study on Effect of Leather Rigidity and Thickness on Drapability of Sheep Garment Leather. *Materials* **2021**, *14*, 4553. [CrossRef]
15. Petrak, S.; Mahnić Naglić, M.; Rogale, D.; Geršak, J. Analysis of Polygonal Computer Model Parameters and Influence on Fabric Drape Simulation. *Materials* **2021**, *14*, 6259. [CrossRef] [PubMed]
16. Rudolf, A.; Stjepanovič, Z.; Cupar, A. Study Regarding the Kinematic 3D Human-Body Model Intended for Simulation of Personalized Clothes for a Sitting Posture. *Materials* **2021**, *14*, 5124. [CrossRef]
17. Bogović, S.; Čorak, A. A New Method for Testing the Breaking Force of a Polylactic Acid-Fabric Joint for the Purpose of Making a Protective Garment. *Materials* **2022**, *15*, 3549. [CrossRef]
18. Glogar, M.; Pušić, T.; Lovreškov, V.; Kaurin, T. Reactive Printing and Wash Fastness of Inherent Flame Retardant Fabrics for Dual Use. *Materials* **2022**, *15*, 4791. [CrossRef]
19. Kumpikaitė, E.; Lapelytė, E.; Petraitienė, S. Method of Predicting the Crimp of Jacquard-Woven Fabrics. *Materials* **2021**, *14*, 5157. [CrossRef]
20. Kumpikaitė, E.; Tautkutė-Stankuvienė, I.; Simanavičius, L.; Petraitienė, S. The Influence of Finishing on the Pilling Resistance of Linen/Silk Woven Fabrics. *Materials* **2021**, *14*, 6787. [CrossRef]
21. Halepoto, H.; Gong, T.; Noor, S.; Memon, H. Bibliometric Analysis of Artificial Intelligence in Textiles. *Materials* **2022**, *15*, 2910. [CrossRef] [PubMed]

Disclaimer/Publisher's Note: The statements, opinions and data contained in all publications are solely those of the individual author(s) and contributor(s) and not of MDPI and/or the editor(s). MDPI and/or the editor(s) disclaim responsibility for any injury to people or property resulting from any ideas, methods, instructions or products referred to in the content.

Review

Microplastics in Wastewater by Washing Polyester Fabrics

Ana Šaravanja, Tanja Pušić and Tihana Dekanić *

Department of Textile Chemistry and Ecology, Faculty of Textile Technology, University of Zagreb, Prilaz Baruna Filipovića 28a, HR-10000 Zagreb, Croatia; ana.saravanja@ttf.unizg.hr (A.Š.); tanja.pusic@ttf.unizg.hr (T.P.)
* Correspondence: tihana.dekanic@ttf.unizg.hr

Abstract: Microplastics have become one of the most serious environmental hazards today, raising fears that concentrations will continue to rise even further in the near future. Micro/nanoparticles are formed when plastic breaks down into tiny fragments due to mechanical or photochemical processes. Microplastics are everywhere, and they have a strong tendency to interact with the ecosystem, putting biogenic fauna and flora at risk. Polyester (PET) and polyamide (PA) are two of the most important synthetic fibres, accounting for about 60% of the total world fibre production. Synthetic fabrics are now widely used for clothing, carpets, and a variety of other products. During the manufacturing or cleaning process, synthetic textiles have the potential to release microplastics into the environment. The focus of this paper is to explore the main potential sources of microplastic pollution in the environment, providing an overview of washable polyester materials.

Keywords: microplastics; wastewater; textiles; polyester; polyester ageing

1. Introduction

Floating microplastics are the most widespread pollutant in the aquatic environment, acting as a contaminant for all aquatic organisms due to their constant deficiencies. Because of their small size, microplastics are consumed in large quantities by aquatic organisms, causing physiological problems. Microplastic ingestion has been related to lower food intake, developmental disorders, and behavioural changes. Data show that nearly 700 aquatic organisms worldwide are threatened by microplastic ingestion [1,2]. A variety of factors influences the biodegradation of microplastics, which is why it is important to understand the properties of the plastics. As plastics continue to disintegrate and defragment, the availability of microplastics will increase. Microplastics also have a tendency to change the density over time and, as a result, float because of biological fouling. The importance of microplastic composition has recently been highlighted in numerous publications. In general, a plastic composition refers to the polymer from which the plastic is made, and the density of microplastics is defined by this composition [1].

All plastic products with a diameter less than 5 millimetres are classified as microplastics [3,4]. Two types of microplastics have been listed in the literature so far: the primary and secondary forms. The primary form of microplastics contains microgranules and can be found in cosmetics, whereas the secondary forms of microplastics are produced by the degradation process of larger plastic parts, e.g., poly(ethylene terephthalate) or PET bottles, or by the abrasion of synthetic textile materials, often called microfibers. However, there are certain inconsistencies in the use of the terms microplastics and microfibers, particularly in the context of textiles. As a result, these terms should not be misunderstood and used together [1,3]. Microplastics enter sewers, seas, and oceans due to human errors, carelessness, and fragmentation when using various materials such as paints, rubber, textiles or other plastics (Figure 1).

Citation: Šaravanja, A.; Pušić, T.; Dekanić, T. Microplastics in Wastewater by Washing Polyester Fabrics. *Materials* **2022**, *15*, 2683. https://doi.org/10.3390/ma15072683

Academic Editor: Gerard Lligadas

Received: 28 February 2022
Accepted: 4 April 2022
Published: 6 April 2022

Publisher's Note: MDPI stays neutral with regard to jurisdictional claims in published maps and institutional affiliations.

Copyright: © 2022 by the authors. Licensee MDPI, Basel, Switzerland. This article is an open access article distributed under the terms and conditions of the Creative Commons Attribution (CC BY) license (https://creativecommons.org/licenses/by/4.0/).

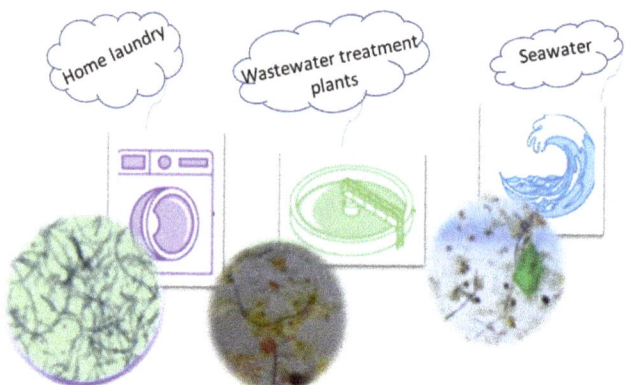

Figure 1. Microfibres detected in laundry effluent, wastewater and seawater [5].

By adding different types of additives to the polymer, they can be modified and their range of application extended. The most common additives are paints, fillers, UV protective agents, modifiers, lubricants, etc. Additives such as flame retardants and plasticizers have been found to be ubiquitous in various production processes and released into water and ground [6]. Regardless of technological advances and changing lifestyles, 2015 data shows that recycling remains a low priority. Only about 18% of PET is recycled, 10% of PE and high-density PE (HDPE), almost 6% of low-density PE (LLDPE), and less than 1% of PP [7].

Plastic additives, such as thermal stabilizers, are often used because they allow high processing temperatures. Plastic additives usually have negative connotations because there is insufficient evidence, as shown by the exposure to bisphenol A (BPA) in polycarbonate products. The EFSA (European Food Safety Authority) stated in 2007 that human exposure to BPA is below the TDI (lifetime exposure). Plasticizers are also found in microplastics, where there is a wide variety, including: adipates, phthalates, trimellitates, etc. [8].

2. Fragments of Plastics and Polymers

The high productivity and extremely slow biotic breakdown of plastics cause them to spread in the environment as a result of adverse wastewater effects. Plastics that enter the aquatic environment and their residues may remain for months, or hundreds or thousands of years. During this time, they are defragmented by mechanical and photochemical processes, resulting in the formation of microplastics (<5 mm) or nanoplastics (<1 µm) [1,9,10]. It should be noted that the term polymer is often used in everyday language in a similar way to plastic, so it is necessary to clarify the basic difference.

A polymer is a large molecule called a macromolecule, which consists of building units, so-called monomers. Macromolecules can be linear, branched, or crosslinked. The main difference between polymers and plastics is that plastics are mixtures/blends of two polymers or a polymer and low-molecular-weight compounds (additives) such as UV or thermal stabilizers, flame retardants, dyes, antioxidants, pigments, antimicrobial agents, lubricants, fillers, and others, according to the final applications [11–14].

Plastics can be used in many applications and various forms: clothing and textiles (PET, PAN, etc.), cookware (Teflon–PTFE), food containers (form HDPE, LDPE, or cups from PS), packaging (bags or bottles from PET), bearings (PA), epoxy glue, isolations (PS or PUR), silicone (heart valves), floor coverings (PVC), etc. [12,13,15–17]. Plastics based on PET, PE, PP and PA are mainly used in the textile industry.

Plastics are mostly derived from petroleum, including the category of polymers containing an ester functional group in each repeating unit of their main chain, but the most commonly used type of polyester in textile use is polyethylene terephthalate (PET), polypropylene, polyethylene, and polyvinyl chloride, of which polypropylene and polyethylene are common and standard products, Table 1.

Table 1. Applications and specific density of different synthetic polymers found in the marine environment [9,18].

Categories	Common Applications	Specific Density [g/cm^3]
Polyethylene (PE-LDPE, LLDPE)	Plastic bags, six-pack rings, bottles	0.91–0.93
Polypropylene (PP)	Rope, bottle caps, netting	0.90–0.92
Foamed polystyrene (PS)	Cups, buoy	0.01–1.05
Polystyrene (PS)	Plastic utensils, food containers, packaging	1.04–1.09
Polyvinyl chloride (PVC)	Bags, tubes	1.16–1.30
Polyamide or nylon (PA)	Ropes	1.13–1.15
Polyethylene terephthalate (PET)	Beverage bottles	1.34–1.39
Polyester resin + fibreglass	Textiles	>1.35
Polycarbonate (PC)	Electronic compounds	1.20–1.22
Cellulose acetate (CA)	Filter cigarettes	1.22–1.24
Polytetrafluoroethylene (PTFE)	Teflon items, tubes	2.10–2.30

Low-density plastics, such as PP and PE, form fragments with a lower density than water and thus float on the water, whereas higher-density polymers, such as PET, form plastics that do not float but already settle depending on their movement and sinking speed. Positively charged plastics usually float on the water surface, but only for a short time prior to becoming contaminated and mixed in with other waste.

One of the most important properties of plastic is its durability. Due to inadequate waste management, plastic pollution occurs on land and in the sea. Larger pieces of plastic are scattered, which is particularly visible in coastal areas where the effects of waves and UV light are strongest. High temperatures combined with intense UV radiation affect the decomposition of plastic, whereas lower temperatures and less UV radiation result in much slower decomposition [19]. Depending on the causes of decomposition, there are several mechanisms: biodegradation, photodegradation, thermal degradation, thermo-oxidative degradation, and hydrolysis [18].

Photodegradation, photo-oxidation, UV decomposition, and oxidative degradation are synonyms for the same process. It is a phenomenon in which UV light in combination with atmospheric oxygen changes the physical and chemical structure of plastics [20]. The rate of degradation can vary depending on the environment and temperature, but photodegradation has been shown to be the most influential factor in the degradation processes of this type [20,21]. UV light affects plastics depending on the conditions present [22]. Ranjan and Goel have shown that the degree of photodegradation varies depending on the environment. The oxygen content of water and oxygen are needed for the degradation process [23].

Biodegradation is carried out by microorganisms under aerobic and anaerobic conditions [24,25]. Microorganisms change the chemical structures, sizes, shapes, and masses of plastics through hydrolysis. Aerobic degradation is a process that uses oxygen as an oxidant and decomposes organic matter to carbon dioxide and water. This process often takes place in nature where oxygen is abundant. Anaerobic decomposition is the process of decomposition in the absence of oxygen. Thermal degradation refers to chemical changes in polymers as a result of elevated temperature [26]. Thermooxidative degradation is a slow oxidative process at moderate temperatures, and hydrolysis means degradation caused by water reaction [18].

Direct plastic contamination is an irreversible process because plastic is not degradable and there is almost no way to collect these microplastic particles in water resources after the plastic has been sprayed into microplastics. A real example of a primary plastic product where plastic directly plays an important role in cleansing the face or skin, besides toothpaste, is exfoliating creams. Here, plastic particles, usually polyethylene (PE) particles, remove dirt from the face. After rinsing the face or body, these particles end up in wastewater and thus in the sewage system. As a result, the United States has banned the use of plastic beads in creams and cosmetics. This has led some industries to list and identify sustainable particles that are certainly biodegradable and are almost equivalent to the microplastic particles found in cosmetics, even in terms of price, solvent resistance, surface shape, uniform size, and mechanical properties. Biodegradable products have been replaced by: natural hard material (human walnut), synthesized bio-based polymers (polylactic acid), and natural polymers (starch, lignin) [27]. There is a growing interest in the development of green plastic using natural resources. Green plastic is divided into two groups: green plastic made from monomers derived from biomass such as vegetable oil, and natural polymers such as chitin or lignin, which have extremely high biocompatibility and biodegradability [13].

Ageing of Polymers

Ageing is a term used in polymer science when the properties of polymers change over a period of time. Previous studies have shown that ageing causes the degradation of polyester at the molecular level. Theoretically, there are two types of ageing mechanisms: physical and chemical. Physical ageing is the most common type of ageing, corresponds to changes that have occurred in the composition (water absorption) and in the chain configuration itself, most often when the intermediate molecules have been modified but without changing the chemical structure of the polymer. The main groups of physical ageing are structural relaxation or a reduction in free volume and physical ageing, which occurs due to solvent penetration or release of plasticizer. Chemical ageing, on the other hand, is the result of reactions with external agents such as water, oxygen, UV radiation, or ionizing radiation. Considering the problem of polyester during washing, the emphasis is on the interaction of polyester with water. Physical ageing in amorphous parts of polymers occurs below the glass transition temperature, T_g. Penetration of water or other small molecules into the interior of the polymer leads to a sharp drop in the glass transition temperature. Water is a polar solvent, and its solubility in polyester is expected to be very low. Water has the ability to react with ester groups, so in this case, it is a type of physical ageing with water. It is a reversible process that is often accompanied by irreversible damage that occurs in the fibre matrix itself or at the fibre interface [28–31].

One of the most important mechanisms for degradation is physical ageing due water or wet ageing. Lemmi et al. studied the nature of physical ageing of polycrystalline PET, varying the temperature and duration of ageing, and then compared how ageing directly affects fibre strength (Figure 2) [32].

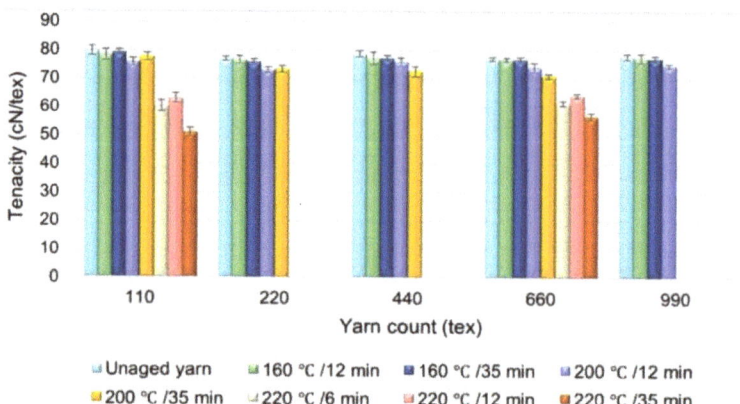

Figure 2. The relationship between the ageing time at ageing temperatures of 160, 200, and 220 °C and yarn strength [32].

The results obtained show that the materials aged for 12 and 35 min at a temperature lower than 160 °C show no difference from the untreated yarns. Materials processed above 160 °C but less than 200 °C show a decrease in strength below 1.11 cN/tex. Polyester samples aged at temperatures of 140 °C and 160 °C for a period of 12 min and 35 min showed almost no difference in strength compared to the untreated samples. However, the strength decreased with increasing temperature, and the toughness also decreased by about 5% at 200 °C and 19% at 220 °C. A sample aged at 220 °C for 35 min has almost 31% lower strength than the untreated polyester.

The parameters of thermal ageing significantly affect the mechanical and physical properties of the samples. In summary, the strength and thermal elongation are inversely proportional to the ageing temperature [32].

3. Textiles—Source of Microplastic Pollution

Textiles are also considered one of the major sources of microplastic pollution. Microplastics from textiles generally have a fibre shape, which is why they are often referred to as microfibres [33,34]. The main sources of microplastic fibres released from textiles are textile manufacturing and industrial and household washing. Cai et al. have shown that, during yarn spinning, microplastic fibres are released five times more from textiles with processed surfaces, such as fleece or plain brushed. Additionally, many studies have provided estimates of microfiber emissions from synthetic textiles during machine washing. Numerous studies and results show that textiles are among the main potential polluters due to the release of fibres, and that one of the reasons for this is the washing process [34–50]. Research has shown that the type of construction itself affects the amount of microfibers released. It would be convenient to use fabrics with a compact weave and as high a density as possible. The first wash releases most of the microfibers, and as the number of washes increases, the release will decrease. It is emphasized that washing parameters such as time and temperature have an influence on the release of microfibers [44]. Some possibilities to avoid this phenomenon are listed in Table 2.

Table 2. Possibility of preventing the release of microfibers [51].

Textile parameters	Type of fibres	Hydrophilic fibres release more fibres than synthetic ones. The strength can also affect tearing
	Yarn properties	Yarns with more twists and longer filaments release less microfibres
	Structure of fabric	Thermally cut fabrics release less than mechanically cut fabrics. The influence of knitted and woven structure is not entirely clear
	Ageing of fabric	The impact is not predictable because the garment does not pass a complete lifecycle
External parameters	Washing machine	Vertical drum machines contribute more to the release than horizontal ones, although this is related to the bath ratio

It has been shown that, among all textile fibres, polyester and cotton are the most widely used, with a total annual demand of 46 million tons, if polyester is included [1,35,52,53].

Polyester fibres belong to a group of fibres composed of macromolecules, synthesized polymers with a linear structure characterized by ester bonds (-CO-O-) linking the constitutional units, after which the whole group is named. Poly(ethylene terephthalate), PET, is the most abundant polymer in the polyester fibre group. Taking into account the data from

2017, PET accounts for 50% of the total man-made fibres produced, and 14% are made from recycled PET [1].

As for its properties, polyester has a density of 1.37–1.45 g/cm^{-3}, sinks very quickly, is non-biodegradable, and shows some weathering resistance. Although polyester is resistant to weathering, the fragmentation mechanisms are not, so photo-oxidation and hydrolysis can occur in marine environments. The change in pH in the ocean can change the chemical equilibrium of the microplastic itself by increasing or decreasing the rate of chemical leaching from the polyester surface, i.e., PET, which is generally considered a safe fibre, may become extremely hazardous in the near future [1,54,55].

Recently, PET polymeric materials and PET fibres, especially in the form of plastic bottles and packaging, have often raised environmental issues related to microplastics. It should be noted that polyester is not harmful to health and the environment, but due to its high presence, i.e., its high volume in waste, low biodegradability, and partial resistance to biological and atmospheric agents, the biggest problem is the release of microplastics into the environment. Due to these characteristics, polyester belongs to the category of environmentally unfriendly materials. Increasing the percentage of recycled PET production is economically and environmentally acceptable due to two segments:

- Creation of added value with low quality materials;
- Reduction in the plastic waste increase [56].

Among the studies reviewed, polyester is the most commonly investigated textile material because of its widespread use in the textile industry [53]. Some other studies deal with other textile materials such as polyamide [57–61], polyacrylic [62], and their blends [63,64]. It has also been found that natural fibres such as cotton minimise the risk of environmental pollution compared to synthetic textiles due to its biodegradability [57,58].

3.1. Textiles—The Release in the Washing Process

It is well-known that, in the washing process, five partners act in synergy: textile material, water, detergent, washing machine and stain. Water serves as a medium for the transport of thermal and mechanical energy, as well as for washing. Various physical and chemical mechanisms occur during textile washing. The main parameters that affect washing performance are mechanics, temperature, chemistry, and time [65–69]. Their joint action is responsible for a satisfactory washing effect (Figure 3).

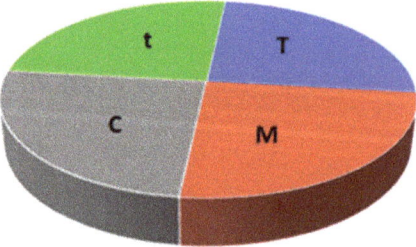

Figure 3. Washing factors according to Sinner's cycle: t—time, T—temperature, M—mechanics, C—chemistry.

Washing in household washing machines increases the mechanical agitation (action), which shortens the time, and the factors in the Sinner's cycle are changed. As far as chemicals are concerned, this group includes detergents that are biodegradable and, therefore, environmentally friendly.

The washing of household textiles has been shown to be an important source of microplastic contamination, although the release of fibres from textiles is not yet fully understood [57]. In some literature sources, the washing process is mentioned as a source of contamination of plastic fibres. These data showed that a single garment can release more than 1900 fibres per wash cycle, and that almost all garments emit more than 100 fibres per litre of wastewater [65]. In addition, studies have shown a release of 0.033–0.039% by weight of fibres from polyester garments after washing, although it should be noted that no experimental conditions are given. In conclusion, home textile washing is a continuous source of microplastic emissions in wastewater, and the relationship between the process of washing, and the release of microplastics is proportional.

As for the release of microplastics during the washing process, it should be noted that this problem applies to a greater extent to synthetic textiles, not so much to natural fibre materials, since natural fibres are biodegradable. The representative of the synthetic fibres is usually the polyester fibre. According to the available literature data, the release of fibres during washing is most often studied at different temperatures, with the addition of different amounts of detergents, and using certain mechanics and time. The results showed that the average size of the released fibre was 11.9 to 17.7 µm in diameter, and the length was 5.0 to 7.8 mm [36,37,62].

Kelly et al. [70] have shown the relationship between the effect of time and washing temperature, confirming that at smaller temperature and time intervals there is no difference, that at 15 and 30 °C for 15 to 30 min there is no effect on the release of microplastics, whereas at higher temperatures from 60 °C, there is a significant increase in the release of microplastics when polyester fabrics are processed. When mentioning and comparing temperatures of 15 and 30 °C, it should be noted that these temperatures fall into the category of "cold" washing. When comparing temperatures, there is a significant increase in the amount of microplastics when washing at 60 °C, but also at less than 60, whereas below 30 °C, there is only a minimal change in the release of microplastics. Therefore, the temperature is not such an important factor affecting the increase in or release of microplastics. Considering the mechanics of washing itself due to a prolonged time cycle of washing, the research found that the release of microplastic would be significantly increased, although it should be noted that there was no significant difference in the release of microplastics at a duration of 8 h. Additionally, the detergent has no effect on the release of microplastics. However, there are some doubts about the effect of detergents. For example, the study shows that steel balls are used with the detergent, which could interact with the detergent as a mechanic when they would enhance the effect of the detergent, thus forcing it into the fabric itself and creating many bubbles. However, this research has not yet been sufficiently explored, but I would like to draw attention to the fact that the safe type of the child itself has some influence on the release of the microplastics [70].

Although there are many publications on the release of microfibres during washing, other possible pathways of microplastic particle release into the environment have only recently been investigated. One of these is drying in tumble dryers [71–73]. Household tumble dryers play an important role in the release of microfibres from textiles into the environment or into the air. This is because the air from the dryer is usually not treated before it is released into the environment, and the microfibres are released directly from the dryer or into the surrounding space or through the ventilation pipe into the environment [71–74]. As the textiles spin in the dryer, the microfibres could be released from the textiles, especially at higher temperatures. The release of microfibres from large commercial dryers is not known but could also be significant and not negligible. Furthermore, if the dryers are not connected to the ventilation system, human microfibres can be released directly from the air of the enclosed space [73,75]. Microplastics have been detected in indoor and outdoor air all over the world. Based on a normal exposure scenario, it has been estimated that a child could ingest more than 900 microplastic particles per year via dust [76].

When investigating the effect of repeated washing cycles on the release of microplastic particles, Sillanpää et al. concluded that the release was reduced by about 90% in the last cycles [35]. On the other hand, Hartline et al. found that older garments released more microplastics during a continuous wash cycle of several hours. Various observations were also made when investigating the effect of detergents on the release of microplastics. For example, Napper and Thompson and De Falco found that the presence of detergents generally leads to an increased release of microplastic particles. Pierc and his colleagues reported that the detergent had no significant effect on the release of microplastics [35,61,62,70,74,77].

The release of textile fibres may also be affected by parameters other than the type of the fabric and the washing and drying process. Some other studies have investigated the effects of water hardness, water softeners, temperature, and the type of washing machine [39,59,61,62,74,77–79]. Zambrano et al. reported that the type of shedding is affected by fibre friction, shape, thickness, stiffness, and abrasion resistance. They linked the stronger shedding of cotton fibres mainly to the stronger hairiness of cotton fabric compared to other fibres studied. The authors found that polyester is one of the materials with the lowest abrasion resistance compared to cellulose fibres. They concluded that fabrics with higher abrasion resistance, higher yarn strength, and lower hairiness are desirable factors to reduce fibre shedding during washing [78].

Among the synthetic materials used, polyester fibres shed more microfibres compared to other synthetic fibres such as acrylic and polyamide. Knitted fabrics were also found to release more fibres. Other researchers suggest that the selection of a suitable spinning method could control the structural compactness and thus solve the problem of loose structures, hairiness, and twisting of the yarn [43,80]. De Falco et al. have found that fabrics release more microfibres than knitwear, depending on the type of fibres used in the textile production. Knitwear is made of filaments, whereas woven fabric is made of double yarn with greater hairiness. As for the surface mass of the textiles, they have shown that the release of fibres is not affected [77].

Further research results show that the number of microfibers released from polyester and cotton fabrics after the first wash varied from 2.1×10^5 to 1.3×10^7 and that the largest number of fibres came from cotton fabrics. Indeed, the results showed that the annual emission of microfibers in Finland was estimated to be 154,000 kg (PES) and 411,000 kg (cotton) [35]. Synthetic waste effluents from washing machines contain released fibres which are then transported to the wastewater and sewage system, and are deposited or floated in them (Figure 4).

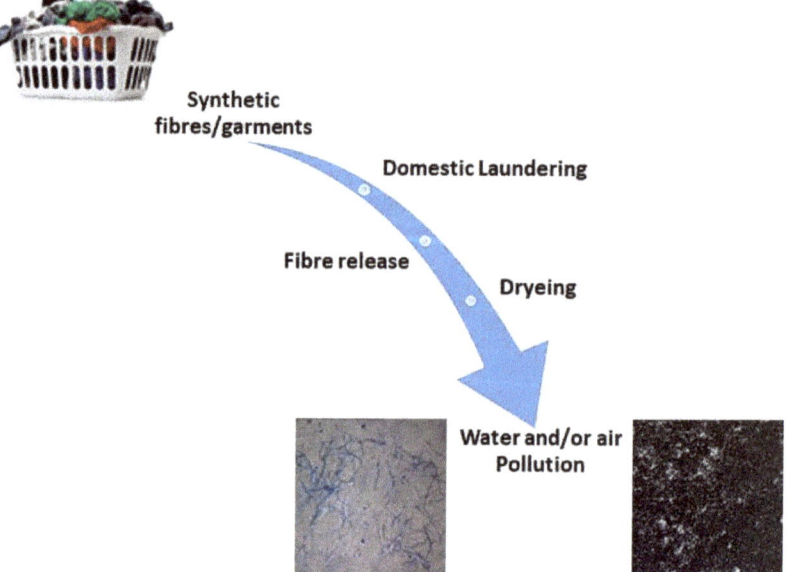

Figure 4. The microfiber shedding mode.

Guo et al. found that polyester microfibers are able to change soil structure and organic matter transfer, which also shows that they can affect not only water resources but also soil function. They wanted to show what effect microfibers have on enzymes by storing soil organic carbon, the micromolecules. They hypothesized that polyester microfibers have a great tendency to alter the overall microbial community in the soil as well as the activation of enzymes and the effect of cycling organic carbon in the soil. The authors concluded that the addition of polyester microfibers had the same effect on the whole soil community, and the activity of lactose and cellulose in the soil was also improved by the addition of microfibers, but the addition of microfibers also reduced the total amount of organic matter in the soil [81].

3.2. Wastewater and Microplastics

Wastewater treatment plants can serve as an entry point for microplastics into the environment. The water treatment process is often influenced by the particle size. Large solid particles are removed from wastewater treatment plants, and thus the concentration of organic matter is reduced [4]. To reduce the amount of particles in wastewater, some wastewater treatment plants use a filter or a membrane bioreactor (MBR). Pretreatment is also a required step in wastewater treatment plants, which is divided into three stages, primary, secondary, and tertiary [6], each involving different procedures:

- mechanical;
- chemical;
- physico-chemical;
- biological; and
- chemico-biological.

Over the past decade, wastewater treatment has been constantly needed to improve wastewater quality. However, the final technologies to improve wastewater quality are not specifically designed to remove microplastics. According to the literature review, new wastewater treatment methods may improve the final step of removing microplastics from wastewater [82]. These newly developed technologies can effectively remove microplastics from wastewater, but they are a very expensive process, which is difficult to install into existing plants and are used only for applications that require high quality standards An example of this is membrane bioreactors, which adsorb only water and small particles after primary and secondary treatment using cross-flow filtration. Another limitation of this technology is the high demand for energy sources, which results in a high cost for the process, making it uneconomical [4].

Thanks to rapid sand filters and discfilters, scientists have achieved the fast removal of microplastics by physical separation [83]. Microplastic can also be isolated during filtration, microfiltration 0.1–1 μm, ultrafiltration 2–100 μm, and nanofiltration 2 nm. Biochar has been used in sand filter systems for filtration efficiency. Microplastic particles with a diameter of up to 10 μm can be removed with an efficiency of 95% [84]. Electrocoagulation, in which coagulants are produced by electric current, is also applicable. In most cases, this involves iron or aluminium electrodes reacting with hydroxide ions formed after electrolysis to form hydroxide coagulants. The microplastic becomes very unstable in the presence of the coagulants formed and remains suspended in the sludge blanket. The results showed that about 90% of microfibers were removed [84] by electrocoagulation.

The number of microplastics can be influenced by lifestyle, population, climate, and seasonal conditions. The fact is that the amount of microplastic is highest in warmer conditions than in colder ones. It has been shown that wastewater samples contain fragments of PP and PE, whereas sludge contains PA, PET, and PS, as denser polymers are deposited during the treatment process [85]. Studies have also shown that a human can ingest between 39,000–520,000 microplastic per year through food and beverages, with levels naturally depending on age and gender [86]. Thanks to the method, it is possible to distinguish microplastics from non-plastic particles on the filter membrane using μ-FTIR [87].

4. Mitigation Measures for Microplastic Contamination

In this context, synthetic textiles stand out and disintegrate, and some measures are proposed to be applied during the production process.

4.1. Prevention

Adjusting the next steps has a significant impact on preventing the formation of microfibers. First of all, it is recommended to use high-quality products, focusing on clothes made of natural fibres. Then, if possible, it is important to avoid the use of mechanically finished fabrics such as fur or fleece. Furthermore, it is recommended to use appropriate chemicals (less alkali detergents—liquid or gel, softener, and other finishing agents) that reduce the release of microfibers. The emphasis is that the main problems include microplastics/microfibres and insufficient regulatory standards [88]. Airing out clothes after each use also reduces the number of washes and thus the release of microfibers [60].

4.2. Measures Proposed for The Production Process

The impacting parameters are: twisting yarn to increase elasticity and resistance, reducing the thickness because this automatically reduces the number of fibres, increasing the fabric density because this increases the strength of the structure, reducing the use of short fibres, and careful choice of textile auxiliaries to prevent friction between fibre–detergent and fibre–fibre, as well as to protect against mechanical stress in the washing machine [43,78,80,89].

The study mentions two biodegradable coatings as protection against physical stress, namely chitosan and pectin. Textile Company Lenzing AG announced that it uses alternatives to biodegradable materials such as Tencel (Lyocel), viscose, and Modal. The functionalization of the protective layer itself is based on the use of pectin, a natural polysaccharide that is extremely cheap and very readily available. It is known as a waste product of sugar and sunflower oil production. The test results of the washed treated fabrics showed a significant reduction in microfibers by up to 90% compared to untreated fabrics and a promising resistance to the washing process as well [90].

The proposed measures cover the exclusive use of an ultrasonic cutter to reduce the amount of precipitated fibres by about 50%, less brushing during processing, the use of recycled PET, and the removal of existing fibres before shipment.

Innovative surface treatment aimed to create a protective layer on the surface to reduce the release of microfibers from the surface. Chitosan is an amorphous solid white, harmless, odourless, and biodegradable biopolymer. It is soluble in acid but not soluble in alkali, water, or any organic solvent. The production of chitosan lacks the desired purity, which is extremely important in the pharmaceutical industry. The chitosan used in the textile industry is of low purity and can be used directly in fibre production. Studies have shown that chitosan can reduce the harmful release of microplastics during the dyeing process [91,92]. Chitosan also has an antibacterial effect and is used in finishing to obtain antibacterial fibres [93–97]. Some studies have shown good resistance to washing and antibacterial activity of chitosan-treated fabrics against *Staphylococcus aureus* in up to five washing cycles [93,94]. The treatment of polyester fabrics with chitosan leads to a significant antistatic effect in addition to its antibacterial properties. Matsukawa et al. treated polyester fabrics with chitosan by hydrolysing its surface with a sodium hydroxide solution to incorporate it into a functional group (-COOH) [23,26]. It is also important to know that polyester can be dyed directly if it has been previously treated with chitosan in an alkaline treatment and impregnated in this bath and is then ready for dyeing. The alkaline treatment improves the adhesion of chitosan to the polyester surface, resulting in a stronger depth of colour [98,99]. It is noted that obtaining better effects depends on the amount of chitosan and its properties, such as the degree of deacetylation or molecular weight. However, the amount of chitosan should be carefully dosed because the higher the viscosity, the more undesirable the effects.

4.3. Mechanical Means of Preventing the Release of Fibres during Use

Cora Ball is made of soft plastic, with pores that can adhere microfibers to its surface in the washing process. Collected microfibers can easily be removed by hand. The results show that the total amount of fibres absorbed by the ball is about 35% in each wash. It is also important that the used ball retains the ability to collect microfibers compared to a newly purchased ball [89].

4.4. Innovative Methods for the Direct Removal of Microparticles from the Aquatic Environment

It is possible for microplastics to degrade through the action of microorganisms that degrade microplastic to biomass, carbon dioxide, methane, water, and other inorganic compounds. Degradation parameters such as environmental parameters, UV radiation, solar radiation, moisture, and the physical and chemical properties of the polymer play an important role. In addition to microorganisms, microplastics can also be degraded by bacteria. Biocatalyst cells, *Comamonas Testosteroni*, are able to degrade PES and thus reduce the release of microplastics. At the beginning of the process, the diameter of PES was 7.30 µm; after treatment in alkaline medium, the diameter decreased to 1.58 µm, which promoted the rapid degradation of PES under alkaline degradation conditions by bacteria, and by biosorption, which allowed the adsorbate to bind to the surface of the adsorbent. Studies have shown that algae and brown algae tend to adsorb microplastics due to the presence of alginic acid in their walls. The functional carboxyl group is present in brown algae, and thanks to this group, the plastic can bind to the adsorbent [100,101].

5. Conclusions

Microplastics have become a serious threat to the environment and human health. Therefore, the gradual release and use of microplastics, which occurs directly in the fibre production itself, must be drastically reduced and be part of a global initiative, even before research studies on the long-term risks and consequences are available. The environmental aspects of microplastics are not sufficiently considered, it is not given enough importance, and most people are not aware of its negative effects. There is a need to develop as many programs as possible to monitor microplastics and thus play a key role in the prevention and management of microplastic pollution. Many countries have not developed a strategic approach to the largest sources of microplastics accumulating in water, nor have they developed processes to clean them up efficiently. There are several research projects that have investigated the impact of microplastics on final waste, as well as the removal of microplastics during each step of the process in wastewater treatment plants. Therefore, it is crucial that the machines are economical, energy efficient, and environmentally friendly to reduce the emission of microplastics as cost-effectively as possible. As far as polyester and the release of microplastics are concerned, the primary objective is to produce polyester with optimal structural parameters and switch to the most recycled production possible. It is necessary to use the most suitable agents that protect the structure of the polymer and prevent the release of microplastics during washing. It is also necessary to optimize the washing and drying processes of synthetic materials, as they are unique and market relevant. Finally, it is obvious that this interdisciplinary topic will continue to be relevant in the future and there is great potential for further research.

Author Contributions: Conceptualization, A.Š., T.D. and T.P.; methodology, T.D. and T.P.; writing—original draft preparation, A.Š.; writing—review and editing, T.D. and T.P. All authors have read and agreed to the published version of the manuscript.

Funding: This research was funded in part by Croatian Science Foundation under the project IP-2020-02-7575, InWaShed-MP and in part by the "Young researchers' career development project—training of doctoral students", DOK-2021-02-6750, of the Croatian Science Foundation.

Institutional Review Board Statement: Not applicable.

Informed Consent Statement: Not applicable.

Data Availability Statement: No new data were created or analyzed in this study. Data sharing is not applicable to this article.

Acknowledgments: A paper recommended by the 14th Scientific and Professional Symposium Textile Science and Economy, The University of Zagreb Faculty of Textile Technology.

Conflicts of Interest: The authors declare no conflict of interest.

References

1. Issac, M.N.; Balasubramanian, K. Effect of microplastics in water and aquatic systems. *Environ. Sci. Pollut. Res.* **2021**, *28*, 19544–19562. [CrossRef] [PubMed]
2. Marn, N.; Jusup, M.; Kooijman, S.; Klanjscek, T. Quantifying impacts of plastic debris on marine wildlife identifies ecological breakpoints. *Ecol. Lett.* **2020**, *23*, 1479–1487. [PubMed]
3. Kolbe, S. Microplastics or microfibers?—Conceptual confusion, Research Institute for Textiles and Clothing (FTB), Niederrhein University of Applied Sciences, Moenchengladbach, Germany. *Tekstil* **2018**, *67*, 235–236.
4. Westphalen, H.; Abdelrasou, A. Challenges and treatment of microplastics in water. In *Water Challenges of an Urbanizing World*; Glavan, M., Ed.; InTech Open: London, UK, 2018; pp. 71–82.
5. Vassilenko, K.; Watkins, M.; Chastain, S.; Posacka, A.; Ross, P.S. Me, My Clothes and the Ocean: The Role of Textiles in Microfiber Pollution, Science Feature. Ocean Wise Conservation Association, Vancouver, QC, Canada, 16p. Available online: https://assets.ctfassets.net/fsquhe7zbn68/4MQ9y89yx4KeyHv9Svynyq/8434de64585e9d2cfbcd3c46627c7a4a/Research_MicrofibersReport_191004-e.pdf (accessed on 26 November 2021).
6. Carr, S.A.; Liu, J.; Tesoro, A.G. Transport and fate of microplastic particles in wastewater treatment plants. *Water Res.* **2016**, *91*, 174–182. [CrossRef]
7. Thiounn, T.; Rhett, C.S. Advances and approaches for chemical recycling of plastic waste. *J. Polym. Sci.* **2020**, *58*, 1347–1364.
8. Andrady, A.L.; Neal, M.A. Applications and societal benefits of plastics, Philosophical transactions of the Royal Society of London. Series B. *Biol. Sci.* **2009**, *364*, 1977–1984. [CrossRef]
9. Espinosa, C.; Esteban, M.Á.; Cuesta, A. Microplastics in aquatic environments and their toxicological implications for fish. In *Toxicology—New Aspects to This Scientific Conundrum*; Larramendy, M.L., Soloneski, S., Eds.; InTech: Rijeka, Croatia, 2016; pp. 113–145.
10. European Parliament's Policy Department for Citizens' Rights and Constitutional Affairs. The Environmental Impacts of Plastics and Micro-Plastics Use, Waste and Pollution: EU and National Measures—Study. 2020. Available online: https://www.europarl.europa.eu/RegData/etudes/STUD/2020/658279/IPOL_STU(2020)658279_EN.pdf (accessed on 28 November 2021).
11. Čatić, I.; Barić, G.; Cvjetičanin, N.; Galić, K.; Godec, D.; Grancarić, A.M.; Katavić, I.; Kovačić, T.; Raos, P.; Rogić, A. Polimeri—Od prapočetka do plastike i elastomera. *Polimeri* **2020**, *31*, 59–70.
12. Van der Vegt, A.K. *From Polymers to Plastics*; Delft University Press, VSSD: Delft, The Netherlands, 2002.
13. Sazali, N.; Ibrahim, H.; Jamaludin, A.S.; Mohamed, M.A.; Salleh, W.N.W.; Abidin, M.N.Z. A short review on polymeric materials concerning degradable polymers. *IOP Conf. Ser. Mater. Sci. Eng.* **2020**, *788*, 012047. [CrossRef]
14. Material Properties of Plastics. Available online: https://application.wiley-vch.de/books/sample/3527409726_c01.pdf (accessed on 16 March 2022).
15. Nmazai, H. Polymers in our daily life. *BioImpacts* **2017**, *7*, 73–74.
16. Maddah, H.A. Polypropylene as a promising plastic: A review. *Am. J. Polym. Sci.* **2016**, *6*, 1–11.
17. Eyerer, P. Plastics: Classification, characterization, and economic data. In *Polymers—Opportunities and Risks I. The Handbook of Environmental Chemistry*; Eyerer, P., Ed.; Springer: Berlin/Heidelberg, Germany, 2010; pp. 1–17.
18. Andrady, A.L. Microplastics in the marine environment. *Mar. Pollut. Bull.* **2011**, *62*, 1596–1605. [CrossRef] [PubMed]
19. Andrady, A.L. Persistence of plastic litter in the oceans. In *Marine Anthropogenic Litter*; Bergmann, M., Gutow, L., Klages, M., Eds.; Springer International Publishing: Cham, Switzerland, 2015; pp. 57–72.
20. Yousif, E.; Haddad, R. Photodegradation and photostabilization of polymers, especially polystyrene: Review. *SpringerPlus* **2013**, *2*, 398. [CrossRef] [PubMed]
21. Lambert, S.; Sinclair, C.J.; Bradley, E.L.; Boxall, A.B.A. Effects of environmental conditions on latex degradation in aquatic systems. *Sci. Total Environ.* **2013**, *447*, 225–234. [CrossRef] [PubMed]
22. Julienne, F.; Delorme, N.; Lagarde, F. From macroplastics to microplastics: Role of water in the fragmentation of polyethylene. *Chemosphere* **2019**, *236*, 124409. [CrossRef] [PubMed]
23. Ranjan, V.P.; Goel, S. Degradation of Low-Density Polyethylene Film Exposed to UV Radiation in Four Environments. *J. Hazard. Toxic Radioact. Waste* **2019**, *23*, 04019015. [CrossRef]
24. Gu, J.D. Microbiological deterioration and degradation of synthetic polymeric materials: Recent research advances. *Int. Biodeterior. Biodegrad.* **2003**, *52*, 69–91. [CrossRef]
25. Hammer, J.; Kraak MH, S.; Parsons, J.R. Plastics in the Marine Environment: The Dark Side of a Modern Gift. *Rev. Environ. Contam. Toxicol.* **2012**, *220*, 192.
26. Faravelli, T.; Pinciroli, M.; Pisano, F.; Bozzano, G.; Dente, M.; Ranzi, E. Thermal degradation of polystyrene. *J. Anal. Appl. Pyrolysis* **2001**, *60*, 103–121. [CrossRef]

27. Ju, S.; Shin, G.; Lee, M.; Koo, J.M.; Jeon, H.; Ok, Y.S.; Hwang, D.S.; Hwang, S.Y.; Oh, D.X.; Park, J. Biodegradable chito-beads replacing non-biodegradable microplastics for cosmetics. *Green Chem.* **2021**, *23*, 6953–6965. [CrossRef]
28. Hunter, L.W.; White, J.W.; Cohen, P.H.; Biermann, P.J. A materials aging problem in theory and practice. *Johns Hopkins APL Tech. Dig.* **2000**, *21*, 575–581.
29. Richaud, E.; Verdu, J. Aging behavior and modeling studies of unsaturated polyester resin and unsaturated polyester resin-based blends. In *Unsaturated Polyester Resins: Fundamentals, Design, Fabrication, and Applications*; Elsevier: Amsterdam, The Netherlands, 2019; pp. 199–231. Available online: https://hal.archives-ouvertes.fr/hal-02568819/document (accessed on 28 November 2021).
30. Salopek Čubrić, I.; Čubrić, G.; Potočić Matković, V.M. Behavior of Polymer Materials Exposed to Aging in the Swimming Pool: Focus on Properties That Assure Comfort and Durability. *Polymers* **2021**, *13*, 2414. [CrossRef] [PubMed]
31. White, J.R. Polymer ageing: Physics, chemistry or engineering? Time to reflect. *Comptes Rendus Chim.* **2006**, *9*, 1396–1408. [CrossRef]
32. Lemmi, T.S.; Barburski, M.; Kabziński, A.; Frukacz, K. Effect of thermal aging on the mechanical properties of high tenacity polyester yarn. *Materials* **2021**, *14*, 1666. [CrossRef] [PubMed]
33. Roos, S.; Levenstam Arturin, O.; Hanning, A.-C. Microplastics Shedding from Polyester Fabrics, Mistra Future Fashion Report No 2017:1, SEREA. 2017. Available online: http://mistrafuturefashion.com/wp-content/uploads/2017/06/MFF-Report-Microplastics.pdf (accessed on 26 November 2021).
34. Manshoven, S.; Smeets, A.; Malarciuc, C.; Tenhunen, A.; Mortensen, L.F. Microplastic Pollution from Textile Consumption in Europe, Eionet Report—ETC/CE 2022/1. Available online: https://www.eionet.europa.eu/etcs/etc-ce/products/etc-ce-products/etc-ce-report-1-2022-microplastic-pollution-from-textile-consumption-in-europe (accessed on 26 November 2021).
35. Sillanpää, M.; Sainio, P. Release of polyester and cotton fibers from textiles in machine washings. *Environ. Sci. Pollut. Res.* **2017**, *24*, 19313–19321. [CrossRef] [PubMed]
36. De Falco, F.; Cocca, M.C.; Avella, M.; Thompson, R.C. Microfiber Release to Water, Via Laundering, and to Air, via Everyday Use: A Comparison between Polyester Clothing with Differing Textile Parameters. *Environ. Sci. Technol.* **2020**, *54*, 3288–3296. [CrossRef] [PubMed]
37. Galvão, A.; Aleixo, M.; De Pablo, H.; Lopes, C.; Raimundo, J. Microplastics in wastewater: Microfiber emissions from common household laundry. *Environ. Sci. Pollut. Res.* **2020**, *27*, 26643–26649. [CrossRef] [PubMed]
38. Gaylarde, C.; Baptista-Neto, J.A.; da Fonseca, E.M. Plastic microfibre pollution: How important is clothes' laundering? *Heliyon* **2021**, *7*, e07105. [CrossRef]
39. Hernandez, E.; Nowack, B.; Mitrano, D. Polyester Textiles as a Source of Microplastics from Households: A Mechanistic Study to Understand Microfiber Release During Washing. *Environ. Sci. Technol.* **2017**, *51*, 7036–7046. [CrossRef]
40. Cai, Y.; Yang, T.; Mitrano, D.M.; Heuberger, M.; Hufenus, R.; Nowack, B. Systematic Study of Microplastic Fiber Release from 12 Different Polyester Textiles during Washing. *Environ. Sci. Technol.* **2020**, *54*, 4847–4855. [CrossRef]
41. De Falco, F.; Di Pace, E.; Cocca, M.; Avella, M. The contribution of washing processes of synthetic clothes to microplastic pollution. *Sci. Rep.* **2019**, *9*, 6633. [CrossRef]
42. Volgare, M.; De Falco, F.; Avolio, R.; Castaldo, R.; Errico, M.E.; Gentile, G.; Ambrogi, V.; Cocca, M. Washing load influences the microplastic release from polyester fabrics by affecting wettability and mechanical stress. *Sci. Rep.* **2021**, *11*, 19479. [CrossRef] [PubMed]
43. Rathinamoorthy, R.; Balasaraswathi, S.R. A review of the current status of microfiber pollution research in textiles. *Int. J. Cloth. Sci. Technol.* **2021**, *33*, 364–387. [CrossRef]
44. Choi, S.; Kwon, M.; Park, M.-J.; Kim, J. Characterization of Microplastics Released Based on Polyester Fabric Construction during Washing and Drying. *Polymers* **2021**, *13*, 4277. [CrossRef]
45. Čurlin, M.; Pušić, T.; Vojnović, B.; Dimitrov, N. Particle Characterization of Washing Process Effluents by Laser Diffraction Technique. *Materials* **2021**, *14*, 7781. [CrossRef] [PubMed]
46. Schöpel, B.; Stamminger, R. A Comprehensive Literature Study on Microfibres from Washing Machines. *Tenside Surfactants Deterg.* **2019**, *56*, 94–104. [CrossRef]
47. Henry, B.; Laitala, K.; Klepp, I.G. Microfibres from apparel and home textiles: Prospects for including microplastics in environmental sustainability assessment. *Sci. Total Environ.* **2019**, *652*, 483–494. [CrossRef]
48. Bayo, J.; Ramos, B.; López-Castellanos, J.; Rojo, D.; Olmos, S. Lack of Evidence for Microplastic Contamination from Water-Soluble Detergent Capsules. *Microplastics* **2022**, *1*, 121–140. [CrossRef]
49. Haap, J.; Classen, E.; Beringer, J.; Mecheels, S.; Gutmann, J.S. Microplastic Fibers Released by Textile Laundry: A New Analytical Approach for the Determination of Fibers in Effluents. *Water* **2019**, *11*, 2088. [CrossRef]
50. Luogo, B.D.P.; Salim, T.; Zhang, W.; Hartmann, N.B.; Malpei, F.; Candelario, V.M. Reuse of Water in Laundry Applications with Micro- and Ultrafiltration Ceramic Membrane. *Membranes* **2022**, *12*, 223. [CrossRef]
51. Palacios-Mateo, C.; van der Meer, Y.; Seide, G. Analysis of the polyester clothing value chain to identify key intervention points for sustainability. *Environ. Sci. Eur.* **2021**, *33*, 2. [CrossRef]
52. Cai, Y.; Mitrano, D.M.; Heuberger, M.; Hufenus, R.; Nowack, B. The origin of microplastic fiber in polyester textiles: The textile production process matters. *J. Clean. Prod.* **2020**, *267*, 121970. [CrossRef]
53. Carmichael, A. Man-made fibers continue to grow. *Text. World* **2015**, *165*, 2588–2597.

54. Weber, A.; Scherer, C.; Brennholt, N.; Reifferscheid, G.; Wagner, M. PET microplastics do not negatively affect the survival, development, metabolism and feeding activity of the freshwater invertebrate Gammarus pule. *Environ. Pollut.* **2018**, *234*, 181–189. [CrossRef] [PubMed]
55. Piccardo, M.; Provenza, F.; Grazioli, E.; Cavallo, A.; Terlitti, A.; Renzi, M. PET microplastics toxicity on marine key species is influenced by pH, particle size and food variations. *Sci. Total Environ.* **2020**, *715*, 136947. [CrossRef]
56. Čorak, I.; Pušić, T.; Tarbuk, A. Enzimi za hidrolizu poliestera. *Tekstil* **2019**, *68*, 142–151.
57. Kärkkäinen, N.; Sillanpää, M. Quantification of different microplastic fibres discharged from textiles in machine wash and tumble drying. *Environ. Sci. Pollut. Res.* **2021**, *28*, 16253–16263. [CrossRef]
58. Cesa, F.S.; Turra, A.; Checon, H.H.; Leonardi, B.; Baruque-Ramos, J. Laundering and textile parameters influence fibers release in household washings. *Environ. Pollut.* **2020**, *257*, 113553. [CrossRef]
59. Yang, L.; Qiao, F.; Lei, K.; Li, H.; Kang, Y.; Cui, S.; An, L. Microfiber release from different fabrics during washing. *Environ. Pollut.* **2019**, *249*, 136–143. [CrossRef]
60. Carney Almroth, B.; Åström, L.; Roslund, S.; Petersson, H.; Johansson, M.; Persson, N. Quantifying shedding of synthetic fibers from textiles; a source of microplastics released into the environment. *Environ. Sci. Pollut. Res.* **2018**, *25*, 1191–1199. [CrossRef]
61. Hartline, N.L.; Bruce, N.J.; Karba, S.N.; Ruff, E.O.; Sonar, S.U.; Holden, P.A. Microfiber masses recovered from conventional machine washing of new or aged garments. *Environ. Sci. Technol.* **2016**, *50*, 11532–11538. [CrossRef]
62. Napper, I.E.; Thompson, R.C. Release of synthetic microplastic plastic fibres from domestic washing machines: Effects of fabric type and washing conditions. *Mar. Pollut. Bull.* **2016**, *112*, 39–45. [CrossRef] [PubMed]
63. Corami, F.; Rosso, B.; Bravo, B.; Gambaro, A.; Barbante, C. A novel method for purification, quantitative analysis and characterization of microplastic fibers using Micro-FTIR. *Chemosphere* **2020**, *238*, 124564. [CrossRef] [PubMed]
64. Belzagui, F.; Crespi, M.; Álvarez, A.; Gutiérrez-Bouzán, C.; Vilaseca, M. Microplastics' emissions: Microfibers' detachment from textile garments. *Environ. Pollut.* **2019**, *248*, 1028–1035. [CrossRef] [PubMed]
65. Soljačić, I.; Pušić, T. *Njega Tekstila, Dio 1: Čišćenje u Vodenim Medijima, Zagreb, Tekstilno-Tehnološki Fakultet*; Sveučilišta u Zagrebu: Zagreb, Croatia, 2005.
66. Sinner, H. *Ueber das Waschen mit Haushaltwaschmaschinen: In Welchem Umfange Erleichtern Haushaltwaschmaschinen und—Geraete das Waeschewaschen im Haushalt?* Haus und Heim: Hamburg, Germany, 1960.
67. Bao, W.; Gong, R.H.; Ding, X.; Xue, Y.; Li, P.; Fan, W. Optimizing a laundering program for textiles in a front-loading washing machine and saving energy. *J. Clean. Prod.* **2017**, *148*, 415–421. [CrossRef]
68. Konstadinos Abeliotis, K.; Amberg, C.; Candan, C.; Ferri, A.; Osset, M.; Owens, J.; Stamminger, R. Trends in laundry by 2030. *Househ. Pers. Care Today* **2015**, *10*, 22–28.
69. Alfieri, F.; Cordella, M.; Staminger, R.; Bues, A. Durability Assessment of Products: Analysis and Testing of Washing Machines, JCR Technical Reports. Available online: https://publications.jrc.ec.europa.eu/repository/bitstream/JRC114329/jrc114329_task_3_durability_final_v3.0.pdf (accessed on 16 March 2022).
70. Kelly, M.R.; Lant, N.J.; Kurr, M.; Burgess, J.G. Importance of water-volume on the release of microplastic fibers from laundry. *Environ. Sci. Technol.* **2019**, *20*, 11735–11744. [CrossRef]
71. O'Brien, S.; Okoffo, E.D.; O'Brien, J.W.; Ribeiro, F.; Wang, X.; Wright, S.L.; Samanipour, S.; Rauert, C.; Alajo Toapanta, T.Y.; Albarracin, R.; et al. Airborne emissions of microplastic fibres from domestic laundry dryers. *Sci. Total Environ.* **2020**, *747*, 141175. [CrossRef]
72. Kapp, K.J.; Miller, R.Z. Electric clothes dryers: An underestimated source of microfiber pollution. *PLoS ONE* **2020**, *15*, 1–17.
73. Danyang, T.; Zhang, K.; Xu, S.; Lin, H.; Liu, Y.; Kang, J.; Yim, T.; Giesy, J.P.; Leung, K.M.Y. Microfibers Released into the Air from a Household Tumble Dryer. *Environ. Sci. Technol. Lett.* **2022**, *9*, 120–126.
74. Pirc, U.; Vidmar, M.; Mozer, A.; Kržan, A. Emissions of Microplastic Fibers from Microfiber Fleece during Domestic Washing. *Environ. Sci. Pollut. Res.* **2016**, *23*, 22206–22211. [CrossRef]
75. Zhang, Q.; Xu, E.G.; Li, J.; Chen, Q.; Ma, L.; Zeng, E.Y.; Shi, H. A Review of Microplastics in Table Salt, Drinking Water, and Air: Direct Human Exposure. *Environ. Sci. Technol.* **2020**, *54*, 3740–3751. [CrossRef] [PubMed]
76. Abbasi, S.; Keshavarzi, B.; Moore, F.; Turner, A.; Kelly, F.J.; Dominguez, A.O.; Jaafarzadeh, N. Distribution and potential health impacts of microplastics and microrubbers in air and street dusts from Asaluyeh County, Iran. *Environ. Pollut.* **2019**, *244*, 153–164. [CrossRef] [PubMed]
77. De Falco, F.; Gullo, M.P.; Gentile, G.; Di Pace, E.; Cocca, M.; Gelabert, L.; Brouta-Agnésa, M.; Rovira, A.; Escudero, R.; Villalba, R.; et al. Evaluation of microplastic release caused by textile washing processes of synthetic fabrics. *Environ. Pollut.* **2018**, *236*, 916–925. [CrossRef] [PubMed]
78. Zambrano, M.C.; Pawlak, J.J.; Daystar, J.; Ankeny, M.; Cheng, J.J.; Venditti, R.A. Microfibers generated from the laundering of cotton, rayon and polyester based fabrics and their aquatic biodegradation. *Mar. Pollut. Bull.* **2019**, *142*, 394–407. [CrossRef]
79. Browne, M.A.; Crump, P.; Niven, S.J.; Teuten, E.; Tonkin, A.; Galloway, T.; Thompson, R. Accumulation of microplastic on shorelines worldwide: Sources and sinks. *Environ. Sci. Technol.* **2011**, *45*, 9175–9179. [CrossRef]
80. Vassilenko, E.; Watkins, M.; Chastain, S.; Mertens, J.; Posacka, A.M.; Patankar, S.; Ross, P.S. Domestic laundry and microfiber pollution: Exploring fiber shedding from consumer apparel textiles. *PLoS ONE* **2020**, *16*, e0250346. [CrossRef]
81. Guo, Q.Q.; Xiao, M.R.; Ma, Y.; Niu, H.; Zhang, G.S. Polyester microfiber and natural organic matter impact microbial communities, carbon-degraded enzymes, and carbon accumulation in a clayey soil. *J. Hazard. Mater.* **2021**, *405*, 124701. [CrossRef]

82. Talvitie, J.; Mikola, A.; Koistinen, A.; Setälä, O. Solutions to microplastic pollution—Removal of microplastics from wastewater effluent with advanced wastewater treatment technologies. *Water Res.* **2017**, *123*, 401–407.
83. Sol, D.; Laca, A.; Laca, A.; Diaz, M. Microplastics in Wastewater and Drinking Water Treatment Plants: Occurrence and Removal of Microfibres. *Appl. Sci.* **2021**, *11*, 10109. [CrossRef]
84. Singh, S.; Madhanraj, K.; Vishal, D. Removal of microplastics from wastewater: Available techniques and way forward. *Water Sci. Technol.* **2021**, *84*, 3689–3704. [CrossRef]
85. Menéndez-Manjón, A.; Martínez-Díez, R.; Sol, D.; Laca, A.; Laca, A.; Rancaño, A.; Díaz, M. Long-Term Occurrence and Fate of Microplastics in WWTPs: A Case Study in Southwest Europe. *Appl. Sci.* **2022**, *12*, 2133. [CrossRef]
86. Liu, F.; Nord, N.B.; Baster, K.; Vollertsen, J. Microplastics removal from treated wastewater by a biofilter. *Water* **2020**, *12*, 1085. [CrossRef]
87. Zhang, Y.; Wang, H.; Xu, J.; Lu, M.; Wang, Z.; Zhang, Y. Occurrence and Characteristics of Microplastics in a Wastewater Treatment Plant. *Bull. Environ. Contam. Toxicol.* **2021**, *107*, 677–683. [CrossRef] [PubMed]
88. Ramasamy, R.; Subramanian, R.B. Synthetic textile and microfiber pollution: A review on mitigation strategies. *Environ. Sci. Pollut. Res.* **2021**, *28*, 41596–41611. [CrossRef]
89. Anis, A.; Classon, S. Analysis of Microplastic Prevention Methods from Synthetic Textiles. iGEM LUND 2017. Available online: http://2017.igem.org/wiki/images/2/2c/T--Lund---Analysis_of_Microplastic_Prevention_Methods_from_Synthetic_Textiles.pdf (accessed on 16 March 2022).
90. De Falco, F.; Gentile, G.; Avolio, R.; Errico, M.E.; Di Pace, E.; Ambrogi, V.; Avella, M.; Cocca, M. Pectin based finishing to mitigate the impact of microplastics released by polyamide fabrics. *Carbohydr. Polym.* **2018**, *198*, 175–180. [CrossRef]
91. Bhavsar, P.S.; Fontana, D.G.; Zoccola, M. Sustainable Superheated Water Hydrolysis of Black Soldier Fly Exuviae for Chitin Extraction and Use of the Obtained Chitosan in the Textile Field. *ACS Omega* **2021**, *6*, 8884–8893. [CrossRef]
92. Matsukawa, S.; Kasai, M.; Mizuta, Y. Modification of Polyester Fabrics Using Chitosan. *Sen'i Gakkaishi* **1995**, *51*, 17–22. [CrossRef]
93. Flinčec Grgac, S.; Tarbuk, A.; Dekanić, T.; Sujka, W.; Draczyński, Z. The Chitosan Implementation into Cotton and Polyester/Cotton Blend Fabrics. *Materials* **2020**, *13*, 1616. [CrossRef]
94. Korica, M.; Peršin, Z.; Trifunović, S.; Mihajlovski, K.; Nikolić, T.; Maletić, S.; Fras Zemljič, L.; Kostić, M.M. Influence of Different Pretreatments on the Antibacterial Properties of Chitosan Functionalized Viscose Fabric: TEMPO Oxidation and Coating with TEMPO Oxidized Cellulose Nanofibrils. *Materials* **2019**, *26*, 3144. [CrossRef]
95. Zhang, Z.-T.; Chen, L.; Ji, J.-M.; Huang, Y.-L.; Chen, D.-H. Antibacterial Properties of Cotton Fabrics Treated with Chitosan. *Text. Res. J.* **2003**, *73*, 1103–1106. [CrossRef]
96. Yilmaz Atay, H. Antibacterial activity of chitosan-based systems. In *Functional Chitosan: Drug Delivery and Biomedical Applications*; Jana, S., Ed.; Springer: Singapore, 2020; pp. 457–489.
97. Ortega-Ortiz, H.; Gutiérrez-Rodríguez, B.; Cadenas-Pliego, G.; Jimenez, L.I. Antibacterial Activity of Chitosan and the Interpolyelectrolyte Complexes of Poly(acrylic acid)-Chitosan. *Braz. Arch. Biol. Technol.* **2010**, *53*, 623–628. [CrossRef]
98. Walawska, A.; Filipowska, B.; Rybicki, E. Dyeing Polyester and Cotton-Polyester Fabrics by Means of Direct Dyestuffs after Chitosan Treatment. *Fibers Text. East. Eur.* **2003**, *11*, 71–74.
99. Najafzadeh, N.; Habibi, S.; Ghasri, M.A. Dyeing of Polyester with Reactive Dyestuffs Using Nano-Chitosan. *J. Eng. Fibers Fabr.* **2018**, *13*, 47–51. [CrossRef]
100. Dey, T.K.; Md, U.; Mamun, J. Detection and removal of microplastics in wastewater: Evolution and impact. *Environ. Sci. Pollut. Res.* **2021**, *28*, 16925–16947. [CrossRef] [PubMed]
101. Gong, J.; Kong, T.; Li, Y.; Li, Q.; Li, Z.; Zhang, J. Biodegradation of Microplastic Derived from Poly(ethylene terephthalate) with Bacterial Whole-Cell Biocatalysts. *Polymers* **2018**, *10*, 1326. [CrossRef] [PubMed]

Article

Electromagnetic Shielding Properties of Knitted Fabric Made from Polyamide Threads Coated with Silver

Tanja Pušić [1], Bosiljka Šaravanja [2,*] and Krešimir Malarić [3]

[1] Department of Textile Chemistry and Ecology, Faculty of Textile Technology, University of Zagreb, 10000 Zagreb, Croatia; tanja.pusic@ttf.unizg.hr
[2] Department of Clothing Technology, Faculty of Textile Technology, University of Zagreb, 10000 Zagreb, Croatia
[3] Department of Communications and Space Technologies, Faculty of Electrical Engineering and Computing, University of Zagreb, 10000 Zagreb, Croatia; kresimir.malaric@fer.hr
* Correspondence: bosiljka.saravanja@ttf.hr

Abstract: This paper investigates a textile material of low surface mass for its protection against electromagnetic radiation (EMR), which is suitable for composite structures of garments, and for technical and interior applications. The shielding effectiveness against EMR of fabric knitted from polyamide threads coated with silver, measured in the frequency range of 0.9 GHz to 2.4 GHz, indicated a high degree of protection. The key contribution of the paper is the evaluation of the stability of the shielding properties against EM radiation after applying apolar and polar solvents, in synergy with the cyclic process parameters of wet and dry cleaning. The results of the study confirmed the decline in the shielding effectiveness after successive cycles of material treatment with dry and wet cleaning. The effect of wet cleaning in relation to dry cleaning is more apparent, which is due to the damage of the silver coating on the polyamide threads in the knitted fabric.

Keywords: electroconductive material; polyamide; silver; shielding effectiveness; wet cleaning; dry cleaning

---

1. Introduction

The increased awareness of EMR has led to the worldwide introduction of new regulations for manufacturers of electrical and electronic devices, which must now comply with the electromagnetic compatibility requirements (EMC requirements). The need to set limits for the EM radiation of electrical and electrical devices (mobile phones, microwave ovens, signals of 'radar' communication, radio transmitters etc.) that radiate EM energy in different frequency ranges aims to minimise the possibility of interference with radio and wired communications. The lifespan and efficiency of electronic devices can be increased by their protection against electromagnetic interference [1–6].

Figure 1 schematically shows the propagation of the signal through a layer of material with EMR protection properties. When EM rays pass through a medium or material, they interact with the molecules of the material; this phenomenon of interaction can be divided into three phases:

- absorption,
- reflection,
- secondary reflection.

When they hit the surface of a material, EM rays cause the charge in the material to oscillate. This forced oscillation of the charge acts as an antenna and results in reflection, whereas the other part is converted into thermal energy due to the oscillation. This kind of signal loss is known as attenuation due to absorption. Thus, the protective property of the material against EMR is based on the reflection against the conductive surface and the absorption in the conductive volume. Part of the wave is reflected, while the rest is transmitted and weakened as it passes through the medium [7].

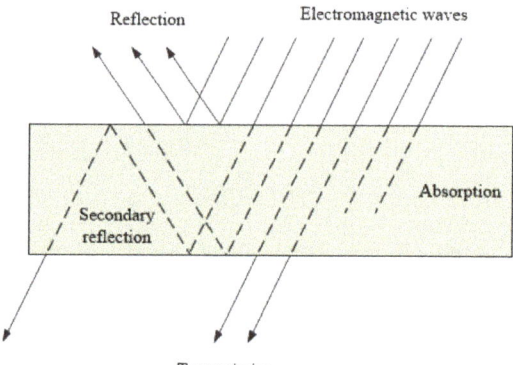

Figure 1. Schematic presentation of the signal propagation through a material with protective properties.

The combined effect of losses through reflection and absorption determines the effectiveness of the protective properties of the material, depending on its electrical and magnetic properties, surface and interior conductivity properties, material thickness, material composition, abrasion, and degree of processing [8].

The ratio of the level of the electric field at a certain distance from the source without protection (shield) and the level of the electric field with protection is defined as the shielding effectiveness (SE). The shielding effectiveness of the conductive barrier SE in dB is the sum of loss of reflection (R), loss of absorption (A), and the loss of secondary reflection (R_r), and is calculated according to Equation (1):

$$SE = R + A + R_r \tag{1}$$

For the purpose of protection against EMR in the electrical and electronic industry, conductive, lightweight and flexible textile structures are produced and developed instead of conductive metals or wire mesh materials. The reduction of the electromagnetic radiation transmission of textile materials can be achieved in various ways, such as by changing the composition [9], structure, or construction [10,11]; by the incorporation of conductive particles into the fibres, or metal threads and foils into the yarn [12–14]; or by the use of metal coatings [6,15], morphology [16], or conductive paints, pigments and varnishes [17].

Numerous studies of such products have been carried out, in which the different construction and finish parameters have varied. The results of measuring the protective properties of materials against EMR not only depend on the material properties but also on the sample size, measurement setup, and EMR source [1,18–22]. Since materials are exposed to various cyclic mechanical stresses, and chemical and atmospheric influences, it is important to monitor the durability of the protective properties under controlled conditions.

This paper deals with SE protective material made of silver-coated polyamide yarn, which—as a light and transparent structure—is suitable for applications in the composite structures of garments, and for technical and interior applications.

The functional material was analysed before and after the cyclic treatment in apolar and polar solvents, with process parameters of wet and dry cleaning. The influence of the solvents and process parameters on the changes on the surface of the material were analysed by scanning electron microscopy (SEM), while the protective properties of EMR were monitored by testing the properties of the shielding on the frequencies of 0.9 GHz, 1.8 GHz, 2.1 GHz, and 2.4 GHz.

2. Materials and Methods

The specifications of the shielding electrically-conductive knitted fabric made of polyamide (PA) yarn coated with silver (Ag) are presented in Table 1.

Table 1. Specifications of the shielding conductive knitted fabric (PA/Ag).

Composition PA/Ag (%)		80/20
Mass per unit area (g/m^2)		35.8
Density (course/wale)/100 mm		150/125
	digital microsope images	
	magnification 50×	magnification 250×
Charmeuse knitted structure		

This functional knitted fabric can either be incorporated as a functional interlining in clothing or used to make children's clothing due to its soft touch and antimicrobial properties, made possible by the silver. The non-stick type of interlining with protective properties against EM radiation is put between the base material and the lining, thus forming part of the composite structure of the garment. Garments are exposed to various mechanical and physicochemical influences, which makes it necessary to objectively evaluate the protective properties of their materials or composite structures before and after exposure to different frequencies. This is an important factor for the assessment of the service life of a garment with added value, which in this analysis is the protection against EMR (electromagnetic radiation).

The functional material (PA/Ag) is exposed to a cyclic treatment with a polar solvent (water) in wet cleaning, and an apolar solvent (perchloroethylene) in dry cleaning. These physicochemical processes were conducted through the synergy of solvents and the process parameters of the Sinner's circle: chemistry, mechanical agitation, temperature, and time [23]. Wet cleaning (W) is an eco-friendly and under-researched process for SE textiles, which is conducted in water at a low temperature, with low mechanical agitation applying special hypoallergenic detergents and protective additives which reduce the swelling of fibres in water [24]. Dry cleaning (P) is a conventional process with excellent cleaning features, and is a promising basis for the retention of original material properties in perchloroethylene. PA/Ag knitted fabric with dimensions of 1 m × 1 m was treated with perchloroethylene 10 times, according to the norm EN ISO 3175-2, while the treatment with water was carried out according to the norm EN ISO 3175-3. The detailed specifications of the Sinner's circle parameters in these processes are described in a piece of previously-published research [25].

2.1. Scanning Electron Microscopy (SEM)

The surface of the PA/Ag fabric was analysed before and after the cyclic treatment with an apolar solvent in dry cleaning (P) and a polar solvent in wet cleaning (W) under Sinner's circle parameters; the samples were observed after the 1st, 3rd, 5th, 7th and 10th cycle. Despite the silver content in the PA/Ag fabric, all of the samples were coated with gold and palladium for 90 s using Emitech Mini sputter coater SC7620 (Quorum Technologies, Ashford, Kent, UK)). The observation of the samples' surface was performed with the SE detector of the scanning electron microscope FE-SEM, MIRAIILMU, Tescan, Czech Republic, under a magnification of 500×.

2.2. Measuring the Shielding Effectiveness (SE) for Microwave Radiation

The shield properties of the tested samples were investigated using a method described in detail elsewhere [25], under the following working conditions:

- temperature 23 ± 1 °C,
- relative humidity 50 ± 10%.

According to the recommendations of the IEE-STD 299-97 [26], MIL STD 285 [27], and ASTM D-4935-89 [28], a measurement setup was designed and installed (Figures 2 and 3), consisting of:

- a measuring instrument: NARDA SRM 3000,
- an HP 8350 B signal generator,
- an IEV horn antenna: Industrija za elektrozveze (Telecommunication Industry), Ljubljana, Type A12,
- a wooden frame, in which a sample of PA/Ag material of 1 m × 1 m was placed.

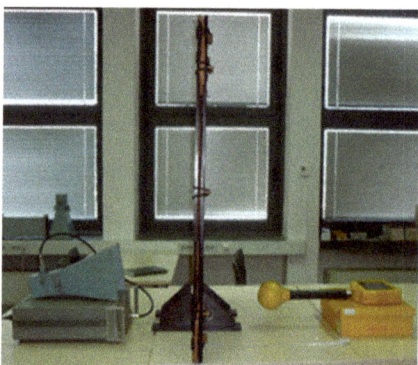

Figure 2. Measurement setup for the SE protective properties of textiles.

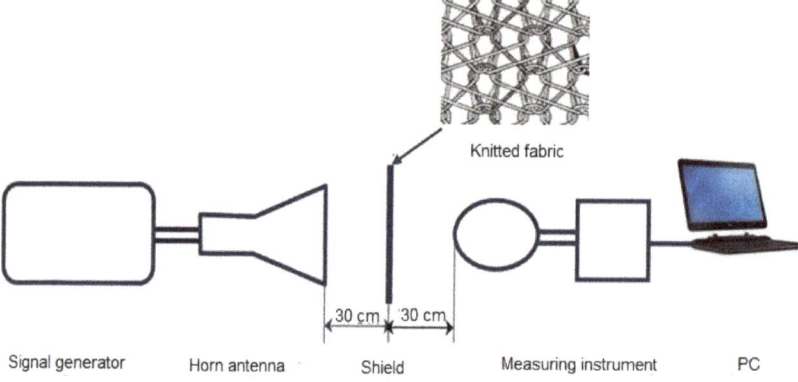

Figure 3. Schematic representation of the measurement setup.

Figures 2 and 3 show the measurement setup of the shield performance test. The signal generator was computer controlled, and provided frequencies of 0.9 GHz, 1.8 GHz, 2.1 GHz, and 2.4 GHz. The generator was connected to the horn/funnel antenna with a coaxial cable (for 900 MHz, a dipole antenna is used). The wooden shield was placed 30 cm away from the antenna and the measuring instrument: a spectrum analyser with a broadband antenna.

The EM protection factor was determined as the ratio between the EM field intensity (E_0) measured without the fabric and the EM field intensity (E_1) with the material placed between the radiation source and the measuring device.

The shielding effectiveness SE (dB) was calculated according to the following Equation (2):

$$SE = 20 \log \frac{E_0}{E_1} \qquad (2)$$

where:
- E_0 is the field level without protection (shield),
- E_1 is the field level with protection (shield).

The change in the shielding effectiveness of the PA/Ag knitted fabrics after the 1st, 3rd, 5th, 7th, and 10th cycles of dry and wet cleaning is expressed using Equations (3) and (4):

$$dSE = SE_0 - SE_P \qquad (3)$$

$$dSE = SE_0 - SE_W \qquad (4)$$

where:
- SE_0 represents the initial shielding effectiveness of the PA/Ag knitted fabric,
- SE_P represents the shielding effectiveness of the PA/Ag knitted fabric after the 1st, 3rd, 5th, 7th, and 10th dry cleaning cycles,
- SE_W represents the shielding effectiveness of the PA/Ag knitted fabric after the 1st, 3rd, 5th, 7th, and 10th wet cleaning cycles.

3. Results and Discussion

The cyclic exposure of the PA/Ag material to solvents in synergy with the process parameters led to a change in the material thickness tested according to EN ISO 5084: 2003, as shown in Table 2.

Table 2. Thickness of the PA/Ag fabric before and after 10 cycles of dry (P) and wet (W) cleaning.

PA/Ag Fabric	Thickness (mm)
Untreated	0.150
10 cycles treatment with P	0.162
10 cycles treatment with W	0.165

Due to the presence of amide bonds in the macromolecules, PA fibres can form hydrogen bonds, owing to which they have a better ability to absorb moisture [29] (compared to some hydrophobic polymers), which implies the possible influence of the polar solvent, e.g., water (W). However, the results in the table indicate a slight increase in the thickness of the material in the wet cleaning (W) and dry cleaning (P) compared to the untreated fabric. Expressed in %, the variability of the fabric thickness after 10 dry cleaning cycles is 0.5 times higher than the variability of the fabric thickness after wet cleaning. The slight shrinkage of the SE fabric in the dry cleaning process can be explained by the presence of a small quantity of water in the system, and subsequent drying.

The photograph, shown in Figure 4, of the surface of the conductive untreated PA/Ag sample shows a uniform coating of silver on the polyamide filament. The synergistic influence of the solvent and other process parameters was achieved through the characterisation of the surface of the PA/Ag material by the scanning electron microscope before and after the 1st, 3rd, 5th, 7th, and 10th processing cycle, under a magnification of 500× (Figure 5).

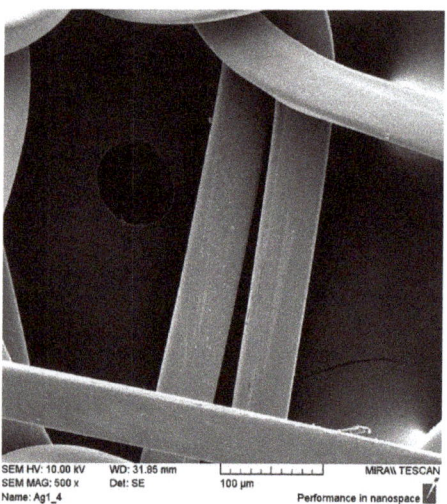

Figure 4. SEM micrograph of the conductive untreated PA/Ag sample under a magnification of 500×.

The influence of the apolar solvent (P) in synergy with the dry cleaning process parameters could be seen after five cycles, which can be attributed to the successive cycles. With the increase in the number of cycles on the knitted fabric, greater longitudinal damage of the silver coating on the threads was noticed, which intensified in the 10th cycle. The irregular shape of the damage and the appearance of ruptures on the silver coating indicate a more intense influence of mechanics as a process factor, which led to fractures of the material. Such local damage excludes the influence of solvents, which would act more evenly over the entire surface. The changes in the surface of the PA/Ag fabric under the influence of the polar solvent (W) and wet cleaning process parameters were visible after the 3rd cycle. The change dynamics were more intense compared to the dry cleaning (P). Additionally, irregular incrustations could be seen on the sample surface after 10 cycles of wet cleaning (W_10), which indicate the interaction of some of the substances in the process. The SEM images indicate that the polar solvent, in synergy with the process parameters of wet cleaning, caused a higher degree of longitudinal and irregular local damage to the PA/Ag fabric compared to the apolar solvent and the process parameters of dry cleaning.

The examination of the durability of the protective properties of elastic yarns with metal coatings has confirmed that the changes in the conductive properties during washing depend on the type of coating. The obtained results deviate from the research conducted on silver-coated textiles after washing in 25 cycles; the silver coating remained almost unchanged [30].

The polyamide, as a pure polymer, exhibits non-conductive properties, while the coating with Ag enhanced the electrical conductivity of the material and increased its shielding effectiveness [31].

The shielding effectiveness (SE) of the face and reverse side of the PA/Ag fabric before the solvent treatment with process parameters at frequencies of 0.9 GHz, 1.8 GHz, 2.1 GHz, and 2.4 GHz are shown in Figure 6.

The protective properties of the face and reverse side of the untreated PA/Ag samples at all of the frequencies are almost identical, as shown in Figure 6. The highest degree of protection was obtained at 2.4 GHz (24.1 dB), while the lowest degree of protection was achieved at 0.9 GHz (SE = 14.8 dB). Despite the difference of almost 10 units, the achieved degree of protection >10 dB represents an acceptable degree of protection [31].

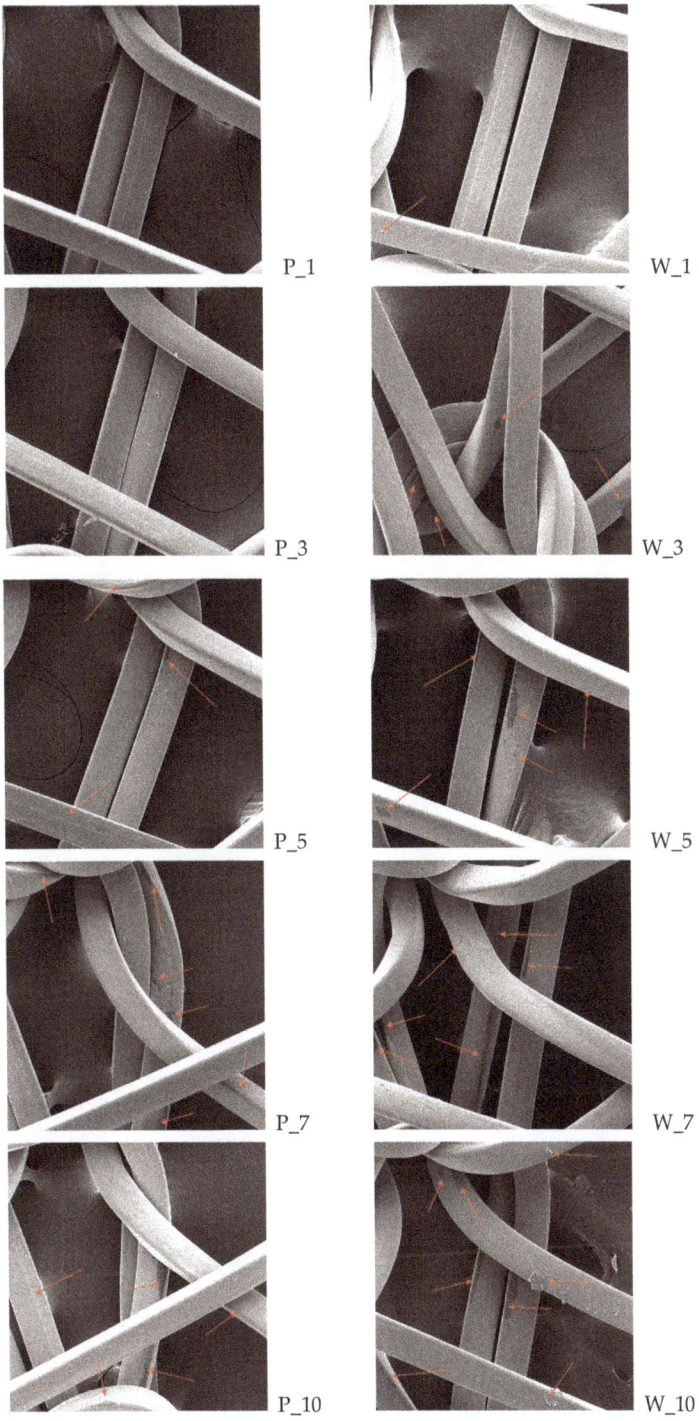

Figure 5. SEM micrographs of the PA/Ag samples before and after the 1st, 3rd, 5th, 7th and 10th treatment cycles of dry (P) and wet cleaning (W), under a magnification of 500×.

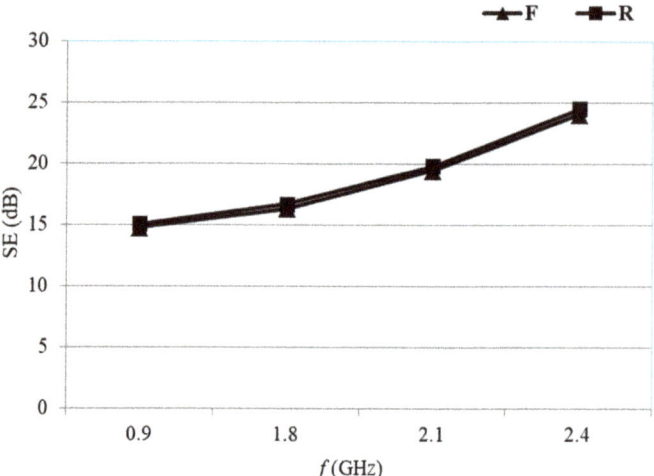

Figure 6. Initial SE of the face (F) and reverse (R) sides of the PA/Ag fabric at frequencies of 0.9 GHz, 1.8 GHz, 2.1 GHz, and 2.4 GHz.

The first cycle of the PA/Ag fabric treatment with the dry cleaning (P) and wet cleaning (W) reduced the degree of protection at 0.9 GHz. The wet cleaning (W) had a stronger influence compared to the dry cleaning (P), while the largest difference of SE properties was confirmed after 3 cycles. The almost linear and parallel decline of the SE properties continues after the 5th, 7th, and 10th cycle (Figure 7).

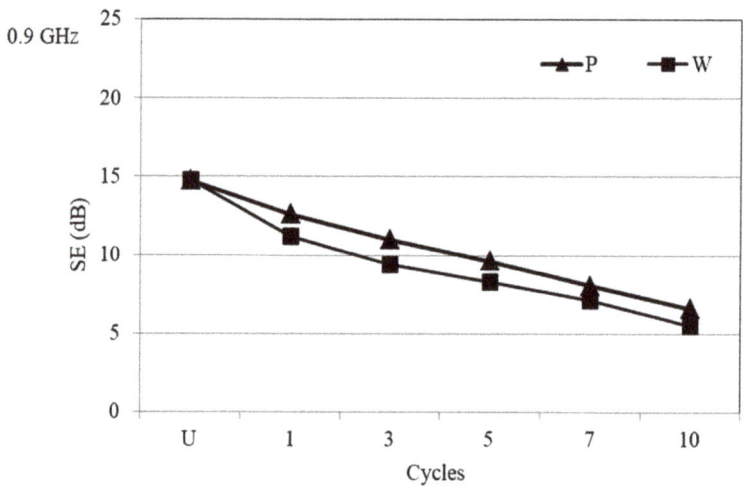

Figure 7. SE of the PA/Ag fabric before and after 10 cycles of treatment with P and W at the frequency of 0.9 GHz.

The first cycle of the PA/Ag fabric treatment with the dry cleaning (P) and wet cleaning (W) solvents, in synergy with process parameters, reduced the degree of protection at 1.8 GHz. The wet cleaning had a stronger influence compared to the dry cleaning; the largest difference of SE properties was found after the 7th cycle (Figure 8).

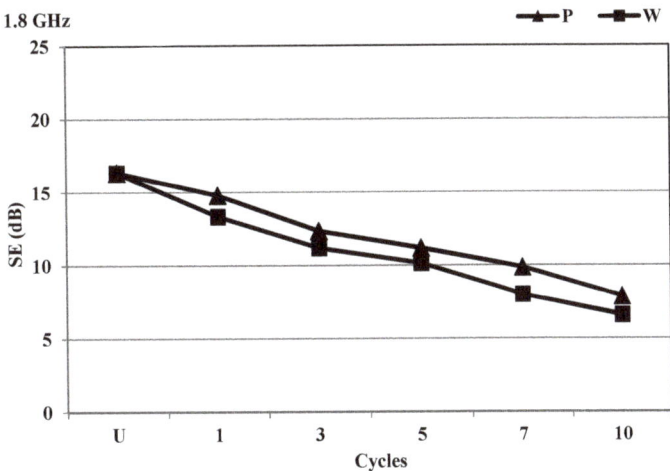

Figure 8. SE of the PA/Ag fabric before and after 10 cycles of treatment with P and W at the frequency of 1.8 GHz.

The first cycle of the PA/Ag fabric dry cleaning and wet cleaning reduced the degree of protection at 2.1 GHz. The wet cleaning had a stronger influence compared to the dry cleaning. The largest difference in SE properties was found after the 1st cycle, and the almost linear and parallel decline in the SE properties continued after the 3rd, 5th, 7th, and 10th cycle (Figure 9).

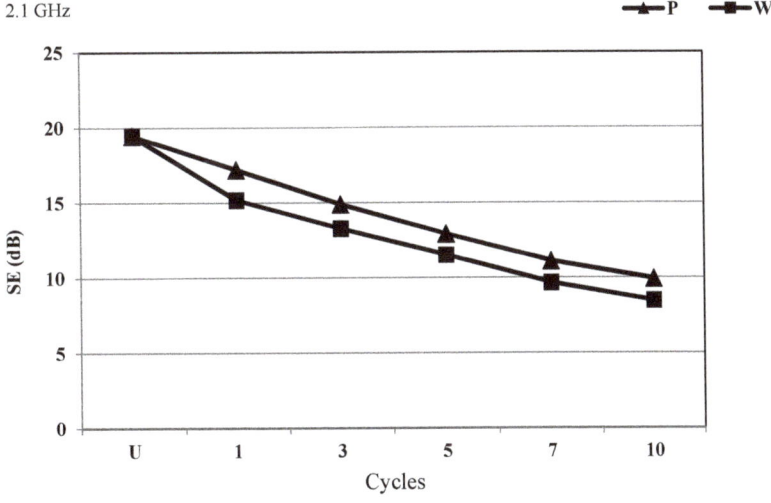

Figure 9. SE of the PA/Ag fabric before and after 10 cycles of treatment with P and W at the frequency of 2.1 GHz.

Textile materials characterized by a shielding effectiveness (SE) of >20 dB are acceptable for industrial applications [31], meaning that untreated PA/Ag fabric possesses an appropriate SE at a frequency of 2.4 GHz. Figure 10 indicates the better preservation of SE in dry cleaning than in wet cleaning. The numerical differences in the SE values of the PA/Ag fabric due to repeated processing cycles are shown in Table 3.

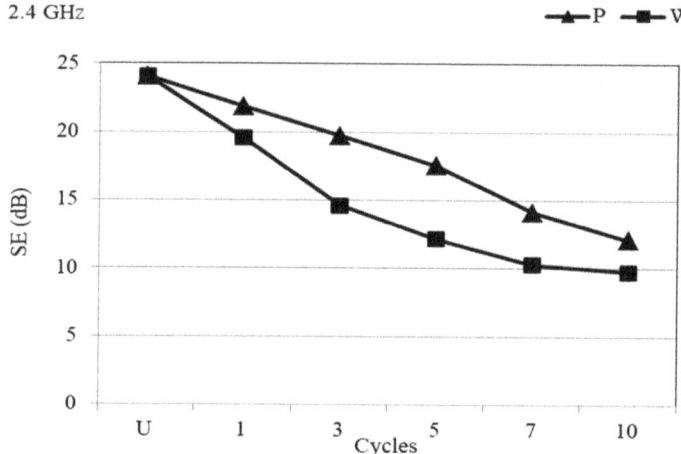

Figure 10. SE of the PA/Ag fabric before and after 10 cycles of treatment with P and W at the frequency of 2.4 GHz.

The SE differences of the treated PA/Ag fabric caused by the physicochemical influence of dry and wet cleaning in relation to the untreated fabrics are shown in Tables 3 and 4.

Table 3. Difference in the shielding effectiveness of the PA/Ag fabric after the impact of dry cleaning (P).

	Dry Cleaning Cycles				
f (GHz)	dSE (dB)				
	P_1	P_3	P_5	P_7	P_10
0.9	2.2	3.8	5.2	6.7	8.2
1.8	1.5	4.0	5.2	6.5	8.5
2.1	2.3	4.6	6.5	8.4	9.5
2.4	2.2	4.3	6.5	9.9	11.9

Table 4. Difference in the shielding effectiveness of the PA/Ag fabric after the impact of wet cleaning (W).

	Wet Cleaning Cycles				
f (GHz)	dSE (dB)				
	W_1	W_3	W_5	W_7	W_10
0.9	3.6	1.8	6.4	7.6	9.3
1.8	4.3	6.2	7.9	9.8	10.9
2.1	4.3	6.2	7.9	9.8	10.9
2.4	4.5	9.5	11.9	13.8	14.3

Based on the obtained efficiency differences (dSE) at all frequencies, a more progressive influence of the wet cleaning on the decrease of the SE value compared to the dry cleaning could be clearly noticed. The largest efficiency differences were found at 2.4 GHz. The impact of the process parameters in wet and dry cleaning on dSE correlated well with the surface observation of the SEM images.

The initial surface damage of PA/Ag fabrics is observed after the 5th dry cleaning. The obtained dSE values of the PA/Ag fabric through five cycles of dry cleaning at all frequencies are almost equal. The differences in dSE between the lower and higher frequencies are noticeable after the 7th and 10th cycles of dry cleaning.

The subsequent cycles of wet cleaning affect larger differences at all frequencies compared with the 1st cycle. The initial surface damage of the Ag layer on the PA fabric observed after the 3rd wet cleaning cycle caused a great decrease in the dSE value, especially at a frequency of 2.4 GHz.

It has been confirmed that the damage of the fabric surface layer in the wet cleaning process affects the decrease in SE values. The obtained results for the samples treated in the wet cleaning process were not in accordance with the results presented in [32], in which the Electromagnetic Shielding Effectiveness (EMSE) values in a low and medium frequency range (0.75 GHz to 3.0 GHz) were attributed to changes in the fabric structure after five cycles of the washing process.

4. Conclusions

Polyamide knitted fabric made of silver-coated thread possesses an optimal electromagnetic shielding effectiveness in the frequency range from 0.8 GHz to 2.4. GHz. The protective factor, minimal weight and thickness are promising characteristics for clothing, interior, and technical applications. The initial SE properties were changed under repeated cycles of dry and wet cleaning. The increased number of wet and dry cleaning cycles caused a linear drop of the SE values at 0.9 GHz, 1.8 GHz, 2.1 GHz, and 2.4 GHz. The SEM images indicated the damage of the silver coating on the polyamide yarn. The degradation was more noticeable after the wet cleaning than the dry cleaning.

Author Contributions: Research concept, T.P., B.Š. and K.M.; Figure 1, K.M.; measurement setup, K.M., methodology, B.Š. and K.M.; results design, T.P., B.Š. and K.M.; discussion, T.P., B.Š. and K.M.; preparation of paper for submission, T.P., B.Š. and K.M.; project leader: K.M. All authors have read and agreed to the published version of the manuscript.

Funding: The research was fully supported by the Croatian Science Foundation under the project HRZZ-IP-2018-01-7028 'Shielding from electromagnetic fields with electrically conductive textile materials (SEMECTEX)'.

Institutional Review Board Statement: Not applicable.

Informed Consent Statement: Not applicable.

Data Availability Statement: Data available in a publicly accessible repository.

Acknowledgments: The authors thank Zorana Kovačević, for the microscopic analysis preformed on the FE-SEM, MIRAIILMU, Tescan.

Conflicts of Interest: The authors declare no conflict of interest.

References

1. Ozen, M.S.; Sancak, E.; Beyit, A.; Usta, I.; Akalin, M. Investigation of electromagnetic shielding properties of needle-punched nonwoven fabrics with stainless steel and polyester fiber. *Text. Res. J.* **2013**, *83*, 849–858. [CrossRef]
2. Chen, H.; Lee, K.; Lin, J.; Koch, M. Comparison of Electromagnetic Shielding Effectiveness Properties of Diverse Conductive Textiles via Various Measurement Techniques. *J. Mater. Process. Technol.* **2007**, *192–193*, 549–554. [CrossRef]
3. Wang, X.; Liu, Z.; Jiao, M. Computation Model of Shielding Effectiveness of Symmetric Partial for Anti-Electromagnetic Radiation Garment. *Prog. Electromagn. Res. B* **2013**, *47*, 19–35. [CrossRef]
4. Kim, H.-R.; Fujimori, K.; Kim, B.-S.; Kim, I.-S. Lightweight Nanofibrous EMI Shielding Nanowebs Prepared by Electrospinning and Metallization. *Compos. Sci. Technol.* **2012**, *72*, 1233–1239. [CrossRef]
5. Prudnik, A.M.; Zamastotsky, Y.; Siarheyev, V.; Siuborov, V.; Stankevich, E.; Pobol, I. Electromagnetic Interference Shielding Properties of the Cu, Ti and Cr Coatings Deposited by Arc—PVD on Textile Materials. *Electr. Rev.* **2012**, *6*, 81–83.
6. Sonehara, M.; Noguchi, S.; Kurashina, T.; Sato, T.; Yamasawa, K.; Miura, Y. Development of an Electromagnetic Wave Shielding Textile by Electroless Ni-Based Alloy Plating. *IEEE Trans. Magn.* **2009**, *45*, 4173–4175. [CrossRef]
7. Maity, S.; Singha, K.; Debnath, P.; Singha, M. Textiles in Electromagnetic Radiation Protection. *J. Saf. Eng.* **2013**, *2*, 11–19.
8. Roh, J.-S.; Chi, Y.-S.; Kang, T.J.; Nam, S.-W. Electromagnetic shielding effectiveness of multifunctional metal composite Fabrics. *Text. Res. J.* **2008**, *78*, 825–835. [CrossRef]
9. Tunakova, V.; Tunak, M.; Tesinova, P.; Seidlova, M.; Prochazka, J. Fashion clothing with electromagnetic radiation protection: Aesthetic properties and performance. *Text. Res. J.* **2020**, *90*, 2504–2521. [CrossRef]

10. Perumalraj, R.; Dasaradan, B.S.; Anbarasu, R.; Arokiaraj, P.; Harish, S.L. Electromagnetic Shielding Effectiveness of Copper Core Woven fabrics. *J. Text. Inst.* **2009**, *100*, 512–524. [CrossRef]
11. Das, A.; Kothari, V.K.; Kothari, A.; Kumar, A. Effect of Various Parameters on Electromagnetic Shielding Effectivenes of Textile Fabrics. *Indian J. Fibers Text. Res.* **2009**, *34*, 144–148.
12. Usta, I.; Sancak, E.; Yüksek, M.; Beyit, A.; Atak, O.; Senyurek, V. Effect of Metal Filament Wire Containing Knitting Fabrics on Electromagnetic Shielding Effectiveness (EMSE). In *Book of Proceedings of the 5th International Textile, Clothing & Design Conference 2010— Magic World of Textile*; Dragčević, Z., Hursa Šajatović, A., Vujasinović, E., Eds.; University of Zagreb Faculty of Textile Technology: Zagreb, Croatia, 2010; pp. 271–276.
13. Cheng, K.B.; Ramakrishna, S.; Lee, K.C. Development of Conductive Knitted Fabric Reinforced Thermoplastic Composites for Electromagnetic Shielding Applications. *J. Thermoplast. Compos. Mater.* **2000**, *13*, 378–399. [CrossRef]
14. Cheng, K.B.; Ramakrishna, S.; Lee, K.C. Electromagnetic Shielding Effectiveness of Copper Glass Fiber Knitted Fabric Reinforced Polypropylene Composites—Part A. *Composites* **2000**, *31*, 1039–1045. [CrossRef]
15. Koprowska, J.; Pietranik, M.; Stawski, W. New Type of Textiles with Shielding Properties. *Fibres Text. East. Eur.* **2004**, *12*, 39–42.
16. Bergman, A.; Michielssen, E.; Taub, A. Comparison of Experimental and Modeled EMI Shielding Properties of Periodic PorousxGNP/PLA Composites. *Polymers* **2019**, *11*, 1233. [CrossRef] [PubMed]
17. Safarova, V.; Militky, J. Multifunctional Metal Composite Textile Shields Against Electromagnetic Radiation—Effect of Various Parameters on Electromagnetic Shielding Effectiveness. *Polym. Compos.* **2015**, *38*, 309–323. [CrossRef]
18. Mistik İlker, S.; Sancak, E.; Ovali, S.; Akalin, M. Investigation of Electromagnetic Shielding Properties of Boron, Carbon and Boron–Carbon Fibre Hybrid Woven Fabrics and their Polymer Composites. *J. Electromagn. Waves Appl.* **2017**, *31*, 1289–1303. [CrossRef]
19. Rajendrakumar, K.; Thilagavathi, G. Electromagnetic Shielding Effectiveness of Copper/PET Composite Yarn Fabrics. *Indian J. FibreText. Res.* **2012**, *37*, 133–137.
20. Ortlek, H.G.; Kilic, G. Electromagnetic Shielding Characteristics of Different Fabrics Knitted from Yarns Containing Stainless Steel Wire. *Ind. Text.* **2011**, *62*, 304–308.
21. Özkan, İ. Investigation on the electromagnetic shielding performance of copper plate and copper composite fabrics: A comparative study. *Tekst. Ve Konfeksiyon* **2020**, *30*, 156–162. [CrossRef]
22. Neruda, M.; Vojtech, L. Electromagnetic Shielding Effectiveness of Woven Fabrics with High Electrical Conductivity: Complete Derivation and Verification of Analytical Model. *Materials* **2018**, *11*, 1657. [CrossRef] [PubMed]
23. Hauthal, H.G. Washing at Low Temperatures–Saving Energy, Ensuring Hygiene. *Tenside Surfactants Deterg.* **2012**, *49*, 171–177. [CrossRef]
24. Lavastir, G.P. Wet cleaning. *Detergo* **2007**, *5*, 28–37.
25. Šaravanja, B.; Pušić, T.; Malarić, K.; Hursa Šajatović, A. Durability of Shield Effectiveness of Copper-Coated Interlining Fabrics. *Tekstilec* **2021**, *64*, 25–31. [CrossRef]
26. EEE STD 299 Standard Method for Measuring the Effectiveness of Electromagnetic Shielding Enclosures. 2006. 299. Available online: https://standards.ieee.org/standard/299-2006.html (accessed on 5 February 2021).
27. MIL-STD-285, Military Standard: Attenuation Measurements for Enclosures, Electromagnetic Shielding. 1956. Available online: http://everyspec.com/MIL-STD/MIL-STD-0100-0299/MIL-STD-285_25102/ (accessed on 5 February 2021).
28. ASTM D-4935-89 Standard Test Method for Measuring the Electromagnetic Shielding Effectiveness of Planar Materials. 1999. Available online: https://www.astm.org/Standards/D4935.htm (accessed on 5 February 2021).
29. Čunko, R.; Andrasy, M. *Vlakna*; Zrinski d.d.: Čakovec, Croatia, 2005; ISBN 978-953-1-550895.
30. Schwarz, A.; Kazani, I.; Cuny, L.; Hertleer, C.; Ghekiere, F.; De Clercq, G.; De Mey, G.; Van Langenhove, L. Electro-conductive and elastic hybrid yarns—The effects of stretching, cyclic straining and washing on their electro-conductive properties. *Mater. Des.* **2011**, *32*, 4247–4256. [CrossRef]
31. Afilipipoaei, C.; Teodorescu-Draghicescu, H. A Review over Electromagnetic Shielding Effectiveness of Composite Material. *Proceedings* **2020**, *62*, 3023. [CrossRef]
32. Kayacan, O. The effect of washing processes on the electromagnetic shielding of knitted fabrics. *Tekst. Ve Konfeksiyon* **2014**, *24*, 356–362.

Moisture Vapor Permeability and Thermal Wear Comfort of Ecofriendly Fiber-Embedded Woven Fabrics for High-Performance Clothing

Hyun-Ah Kim

Korea Research Institute for Fashion Industry, 45-26, Palgong-ro, Dong-gu, Daegu 41028, Korea; ktufl@krifi.re.kr

Abstract: This study examined the moisture vapor permeability and thermal wear comfort of ecofriendly fiber-embedded woven fabrics in terms of the yarn structure and the constituent fiber characteristics according to two measuring methods. The moisture vapor permeability measured using the upright cup ($CaCl_2$) method (JIS L 1099A-1) was primarily dependent on the hygroscopicity of the ecofriendly constituent fibers in the yarns and partly influenced by the pore size in the fabric because of the yarn structure. On the other hand, the moisture vapor resistance measured using the sweating guarded hot plate method (ISO 11092) was governed mainly by the fabric pore size and partly by the hygroscopicity of the constituent ecofriendly fibers. The difference between the two measuring methods was attributed to the different mechanisms in the measuring method. The thermal conductivity as a measure of the thermal wear comfort of the composite yarn fabrics was governed primarily by the pore size in the fabric and partly by the thermal characteristics of the constituent fibers in the yarns. Lastly, considering market applications, the Coolmax®/Tencel sheath/core fabric appears useful for winter warm feeling clothing because of its the good breathability with low thermal conductivity. The bamboo and Coolmax®/bamboo fabrics are suitable for summer clothing with a cool feel because of their high thermal conductivity with good breathability. Overall, ecofriendly fibers (bamboo and Tencel) are of practical use for marketing environmentallyfriendly high-performance clothing.

Keywords: moisture vapor permeability; ecofriendly fiber; Tencel; bamboo; KES-F7; pore diameter

1. Introduction

The environmental impact of human beings has taken various forms, some familiar and others not generally recognized. The former includes energy consumption and pollution, together with global warming, melting icecaps, rising sea levels, and increasing frequency of adverse weather conditions [1] (p. 171). Albeit a minor contributor, the textile industry is exerting some impact, and the contribution of synthetic fibers such as PET and nylon must be taken into account. The term ecofriendly has been coined to define a process that is effective without harming the environment. Environmentally friendly fiber materials in textiles are divided into three areas: (organic) natural fibers, biodegradable synthetic fibers, and recycled fibers. An organic natural fiber implies organic cotton and bamboo, as well as Tencel as a regenerated cellulose fiber. Bamboo fibers made from bamboo pulp have a noncircular cross-section and impart good wear comfort with superior absorption and breathability while wearing clothing. It has 100% biodegradable characteristics (decaying after 3 or 4 months in the soil) and does not cause environmental pollution [2]. Tencel fibers developed by Lenzing AG in Austria originated from wood. The biobased Tencel (brand name of Lyocell) fibers are certified as compostable and biodegradable, and they can degrade after 3–4 months in the soil [2]. Biodegradable synthetic fibers include polylactic acid (PLA). PLA made from corn starch degrades after 3–5 years in the atmosphere and degrades after 2–4 months in landfill [2]. Recycled fibers are commercialized from recycled PET bottle and ECO CIRCLE® developed by Teijin Fibers Limited in Japan.

PET is manufactured from petrochemicals and will not decompose naturally. One manner in which this problem is tackled is through recycling [2]. The consumption of synthetic fibers such as PET, PP, and nylon is growing steadily every year with the appearance of highly functional new synthetic fibers. Two concerns in the textile industry are how the consumption of synthetic fibers can be reduced by substituting them with ecofriendly fibers with improved moisture transmission while wearing clothing, and how they can be achieved using various yarn manufacturing technology.

Moisture transmission through textile materials is divided into two areas: moisture liquid transmission and moisture vapor transmission. Moisture liquid transmission involves a two-stage process: initially wetting and then wicking. In contrast, moisture vapor transmission is governed by the diffusion of moisture vapor through the inter-yarn and inter-fiber air spaces of the fabrics, called breathability. The moisture vapor transmission behavior of fabric materials plays a vital role in maintaining the clothing wear comfort. In particular, breathable fabrics that maintain high breathability with good perspiration absorption and fast-drying properties are needed in sportswear, work wear, and various types of protective clothing. In addition, high breathability in clothing allows the human body to provide cooling due to perspiration and evaporation. Moreover, minimizing sweat build-up in clothing is also important in cold environments. On the other hand, when examining the mechanism of moisture vapor transmission behavior from a human body wearing clothing, the first behavior is the diffusion of moisture vapor by sweating through the air spaces in the fabric. The drying of moisture and the moisture vapor absorbed by perspiration from the human body coincides, which is considered the second type of moisture vapor transmission behavior but is slightly complicated.

Therefore, many studies [3–7] related to the moisture vapor permeability (MVP) of various fabrics have been carried out. They reported the breathable characteristics of fabrics according to the fiber materials and fabric structural parameters with various measuring methods of breathability. Rego et al. [3] examined the thermo-physiological wear comfort of cotton/polyester fabrics using a sweating guarded hot plate method. Gorjanc et al. [4] used a water cup measuring method to examine the effects of the fabric structural parameters on the thermal and moisture vapor resistance of cotton fabrics. Lee et al. [5] reported the effects of fiber materials and fabric structural parameters on the MVP using statistical modeling. Kim et al. [6] examined the relationship between the clothing performance of fabrics made from artificial and natural fibers and the dynamic moisture vapor transfer in a microclimate. Cubric et al. [7] explored the technical parameters affecting the moisture vapor resistance of knitted fabrics using a sweating guarded hot plate and a thermal manikin. Thus far, most studies have used traditional staple yarns and fabrics made from natural fibers and blended yarns using different methods to measure the MVP.

On the other hand, some studies [8–14] on the moisture vapor transmission of water proof breathable fabrics have been performed under various conditions, such as steady-state, rainy, windy, and rainy and windy. In particular, Ruckman and coworkers [11–14] examined the condensation phenomena of waterproof breathable fabrics. Yoo and coworkers [15,16] analyzed the condensation of the inner surface of the fabrics at ambient temperatures below 0 °C in cold weather. The textile materials used in previous studies were divided into two areas: one on traditional fabrics using natural fibers, such as wool, cotton, and their blended fibers, and the other on waterproof breathable fabrics made from synthetic filaments, such as nylon and polyester. Moreover, the method for measuring breathability in each study was different, making it difficult to compare the breathability of waterproof breathable fabrics.

In particular, many studies [17–29] have examined breathability characteristics according to the measuring method. Lomax [17] reported a difference in the MVP between ISO 11092 and BS 7209 methods using coated nylon and PET fabrics. Salz [18] pointed out that the moisture vapor transmission rates are often difficult to compare because of various test methods. He developed a laboratory method for measuring the MVP using a heated cup method combined with an artificial rain condition. Yoo et al. [19] explored simulated heat and moisture transport characteristics in fabrics and garments determined using a vertical plated sweating skin model. Several studies [20–29] compared the performance of moisture vapor transmission of waterproof breathable fabrics using different measuring methods. Gibson et al. [20–22] examined the parameters affecting the steady-state heat and moisture vapor transmission measurements for clothing using a hydrophobic and hydrophilic membrane laminated with two- and three-layer fabrics. They used the most common techniques, such as a sweating guarded hot plate and the cup-type method to measure moisture vapor resistance and moisture vapor transmission rate.

Dolhan [23] reported a correlation between the upright cup and Canadian control dish methods when comparing the measuring apparatus for the moisture vapor resistance. Congalton [24] examined the heat and moisture transport of clothing ensembles and reported a strong correlation between the Hohenstein measuring method (ISO 11092) and the evaporative dish method (BS 7209), which was in contrast to Lomax [17], who reported an inverse correlation between the two measuring methods. Gretton et al. [25] reported a linear correlation between moisture vapor resistance measured using the Gore-modified desiccant method and the MVP index measured by BS 7209. McCullogh et al. [26] examined the correlation among five measuring methods of breathability using 26 waterproof breathable fabrics. They reported that the upright cup method showed the lowest water vapor transmission rate, followed by the dynamic cell method, inverted cup method, and desiccant inverted cup method. In addition, the correlation coefficient between the sweating guarded hot plate and desiccant inverted cup methods showed a high inverse correlation.

Huang [27] and Huang and coworkers [28,29] examined the factors affecting the moisture vapor resistance obtained using the ISO 11092 method and compared them with the existing water vapor transmittance method. Many studies carried out thus far have reported that the breathability of the fabrics differed according to the fiber materials, fabric structural factors, and surface modification method, such as coating and laminating, as well as the measuring method. Measuring the breathability of fabrics can be achieved using two methods. The first is to measure the water vapor transmittance rate (WVTR), which is a simple method used for quality control and marketing purposes. The second is to measure the moisture vapor resistance using the wet thermal resistance method, which is more precise (accurate) and used mainly for fabric development and research. On the other hand, many studies [30–36] focusing on improving wear comfort by sweating focused on the yarn and fabric manufacturing technologies by combining hydrophobic and hydrophilic yarns. Several wear comforts of woven fabrics made from various yarns were examined using a different yarn structure and various constituent fiber characteristics, such as Coolmax®, Tencel, Bamboo, and other ecofriendly fibers [37–43].

Despite these studies, there are few reports on the MVP (breathability) of the woven fabrics made from composite yarns, such as siro, siro-fil, and sheath/core yarns, using Coolmax®, Tencel, bamboo, PET, and polypropylene (PP) filaments. This study used bamboo and Tencel fibers to produce ecofriendly yarns with PET and PP filaments using siro and sheath/core spinning systems. The PET and PP filaments in the sheath/core or siro-fil yarn structures play a very important role in passing water and moisture vapor as drainage in the yarns and fabrics. Accordingly, PET and PP filaments were used to enhance the wear comfort characteristics with superior water absorption and moisture vapor permeability, even though they are not ecofriendly fibers.

There are no reports on the difference in breathability characteristics according to the yarn structure and the measuring method of breathability. Therefore, the main concern of this study is how the breathability of different composite yarn fabrics is influenced by the yarn structures and constituent fiber characteristics, and how it is associated with the thermal and absorption properties of the fabrics regarding the pore size of the fabrics according to the measuring method. Accordingly, in this study, two types of breathability, i.e., WVTR and moisture vapor resistance, by the wet thermal transmission of 15 fabric specimens made from different composite yarns were measured and compared in terms of the yarn structure and constituent fiber characteristics according to the two measuring methods. In addition, the moisture vapor transmission characteristics were compared with the thermal conductivity of the fabric specimens measured using the KES-F7 in terms of the pore size of the fabric and the thermal characteristics of constituent fibers.

2. Materials and Methods

2.1. Yarn Preparation

Coolmax® (Dupont, Torrance, CA, USA) and Tencel (Gemeindeverwaltung, Lenzing, Austria) as ecofriendly fibers are used widely in the textile market. Various composite yarns have been made and commercialized using hi-multi PET and PP filaments in the functional and work wear market. In particular, a novel polypropylene (PP) filament with good absorption properties was applied to achieve clothing with thermo-physiological comfort [1–3]. On the other hand, few studies have investigated the moisture and heat transport of Tencel/Coolmax®-incorporated yarns and their fabrics according to the yarn structure with their measuring method. In this study, seven types of composite yarn were spun using ring, siro, and sheath/core spinning systems. Two existing hi-multi PET (75d/144f) and PP (100d/48f) filaments were used. Table 1 provides details of the yarn specimens. Sheath/core composite yarn specimens (no. (1), 147.5 dtex) were made by feeding PP DTY (draw-textured yarn) filament (30d/24f) between the double roving of Tencel. The siro-fil composite yarn (no. (2), 147.5 dtex) was prepared by replacing one of the siro components with a PET DTY filament (55d/216f) inserted at the back of the front rollers on the siro-spinning machine. The siro-spun yarn specimen (no. (3), 147.5 dtex) was made using Tencel roving on the siro-spinning frame. These were used as warp yarns. Sheath/core composite yarn specimen (no. (4), 196.7 dtex) was made by feeding Coolmax® filament (50d/36f) between the double roving of Tencel. A ring-spun yarn specimen (no. (5), 196.7 dtex) was prepared using bamboo and Coolmax® rovings on a ring spinning frame (Zinzer MAT 670, Krefeld, Germany). The siro-fil composite yarn specimen (no. (6), 196.7 dtex) was spun by replacing one of the siro components with a PET DTY filament (55d/216f) inserted at the back of the front rollers on the siro-spinning frame. The bamboo spun yarn specimen (no. (7), 196.7 dtex) was made using bamboo roving. In addition, the hi-multi PET filament (no. (8), 83.3 dtex) and PP filament (no. (9), 111.1 dtex) were prepared as existing commercial yarns. These were used as weft yarns. Each warp yarn specimen was prepared as a 147.5 dtex and a 196.7 dtex for each weft yarn specimen by regulating the draft ratio on the ring (siro) and sheath/core spinning frames. Table 1 lists their twist multiplier (TM), spindle rpm, and blend ratio.

Table 2 presents a schematic diagram of the yarn specimens in this study, which was drawn with reference to the images of yarn cross-sections obtained by SEM and optical microscopy. As shown in Table 2, the yarn structure of the sheath/core composite yarns (no. (1) and (4)) was composed of filaments in the core and staple fibers in the sheath, respectively. The siro-fil yarns (no. (2) and (6)) were drawn as a side-by-side cross-section, which was assumed to be formed as two parts caused by the centrifugal force by the traveler rotation on the ring frame. The siro-spun yarn (no. (3)) was twisted using two Tencel rovings on the siro-spinning frame and had a compact yarn structure with uniformly distributed Tencel fibers in the yarn cross-section. The Coolmax®/bamboo spun yarn (no. (5)) was twisted by a traveler on the ring-spinning system and had a relatively bulky yarn structure with noncircular cross-sections of Coolmax® and bamboo fibers. The bamboo spun yarn (no. (7)) had a slightly compact yarn structure with bamboo fibers distributed uniformly in the yarn cross-section. PET and PP filaments (no. (8) and (9)) were composed of parallel filament bundles and had yarn structures with many pores in the yarn cross-section.

Table 1. Details of the yarn specimens used in this study.

Yarn No.	Yarn Type	BlendRatio (%)	Spinning Method	Yarn No. (dtex)	Twist TM tpi/√Ne	Twist Spindle (rpm)	Fiber (Filament) Used	Period of Biological Decay in Soil (Month)
(1)	PP/Tencel S/C	PP: 39.3/ T: 60.7	Sheath/core	147.5	4.53	7000	PP DTY 30d/24f and Tencel S/F	PP: no decay T: 3-4
(2)	PET/Tencel Siro-fil	P: 44.4/ T: 55.6	Siro-fil	147.5	4.12	9000	PET DTY 55d/216f and Tencel S/F	P: no decay T: 3-4
(3)	Tencel Siro-spun	T: 100	Siro-spun	147.5	4.42	11,000	Tencel S/F	T: 3-4
(4)	Coolmax/TencelS/C	C: 39.3 T: 60.7	Sheath/core	196.7	4.34	9000	Coolmax 50d/36f and Tencel S/F	C: no decay T: 3-4
(5)	Coolmax/ BambooSpun	C: 48.6/ B: 51.4	Ring-spun	196.7	3.82	12,000	Coolmax/bamboo S/F	C: no decay B: 3-4
(6)	PET/Tencel Siro-fil	P: 44.4/ T:55.6	Siro-fil	196.7	4.12	9000	PET DTY 55d-216f and Tencel S/F	P: no decay T: 3-4
(7)	Bamboo spun	B: 100	Ring-spun	196.7	3.82	12,000	Bamboo S/F	B: 3-4
(8)	* Hi-multi PET	P: 100	-	83.3	-	-	PET DTY 75d/144f	no decay
(9)	* PP filament	PP: 100	-	111.1	-	-	PP DTY 100d/48f	no decay

Note: S/F: staple fiber, T: Tencel, *: existing filament. PP: polypropylene, C: Coolmax. P: PET, B: bamboo, DTY: draw textured yarn.

Table 2. Composite warp and weft yarn models [44].

Model			
Spec.	PP DTY 30d/24f + Tencel sheath/core yarn	PET DTY 55d/216f + Tencel siro-fil for yarn	Tencel + Tencel staple fibres siro-spun yarns
Yarn specimens	PP/Tencel S/C 147.5 dtex No (1)	PET/Tencel siro-fil 147.5 dtex No (2)	Tencel siro-spun 147.5 dtex No (3)
Model			
Spec.	Coolmax 50d/36f+ Tencel sheath/core yarn	Coolmax/bamboo staple fibers ring-spun yarns	PET DTY 55d/216f + Tencel siro-fil yarn
Yarn specimens	Coolmax/Tencel S/C 196.7 dtex No (4)	Coolmax/bamboo spun yarn 196.7 dtex No (5)	PET/Tencel siro-fil 196.7 dtex No (6)

Table 2. *Cont.*

Model	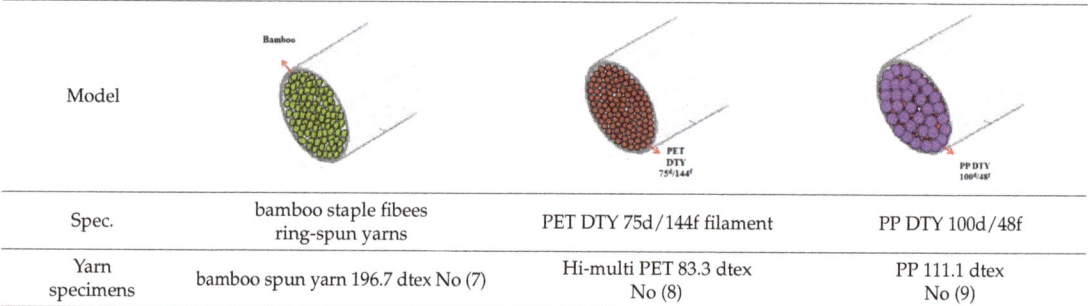		
Spec.	bamboo staple fibees ring-spun yarns	PET DTY 75d/144f filament	PP DTY 100d/48f
Yarn specimens	bamboo spun yarn 196.7 dtex No (7)	Hi-multi PET 83.3 dtex No (8)	PP 111.1 dtex No (9)

2.2. Fabric Preparation

Fifteen types of fabric specimens were woven on a rapier loom (GTX 4-R, Picanol, Belgium), which was divided into three groups according to the warp beams. Three types of warp beams were prepared on a single warping machine (ROM2, Karl Mayer, Germany) using yarn specimens (1), (2), and (3) (Table 1), with which three groups of fabric specimens were woven using five types of weft yarn specimens ((4) to (8) in Table 1). Yarn specimen (9) was alternatively inserted as second weft yarn. Table 3 lists these 15 fabric specimens. Group A was composed of five different fabric specimens ((1) to (5)) using five weft yarn specimens ((4) to (8) in Table 1), with the first warp beam made from PP/Tencel core/sheath yarn (no. (1) in Table 1). Group B was composed of five types of fabric specimens ((6) to (10)) using the same five weft yarn specimens ((4) to (8) in Table 1) with the second warp beam made from PET/Tencel siro-fil yarn (no. (2) in Table 1). Group C was composed of five fabric specimens ((11) to (15)) using the same five weft yarn specimens ((4) to (8) in Table 1) with the third warp beam made from Tencel siro-spun yarn (no. (3) in Table 1). PP DTY (111.1 dtex/48f) was inserted alternatively for all fabric specimens as a second weft yarn for a plain weave pattern.

The warp density of all fabric specimens was 36.0 ends/cm and 24.6 picks/cm for the weft. The fabric weight was calculated using yarn linear density and fabric density of the fabric specimen. The fabric thickness was measured at a pressure of 2 gf/cm^2 using a FAST-1 compression meter [42]. Fifteen types of gray fabric specimens, 20 m long each, were prepared and followed by dyeing and finishing processes. Gray fabric specimens were scoured on a CPB scouring machine (BPB, Kuester, Germany) and washed at a speed of 50 m/min on a continuous drying machine (Extra-CTA 2400, Benninger, Switzerland). A preset was done at 40 m/min at 150 °C. Dyeing was performed on a rapid dyeing machine (Cut-MF-1, Hisaka work Ltd., Osaka, Japan) at 120 °C for 60 min, followed by a drying treatment on a continuous dryer machine (Shrink dryer, Ilsung Ltd. Co., Seoul, Korea).The final setting was performed on a stenter machine (Sun-super, Ilsung Ltd. Co., Seoul, Korea) at a speed of 50 m/min at 130 °C.

Table 3. Specification of the fabric specimens.

Group	Fabric Specimen No.	Warp Yarn	Weft Yarn		Fabric Density (Ends, Picks/cm)		Weight (g/y)	Thickness (10^{-3} m)
			Yarn 1	Yarn 2	Wp	Wf		
A	1	PP/Tencel Sheath/core (147.5 dtex)	Coolmax/Tencel S/C (196.7 dtex)	PP (111.1 dtex)	36.0	24.6	162	0.368
	2		Coolmax/bamboo spun (196.7 dtex)				162	0.345
	3		PET/Tencel Siro-fil (196.7 dtex)				162	0.341
	4		Bamboo spun (196.7 dtex)				162	0.364
	5		Hi-multi PET (83.3 dtex)				137	0.352
B	6	PET/Tencel Siro-fil (147.5 dtex)	Coolmax/Tencel S/C (196.7 dtex)	PP (111.1 dtex)	36.0	24.6	160	0.396
	7		Coolmax/bamboo spun (196.7 dtex)				161	0.337
	8		PET Tencel Siro-fil (196.7 dtex)				161	0.345
	9		Bamboo spun (196.7 dtex)				161	0.345
	10		Hi-multi PET (83.3 dtex)				137	0.294
C	11	Tencel Siro-spun (147.5 dtex)	Coolmax/Tencel S/C (196.7 dtex)	PP (111.1 dtex)	36.0	24.6	158	0.380
	12		Coolmax/bamboo spun (196.7 dtex)				160	0.356
	13		PET Tencel Siro-fil (196.7 dtex)				160	0.345
	14		Bamboo spun (196.7 dtex)				161	0.349
	15		Hi-multi PET (83.3 dtex)				133	0.301

2.3. Measurement of Pore Size of the Fabric Specimens

The moisture and heat transport of woven fabrics were strongly dependent on the constituent fiber characteristics, pore size, and fabric structural parameters [45–49]. Kim and Kim [50] reported that the thermal conductivity, drying rate, and air permeability of hollow filament-embedded woven fabrics were strongly dependent on the porosity and pore size of the fabrics. In this study, the primary concern of pore size measurements was how the pore size is influenced by the constituent yarn structure and how it affects the moisture vapor transport of fabrics according to the measuring method. The pore diameter (D, μm) was measured using a capillary flow porometer (CFP-1200 AE PMI Co., Ithaca, NY, USA) according to the ASTM measuring method. Figure 1 presents the porometer used in this study.

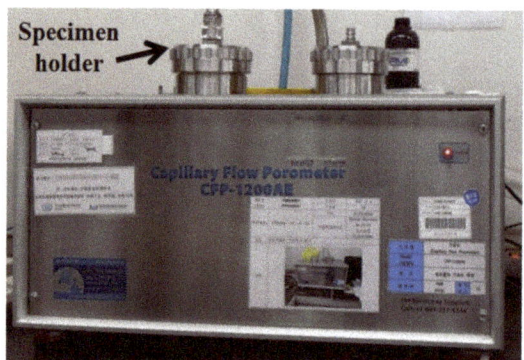

Figure 1. Capillary flow porometer machine.

A fabric specimen 47 mm in diameter was placed in the specimen holder shown in Figure 1. The specimen holder was closed, and slight gas pressure was applied to eliminate the possible liquid backflow. The gas pressure was increased slowly. Finally, the lowest pressure at which a steady stream of bubbles rises from the central area of the liquid reservoir was recorded. The maximum pore diameter (D) was calculated using Equation (1) from the median value of the graph between the airflow and pressure.

$$D = C\, Y/p, \tag{1}$$

where D, Y, and p are the maximum pore diameter (μm), the surface tension of the liquor (dynes/cm), and pressure (psi); C = 0.415 when p is in psi units. The yarn and fabric cross-sections were measured by field-emission scanning electron microscopy (FE-SEM, S-4100, Hitachi Co., Omori, Japan) and optical microscopy (i-Camscope-305A, Seoul, Korea).

2.4. Measurement of the WVTR of the Fabric Specimens

The resistance to moisture vapor diffusion (i.e., moisture vapor resistance) depends mainly on the air permeability of the fabric and indicates its ability to transfer perspiration vapor leaving of human skin. In this study, the main concerns of moisture vapor transmission measurements are how the moisture vapor resistance is associated with the absorption rate of the fabrics according to the constituent yarn structure (porosity) in the fabric, and how the thermal conductivity of the fabric is related to the moisture vapor resistance and is dependent on the yarn structure and thermal conductivity of the constituent fibers. The WVTR (g/m^2·h) was measured using the JIS L 1099A-1, which was based on BS7209, similar to the upright cup method, as shown in Figure 2a. Five aluminum cups, 6 cm in diameter and 2.5 cm in height, were prepared, and the cup inside was heated to 40 °C by heating in an air-conditioned room (container), then filled with 33 g of CaCl$_2$ as a desiccating agent. Five fabric specimens, 7 cm in diameter, were prepared and conditioned at 20 ± 2 °C and an RH of 65 ± 2% for 24 h. The fabric specimen was laid 3 mm away from CaCl$_2$ in an aluminum cup; its surface was faced toward the CaCl$_2$ in the cup. Packing rubber with a circular covering was clamped and sealed over the fabric specimen on the mouth of the aluminum cup to prevent the leakage of moisture vapor between the fabric specimens and the aluminum cup. Five aluminum cup assemblies with fabric specimens were placed in the conditioning room at 40 ± 2 °C and 90 ± 5% RH for 1 h. The water vapor transmission rate was calculated using Equation (2).

$$\text{WVTR (g/m}^2\cdot\text{h)} = 10(W_2 - W_1)/(A \times t), \tag{2}$$

where, WVTR is water vapor transmission rate (g/m²·h), W_2 is the mass of the fabric specimen (mg) after the test, W_1 is the mass of the fabric specimen (mg) before the test, A is the specimen area (cm²), and t is the testing time (h).

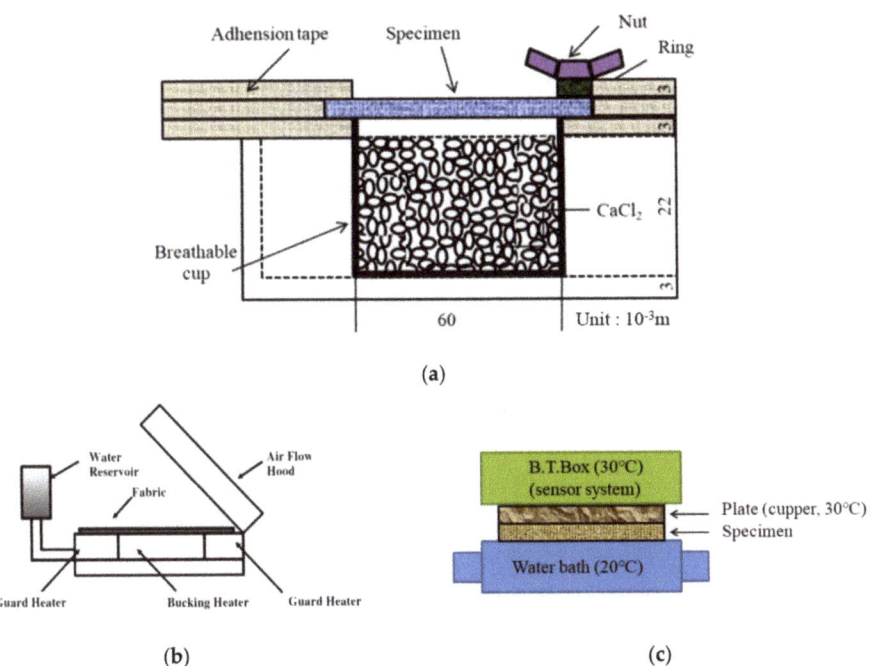

Figure 2. Schematic diagram of the measuring equipment of (**a**) upright cup method (JIS L 1099A-1), (**b**) sweating guarded hot plate method (ISO 11092), and (**c**) KES-F7 system [44].

2.5. Measurement of Moisture Vapor Resistance of the Fabric Specimens

Moisture vapor resistance (R_{ef}, m²Pa/W) of the fabric was measured using a sweating guarded hot plate (Therm DAC, London, UK) according to the ISO 11092 method. Figure 2b presents a schematic diagram of this apparatus. A fabric specimen, 50.8 cm × 50.8 cm in size, was prepared and conditioned in a standard atmosphere with an RH of 65% and a temperature of 20 °C. The specimen was placed over the PTFE membrane on perforated metal on a hot plate, which was used to prevent water on the perforated metal of the hot plate from wetting the fabric specimen. The temperatures of the guarded hot plate and air in the chamber were kept at 35 ± 0.5 °C (i.e., the temperature of human skin) with an RH of 40% and an air speed of 1 m/s. The moisture vapor resistance (R_{ef}) of the fabric was determined by measuring the evaporative heat loss (H) under the steady-state condition, using Equations (3) and (4).

$$R_{e,t} = \frac{(p_s - p_a)A}{H}, \tag{3}$$

where $R_{e,t}$ is the total resistance to evaporative heat transfer provided by the fabric system and air layer (m²·Pa·W⁻¹), A is the area of the plate test section (m²), p_s is the water vapor pressure at the plate surface (Pa), p_a is the moisture vapor pressure in the air (Pa), and H is the power input (heat loss)(W).

$$R_{e,f} = R_{e,t} - R_{e,a}, \tag{4}$$

where $R_{e,f}$ is the resistance to evaporative heat transfer provided by the fabric (i.e., moisture vapor resistance of fabric), and $R_{e,a}$ is the resistance measured for the air layer and liquid barrier. The arithmetic mean of five readings from each fabric specimen was calculated.

2.6. Measurement of the Thermal Conductivity of the Fabric Specimens

Thermal transmission through textile materials is divided into two methods: wet and dry heat transmission. Moisture vapor transmission by wet heat transport is governed by diffusion and convection, whereas dry heat transport occurs through conduction, convection, and radiation from the human body to the atmosphere. Moisture vapor resistance measurements using the ISO 11092 method were assessed using the principle of wet heat transmission, which means the movement of wet heat particles evaporated by perspiration sweated from human skin. In this study, one concern of the thermal transport measurement was how the wet heat transmission related to the moisture vapor resistance is associated with the dry heat transmission, and how they are influenced by the yarn structure and measuring method. Accordingly, the thermal conductivity of the fabric specimens as a measure of dry heat transport was measured to determine how it is influenced by the constituent yarn structure and then how it affects the moisture vapor resistance of the fabric specimens. The thermal conductivity of the fabric specimens was measured using the KES-F7 system (Kato Tech. Co., Ltd., Kyoto, Japan), of which a schematic diagram is shown in Figure 2c. First, the B.T. Box temperature was set to 30 °C, and water was circulated at a constant temperature of 20 °C in a water bath. A fabric specimen was placed on the water bath. Heat flowed from a high temperature (B.T. Box 30 °C) to a low temperature (water bath, 20 °C) in the apparatus through a plate and specimen. The B.T. Box (composed of an electrical system equipped with temperature sensors) then measured the heat loss emanating from the plate as watts (W) from the change in electrical voltage. The heat loss (W/10^{-4}m^2) with the fabric specimen placed on the water bath is H in Equation (5). The thermal conductivity (K) was calculated using Equation (5) as follows:

$$K = \frac{H}{t} \times \frac{D}{A \cdot T}, \quad (5)$$

where, K, H, and D are the thermal conductivity (W/10^{-2} m·°C), dry heat loss (W/10^{-4} m^2), and fabric thickness (10^{-2} m), respectively. A and t are the fabric area (10^{-4} m^2) and time (h), respectively. ΔT is the temperature difference (°C).

2.7. Measurement of the Absorption Rate of the Fabric Specimens

The moisture vapor particles sweated from the human body move throughout the fabric and are partly adsorbed and wetted, after which drying will occur. The absorption rate of the fabric specimens was measured using a drying rate measuring apparatus (IT-ACD, INTEC Co. Ltd., Tokyo, Japan), as shown in Figure 3. A square fabric specimen (40 cm × 40 cm) was conditioned at 20 °C and 65% RH in the conditioning room (JIS L 1096, 2010), and the initial mass (m_1) was then measured. A square fabric specimen was submerged in distilled water for 20 min at 27 ± 2 °C in a water bath. The specimen in the distilled water bath was passed through a mangle at 25 cm/s, and the mass (m_2) was weighed. The absorption rate (%) of each fabric specimen was calculated using Equation (6).

$$R\ (\%) = (m_2 - m_1)/m_1, \quad (6)$$

where R is the absorption rate of fabric (%), m_1 is the initial mass (g) of the fabric specimens, and m_2 is fabric mass (g) after passing through the mangle.

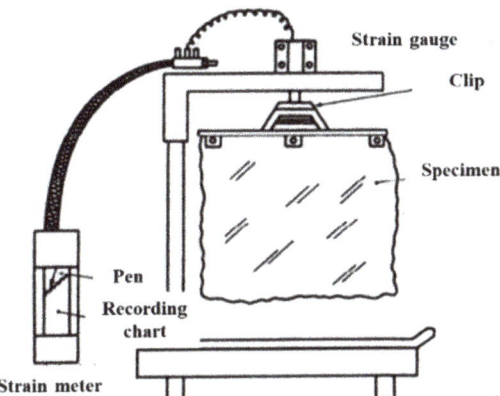

Figure 3. Schematic diagram of measuring equipment of the absorption rate.

3. Results and Discussion

3.1. Pore Size of the Fabric Specimens with SEM Images of the Cross-Sections of Yarns

The fabric porosity affecting clothing wear comfort is divided into two types: micro and macro porosity [51,52]. In particular, the moisture vapor and heat permeabilities are incorporated with both micro and macro porosities. In this study, the calculated porosity does not apply because the fabric specimens prepared in this study were made from the same yarn count and fabric sett, which means that the calculated porosity may not be available to examine the difference in the MVP among the fabric specimens prepared using the same yarn count and fabric sett. Therefore, in this study, the pore diameter was considered a measure to examine the WVP according to the yarn structure and measuring method. Table 3 lists the measured physical properties of the fabric specimens with the measured pore diameters. ANOVA (F-test) was carried out to verify the statistical significance of the experimental data shown in Table 4. ANOVA was performed between the mean value of the physical properties of each specimen (five specimens) in each group with the 95% confidence limit (5% significance level). Table 5 lists the ANOVA analysis of the physical properties of 15 fabric specimens. As shown in Table 5, the significance test between each mean pore diameter among the five specimens in each group A, B, and C was statistically significant, as F_0 (V/Ve) > F (4, 20, 0.95) and $p < 0.05$. Similarly, WVTR, R_{ef}, and thermal conductivity were statistically significant, as F_0 (V/Ve) > F (4, 20, 0.95) and $p < 0.05$, as shown in Table 5. Figure 4 presents a diagram of the pore diameters (mean) with the deviation of the fabric specimens listed in Table 4. The deviation in Table 4 denotes the difference between maximum and minimum values of five experimental data of each specimen.

Table 4. Physical properties of the fabric specimens.

Group	Fabric Specimen No.	Pore Diameter D (μm)		Moisture Vapor Permeability				Thermal Conductivity		Absorption Rate R (%)
				Water Vapor Transmission Rate WVTR (g/m²·h)		Moisture Vapor Resistance R_{ef} (m²·Pa/W)		K (m²·Pa/W)		
		Mean	Dev.	Mean	Dev.	Mean	Dev.	Mean	Dev.(10^{-3})	Mean
A	1	3.82	0.100	425.4	11.1	1.91	0.103	0.0445	1.66	28.3
	2	3.19	0.088	413.2	10.0	2.80	0.109	0.0458	1.67	27.2
	3	2.85	0.112	410.6	9.2	3.42	0.118	0.0489	1.64	26.4
	4	2.98	0.110	430.2	11.1	2.91	0.101	0.0473	1.49	30.2
	5	3.56	0.124	385.5	10.0	1.74	0.107	0.0423	1.74	26.3
B	6	3.20	0.104	348.2	10.1	1.56	0.101	0.0439	1.57	30.2
	7	2.85	0.115	340.5	10.0	2.85	0.109	0.0437	1.67	28.5
	8	2.30	0.096	337.6	7.1	3.24	0.107	0.0481	1.58	28.1
	9	2.70	0.121	375.4	9.0	2.88	0.101	0.0453	1.67	32.4
	10	3.11	0.116	326.2	9.1	1.34	0.109	0.0395	1.64	26.2
C	11	3.45	0.100	369.8	11.0	2.12	0.118	0.0453	1.80	34.4
	12	2.95	0.116	358.6	11.1	2.43	0.103	0.0461	1.81	34.2
	13	2.55	0.106	355.4	10.3	3.32	0.101	0.0478	1.80	34.0
	14	2.76	0.094	378.5	9.1	2.86	0.118	0.0472	1.54	36.7
	15	3.25	0.111	337.7	11.2	1.78	0.101	0.0421	1.45	32.1

Note: dev = max − min.

Table 5. ANOVA analysis of the fabric physical properties.

Physical Properties		F-Value(F_0)	F(4, 20, 0.95)	p-Value
Pore diameter	Group A	318.0	2.87	8.57×10^{-18} ($p < 0.05$)
	Group B	221.9	2.87	2.91×10^{-16} ($p < 0.05$)
	Group C	279.3	2.87	3.06×10^{-17} ($p < 0.05$)
WVTR	Group A	83.7	2.87	3.36×10^{-12} ($p < 0.05$)
	Group B	129.7	2.87	5.29×10^{-14} ($p < 0.05$)
	Group C	67.4	2.87	2.54×10^{-11} ($p < 0.05$)
Ref	Group A	1119.5	2.87	3.31×10^{-23} ($p < 0.05$)
	Group B	1971.7	2.87	1.18×10^{-25} ($p < 0.05$)
	Group C	872.7	2.87	3.94×10^{-22} ($p < 0.05$)
K	Group A	64.4	2.87	3.90×10^{-11} ($p < 0.05$)
	Group B	124.8	2.87	7.66×10^{-14} ($p < 0.05$)
	Group C	54.7	2.87	1.72×10^{-10} ($p < 0.05$)

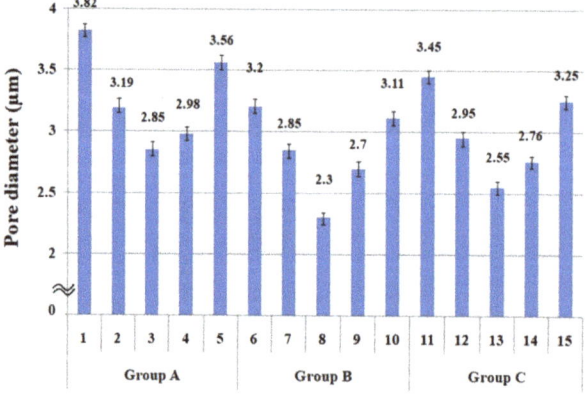

Figure 4. Diagram of pore diameters of the fabric specimens.

As shown in Figure 4, of the three fabric specimen groups (A, B, and C), the pore diameters of the fabric specimens in group A ((1) to (5)) showed higher values than those in groups B ((6) to (10)) and C ((11) to (15)). This suggests that the pore size of the PP/Tencel sheath/core yarns used as a warp yarn of group A was larger than that of the PET/Tencel siro-fil and Tencel siro-spun yarns used as the warp yarns of groups B and C. Hence, the sheath/core yarn has larger pores and voids. By contrast, the siro-fil and siro-spun yarns have compact yarn structures, resulting in relatively small pore diameters in the yarns, even though the fabric specimens were made from the same yarn count and fabric density. These were verified by SEM and optical microscopy of the constituent warp yarns used in the fabric specimens. Table 6 presents SEM and optical microscopy images of the yarn specimens shown in Table 1. As shown in yarn specimen (1), which was used as a warp yarn of fabric specimens (1) to (5) (group A), many air voids were observed in the sheath and core. Round-shaped capillary channels at the border between the Tencel fibers and the filaments in the core of the PP/Tencel sheath/core yarn were found, resulting in a higher pore diameter of fabric group A ((1) to (5)) than fabric groups B ((6) to (10)) and C ((11) to (15)). On the other hand, as shown in yarn specimen (2) in Table 6, a compact yarn cross-section was observed in the PET/Tencel siro-fil yarn, resulting in a smaller pore diameter of the fabric specimen group B ((6) to (10)). In yarn specimen (3) in Table 6, small air voids were observed in the Tencel siro-spun yarn, resulting in relatively small pore diameters of the fabric specimen group C ((11) to (15)). Regarding the pore size of the fabric specimens according to the yarn structure in the weft direction, of the five types of fabric specimens ((1) to (5), (6) to (10), and (11) to (15)), as shown in Figure 4, the Coolmax®/Tencel sheath/core fabric (specimen (1)) and Coolmax®/bamboo spun fabric (specimen (2)) exhibited larger pore diameters than the PET/Tencel siro-fil fabric (specimen (3)). By contrast, the pore diameter of the siro-fil fabric (specimen (3)) was smaller than that of the bamboo spun fabric (specimen (4)) and hi-multi PET filament fabric (specimen (5)). These results suggest that sheath/core and spun yarn fabrics have larger pores in the fabrics, whereas siro-fil has a compact yarn structure, which results in small pores in the fabric. Relatively large air voids in the Coolmax®/bamboo spun yarns observed in yarn specimen ((5)) in Table 6 were noted in the weft direction in fabric specimen (2), resulting in relatively large pore diameters for fabric specimens (2), (7), and (12), as shown in Table 4 and Figure 4. On the other hand, a compact yarn cross-section was observed in yarn specimen (6) in Table 6, resulting in small pore diameters for fabric specimens (3), (8), and (13), as shown in Table 4 and Figure 4. Similar to yarn specimen (5), yarn specimen (7) showed a relatively compact yarn cross-section, as shown in Table 6. The pore sizes in yarn specimen (7) were smaller than those of yarn specimen (5), resulting in smaller pore diameters of fabric specimens (4), (9), and (14) than those of fabric specimens (2), (7), and (12), as shown in Table 4 and Figure 4. Regarding the hi-multi PET filament of the yarn specimen (8), many air voids were observed in the SEM and optical microscopy images of the yarn cross-section. In addition, non-twisted parallel filament bundles were observed in the SEM image of the yarn surface shown in Table 6, which produced fine capillary channels along the filament bundles with many pores in the yarn cross-section, as shown in the SEM and optical microscopy images of yarn specimen (8) in Table 6, resulting in a high pore diameter of fabric specimens (5), (10), and (15) compared to siro-fil and spun yarn fabrics made from yarn specimens (5), (6), and (7) in Table 6. According to previous studies [45,46], the MVP is incorporated with micro and macro porosities, and fine voids cause microporosity among the fibers in the yarns. By contrast, macro porosity is produced from the void spaces among the threads in the fabric. The pore diameters measured in this study were assessed as a measure of the fabric porosity considering both micro voids among the fibers in the yarns and macro voids among the threads in the fabrics. Therefore, the MVP was examined in relation to the pore diameter measured from SEM images of the air void and capillary channels formed according to the different yarn structures. In addition, the MVP of the fabrics made from the different yarn structures was compared and discussed with its two methods (WVTR and R_{ef}) for measuring the breathability, as shown in the next section.

Table 6. SEM images of cross-sections (×500) and surfaces (×150) of the yarns and optical microscopy (×300) [53].

Yarn Specimen No	Yarns	SEM (Cross-Section)	Optical Microscopy (Cross-Section)	SEM (Surface)
(1)	PP/Tencel Sheath/core			
(2)	PET/Tencel Siro-fil			
(3)	Tencel Siro-spun			
(4)	Coolmax/Tencel Sheath/core			
(5)	Coolmax/Bamboo Spunyarn			
(6)	PET/Tencel Siro-fil			

Table 6. Cont.

Yarn Specimen No	Yarns	SEM (Cross-Section)	Optical Microscopy (Cross-Section)	SEM (Surface)
(7)	Bamboo Spunyarn			
(8)	Hi-multi PET 75d/144f			
(9)	PP DTY 100d/48f			

3.2. WVTR of the Fabric Specimens Using Upright Cup Method

Figure 5 presents the WVTR of the 15 fabric specimens. The mean value of the five specimens for the WVTR in groups A, B, and C was statistically significant, as shown in Table 5. The mean values of the 15 specimens were plotted with the maximum and minimum values of the five experimental data of each specimen, respectively. A comparison of the WVTR according to the warp yarn structure (i.e., groups A, B, and C) revealed the WVTR of group A to be higher than that of groups B and C because of the larger pore diameter by the warp yarn (PP/Tencel sheath/core) of group A than groups B (PET/Tencel siro-fil) and C (Tencel siro-spun), as shown in Figure 4. This was verified by SEM images (Table 6), i.e., larger pores with a capillary channel between PP filament in core and Tencel fibers in the sheath were observed in the yarn specimen (1) (a warp yarn of group A fabric specimens), which resulted in a higher WVTR of group A fabrics. The WVTR of group C was slightly higher than group B because of the larger pore diameter by the warp yarn (Tencel siro-spun) of group C than group B (PET/Tencel siro-fil), as shown in Figure 4. These results were consistent with Fohr et al. [54], who reported that the WVTR was strongly dependent on the porosity and diffusion characteristics of the moisture vapor particles, which is in agreement with the current findings.

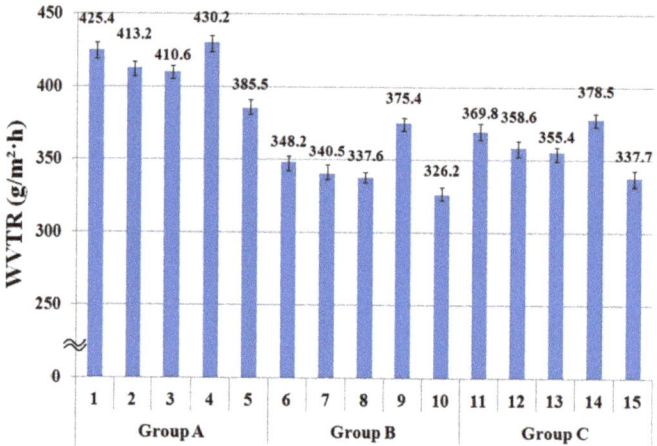

Figure 5. WVTR of the specimens using the upright cup method.

The WVTR of the fabric specimens was examined according to the yarn structure in the weft direction. As shown in Figure 5, of five fabric specimens ((1) to (5)) in group A, the WVTR of fabric specimen (1) was higher than that of specimens (2) and (3). That of fabric specimen (2) was slightly higher than that of specimen (3), because of the larger pore diameter by the weft yarn (Coolmax/Tencel sheath/core yarn) of fabric specimen (1) than fabric specimen (2) (Coolmax/bamboo spun yarn) and 3 (PET/Tencel siro-fil yarn), and of fabric specimen (2) than (3). Similar results were shown in groups B ((6) to (10)) and C ((11) to (15)). On the other hand, the WVTR of fabric specimen (4) (as well as (9) and (14)) inserted with bamboo spun yarn was the highest compared to the other fabric specimens, which was attributed to the high absorption rate and the diffusivity of bamboo spun yarn fabric with the appropriate pores in the yarns. As shown in Table 3, of the 15 fabric specimens, specimens (4), (9), and (14) composed of bamboo spun yarns in the weft direction exhibited the highest absorption rate, which was partly responsible for the highest WVTR of the fabrics. These results are in accordance with previous studies [55,56] reporting that the water vapor transmission of hygroscopic fibrous materials was higher than that of the materials that do not absorb moisture. This reduces the moisture built up in the microclimate, enhancing moisture vapor transmission from human skin to the environment. According to Li et al. [57], an increase in the WVTR by the absorption of moisture vapor is mainly because the heat of sorption increases the temperature of the fibrous assemblies, which in turn affects the moisture vapor transmission rate. The effect of water vapor absorption on the WVTR can be explained in fabric specimens (5), (10), and (15). As shown in Figure 5, the WVTR of fabric specimens (5), (10), and (15) showed the lowest value compared to the other specimens. This is because the PET 75d/144f filaments inserted in fabric specimens (5), (10), and (15) are hydrophobic and do not absorb moisture, resulting in a low WVTR. Summarizing the WVTR measured by the upright cup method according to the yarn structures in the warp and weft directions, the WVTR of the fabric specimens divided into groups A, B, and C was dependent on the pore diameter of the fabric, i.e., the WVTR of group A fabrics (specimens (1) to (5)) was higher than that of group B (specimens (6) to (10)) and C (specimens (11) to (15)) fabrics, which was attributed to the larger pore diameter of the group A fabrics. On the other hand, a study of the WVTR of fabric specimens according to yarn structure in the weft direction revealed the WVTR to be primarily dependent on the absorption rate of the constituent fibers in the fabric and partly on the pore size of the fabric. The fabric specimens composed of bamboo fibers with hygroscopic characteristics in the weft direction (i.e., high absorption rate) exhibited the highest WVTR. In contrast, the fabric specimens composed of hydrophobic PET filaments showed the lowest WVTR. Hence, in this study, the WVTR of the fabric measured using the upright cup method according to the weft yarn structure and fiber characteristics was strongly dependent on the hygroscopicity of the constituent fibers. On the other hand, the WVTR of the fabric according to the warp yarn structure was governed by the pore diameter of the fabric, i.e., dependent on the warp yarn structure.

3.3. Moisture Vapor Resistance of the Fabric Specimens by ISO 11092 Method

The ISO 11092 method uses a sweating guarded hot plate apparatus to simulate moisture transport through the textile when worn next to the human skin. This model measures the moisture vapor resistance of the fabric by measuring the evaporative heat loss in the steady state. Its measuring mechanism is different from the upright cup method. Figure 6 shows the moisture vapor resistance (R_{ef}) of the fabric specimens. The mean value of the moisture vapor resistance of the 15 fabric specimens was statistically significant, as shown in Table 5.

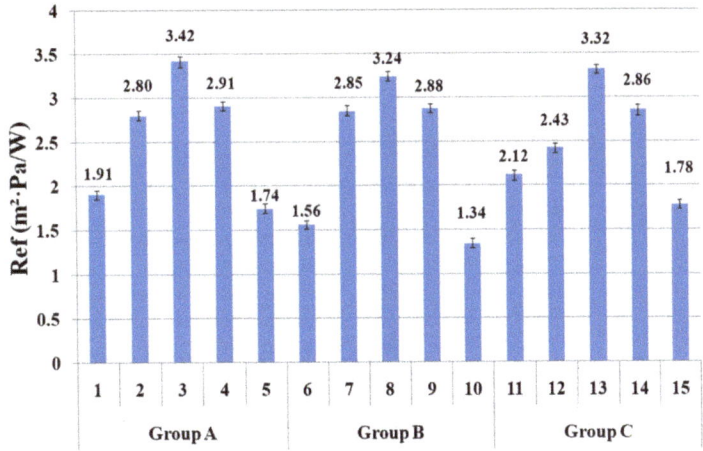

Figure 6. R_{ef} of the specimens using the ISO 11092 method.

The moisture vapor resistance of the 15 fabric specimens according to the weft yarn structure showed a distinctive result, and a similar trend among fabric groups A, B, and C was observed, i.e., proportional to the pore diameters. As shown in Figure 6, fabric specimens (1) and (5) in group A, (6) and (10) in group B, and (11) and (15) in group C exhibited low moisture vapor resistance, i.e., superior breathability to other fabric specimens. These results were attributed to the larger pore sizes (specimens (1), (5), (6), (10), (11), and (15) in Figure 4) in the fabrics depending on the weft yarn structure. Hence, the superior moisture vapor transmittance of fabric specimens (1), (6), and (11) was due to the following: high pore diameters with the capillary channels (yarn specimen (4) in Table 6) between the Coolmax® noncircular filaments in the core and Tencel fibers in the sheath of the yarns, and the fine capillary channels (yarn specimen (8) in Table 6) between the hi-multi PET filaments of fabric specimens (5), (10), and (15), which enable more moisture vapor to be transmitted from the fabric toward the outside. On the other hand, fabric specimens (3), (8), and (13) showed the highest moisture vapor resistance, i.e., inferior breathability to that of the other fabric specimens. This was attributed to the low pore diameters (specimens (3), (8), and (13) in Figure 4) in the fabrics because of the compact yarn structure of the PET/Tencel siro-fil yarns (yarn specimen (6) in Table 6), which prevents moisture vapor from being pushed by the moisture vapor pressure. In addition, the moisture vapor resistance of the fabric specimens composed of spun yarns in the weft direction ((2) and (4) in group A, (7) and (9) in group B, and (12) and (14) in group C) was higher, i.e., showed inferior breathability to that of the fabric specimens with the sheath/core and hi-multi PET filament. This was attributed to the smaller pore diameters (Figure 4) of the spun yarn fabrics than the sheath/core and hi-multi PET fabrics and partly to the higher hygroscopicity of the bamboo fibers. According to previous studies [1,58], the correlation between the diffusion process and moisture vapor transmission can be explained by the swelling of the fibers due to the affinity of the hydrophilic fiber molecules. Hence, as moisture vapor diffuses through the fibers in the fabric, it is absorbed by the fibers, causing fiber swelling and reducing the size of the air (void) spaces between the fibers. This delays the diffusion process, which reduces the rate of moisture vapor particle movements [58]. This explains why hygroscopic spun yarn fabrics (specimens (2) and (4) in group A, (7) and (9) in group B, and (12) and (14) in group C) exhibited higher moisture vapor resistance than the filament fabrics (specimens (5), (10), and (15)). This explains why the R_{ef} measured by the sweating guarded hot plate method differs from the WVTR measured by the upright cup method, which is due to the difference of mechanism between the two measuring methods, i.e., the transmission of moisture vapor by forced convection due to $CaCl_2$ in the upright cup and the diffusion process by the free convection of wet heat particles in a sweating guarded hot plate apparatus. Furthermore, these results indicate that the sweating guarded hot plate method is appropriate to measure the breathability of coated (or laminated) nylon (or PET) fabrics, whereas the upright cup method is suitable for non coated ordinary fabrics.

3.4. Thermal Conductivity of the Fabric Specimens

The sweating guarded hot plate (ISO 11092 method) applies the transmission (diffusion, or convection) of wet thermal particles to measure the moisture vapor resistance, which is similar to the transmission by the conduction of dry thermal particles. Understanding how the moisture vapor resistance of the fabrics measured using the diffusion of wet thermal particles (sweating) is associated with the thermal resistance (conductivity) measured by the conduction of dry thermal particles is very important for examining dry and wet thermal wear comforts and their correlations according to the yarn structure of the fabric and the thermal characteristics of constituent fibers. Figure 7 presents the thermal conductivity of 15 fabric specimens. The mean value of the thermal conductivity of the 15 fabric specimens was statistically significant, as shown in Table 5.

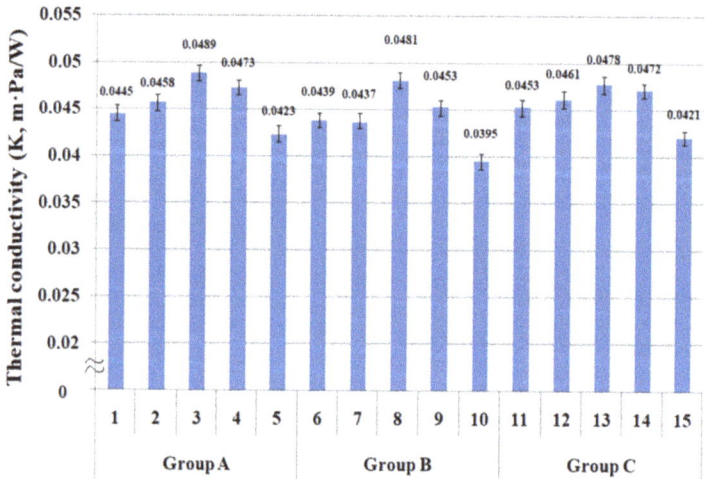

Figure 7. Thermal conductivity of the fabric specimens.

As shown in Figure 7, the thermal conductivity of fabric specimens ((3), (8), and (13)) with PET/Tencel siro-fil in the weft direction was the highest, followed by the fabric specimens ((4), (9), and (14)) composed of bamboo spun yarns. These fabrics have smaller pore diameters (Figure 4) than the other fabrics. Small air voids in the low porosity fabric due to the compact yarn structure of the PET/Tencel siro-fil and bamboo spun yarns cannot entrap the neighboring air in the compact yarns and their fabrics. This enables easy heat conduction from the inner layer to the outer one of the fabrics, resulting in higher thermal conductivity than the other fabrics. On the other hand, fabric specimens (5), (10), and (15) in groups A, B, and C composed of the hi-multi PET in the weft direction showed the lowest thermal conductivity because of the larger pore diameter in the filament bundles (Figure 4), which entraps the neighboring air and prevents heat flow from the inner layer to the outer layer of the fabrics, resulting in lower thermal conductivity than the other fabrics. The lower thermal conductivity of these fabrics was attributed partly to the lower thermal conductivity of the PET filament than that of the Tencel and bamboo fibers (p. 150) [59,60]. In addition, the thermal conductivity of the Coolmax®/Tencel sheath/core fabrics (specimens (1), (6), and (11)) was lower than that of the Coolmax®/bamboo spun, bamboo spun, and PET/Tencel siro-fil fabrics (specimens (2) to (4), (7) to (9), and (12) to (14)) in groups A, B, and C, which was also attributed to the larger pore diameters of the sheath/core fabrics than the other fabrics. In particular, fabric specimens (1), (6), and (11) with Coolmax®/Tencel sheath/core yarns in the weft direction exhibited higher thermal conductivity than the hi-multi PET fabric specimens (5), (10), and (15), despite the sheath/core fabrics having much larger pore diameters than the hi-multi PET fabrics. The higher thermal conductivity of the Tencel fibers than the PET filament was considered the cause [59,60], because it may compensate for the thermal conductivity of the two fabrics, resulting in higher thermal conductivity of the Coolmax®/Tencel sheath/core fabrics. Overall, the thermal conductivity of the composite yarn fabrics is influenced mainly by the pore size according to the yarn structure and partly by the thermal characteristics of the constituent fibers. These results are in agreement with previous findings [61–63]. Das et al. [61,62] reported that cotton/acrylic blend fabric exhibited low thermal conductivity because of its high porosity. Das and Istiaque [63] published a similar result for the hollow yarn fabric. The thermal properties of the composite yarn fabrics were influenced mainly by the fabric porosity; the sheath/core yarn fabric with a noncircular cross-sectional filament exhibited low thermal conductivity. Furthermore, the bulky yarn fabric with noncircular cross-sectional fibers showed low thermal conductivity because of the high porosity and hairy and crimpy constituent fibers. Thus, the thermal conductivity of the composite yarn fabrics was governed partly by the thermal characteristics of the constituent fibers. Finally, considering the market application of ecofriendly fiber-embedded fabrics for high-performance clothing with good wear comfort, the Coolmax®/Tencel sheath/core fabrics are useful for winter clothing with a warm feel due to good breathability with low thermal conductivity. Bamboo and Coolmax®/bamboo spun yarn fabrics are suitable for summer clothing with a cool feeling because of their high thermal conductivity and good breathability.

3.5. Correlation Analysis between Wear Comfort Characteristics of Fabrics

Heat and moisture vapor transmission is essential for characterizing the thermophysiological behavior of fabrics to assess their wear comfort for the design of high-performance clothing. They are characterized by two parameters: resistance to dry heat (inverse of thermal conductivity) and resistance to moisture vapor. Therefore, correlation analysis was carried out to determine the interrelationships between the thermal conductivity and moisture vapor resistance according to the yarn structure and constituent fiber characteristics with the measuring method, as well as the fabric absorption rate. Table 7 lists the correlation coefficient between each parameter. The correlation between the moisture vapor resistance (R_{ef}) and pore diameter was significant at the 99% confidence level. In addition, the correlation between R_{ef} and thermal conductivity (K) was significant at the 99% confidence level. Therefore, the results obtained with significant correlation at the 99% confidence level were graphed. A trend line was added to the graph shown in Figure 8.

Table 7. Correlation coefficient between each parameter of the wear comfort characteristics of the fabrics.

	Pore Diameter (μm)	Water Vapor Transmission Rate (g/m²·h)	Moisture Vapor Resistance (m²·Pa/W)	Thermal Conductivity (m·Pa/W)	Absorption Rate (%)
Pore diameter (D)	1				
Water vapor transmission rate(WVTR)	0.354	1			
Moisture vapor resistance (R_{ef})	−0.734 [a]	0.264	1		
Thermalconductivity (K)	−0.545 [b]	0.412 [b]	0.872 [a]	1	
Absorptionrate (A)	−0.226	−0.159	0.161	0.301	1

Note: [a] significant at the 0.01 level; [b] significant at the 0.05 level.

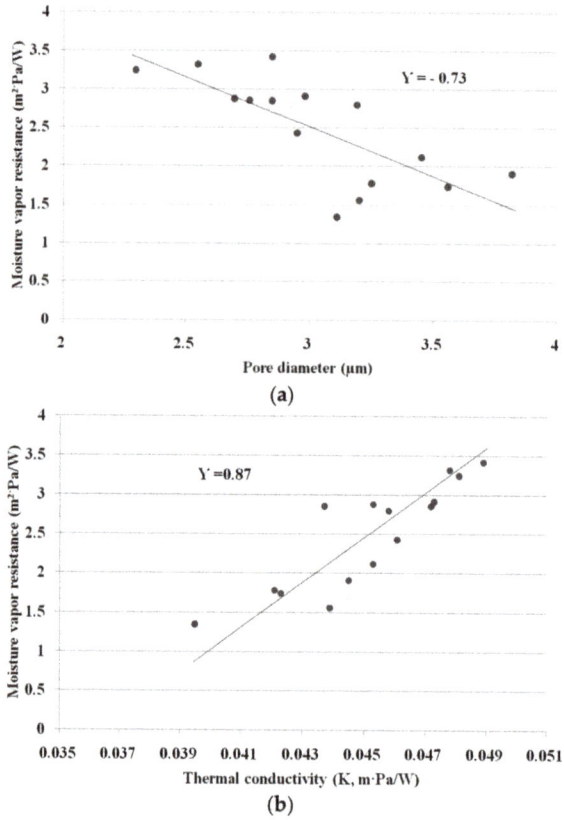

Figure 8. Correlation diagram between each wear comfort characteristic of the fabrics: (**a**) moisture vapor resistance and pore diameter; (**b**) moisture vapor resistance and thermal conductivity.

Figure 8 presents a diagram showing the interrelationships among the moisture vapor resistance, thermal conductivity, and pore diameter of the fabric specimens.

As shown in Table 7 and Figure 8a correlation coefficient between the moisture vapor resistance (R_{ef}) and pore diameter was observed as an inverse correlation (−0.73). Hence, the moisture vapor resistance is influenced by the pore diameters of the fabrics and partly by the hygroscopicity of the constituent fibers, which is consistent with previous results [54–58]. Fohr et al. [54] and Lomax [64] reported that moisture vapor through a textile medium diffuses in two ways: through the air space (pore) and along the fiber itself. In addition, the diffusion rate along the textile material depends on the porosity, and the moisture vapor diffusivity of the fiber is affected by the absorption and its hygroscopicity.

Barnes et al. [55] and Hong et al. [56] reported that the moisture vapor transmission of the hygroscopic fibrous materials was higher than that of non-absorbing materials. On the other hand, the correlation between R_{ef} and WVTR was very low (0.26) because the measuring mechanisms of the two methods were different. The upright cup method by $CaCl_2$ measures the moisture vapor transport by Fick's law, whereas R_{ef} measures the wet thermal transport by the difference in vapor pressure. This is in accordance with McCullogh et al. [26], who reported that the correlations between R_{ef} and the upright and inverted cup methods of WWB (waterproof, windproof, and breathable) shell fabrics made of nylon and PET were very low (−0.59 and 0.1, respectively). The results are in contrast to those by Huang and Qian [29] and Gorjanc et al. [4]. Huang and Qian [29] reported that the correlation coefficients between R_{ef} and the upright and inverted cup methods using ordinary and breathable fabrics were −0.87 and −0.66, respectively. Gorjanc et al. [4] reported that the correlation coefficient between the water cup method and moisture vapor resistance by the Permetest method using cotton and cotton-stretch fabrics was >0.9. Previous findings [4,26,29] suggested that the breathability characteristics according to the constituent yarn structure were dependent on the measuring method. In addition, the correlation coefficient between R_{ef} and thermal conductivity (K) was 0.87, meaning that the K values measured by dry heat transmission and R_{ef} measured by wet heat movement are governed by a similar mechanism: the movement of heat and moisture vapor particles, as shown in Figure 8b. Moreover, they are dependent on the pore size of the fabrics and the hygroscopicity and thermal characteristics of the constituent fibers.

4. Conclusions

The MVP and thermal wear comfort of the ecofriendly fiber-embedded woven fabrics were examined according to the measuring method in relation to their absorption and thermal properties with the constituent yarn structure and fiber characteristics. Fifteen fabric specimens composed of sheath/core, siro-fil, siro-spun, and ring-spun yarns were prepared using bamboo and Tencel as ecofriendly fibers, as well as PP, PET, and Coolmax® as core filaments. The WVTR and moisture vapor resistance (R_{ef}) were measured using the upright cup method by $CaCl_2$ and a sweating guarded hot plate method, respectively. The WVTR measured by the upright cup method was dependent primarily on the hygroscopicity of the eco-friendly constituent fibers (bamboo and Tencel) in the yarns and partly influenced by the pore size in the fabric depending on the yarn structure. Of the 15 fabric specimens, the bamboo spun yarn fabrics exhibited superior WVTR, followed by the Coolmax®/Tencel sheath/core fabrics.

On the other hand, the hi-multi PET fabrics showed inferior WVTR, followed by the PET/Tencel siro-fil fabrics. The moisture vapor resistance (R_{ef}) by the sweating guarded hot plate method was governed mainly by the pore size in the fabric and partly by the hygroscopicity of the constituent ecofriendly fibers. These results suggest that ecofriendly fibers, bamboo, and Tencel can contribute to environmental improvement and wear comfort related to water and moisture vapor transmission. The moisture vapor resistance of hi-multi PET filament fabrics was the lowest, i.e., best, followed by the Coolmax®/Tencel sheath/core fabrics. In contrast, the PET/Tencel siro-fil fabric was the highest, i.e., showed inferior breathability. R_{ef} measured by the sweating guarded hot plate method differed from the WVTR measured by the upright cup method due to the difference in measuring mechanism between the two methods.

The thermal conductivity of the composite yarn fabrics was influenced by pore size in the fabric and the thermal characteristics of the constituent ecofriendly fibers in the yarns. The hi-multi PET filament fabrics exhibited the lowest thermal conductivity, followed by the Coolmax®/Tencel sheath/core fabrics, whereas the PET/Tencel siro-fil fabrics showed the highest thermal conductivity, followed by the bamboo spun yarn fabrics. These results were verified by correlation analysis. The correlation coefficient between the moisture vapor resistance and pore diameter was −0.73. The correlation coefficients between the moisture vapor resistance (R_{ef}) and thermal conductivity (K) and between K and pore diameter were 0.87 and −0.55, respectively. On the other hand, the correlation coefficient between the WVTR and moisture vapor resistance was very low (0.27), which was attributed to the different mechanisms between the two measuring methods, i.e., transmission of moisture vapor by forced convection in an upright cup and the diffusion process via the free convection of wet heat particles in a sweating guarded hot plate apparatus.

Lastly, considering the market application for high-performance ecofriendly clothing with good wear comfort, the Coolmax®/Tencel sheath/core yarn fabrics are useful for winter clothing with a warm feeling due to the good breathability with low thermal conductivity. Bamboo and Coolmax®/bamboo spun yarn fabrics are suitable for summer clothing with a cool feeling because of their high thermal conductivity and good breathability. Although based on the MVP and thermal wear comfort obtained in this study, the market application of Tencel fibers for winter outdoor clothing and bamboo fibers for summer outdoor clothing is of practical use for engineering high-performance fabrics. These results suggest that an increase in the consumption of ecofriendly fibers with a decrease in synthetic fibers can reduce environmental pollution in the textile industry.

Funding: This research is supported by Ministry of SMEs and Startups (Project Number: S240111).

Institutional Review Board Statement: Not applicable.

Informed Consent Statement: Not applicable.

Data Availability Statement: The data presented in this study are available on request from the corresponding author.

Conflicts of Interest: The author declares no conflict of interest.

References

1. Deopura, B.L.; Alagirusamy, R.; Joshi, M.; Gupta, B. *Polyesters and Polyamides*, 1st ed.; Woodhead Publishing Limited: Cambridge, UK, 2008; pp. 171–199.
2. Ozgen, B. New biodegradable fibers, yarn properties and their applications in textiles: A review. *Ind. Text.* **2012**, *63*, 3–7.
3. Rego, J.M.; Verdu, P.; Nieto, J.; Blanes, M. Comfort Analysis of Woven Cotton/Polyester Fabrics Modified with a New Elastic Fiber, Part 2: Detailed Study of Mechanical, Thermo-Physiological and Skin Sensorial Properties. *Text. Res. J.* **2009**, *80*, 206–215. [CrossRef]
4. Gorjanc, D.; Dimitrovski, K.; Bizjak, M. Thermal and water vapor resistance of the elastic and conventional cotton fabrics. *Text. Res. J.* **2012**, *82*, 1498–1506. [CrossRef]
5. Lee, S.; Obendorf, S.K. Statistical modeling of water vapor transport through woven fabrics. *Text. Res. J.* **2012**, *82*, 211–219. [CrossRef]
6. Kim, J.; Spivak, S. Dynamic Moisture Vapor Transfer Through Textiles. *Text. Res. J.* **1994**, *64*, 112–121. [CrossRef]
7. Cubric, I.S.; Skenderi, Z.; Havenith, G. Impact of raw material, yarn and fabric parameters, and finishing on water vapor resistance. *Text. Res. J.* **2013**, *83*, 1215–1228. [CrossRef]
8. Ruckman, J. Water vapour transfer in waterproof breathable fabrics: Part I: Under steady-state conditions. *Int. J. Cloth. Sci. Technol.* **1997**, *9*, 10–22. [CrossRef]
9. Ruckmann, J.E. Water vapour transfer in waterproof breathable fabrics: Part II: Under windy conditions. *Int.J. Cloth. Sci. Technol.* **1997**, *9*, 23–33. [CrossRef]
10. Ruckmann, J.E. Water vapour transfer in waterproof breathable fabrics: Part III: under rainy and windy conditions. *Int.J. Cloth. Sci. Technol.* **1997**, *9*, 141–153. [CrossRef]
11. Ruckman, J.; Murray, R.T.; Choi, H.S. Engineering of clothing systems for improved thermophysiological comfort. *Int. J. Cloth. Sci. Technol.* **1999**, *11*, 37–52. [CrossRef]
12. Ren, Y.J.; Ruckman, J.E. Effect of condensation on water vapour transfer through waterproof breathable fabrics. *J. Coat.Fabr.* **1999**, *29*, 20–36.
13. Ren, Y.J.; Ruckman, J.E. Water Vapour Transfer in Wet Waterproof Breathable Fabrics. *J. Ind. Text.* **2003**, *32*, 165–175. [CrossRef]
14. Ren, Y.J.; Ruckman, J.Y. Condensation in three-layer waterproof breathable fabrics for clothing. *Int. J. Cloth. Sci. Technol.* **2004**, *16*, 335–347. [CrossRef]
15. Yoo, S.J.; Kim, E.A. Wear trial assessment of layer structure effects on vapor permeability and condensation in a cold weather clothing ensemble. *Text. Res. J.* **2012**, *82*, 1079–1091. [CrossRef]
16. Yoo, H.S.; Kim, E.A. Effects of multilayer clothing system array on water vapor transfer and condensation in cold weather clothing ensemble. *Text. Res. J.* **2008**, *78*, 189–197. [CrossRef]
17. Lomax, G.R. Hydrophilic polyurethane coatings. *J. Coat. Fabr.* **1990**, *20*, 88–107. [CrossRef]

18. Salz, P. Testing the Quality of Breathable Textiles. *Performance of Protective Clothing: Second Symposium*; ASTM Special Technical Publication, 989; Mandorf, F.Z., Sagar, R., Bielson, A.P., Eds.; American Society for Testing and Materials: Philadelphia, PA, USA, 1988; p. 295.
19. Yoo, H.; Hu, Y.; Kim, E. Effects of Heat and Moisture Transport in Fabrics and Garments Determined with a Vertical Plate Sweating Skin Model. *Text. Res. J.* **2000**, *70*, 542–549. [CrossRef]
20. Gibson, P. Factors Influencing Steady-State Heat and Water Vapor Transfer Measurements for Clothing Materials. *Text. Res. J.* **1993**, *63*, 749–764. [CrossRef]
21. Gibson, P.; Kendrick, C.; Rivin, D.; Sicuranza, L.; Charmchi, M. An Automated Water Vapor Diffusion Test Method for Fabrics, Laminates, and Films. *J. Coat. Fabr.* **1995**, *24*, 322–345. [CrossRef]
22. Gibson, P.; Kendrick, C.; Rivin, D.; Charmchii, M. An Automated Dynamic Water Vapor Permeation Test Method. In *Performance of Protective Clothing: Sixth Volume*; ASTM International: West Conshohocken, PA, USA, 1997; pp. 93–107. [CrossRef]
23. Dolhan, P.A. A Comparison of Apparatus Used to Measure Water Vapour Resistance. *J. Coat. Fabr.* **1987**, *17*, 96–109. [CrossRef]
24. Congalton, D. Heat and moisture transport through textiles and clothing ensembles utilizing the Hohenstein skin model. *J. Coat. Fabr.* **1999**, *28*, 183–196.
25. Gretton, J.; Brook, D.; Dyson, H.; Harlock, S. A Correlation between Test Methods Used to Measure Moisture Vapour Transmission through Fabrics. *J. Coat. Fabr.* **1996**, *25*, 301–310. [CrossRef]
26. McCullough, E.A.; Kwon, M.; Shim, H. A comparison of standard methods for measuring water vapour permeability of fabrics. *Meas. Sci. Technol.* **2003**, *14*, 1402–1408. [CrossRef]
27. Huang, J. Review of test methods for measuring water vapor transfer properties of fabrics. *Cell.Polym.* **2007**, *26*, 167–191. [CrossRef]
28. Huang, J. Sweating guarded hot plate test method. *Polym. Test.* **2006**, *25*, 709–716. [CrossRef]
29. Huang, J.; Qian, X. Comparison of Test Methods for Measuring Water Vapor Permeability of Fabrics. *Text. Res. J.* **2008**, *78*, 342–352. [CrossRef]
30. Chen, Q.; Fan, J.T.; Sarkar, M.K.; Bal, K. Plant-based biomimetic branching structures in knitted fabrics for improved comfort-related properties. *Text. Res. J.* **2011**, *81*, 1039–1048. [CrossRef]
31. Chen, Q.; Fan, J.; Sarkar, M.; Jiang, G. Biomimetics of Plant Structure in Knitted Fabrics to Improve the Liquid Water Transport Properties. *Text. Res. J.* **2009**, *80*, 568–576. [CrossRef]
32. Okada, H. Sweat-Absorbent Textile Fabric. U.S. Patent 4,530,873A, 1985.
33. Strauss, I.; Rankin, S.A., Jr. Fabric for recreational clothing. U.S. Patent 5,050,406A, 1991.
34. Lee, Y.K. Method for making fabric with excellent water transition ability. U.S. Patent 6,381,994B1, 2002.
35. Yeh, P. Fabric for moisture management. U.S. Patent 6,509,285B1, 2003.
36. Burrow, T.R.; Firgo, H. Wicking fabric and garment made therefrom. U.S. Patent 8127575B2, 2012.
37. Kim, H.A.; Kim, S.J. Physical Properties and Wear Comfort of Bio-Fiber-Embedded Yarns and their Knitted Fabrics According to Yarn Structures. *Autex Res. J.* **2019**, *19*, 279–287. [CrossRef]
38. Kim, H.A. Physical Property of PTT/Wool/Modal Air Vortex Yarns for High Emotional Garment. *J. Korean Soc. Cloth. Text.* **2015**, *39*, 877–884. [CrossRef]
39. Kim, H.A.; Kim, S.J. Flame retardant, anti-static and wear comfort properties of modacrylic/Excel®/anti-static PET blend yarns and their knitted fabrics. *J. Text. Inst.* **2019**, *110*, 1318–1328. [CrossRef]
40. Kim, H.A. Tactile hand and wear comfort of flame-retardant rayon/anti-static polyethylene terephthalate imbedded woven fabrics. *Text. Res. J.* **2019**, *89*, 4658–4669. [CrossRef]
41. Kim, H.A.; Kim, S.J. Hand and Wear Comfort of Knitted Fabrics Made of Hemp/Tencel Yarns Applicable to Garment. *Fibers Polym.* **2018**, *19*, 1539–1547. [CrossRef]
42. Kim, H.A.; Kim, S.J. Mechanical Properties of Micro Modal Air Vortex Yarns and the Tactile Wear Comfort of Knitted Fabrics. *Fibers Polym.* **2018**, *19*, 211–218. [CrossRef]
43. Kim, H.A.; Kim, S.J. Flame-Retardant and Wear Comfort Properties of Modacrylic/FR-Rayon/Anti-static PET Blend Yarns and Their Woven Fabrics for Clothing. *Fibers Polym.* **2018**, *19*, 1869–1878. [CrossRef]
44. Kim, H.A. Water/moisture vapor permeabilities and thermal wear comfort of the Coolmax®/bamboo/tencel included PET and PP composite yarns and their woven fabrics. *J. Text. Inst.* **2020**. [CrossRef]
45. Saricam, C.; Kalaoglu, F. Investigation of the wicking and drying behaviour of polyester woven fabrics. *Fibers Text. East. Eur.* **2014**, *22*, 73–78. [CrossRef]
46. Tashkandi, S.; Wang, L.; Kanesalingam, S. An investigation of thermal comfort properties of Abaya woven fabrics. *J. Text. Inst.* **2013**, *104*, 830–837. [CrossRef]
47. Varshney, R.K.; Kothari, V.K.; Dhamija, S. A study on thermophysiological comfort properties of fabrics in relation to constituent fiber fineness and cross-sectional shapes. *J. Text. Inst.* **2010**, *101*, 495–505. [CrossRef]
48. Vimal, J.T.; Murugan, R.; Subramaniam, V. Effect of Weave Parameters on the Air Resistance of Woven Fabrics. *Fibres Text. East. Eur.* **2016**, *24*, 67–72. [CrossRef]
49. Wei, J.; Xu, S.; Liu, H.; Zheng, L.; Qian, Y. Simplified modal for predicting fabric thermal resistance according to its micro-structural parameters. *Fibers Text. East. Eur.* **2015**, *23*, 57–60. [CrossRef]

50. Kim, H.A.; Kim, S.J. Moisture and thermal permeability of the hollow textured PET imbedded woven fabrics for high emotional garments. *Fibers Polym.* **2016**, *17*, 427–438. [CrossRef]
51. Beskisiz, E.; Ucar, N.; Demir, A. The Effects of Super Absorbent Fibers on the Washing, Dry Cleaning and Drying Behavior of Knitted Fabrics. *Text. Res. J.* **2009**, *79*, 1459–1466. [CrossRef]
52. Ucar, N.; Beskisiz, E.; Demir, A. Design of a Novel Filament with Vapor Absorption Capacity Without Creating Any Feeling of Wetness. *Text. Res. J.* **2009**, *79*, 1539–1546. [CrossRef]
53. Kim, H.A. Wear Comfort of Woven Fabrics for Clothing Made from Composite Yarns. *Fibers Polym.* **2021**, *22*, 2344–2353. [CrossRef]
54. Fohr, J.; Couton, D.; Treguier, G. Dynamic Heat and Water Transfer Through Layered Fabrics. *Text. Res. J.* **2002**, *72*, 1–12. [CrossRef]
55. Barnes, J.C.; Holcombe, B.V. Moisture sorption and transport in clothing during wear. *Textile Research Journal* **1996**, *66*, 777–786. [CrossRef]
56. Hong, K.; Hollies, N.R.S.; Spivak, S.M. Dynamic moisture vapour transfer through textile. *Textile Research Journal* **1988**, *68*, 697–706. [CrossRef]
57. Li, Y.; Holcombe, B.V.; Scheider, A.M.; Apcar, F. Mathematical modeling of the coolness to the touch of hygroscopic fabrics. *J. Text. Inst.* **1993**, *84*, 267–273. [CrossRef]
58. Pause, B. Measuring the Water Vapor Permeability of Coated Fabrics and Laminates. *J. Coat. Fabr.* **1996**, *25*, 311–320. [CrossRef]
59. Bona, M. *Textile Quality: Physical Methods of Product and Process Control*; Texilia: Biella, Italy, 1994; pp. 83–150.
60. Lavate, S.S.; Burji, M.C.; Patil, S. Study of yarn and fabric properties produced from modified viscose Tencel, Excel, Modal and Their Comparison against cotton. *Text. Today* **2016**, *9*, 36–42.
61. Das, A.; Kothari, V.K.; Balaji, M. Studies on cotton-acrylic bulked yarns and fabrics. Part I: Yarn characteristics. *J. Text. Inst.* **2007**, *98*, 261–267. [CrossRef]
62. Das, A.; Kothari, V.K.; Balaji, M. Studies on cotton–acrylic bulked yarns and fabrics. Part II: Fabric characteristics. *J. Text. Inst.* **2007**, *98*, 363–376. [CrossRef]
63. Das, A.; Ishtiaque, S.M. Comfort characteristics of fabrics containing twist-less and hollow fibrous assemblies in weft. *J. Text. App. Technol. Manag.* **2004**, *3*, 1–7.
64. Lomax, G.R. The design of waterproof, water vapour-permeable fabrics. *J. Coat. Fabr.* **1985**, *15*, 40–49. [CrossRef]

Article

Study of the Filtration Performance of Multilayer and Multiscale Fibrous Structures

Vânia Pais [1,2,*], Carlos Mota [1,2], João Bessa [1,2], José Guilherme Dias [3], Fernando Cunha [1,2] and Raul Fangueiro [1,2,4]

1. Fibrenamics, Institute of Innovation on Fiber-based Materials and Composites, University of Minho, 4800 Guimarães, Portugal; cmota@tecminho.uminho.pt (C.M.); joaobessa@fibrenamics.com (J.B.); fernandocunha@det.uminho.pt (F.C.); rfangueiro@dem.uminho.pt (R.F.)
2. Centre for Textile Science and Technology (2C2T), University of Minho, 4800 Guimarães, Portugal
3. Poleva—Termoconformados, S.A. Rua da Estrada, 4610 Felgueiras, Portugal; josedias@poleva.pt
4. Department of Mechanical Engineering, University of Minho, 4800 Guimarães, Portugal
* Correspondence: vaniapais@fibrenamics.com

Abstract: As the incidence of small-diameter particles in the air has increased in recent decades, the development of efficient filtration systems is both urgent and necessary. Nanotechnology, more precisely, electrospun nanofibres, has been identified as a potential solution for this issue, since it allows for the production of membranes with high rates of fibres per unit area, increasing the probability of nanoparticle collision and consequent retention. In the present study, the electrospinning technique of polyamide nanofibre production was optimized with the variation of parameters such as polymer concentration, flow rate and needle diameter. The optimized polyamide nanofibres were combined with polypropylene and polyester microfibres to construct a multilayer and multiscale system with an increased filtration efficiency. We observed that the penetration value of the multilayer system with a PA membrane in the composition, produced for 20 min in the electrospinning, is 2.7 times smaller than the penetration value of the system with the absence of micro and nano fibers.

Keywords: electrospinning; nanofibres; filtration; particles retention; multilayer systems

Citation: Pais, V.; Mota, C.; Bessa, J.; Dias, J.G.; Cunha, F.; Fangueiro, R. Study of the Filtration Performance of Multilayer and Multiscale Fibrous Structures. *Materials* **2021**, *14*, 7147. https://doi.org/10.3390/ma14237147

Academic Editor: Dubravko Rogale

Received: 11 October 2021
Accepted: 19 November 2021
Published: 24 November 2021

Publisher's Note: MDPI stays neutral with regard to jurisdictional claims in published maps and institutional affiliations.

Copyright: © 2021 by the authors. Licensee MDPI, Basel, Switzerland. This article is an open access article distributed under the terms and conditions of the Creative Commons Attribution (CC BY) license (https://creativecommons.org/licenses/by/4.0/).

1. Introduction

The incidence of particles with a small diameter (lower than 2.5 µm) in the air has risen in recent decades [1]. The rapid growth of urbanization and industrialization has led to a release of small particles to the atmosphere, such as solid particles and liquid droplets, which is concerning. Particles with diameters smaller than 2.5 µm may cause considerable damage due to their ability to penetrate the human bronchi, lungs and even the extrapulmonary organs. Furthermore, these particles can be linked to bacteria or viruses and cause serious human health problems due to the development of acute and chronic diseases [2,3]. Developing a solution to this problem is extremely important, and certain approaches such as filtration membranes have been identified as useful. These filter membranes can be applied in various products, such as face masks and NBC suits (protection against nuclear, biological, and chemical warfare agents) [4].

There are several types of particles that require filtering and each one of them has unique properties. The particle's diameter is one such property, and it can span from a few nanometers, as is the case for antibodies and viruses, to microns, such as for pollen [5]. Figure 1 shows the previously mentioned variation in particle size retention. Therefore, since there are different particles with distinct sizes, the filtration process should be optimized, depending on the objectives. Permeability, filtration performance and the uniformity of the structure are the three principal factors to consider when developing or applying a filtration process. Permeability, mostly related to breathability and water vapour transmission, should be optimized to make the structure wearable without compromising the filtration efficiency [4]. Concerning the filtration performance, the filtration theory provides that the efficiency of this process increases with a decrease in the dimension of the fibres that compose the filter. This statement relates to the increase in fibres per unit area, which leads to an increase in the probability of impact between the filter and particles that need to be filtered. The uniformity is related to efficacy [6].

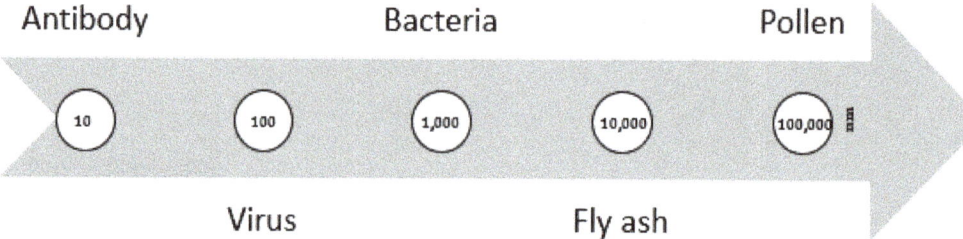

Figure 1. Different particle sizes to be filtered.

The filtration process occurs through different methods, dependent on the size of the particle requires filtering, and the parameter that most influences this. In Table 1 the different mechanisms of filtration are described, corresponding with the particles size being filtered. Larger particles are usually trapped by gravity sedimentation, since the pore sizes of filters are smaller than the particles size, blocking the particles outside the porous structure. Inertial impaction is also a possibility in the retention of larger particles. This mechanism occurs when the particles do not follow the direction of the airflow due to their large inertia. Thus, when associating high speed with larger particles, there is an increase in the probability of collision between particles and fibres. After collision, the particles can adhere to the fibres, and are retained in the filter. The interception mechanism is related to the retention of particles below 0.6 µm and occurs when the particles follow the airflow. Eventually, the particles come into contact with fibres that compose the filter and remain connected by Brownian forces. The efficiency of this process increases with the decrease in particle size. For nano-sized particles (below 0.2 µm), diffusion is the predominant mechanism. The particles do not follow the streamline direction and have a very slow and random movement. At some point, the particles and fibres collide and remain attached. Electrostatic attraction occurs in particles of different dimensions and occurs when the fibres are electrically charged and can capture the particles that are oppositely charged [5,7,8].

Table 1. Mechanisms of filtration and respective particle sizes to be filtered [7].

Mechanisms of Filtration	Size of Particles
Gravity sedimentation	Between 1 and 10 µm
Inertial impaction	Above 0.6 µm
Interception	Below 0.6 µm
Diffusion	Below 0.2 µm
Electrostatic attraction	Charged particles

Concerning the retention of small particles, diffusion is the predominant mechanism of filtration. The diffusion efficiency increases with a decrease in the diameter of the fibres that constitute the filter. However, when the particles have a size of around 0.3 µm, the retention process can be harder to achieve because the diffusion mechanism may not occur. Therefore, to increase retention via interception, a multi-layer approach should be applied [5]. To promote nanoparticles' retention, two factors should be considered: the use of fibres with very small diameters in the filter membrane and the application of a multi-layer system [9–11].

Concerning the relationship between fibres with lower diameters and higher efficiencies, nanofibres have been identified as a solution with great potential. Nanofibre filters have a controllable small diameter, low basis weight, high permeability values, reduced thickness and a porous structure [7,12]. The electrospinning technique is a simple and effective method used to produce fibres at a nanoscale. In this methodology, a high electric field is applied which promotes repulsive interactions among the polymeric solution, and the Taylor cone is formed. When the electrostatic forces overlap the repulsive interactions, a charged jet is ejected from the Taylor cone with a dynamic whipping. Concurrently, the solvent evaporates, and the jet is stretched into fibres with finer diameters that are deposited on a grounded collector [13,14]. The electrospun nanofibre diameters can range from a few nanometers to micrometres. Additionally, this filtration layer can be produced with distinct raw materials and by the application of different parameters, obtaining specific nanofibres with several functionalities. So, the nanoscale diameters and the interconnected porous structure of the electrospun nanofibres make the electrospinning technique a very attractive approach for filtration applications. Furthermore, the static charge, which is a result of the electrospinning process, may remain on the fibres that have been produced and enhance the filtration retention by electrostatic attraction [12,15,16].

The selection of the most suitable polymer to produce the fibres to be applied in particles retention should consider good mechanical properties, hydrophobicity, biocompatibility and compatibility with non-toxic solvents [17–19]. Polyamide (PA) stands out as a potential polymer since it has excellent chemical stability and thermal resistance. It is a synthetic polymer that is biodegradable and biocompatible. Usually, polyamide is dissolved in formic acid, and this combination can be electrospun to efficiently produce nanometric fibres [17,18], as opposed to polycaprolactone (PCL) (for example), which is usually dissolved in chloroform and dimethylformamide (DMF) [11]. According to EU directive 67/548/EEC, DMF is toxic [18].

In this study, electrospun nanofibres were produced with PA polymers to optimize systems with a higher filtration efficiency. Firstly, the parameters' polymers concentration, flow rate and needle's diameter were optimized to obtain fibres with very low diameters and a mat with controlled porosity. The morphology and intrinsic properties of the produced nanofibres were analyzed. In the second part of the study, several combinations of the optimized electrospun PA nanofibres with polypropylene (PP) and polyester (PES) microfibres were analyzed to obtain a multi-layer and multiscale system with a high retention capacity of small particles. The performance of the obtained combinations was evaluated by measuring the filtering material penetration, air permeability and breathing resistance.

2. Materials and Methods

2.1. Materials

PA 6.6 pellets (with a molecular weight of 262.35 g/mol, Tm = 250–260 °C, density = 1.14 g/mL at 25 °C, Sigma Aldrich, St. Louis, MO, USA) were used as a polymeric matrix. The solvent used was formic acid (FA) (98–100%, Fisher Scientific, Leics, UK).

The PP microfibres membrane (weight = 50 g per square meter (gsm), thickness = 360 µm, average fibre diameter = 3.7 µm), applied as a substrate for PA nanofibres, were obtained from Protechnic S.A. (Cernay, France).

2.2. Production of Electrospun PA Membranes

The polymeric solution was optimized after studying different PA concentrations (20% w/v and 25% w/v) to obtain fibres with small diameters and without defects. The solvent applied was FA in 1:1 proportion. The polymeric solution was prepared through the dissolution of PA pellets in FA, for at least 6 h at 30 °C, at constant stirring.

PA nanofibre webs were produced by electrospinning NF-103 from MECC Co., Ltd. (Fukuoka, Japan). The electrospinning parameters were also optimized to obtain fibres with small diameters to increase filtration efficacy. The tested parameters include the flow rate (ranging from 0.4 to 2 mL/h), needle's diameter (0.33, 0.41 and 0.61 mm) and the fibre deposition time (10, 20 and 30 min). The voltage applied was 28 kV and the collector-needle distance was 100 mm. The electrospinning process was conducted at 60% ± 5 RH and 20 °C ± 2. The group's previous studies were consulted to define certain fixed parameters, including voltage, collector-needle distance and the applied solvent [20,21]. The studied conditions and the corresponding produced membranes are presented in Table 2.

Table 2. Operational conditions tested during electrospinning production.

Sample	Solution Parameters		Electrospinning Parameters			
	Concentration (% (w/v))	Solvent	Voltage (kV)	Collector-Needle Distance (mm)	Flow Rate (mL/h)	Needle-Diameter (mm)
A	20	100% FA	28	100	0.4	0.33
B					0.8	
C					1	
D					0.4	0.41
E					0.8	
F					1	
G					1	0.61
H					2	
I	25				0.4	0.41

After optimizing the ideal conditions to produce the nanofibres, deposition was performed over PP microfibres to construct a multilayer and multiscale system with higher performance in terms of small particles retention. Another layer of PP microfibres were added to the PA nanofibres deposited above the PP microfibres. Thus, the filtration layer is composed of 3 layers: PP microfibres, PA nanofibres and PP microfibres. Two different polyester (PES) nonwovens were added to the filtration layer, one for the inner layer—PES IL—and the other for the outer layer—PES OL. A schematic representation of the multilayer system is presented in Figure 2. A total of 4 different combinations were obtained, as represented in Table 3. The difference between the multiple combinations relates to the nanofibre production time, of 0, 10, 20 and 30 min. The membranes were combined through a thermoforming process using a mould with a face-mask shape. However, it is important to mention that the structure can be molded to other shapes for application in other types of products.

2.3. Electrospun PA Membranes Characterization

The morphology of the produced PA nanofibres was investigated using a scanning electron microscope (SEM). The analyses were performed using a NOVA 200 Nano SEM from the FEI Company (Hillsboro, OR, USA). Due to the polymeric nature of the analysed specimens, the samples were vacuum metalized with a thin film of gold-palladium (Au-Pd) before the analysis. The average diameters and the porous distribution of the fibres were calculated by taking measurements from different regions using ImageJ software (1.52a).

The chemical composition and the structural aspects of the electrospun nanofibres were analyzed using Fourier transform infrared spectroscopy (FTIR) coupled with the attenuated reflection (ATR) technique using IRAffinity-1S, SHIMADZU equipment (Kyoto, Japan). All spectra were obtained in the transmittance mode with 45 scans over a wavenumber range of 4000–400 cm^{-1}.

The thermal behaviour of the electrospun nanofibres was assessed via a thermogravimetric analysis (TGA) of the electrospun fibres, which was performed with an STA 700 from HITACHI (Tokyo, Japan). The samples were tested in a temperature range from 25 °C to 500 °C, at a heating rate of 10 °C/min.

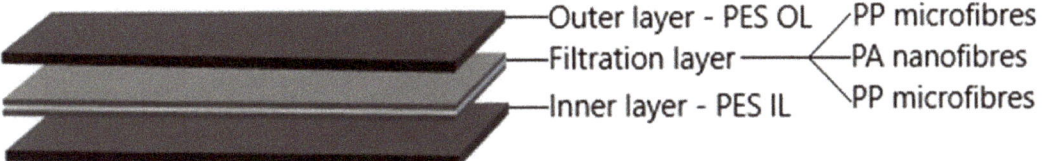

Figure 2. Multilayer system produced.

Table 3. Multilayer systems tested in terms of filtration efficiency.

Reference	Inner Layer	Filtration Layer	Outer Layer	PA Deposition Time
Multilayer_0 min		PP microfibres/PP microfibres (Note: without PA nanofibres)		0 min
Multilayer _10 min	PES IL	PP microfibres/ PA (10 min)/ PP microfibres	PES OL	10 min
Multilayer _20 min		PP microfibres/ PA nanofibres (20 min)/ PP microfibres		20 min
Multilayer _30 min		PP microfibres/ PA nanofibres (30 min)/ PP microfibres		30 min

2.4. Electrospun PA + Nonwoven Fabrics Characterization

The PP microfibres, PES IL and PES OL purchased from Protechnic S.A. (Cernay, France) were also studied. Morphology was analyzed through brightfield microscopy using a Microscope Leica DM750 M (brightfield) (Leica, Wetzlar, Germany) with a coupled camera.

The areal mass of several layers from the multilayer system was calculated by dividing the weight of the sample mat by its effective area. The thicknesses of the different membranes and multilayers systems were measured by analyzing the cross-section at the corresponding SEM images. The areal mass and thickness were obtained by performing measurements on 10 different regions of the samples.

The air permeability, filtration and respiratory evaluations were performed according to the standard EN 149:2001+A1:2009., For the air permeability evaluation, 40 Pa pressure was applied, using an air permeability tester from TEXTEST instruments (Zurich, Switzerland), model FX 3300. The tests related to filtration and respiratory evaluation were performed at Aitex—textile research institute, in Spain. The penetration of sodium chloride aerosol was tested by applying a flow rate of 95 L/min, and the maximum value was registered after 3.5 min of exposure. To perform the respiratory resistance evaluation, the pressure at 3 different conditions was registered, namely, inhalation at 30 L/min, inhalation at 95 L/min and exhalation at 160 L/min.

3. Results

3.1. Electrospun Fibre Characterization

3.1.1. Fourier Transform Infrared Spectroscopy and Thermogravimetric Analysis

The PA nanofibres were produced using an electrospinning technique according to the conditions reported in Table 2. The FTIR spectra of the nanofibres produced with PA polymer are shown in Figure 3. From the figure, it is possible to identify one significant band at 3300 cm^{-1}, typically attributed to the N-H stretching vibration. The asymmetric and symmetric stretching of CH_2 appears at 2931 and 2860 cm^{-1}, respectively. In this spectrum, amide bands of the PA 6.6. appear at 1637 and 1537 cm^{-1}, and the bands located at 1145 cm^{-1} correspond to the CO–CH symmetric bending vibration when combined with CH_2 twisting. The bands at 935 and 688 cm^{-1} are typically attributed to the stretching and bending vibrations of C–C bonds and the band at 580 cm^{-1} may be a result of O=C–N bending. The FTIR spectra obtained agree with those of other authors as a PA 6.6. FTIR spectrum [22–25].

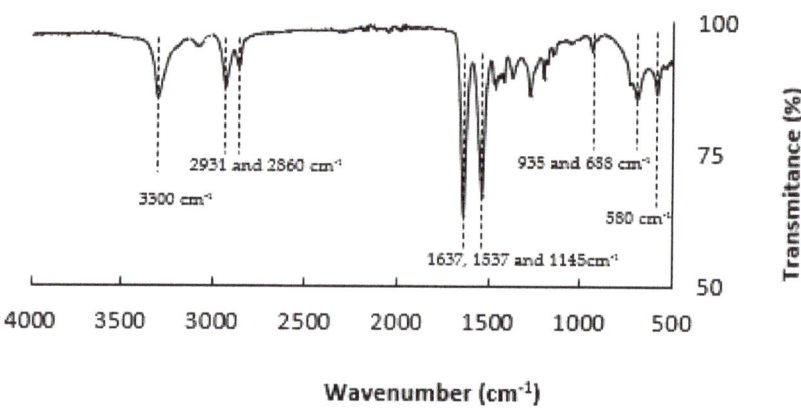

Figure 3. Fourier transform infrared spectra of electrospun PA nanofibres.

3.1.2. Thermogravimetric Analysis

To study the polymer behaviour at different temperatures, a TGA analysis was performed. The results are represented in Figure 4. At the TGA plot, a rapid decline in values starting at 350 °C is visible. No significant changes are observed preceding this temperature value. In this way, the PA polymer is found to have a very resistant thermal profile, that is only able to support temperatures up to approximately 320 °C. After this, a single-step degradation of the electrospun fibres is visible. The obtained results are in accordance with the results of other authors [26–28].

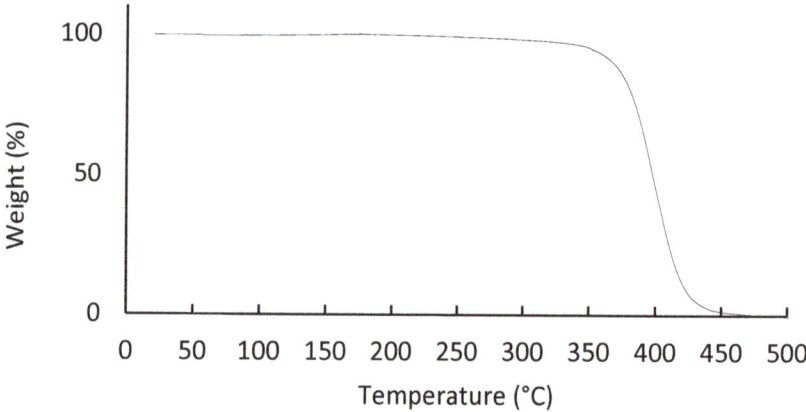

Figure 4. Thermogravimetric analysis curves of electrospun PA nanofibres.

3.1.3. Morphological Analysis

The PA nanofibre morphology was studied using SEM images. The images and the corresponding fibre diameters and porous distribution are represented in Table 4.

Table 4. SEM images of the produced PA nanofibres and corresponding characteristics.

Sample and Production Parameters	SEM Images (×1000)	SEM Images (×5000)	Fiber Diameter ± STDEV (nm)	Porous Distribution (%)
A Flow rate = 0.4 mL/h Needle-diameter = 0.33 mm			302 ± 46	21.66
B Flow rate = 0.8 mL/h Needle-diameter = 0.33 mm			442 ± 170	48.97
C Flow rate = 1 mL/h Needle-diameter = 0.33 mm			402 ± 141	37.50
D Flow rate = 0.4 mL/h Needle-diameter = 0.41 mm			293 ± 60	25.27
E Flow rate = 0.8 mL/h Needle-diameter = 0.41 mm			402 ± 84	22.71
F Flow rate = 1 mL/h Needle-diameter = 0.41 mm			363 ± 80	28.57
G Flow rate = 1 mL/h Needle-diameter = 0.61 mm			399 ± 193	25.09

Table 4. Cont.

Sample and Production Parameters	SEM Images (×1000)	SEM Images (×5000)	Fiber Diameter ± STDEV (nm)	Porous Distribution (%)
H Flow rate = 2 mL/h Needle-diameter = 0.61 mm			356 ± 70	21.11
I Flow rate = 0.4 mL/h Needle-diameter = 0.41 mm 25% (w/v) PA			755 ± 711	56.05

The SEM images reveal that randomly deposited fibres were produced. This feature is important, as filtration aims to produce a structure composed of innumerable pores with very small dimensions. This way, the probability of promoting the retention of nanoparticles is enhanced. The porous distribution, based on size, was measured using imageJ software. The obtained values ranged from 21 to 56%. To promote higher filtration efficiencies, a perfect combination of the fibres diameter and porous distribution should be achieved—a small fibre diameter allows for a lot of tinny pores, but the percentual distribution based on the size of these pores should be as small as possible, to increase the retention of particles probability. The PA nanofibres with the highest polymer concentration have the highest value of porous distribution. Most of the samples obtained percentage distribution values varying from 20 to 30%, thus emphasizing the applicability of PA nanofibres in the field of filtration. It is also important to produce fibre matts with higher densities, to ensure an increase in the collision points' probability. For this purpose, samples A, D, E, F, G and H are identified as viable options. Regarding the diameters of the fibres, samples A and D were found to be significant. Both samples were produced with a flow rate of 0.4 mL/h, with a needle with a diameter of 0.33 mm applied on sample A, and a needle with a diameter of 0.41 mm applied on sample B. Although the two samples have the two lowest mean fibre diameters, sample B had a smaller value—293 nm. Furthermore, samples A and D also have the lowest standard deviation value, which means that the produced fibres are more uniform with each other in terms of size. Sample I was produced using the same ideal conditions as sample B, however, the polymer proportion was higher (25% w/v instead of 20% w/v). When comparing sample I with the others, it is clear that the number of fibres produced per unit area was considerably lower. The average diameters of the produced fibres, as well as the corresponding standard deviation, are higher when compared to the other samples. This result reveals that the application of 20% (w/v) PA is the best approach to produce homogeneous mats of PA nanofibres with smaller diameters.

It can be observed that different parameters in PA-nanofiber production affect the fibre's diameter and porous distribution values, as presented in Figure 5. The values are related to the samples produced with 20% (w/v) PA. By analyzing the bars related to the diameter of the needles between 0.33 mm and 0.41 mm in Figure 5a, it is possible to conclude that nanofibres with lower diameters are always obtained when using a needle with a diameter equal to 0.41 mm. Additionally, by analyzing the flow rate values, it is possible to conclude that generally, lower flow rates promote the production of fibres with smaller diameters. The results of porous distribution are presented in Figure 5b, and it seems that a direct correlation between this value and the nanofiber construction parameters cannot be established. However, lower values for porous distribution are related to smaller fibres diameters, which becomes apparent when looking at both Figure 5a,b. When combining lower flow rates with smaller needle diameters, homogeneous mats of nanofibres (low standard deviation values) with small diameters and a porous distribution are produced. These conditions are optimal filtration purposes [6].

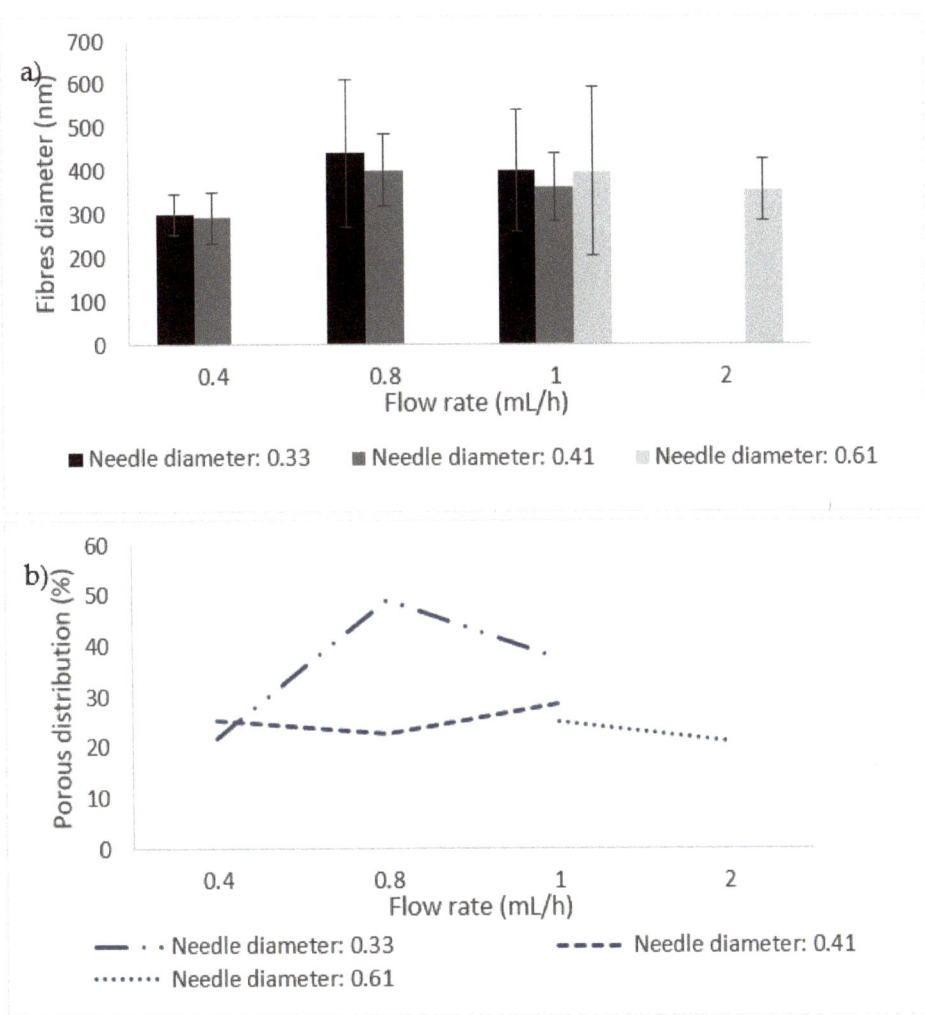

Figure 5. Influence of needle diameter (**a**) and flow rate (**b**) on fibre diameter and porous distribution values.

3.2. Electrospun PA + Non-Woven Fabrics Characterization

The above-mentioned optimal conditions for producing PA nanofibres, using electrospinning methods, were applied in this phase of the study The production and deposition of PA nanofibres onto the PP microfibres were conducted to obtain a multilayer structure for filtration purposes. The PP microfibres were morphologically and physically characterized, and are presented in the microscope images in Figure 6. PP fibres were found to have a random orientation. The average fibre diameter is 3.7 µm (Table 5). This value is much higher than the value obtained for PA nanofibres. Due to the conjugation of these 2 different structures, a multiscale system was constructed, and the filtration efficiency is expected to increase with minimal levels of breathability and/or permeability. The nanofibre membrane is much thinner (ranges from 1.16 to 2.64 µm depending on the production time for PA nanofibres) than the microfibres membrane (368.3 µm), making its application on wearable devices much more attractive. This is one of the greatest advantages of nanoscale, through the addition of a very small amount of nanomaterials it is possible to maximize specific functionalities on a high scale.

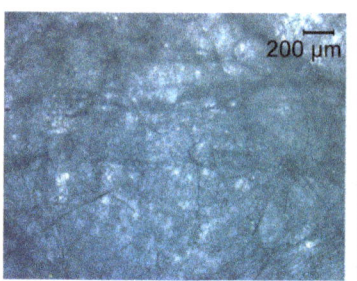 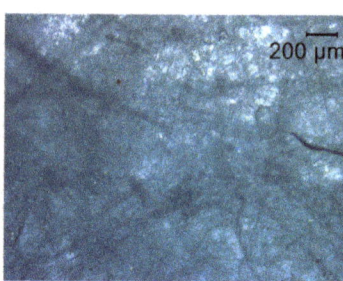 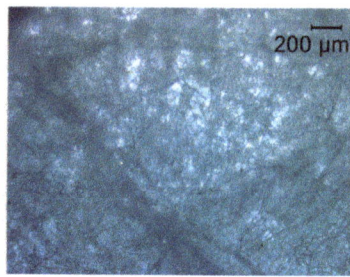

Figure 6. Morphology of (PP) microfibres under an optical microscope.

Table 5. PP microfibres characteristics.

Polymer	Average Fibres Diameter ± STDEV (µm)	Thickness (µm)	Air Permeability (L/m^2/s)	Aerial Mass (gsm)
PP	3.7 ± 1.5	368.3	102.3	53.5

PA electrospun nanofibres were produced and deposited succesfully. In total, 3 different combinations were obtained, since 3 different deposition times of the PA nanofibres were studied: 10, 20 and 30 min. A control sample consisting of the PP membrane without PA nanofibres was added to the study. By changing the time of nanofibre production, a different aerial mass, thickness and air permeability were obtained. These differences will influence the filtration efficiency. Results are shown in Figure 7. All the parameters corresponded to increasesin the of PA nanofibre production time. In the case of aerial mass results, it is observed that the value increased with a higher deposition time of PA nanofibres. This result was expected since higher deposition times allow for the production of more nanofibres, thereby increasing its density. The sample in which nanofibres were produced for 10 min is very similar to the PP microfibres. The sample in which nanofibres were produced for 20 min has an aerial mass closer to the sample in which nanofibres were produced for 30 min. This result indicates that the deposition time is not uniform, as differences were notices with regard to time. This relationship between higher deposition times and the properties of the produced electrospun nanofibres were also observed by Vrieze et al. The authors stated that after some time of fibre deposition the electric field becomes distorted, causing an inhomogeneous deposition in the collector [29]. The thickness of the samples are very similar to each other. This result is a consequence of the higher thickness of the PP microfibres (368.3 µm) compared to the PA nanofibres (10 min: 1.16 µm; 20 min: 1.52 µm; 30 min: 2.64 µm). So, when combining the structures, the PA nanofibres thickness is hidden by the PP microfibres due to the different thicknesses of the two structures. It was also observed that the PP microfibres have a non-homogeneous mat in terms of thickness, with a high standard deviation value. It is also possible to conclude that the addition of PA nanofibres decreases the permeability values, due to the presence of extra material. Once again, greater differences are observed between the samples for which nanofibres were produced for 10 min and 20 min, and the samples produced during 20 and 30 min. The air-permeability parameter is more affected by the presence of the nanofibre membrane. On one hand, lower air permeability values may increase the filtration efficiency, however, on the other hand, it may compromise the breathability parameter. So, this topic needs to be considered when analyzing the filtration parameter and some optimization may be required to obtain the appropriate equilibrium between these two factors.

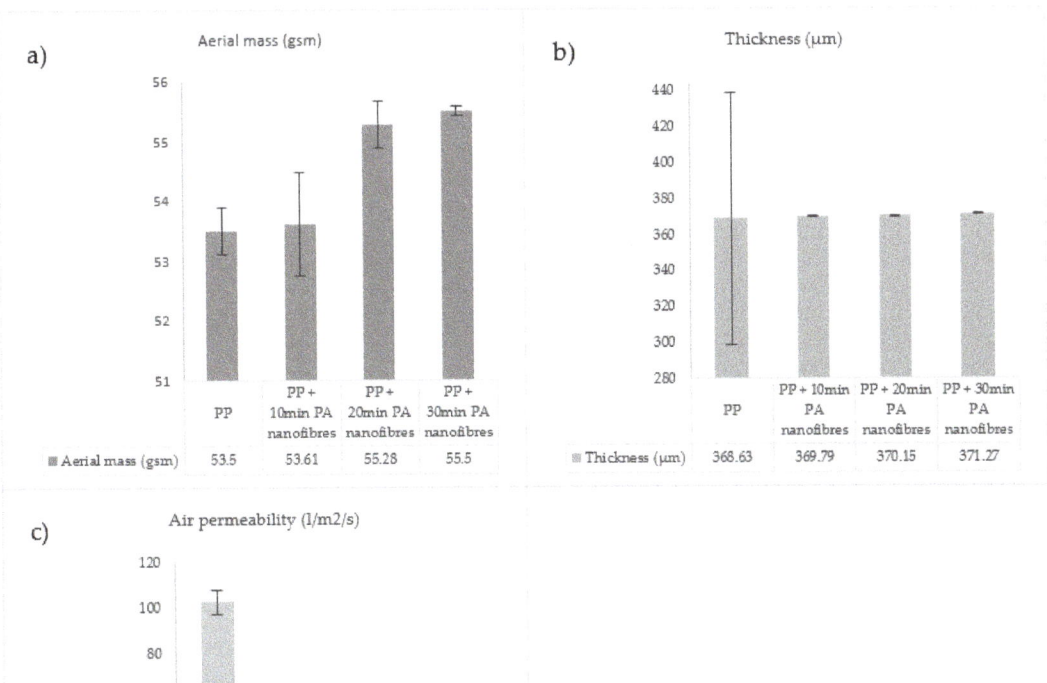

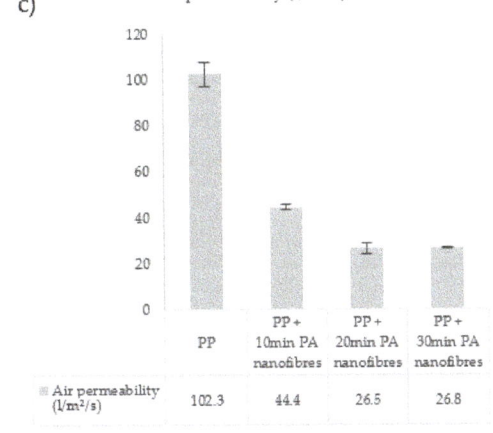

Figure 7. Aerial mass (a), thickness (b) and air permeability (c) of PP and PP + PA nanofibres combinations.

The SEM images shown in Figure 8 represents three different deposition times of the electrospun nanofibres onto the PP microfibres studied. Due to the different dimensions of PA (nano-scale) and PP (micro-scale) fibres, the two fibres are perfectly distinguishable from each other. From the images provided, it is possible to observe that two different layers were obtained via the establishment of contact points between the two adjacent layers. It is also clear that the presence of the PP membrane as a substrate during nanofibres production did not affect the electrospinning process, as the nanofibres with the presence of the PP membrane have the same morphology as the ones represented in Table 4. The density of the PA nanofibres increases with the deposition time, observable in the SEM images. This result increases the aerial mass and thickness and decreases the air permeability, as was observed in the results represented in Figure 7.

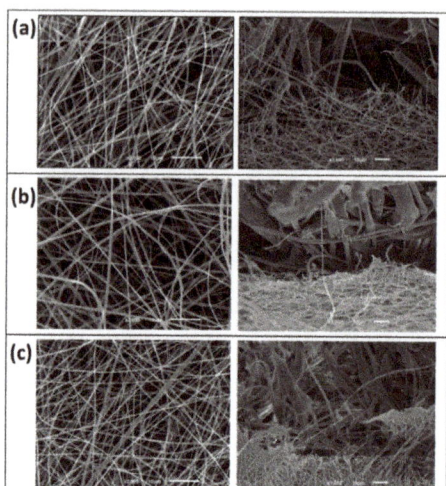

Figure 8. SEM images of the PP microfibres combined with PA nanofibres. The nanofibres were produced for (**a**) 10 min, (**b**) 20 min and (**c**) 30 min.

Since the multilayer filtration system is intended as personal protective equipment, it will probably have direct contact with skin on one side and with the external environment on the other side. For this reason, two extra layers were added, one as an inner layer and the other as an outer layer, to promote higher comfort levels and mechanical support, respectively. These two layers were composed of polyester (PES) polymer, for which microscopic images are represented in Figure 9. The PES IL is denser since it has been selected due to softness. The PES OL will be in contact with the environment and was chosen due to its higher rigidity to provide protection from possible dangers in the external environment and to support the entire structure. Therefore, a multilayer system with three different layers—inner layer (PES IL); filtration layer (PP microfibres + PA nanofibres + PP microfibres); outer layer (PES OL)—were defined. At this point, 4 different combinations were tested: the inner and outer layer remained the same throughout, the filtration layer changed because of the three different deposition times of the studied electrospun nanofibres. The several combinations studied are represented in Table 3. The membranes were combined using a thermoforming process and moulded using a mask shape mould.

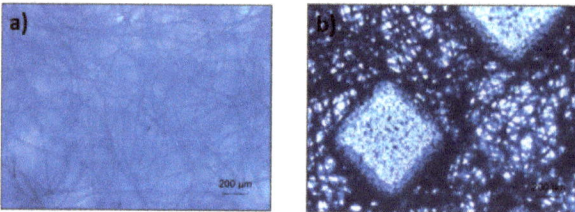

Figure 9. Morphology of PES nonwoven for (**a**) inner and (**b**) outer layer under an optical microscope.

The multilayer system has been optimized for application in filtration systems, to promote micro and nanoparticle retention. More specifically, it can be applied in products such as suits or respiratory masks. The filtration and respiratory parameters were analyzed in relation to the intended purpose of the filtration system. The results are represented in Table 6. The filtration topic was evaluated by the penetration of sodium chloride after 3.5 min of exposure, and it was observed that the multilayer system without the presence of PA nanofibres achieved much higher values (15.39%) than the systems with PA nanofibres

(7.83, 5.9 and 7.11%). This result highlights that the presence of nanofibres promotes smaller penetration rates, and, consequently higher retention rates, thereby increasing the filtration efficiency. Therefore, the results show that the penetration value of the multilayer system, composed of a PA membrane, produced for 20 min, is 2.7 times lower than the penetration value of the system with the absence of PA nanofibres. It is also important to emphasize that the amount of PA present in the system is very small compared to the remaining layers that make up the entire system. By comparing the three multilayer systems, composed with PA nanofibres, the system composed of nanofibres produced for a smaller deposition time—10 min—was identified as having a higher rate of particle penetration and a lower efficiency in terms of filtration. The system in which nanofibres had been produced for 20 min resulted in smaller penetration rates with higher filtration capabilities. The results represented in Figure 7, related to aerial mass, thickness and air permeability reveal that the multilayer systems in which PA nanofibres were produced for 20 and 30 min are very similar. The deposition time can only be increased to a certain threshold, due to electric field distortion occuring in nanofibre production, thereby changing the deposition profile, an effect that has previously been described by Vrieze et al. [29]. So, the different values shown in Table 6 for the multilayer systems composed of nanofibres produced during 20 and 30 min are probably related to the lower homogeneity of the PA nanofibres produced for 30 min, creating points of lower density and decreasing the filtration efficiency.

Table 6. Filtration and respiratory evaluation of the four different multilayer systems studied.

Sample:	Multilayer _0 min	Multilayer _10 min	Multilayer _20 min	Multilayer _30 min
Test	Result			
Penetration of the filter material with sodium chloride after 3.5 min of exposure (%)	15.39	7.83	5.90	7.11
Respiratory resistance (mbar): Inhalation at 30 L/min	0.37	0.66	0.65	0.66
Respiratory resistance (mbar): Inhalation at 95 L/min	1.53	2.23	2.33	2.34
Respiratory resistance (mbar): Exhalation at 160 L/min	2.36	3.73	3.62	4.09

The last three rows of Table 6 are related to the respiratory parameter and the evaluation is performed by analyzing the resistance in millibar (mbar). The lower the resistance values, the greater the breathing facility. The system without the presence of nanofibres has the smallest resistance values, both in terms of exhalation and in terms of inhalation. This result is logical since the combination has one missing layer compared to the other systems. Additionally, the higher value may not be problematic, if the values are under the limits defined by the legislation. These limit values change according to the location in which the product is distributed and also depending on the type of product. For example, in Europe, the legislation that is applied is the RfU PPE-R/02.075.02—certification of filtering half mask against SARS-CoV-2. This legislation defines that when analyzing the respiratory resistance of an FFP2 mask, in inhalation at 30 m/min, the maximum value permitted is 0.7 mbar and at 95 L/min the maximum value permitted is 2.4 mbar. Additionally, when evaluating the exhalation at 160 L/min the value must not exceed 3.0 mbar [30]. In the results obtained within this work, the respiratory resistance at inhalation achieved values that are under values specified by the legislation. In the exhalation evaluation, it was observed that the masks composed of nanofibres membranes produced during 20 and 30 min have a respiratory resistance value that is slightly higher than defined by legislation. Therefore, for this specific application, this aspect should be optimized. Optimization can be performed by changing the electrospinning parameters, such as the production time or the flow rate, or by changing the thermoforming parameters, such as the temperature or the time of forming. When comparing the three multilayers composed with PA nanofibres, very similar results are obtained to the three systems at each respiratory resistance condition. The value that differs the most is the exhalation evaluation at the multilayer system with PA nanofibres produced for 30 min. This value is slightly higher and is related to the presence of more material. At this point, the differences previously observed for 20 and 30 min are no longer observable because the respiratory parameters are affected by the entire structure.

4. Conclusions

Multilayer systems composed of PA nanofibres at the filtration layer were produced. Within this study, different polymer concentrations, flow rates and needle diameters, applied during electrospinning technique, were tested to define suitable conditions for filtration. It was concluded that lower flow rates, combined with smaller needle diameters, enhance the production of nanofibres with smaller diameters. In the second section of the study, different multilayer systems, with PA nanofibres located at the filtration layer, were studied. The PA nanofibres between the different multilayer systems had different production times which led to membranes with a distinct aerial mass, thickness and air permeability. The three-pointed properties affect the filtration efficiency. It was observed that the ideal deposition time for the PA nanofibres is 20 min. The nanofibres, together with the substrate layer—PP microfibres—have an aerial mass equal to 55.28 gsm, a thickness of 370.15 μm and an air permeability of 26.5 $L/m^2/s$. For lower values, the filtration performance was compromised due to a lack of sufficient material to prevent the penetration of the particles. On the other hand, for higher time periods of nanofibre production, it was observed that the properties of the samples do not increase proportionally because the produced nanofibre mats suffer some distortion during the electrospinning technique. The most important conclusion taken from this study is the high potential of nanotechnology, more precisely of electrospun nanofibres, since, with the addition of a very tinny layer of nanofibres, the penetration values decreased 2.7 times when compared with membranes without electrospun nanofibres.

Author Contributions: Conceptualization, V.P. and J.G.D.; methodology, V.P.; software, V.P.; validation, R.F.; investigation, V.P.; resources, J.G.D.; data curation, V.P.; writing—original draft preparation, V.P.; writing—review and editing, V.P. and R.F.; supervision, J.B. and F.C.; project administration, C.M.; funding acquisition, J.G.D. All authors have read and agreed to the published version of the manuscript.

Funding: This research was funded by Portugal 2020 through project nº POCI-01-02B7-FEDER-048171, NanoMask—"Desenvolvimento de máscaras de proteção do tipo FFP2 termoconformadas, com gradiente de filtração (nano e micro), para contexto profissional".

Institutional Review Board Statement: Not applicable.

Informed Consent Statement: Not applicable.

Data Availability Statement: Not applicable.

Conflicts of Interest: The authors declare no conflict of interest.

References

1. Harrison, R.M.; Yin, J. Particulate matter in the atmosphere: Which particle properties are important for its effects on health? *Sci. Total Environ.* **2000**, *249*, 85–101. [CrossRef]
2. Li, H.; Wang, Z.; Zhang, H.; Pan, Z. Nanoporous PLA/(Chitosan Nanoparticle) Composite Fibrous Membranes with Excellent Air Filtration and Antibacterial Performance. *Polymers* **2018**, *10*, 1085. [CrossRef] [PubMed]
3. Park, K.; Kang, S.; Park, J.-W.; Hwang, J. Fabrication of silver nanowire coated fibrous air filter medium via a two-step process of electrospinning and electrospray for anti-bioaerosol treatment. *J. Hazard. Mater.* **2021**, *411*, 125043. [CrossRef] [PubMed]
4. Gorji, M.; Bagherzadeh, R.; Fashandi, H. Electrospun nanofibers in protective clothing. In *Electrospun Nanofibers*; Elsevier Ltd.: Amsterdam, The Netherlands, 2017; pp. 571–598.
5. Tebyetekerwa, M.; Xu, Z.; Yang, S.; Ramakrishna, S. Electrospun Nanofibers-Based Face Masks. *Adv. Fiber Mater.* **2020**, *2*, 161–166. [CrossRef]
6. Selvam, A.K.; Nallathambi, G. Polyacrylonitrile/silver nanoparticle electrospun nanocomposite matrix for bacterial filtration. *Fibers Polym.* **2015**, *16*, 1327–1335. [CrossRef]
7. Tcharkhtchi, A.; Abbasnezhad, N.; Seydani, M.Z.; Zirak, N.; Farzaneh, S.; Shirinbayan, M. An overview of filtration efficiency through the masks: Mechanisms of the aerosols penetration. *Bioact. Mater.* **2020**, *6*, 106–122. [CrossRef] [PubMed]

8. Adanur, S.; Jayswal, A. Filtration mechanisms and manufacturing methods of face masks: An overview. *J. Ind. Text.* **2020**, 1–35. [CrossRef]
9. Srikrishnarka, P.; Kumar, V.; Ahuja, T.; Subramanian, V.; Selvam, A.K.; Bose, P.; Jenifer, S.K.; Mahendranath, A.; Ganayee, M.A.; Nagarajan, R.; et al. Enhanced Capture of Particulate Matter by Molecularly Charged Electrospun Nanofibers. *ACS Sustain. Chem. Eng.* **2020**, *8*, 7762–7773. [CrossRef]
10. Barhate, R.S.; Ramakrishna, S. Nanofibrous filtering media: Filtration problems and solutions from tiny materials. *J. Membr. Sci.* **2007**, *296*, 1–8. [CrossRef]
11. Bazgir, M.; Zhang, W.; Zhang, X.; Elies, J.; Saeinasab, M.; Coates, P.; Youseffi, M.; Sefat, F. Fabrication and Characterization of PCL/PLGA Coaxial and Bilayer Fibrous Scaffolds for Tissue Engineering. *Materials* **2021**, *14*, 6295. [CrossRef] [PubMed]
12. Khandaker, M.; Progri, H.; Arasu, D.; Nikfarjam, S.; Shamim, N. Use of Polycaprolactone Electrospun Nanofiber Mesh in a Face Mask. *Materials* **2021**, *14*, 4272. [CrossRef] [PubMed]
13. Qin, X.; Subianto, S. *Electrospun Nanofibers for Filtration Applications*; Elsevier Ltd.: Amsterdam, The Netherlands, 2017; pp. 449–466. [CrossRef]
14. Xue, J.; Wu, T.; Dai, Y.; Xia, Y. Electrospinning and electrospun nanofibers: Methods, materials, and applications. *Chem. Rev.* **2019**, *119*, 5298–5415. [CrossRef] [PubMed]
15. Yang, Y.; Li, B.; Chen, Z.; Sui, N.; Chen, Z.; Xu, T.; Li, Y.; Fu, R.; Jing, Y. Sound insulation of multi-layer glass-fiber felts: Role of morphology. *Text. Res. J.* **2016**, *87*, 261–269. [CrossRef]
16. Wu, S.; Wang, B.; Zheng, G.; Liu, S.; Dai, K.; Liu, C.; Shen, C. Preparation and characterization of macroscopically electrospun polyamide 66 nanofiber bundles. *Mater. Lett.* **2014**, *124*, 77–80. [CrossRef]
17. Heikkilä, P.; Taipale, A.; Lehtimäki, M.; Harlin, A. Electrospinning of polyamides with different chain compositions for filtration application. *Polym. Eng. Sci.* **2008**, *48*, 1168–1176. [CrossRef]
18. Matulevicius, J.; Kliucininkas, L.; Martuzevicius, D.; Krugly, E.; Tichonovas, M.; Baltrusaitis, J. Design and Characterization of Electrospun Polyamide Nanofiber Media for Air Filtration Applications. *J. Nanomater.* **2014**, *2014*, 859656. [CrossRef]
19. Yun, K.M.; Hogan, C.J.; Matsubayashi, Y.; Kawabe, M.; Iskandar, F.; Okuyama, K. Nanoparticle filtration by electrospun polymer fibers. *Chem. Eng. Sci.* **2007**, *62*, 4751–4759. [CrossRef]
20. Pais, V.; Navarro, M.; Guise, C.; Martins, R.; Fangueiro, R. Hydrophobic performance of electrospun fibers functionalized with TiO_2 nanoparticles. *Text. Res. J.* **2021**. [CrossRef]
21. Francavilla, P.; Ferreira, D.P.; Araújo, J.C.; Fangueiro, R. Smart Fibrous Structures Produced by Electrospinning Using the Combined Effect of PCL/Graphene Nanoplatelets. *Appl. Sci.* **2021**, *11*, 1124. [CrossRef]
22. Zarshenas, K.; Raisi, A.; Aroujalian, A. Surface modification of polyamide composite membranes by corona air plasma for gas separation applications. *RSC Adv.* **2015**, *5*, 19760–19772. [CrossRef]
23. Charles, J.; Ramkumaar, G.R.; Azhagiri, S.; Gunasekaran, S. FTIR and Thermal Studies on Nylon-66 and 30% Glass Fibre Reinforced Nylon-66. *E-J. Chem.* **2009**, *6*, 23–33. [CrossRef]
24. Kang, E.; Kim, M.; Oh, J.S.; Park, D.W.; Shim, S.E. Electrospun BMIMPF6/nylon 6,6 nanofiber chemiresistors as organic vapour sensors. *Macromol. Res.* **2012**, *20*, 372–378. [CrossRef]
25. Díaz-Alejo, L.A.; Menchaca-Campos, C.; Chavarín, J.U.; Sosa-Fonseca, R.; García-Sánchez, M.A. Effects of the Addition of *Ortho*- and *Para*-NH_2 Substituted Tetraphenylporphyrins on the Structure of Nylon 66. *Int. J. Polym. Sci.* **2013**, *2013*, 323854. [CrossRef]
26. Biglari, M.J.; Rahbar, R.S.; Shabanian, M.; Khonakdar, H.A. Novel composite nanofibers based on polyamide 66/graphene oxide-grafted aliphatic- aromatic polyamide: Preparation and characterization. *Polym. Technol. Mater.* **2018**, *58*, 879–888. [CrossRef]
27. Nguyen, T.N.M.; Moon, J.; Kim, J.J. Microstructure and mechanical properties of hardened cement paste including Nylon 66 nanofibers. *Constr. Build. Mater.* **2020**, *232*, 117134. [CrossRef]
28. Xiao, L.; Xu, L.; Yang, Y.; Zhang, S.; Huang, Y.; Bielawski, C.W.; Geng, J. Core–Shell Structured Polyamide 66 Nanofibers with Enhanced Flame Retardancy. *ACS Omega* **2017**, *2*, 2665–2671. [CrossRef] [PubMed]
29. De Vrieze, S.; Westbroek, P.; Van Camp, T.; van Langenhove, L. Electrospinning of chitosan nanofibrous structures: Feasibility study. *J. Mater. Sci.* **2007**, *42*, 8029–8034. [CrossRef]
30. C.R.E. 2020/403. Co-Ordination of Notified Bodies PPE Regulation 2016/425. 2020. Available online: https://eur-lex.europa.eu/legal-content/EN/TXT/?uri=CELEX:32016R0425 (accessed on 18 November 2021).

Article

Mechanical and Thermal Properties of Polyurethane Materials and Inflated Insulation Chambers

Goran Čubrić [1], Ivana Salopek Čubrić [2,*], Dubravko Rogale [1] and Snježana Firšt Rogale [1]

- [1] Department of Clothing Technology, Faculty of Textile Technology, University of Zagreb, 10 000 Zagreb, Croatia; goran.cubric@ttf.unizg.hr (G.Č.); dubravko.rogale@ttf.unizg.hr (D.R.); sfrogale@ttf.unizg.hr (S.F.R.)
- [2] Department of Textile Design and Management, Faculty of Textile Technology, University of Zagreb, 10 000 Zagreb, Croatia
- * Correspondence: ivana.salopek@ttf.unizg.hr

Abstract: Evaluating mechanical and thermal characteristics of garment systems or their segments is important in an attempt to provide optimal or at least satisfying levels of comfort and safety, especially in the cold environment. The target groups of users may be athletes engaged in typical sports that are trained in the cold, as well as football players that play matches and train outdoors during the winter season. Previous studies indicated an option to substitute the inner layers of an intelligent garment with polyurethane inflated chambers (PIC) to increase and regulate thermal insulation. In this paper, the authors investigate the mechanical properties of polyurethane material with and without ultrasonic joints. Furthermore, they investigate the potential of designed PICs in terms of efficiency and interdependence of air pressure and heat resistance. The results indicated that an inflated PIC with four diagonal ultrasonic joints has the highest ability to maintain the optimal thermal properties of an intelligent clothing system. The influence of direction and number of ultrasonic joints on the mechanical properties of polyurethane material is confirmed, especially in terms of compression resilience and tensile energy.

Keywords: polyurethane; material; clothing; mechanical properties; heat resistance

1. Introduction

The comfort of textile and clothing has been the focus of number investigations, especially the aspects of thermophysiological [1–5] and sensorial comfort [6–9]. A number of parameters affect the establishment of comfort. According to the comfort equations, parameters are dominantly related to the parameters of body, environment, and clothing. More specifically, the parameters are as follows:

a. Parameters of the body: skin wetness (w), skin temperature (t_{sk}), DuBois area (A_{Du}), the part of skin included in the transfer by radiation (A_r/A_{Du}), and skin emissivity (ε_{sk});
b. Parameters of the environment: radiant temperature (t_r), air velocity (v), air temperature (t_a), air pressure (p_a), and pressure on the skin ($p_{sk,s}$);
c. Parameters of textile/clothing: clothing insulation (I_{cl}), clothing area factor (f_{cl}), and clothing temperature (t_{cl}).

Furthermore, there are additional parameters of textile materials that affect thermophysiological comfort. Evaluating the thermal properties of garments is crucial for the establishment of optimal or at least satisfying levels of wear comfort, especially when it comes to protective clothing. The evaluation is important for the estimation of the potential heat stress in cold or extremely cold environmental conditions. One of the ongoing missions of researchers is to establish an optimal combination of comfort and thermal insulation to improve garment functionality and sustainability in work and leisure. Clothing has a substantial effect on the thermoregulation of the human body, especially in extreme climatic conditions. Therefore, the thermal comfort of clothing should be precisely defined to ensure optimal use [10]. Clothing with inadequate thermal properties not only diminishes a person's satisfaction but can also affect their physical performance and may even affect health. Clothing needs to provide optimal insulation in the cold to protect the body from hypothermia. Furthermore, it needs to facilitate the optimal transfer of produced sweat through the garment and the whole clothing system [11].

It is well known that wool is a good regulator in terms of temperature and humidity. Having this in mind, a group of researchers focused on the comparison of outdoor jackets with two different battings in the cold environment, specifically, air temperature of $-5\,^\circ$C and 43% relative humidity [12]. The first batting was made of sheep wool, while the second was made of polyester microfiber. The outcomes of the experiment indicated that the wool batting reduces the chill effect. At the same time, this batting caused an undesirable accumulation of moisture in the jacket, which led to a longer drying rate. Another group of scientists focused on the investigation of the transfer abilities (both heat and mass transport) of single layers through a multilayered system [13]. They measured six characteristics important for the classification of transfer properties. The results showed the heat transfer of the whole system depends on the transfer measured through a single layer. The results of another study indicated a significant influence of fabric finishing on the preservation of heat between inner layers [14]. The performance of multilayer systems changes when the clothing is exposed to various ambient conditions. To investigate this effect, a system consisting of four layers was exposed to seven significantly different ambient conditions, ranging from extremely hot to cold [15]. The authors concluded that exposure to different environmental conditions affects the distribution of moisture accumulated within the layers of the system. In cold, approximately 80% of the moisture was accumulated in the zone of the thermic liner. To assure satisfactory thermal insulation of a garment and increase the protection against cold, inner layers may be replaced by different inflated chambers. The garment can be further improved by the incorporation of activated polyurethane chambers [16]. Further investigation showed that the insulation of a completely activated chamber increases significantly (2.7×) when compared to the chamber before the activation [17]. Over time, the properties of polyurethane material change due to aging. The results of a study focusing on the change in thermal properties of polyurethane-coated material due to natural weathering indicated that, after 3 months of exposure to cold conditions, the thermal resistance decreased by 25% [18].

In this paper, the authors focus on a further investigation of the potential of polyurethane material and polyurethane inflated chambers. The intention is to investigate the mechanical and thermal properties of polyurethane material to design different types of polyurethane inflated chambers, as well as to investigate their efficiency and the interdependence of air pressure and heat resistance.

2. Research Methodology

Intelligent thermally adaptive clothes (ITAC) are developed at the Faculty of Textile Technology University of Zagreb [19]. The ITAC activity is based on the integration of air pressure in thermal expanding insulation chambers (TEICs) and sensors for indoor (clothing microclimate) and outdoor (carrier environment) temperature, using a microcontroller with an algorithm and actuators to control the TEIC. When the wearer carries the ITAC, the microcontroller system monitors changes in indoor and outdoor depending on the wearer's physical activity, compares them with the achieved thermal protection, and makes autonomous decisions on the required degree of thermal protection.

The ITAC has a feature of intelligent clothes because the built-in components automatically follow changes in the clothing microclimate and carrier environment, assess the substantial and necessary state of thermal protection, make decisions, and independently perform the adaptation of thermal insulation to the level ensuring constant thermal comfort. The target groups of users may be sportsmen and recreational athletes engaged in different sports that are trained in the cold, such as skiers, hikers, and sailors, as well as football players that play matches and train outdoors during the winter season. Furthermore, targeted groups include other people doing activities in cold weather conditions [16].

TEIC made of polyurethane foil can replace multiple thermal insulation layers of textile materials. Conventional textile materials have a fixed and constant value of thermal insulation, while TEIC made of highly elastic polyurethane foil can change the value of thermal insulation depending on its filling with air, i.e., thickness. The functional dependencies of changes in the thickness of the TEIC on the air pressure and in the thermal resistance depending on the thickness of the TEIC were investigated. The TEIC was made and integrated into an ITAC and filled with an air pressure of 0–70 mbar, whereby TEIC thicknesses of 0–25 mm were measured. A thermal resistance of 0.188–0.502 $m^2 \cdot K \cdot W^{-1}$ was measured on the thermal manikin. The ratio of thermal insulation of the maximum activated TEIC (thickness 25 mm, air pressure 70 mbar) was almost three times higher than that of the nonactivated TEIC (thickness 0 mm, air pressure 0 mbar).

Therefore, the level of thermal protection depends on the thickness of the TEIC, because there is more air, representing a good insulator, in the thicker chambers. This is the reason why the TEIC was made from airtight highly elastic polyurethane foil. [17].

The TEIC was made of highly elastic polyurethane foil designated as Walopur 4201AU made by Bayer Epurex Films GmbH, Germany. It is characterized by a material density of 1.15 g/cm^3, a softening point of 140 to 150 °C, a thickness of 0.196 mm, and very high elongation at breaking force, amounting to 550%. Moreover, the material is highly ultraviolet (UV)-resistant and hydrolytically stable, while it has good properties according to thermal and ultrasonic methods, as well as good microbiological stability, which is important for its incorporation into the clothes. The measuring samples of the TEIC were joined using an ultrasonic welding machine.

The research methodology of this paper includes a definition of the selected polyurethane material, measurement of mechanical properties of the polyurethane materials, design of polyurethane insulation chambers (PICs), and a measurement of the heat resistance of inflated polyurethane insulation chambers (PICs). A PIC is an experimental laboratory sample of polyurethane material joined with an ultrasonic machine. Into such a chamber, the air is blown. The blown air provides thermal insulation.

2.1. Materials and Joining Technique

For the investigation, we selected polyurethane (PU) foil. In Figure 1, microscopic images of the selected foil under a magnification of 50× and 200×, taken by a DINO-Lite microscope, are given. The surface mass of the selected foil was 232 g·m^{-2} with a thickness of 0.25 mm [20].

 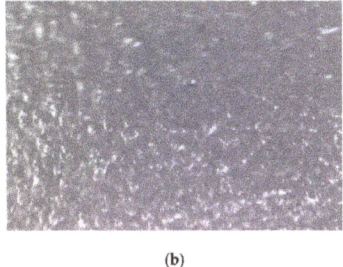

(a) (b)

Figure 1. Microscopic image of polyurethane foil: (**a**) under a magnification of 50×; (**b**) under a magnification of 200×.

For the preparation of polyurethane material with ultrasonic joints and PICs, an ultrasonic joining machine was used. The machine for ultrasonic joining (Figure 2) had a lower engraving counter-roll, 8 mm wide (the joint was in the form of three lines of the same width). The distance between the lower engraving counter and the sonotrode was set to 0.28 mm. Additionally, the speed of welding was set to 3 m·min^{-1}, and the welding energy was 285 W.

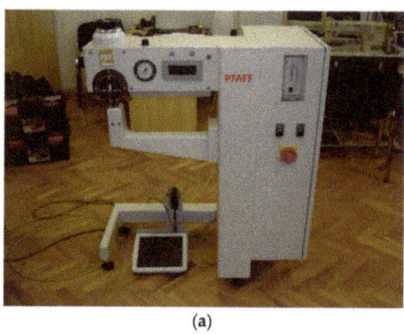
(a)

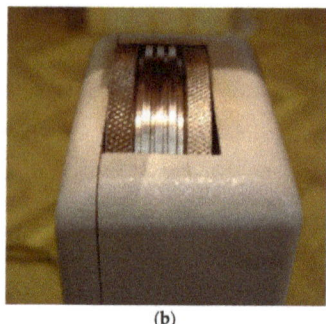
(b)

Figure 2. The machine for ultrasonic joining: (**a**) the machine; (**b**) lower engraving counter-roll.

2.2. Measurement of Mechanical Properties of Polyurethane Materials

For the investigation of mechanical properties, six samples consisting of two layers of polyurethane material were prepared. Five of them were joined ultrasonically. For further reference, prepared samples are designated as follows: PU–number of joints (1, 2, or 4)–the direction of joints (x, y, or d), where the "y" direction of joints corresponds to the direction of extrusion of the PU foil, the "x" direction corresponds to the direction perpendicular to the extrusion direction, and the "d" direction refers to the diagonal direction of joints.

The mechanical properties of polyurethane materials were investigated using a Kawabata Evaluation System (KES). Before testing on the KES system, all samples were conditioned for 24 h (conditions: temperature (T) = 20 ± 2 °C; relative humidity (RH) = 65% ± 2%). All polyurethane materials were investigated in terms of tensile energy (WT), extension (EMT), bending stiffness (B), and shear stiffness (G). The polyurethane material without an ultrasonic joint was additionally tested for compressional energy (WC), compressional resilience (RC), coefficient of friction (MIU), and surface roughness (SMD). A detailed description of measuring techniques was given in previously published papers [20,21].

2.3. Design of PICs

All designed PICs consisted of two layers of polyurethane material. PICs were joined ultrasonically. The dimensions of the chambers were 290 × 290 mm, while the dimensions into which the air was blown were 220 × 220 mm. Table 1 shows the PIC dimensions, along with a detailed description.

Adequate measuring equipment was designed for inflating air into the PICs (Figure 3). It consisted of a mini compressor, a manometer with a pressure sensor, a shut-off valve, and the PIC. The mini compressor inflated air into the PIC. When appropriate pressure was reached within the PIC, the mini compressor stopped inflating air. The shut-off valve prevented the exit of air from the PIC.

Table 1. Design of polyurethane insulation chambers (PICs). ID, identifier.

No.	PIC ID	Dimension	Description
1	C-0	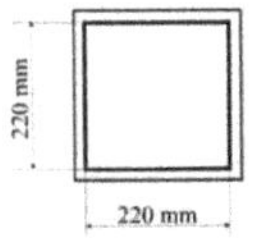 220 mm × 220 mm	- joint type: ultrasonic - number of joints: 0 - position of joints: n/a - length of joint: n/a
2	C-1V	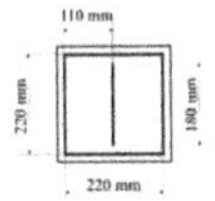 220 mm × 220 mm, 110 mm, 180 mm	- joint type: ultrasonic - number of joints: 1 - position of joints: in the middle - length of joint: 180 mm
3	C-2V	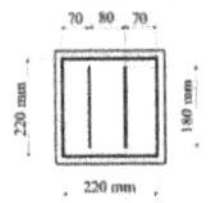 220 mm × 220 mm, 70/80/70, 180 mm	- joint type: ultrasonic - number of joints: 2 - position of joints: two joints with a spacing of 80 mm - length of joint: 180 mm
4	C-3V	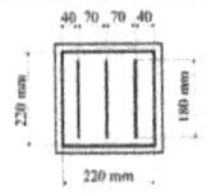 220 mm × 220 mm, 40/70/70/40, 180 mm	- joint type: ultrasonic - number of joints: 3 - position of joints: three joints with a spacing of 70 mm - length of joint: 180 mm
5	C-4V	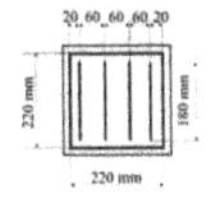 220 mm × 220 mm, 20/60/60/60/20, 180 mm	- joint type: ultrasonic - number of joints: 4 - position of joints: four joints with a spacing of 60 mm - length of joint: 180 mm
6	C-4D	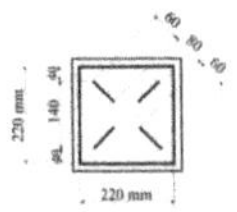 220 mm × 220 mm, 60/60, 140	- joint type: ultrasonic - number of joints: 4 - position of joints: diagonal with break - length of joint: 60 mm

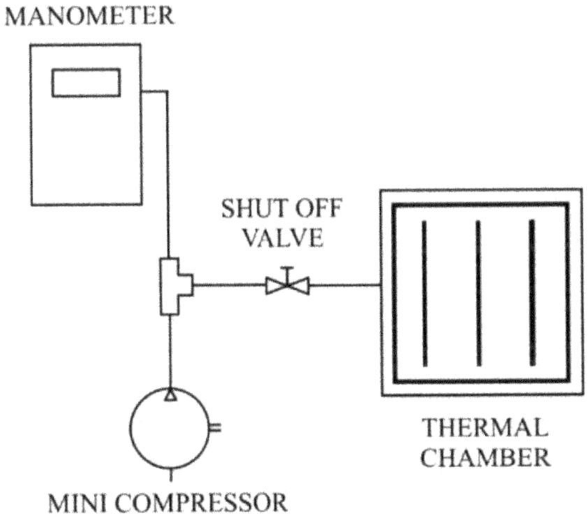

Figure 3. Measuring equipment.

2.4. Measurement of Thermal Properties of PICs

For measurement of the heat resistance of designed PICs, a sweating guarded hotplate (model SGHP-8.2) was used. The device simulates the processes of heat and moisture transfer that occur next to the human skin (Figure 4). The measuring device consists of a heated plate with a temperature that matches the skin temperature [22,23].

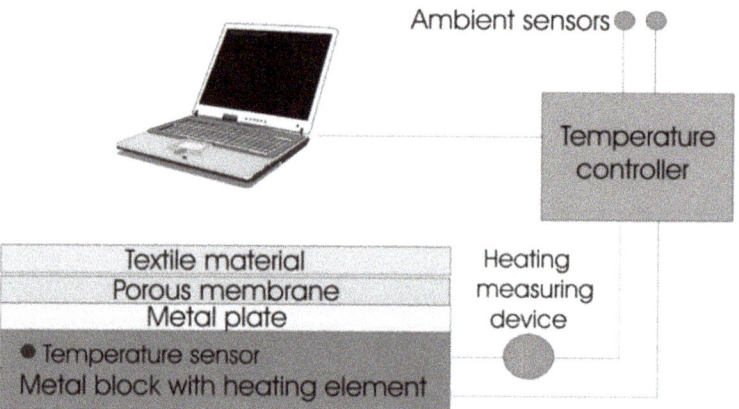

Figure 4. Principle of sweating guarded hotplate (SGHP) device.

The heat resistance was calculated using the following equation:

$$R_{ct} = \frac{(T_s - T_a)}{\frac{H}{A}} - R_{ct0}, \tag{1}$$

where R_{ct} is the heat resistance of a sample (m²·K·W⁻¹), T_s is the hotplate surface temperature (°C), T_a is the air temperature (°C), H/A is the heat flux in the observed zone (W·m⁻²), and R_{ct0} is the heat resistance of the plate (m²·K·W⁻¹).

For measurement of the heat resistance, noninflated PICs and PICs inflated at an air pressure of 10, 20, 30, 40, and 50 mbar were prepared.

Testing was conducted at temperatures of 20 °C, 15 °C, and 10 °C. During the testing, the humidity was set to 65% ± 1%. The air velocity was 1 m·s⁻¹.

3. Results and Discussion

The results include the measured mechanical properties of polyurethane materials and thermal properties of designed PICs.

3.1. Measured Mechanical Properties of Polyurethane Materials

In Table 2 and Figures 5 and 6, results of the measured mechanical properties (WC, RC, MIU, and SMD) of polyurethane materials without ultrasonic joints are given. In Figures 7–10, a comparison of the measured mechanical properties (WT, EMT, B, and G) of polyurethane materials with and without ultrasonic joints is given.

Table 2. Mechanical properties of polyurethane material without ultrasonic joint. WC, compressional energy; RC, compressional resilience; MIU, coefficient of friction; SMD, surface roughness.

Property	Index	Average Value	Coefficient of Variation, %
Compressional energy, N·m⁻¹	WC	0.046	13.50
Compressional resilience, %	RC	81.02	11.13
Coefficient of friction, dimensionless	MIU	1.064	5.71
Surface roughness, μm	SMD	0.221	3.43

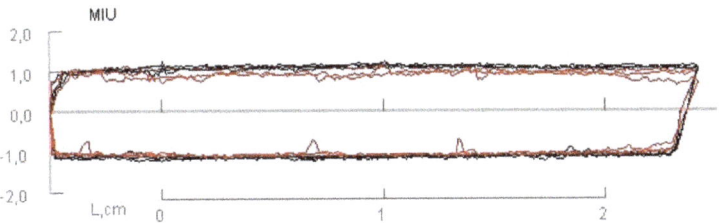

Figure 5. Coefficient of friction of polyurethane material without ultrasonic joint.

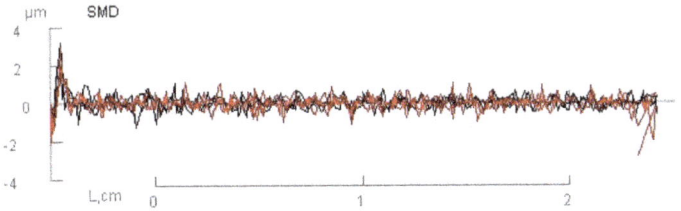

Figure 6. The surface roughness of polyurethane material without ultrasonic joint.

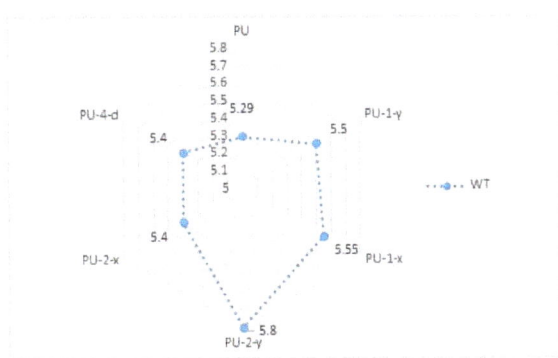

Figure 7. Tensile energy (WT) of polyurethane materials.

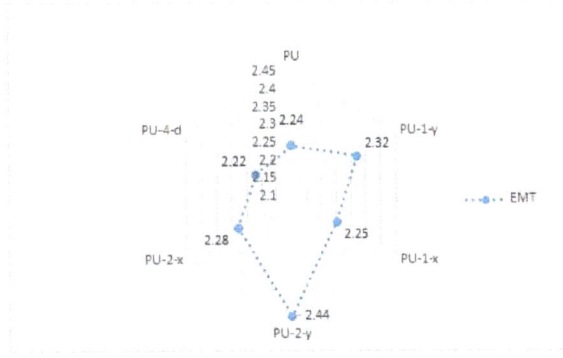

Figure 8. The extension (EMT) of polyurethane materials.

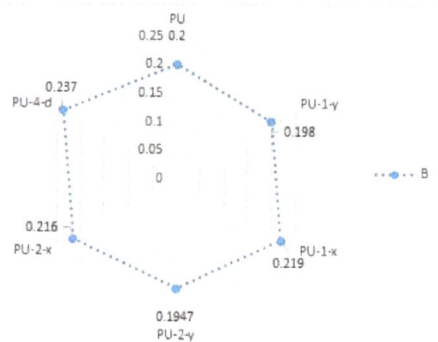

Figure 9. Bending stiffness (B) of polyurethane materials.

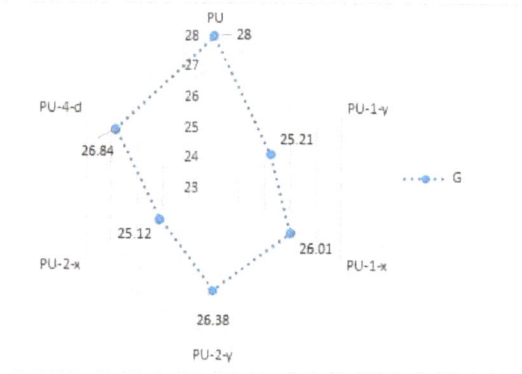

Figure 10. Shear stiffness (G) of polyurethane materials.

As can be seen from Table 2, the compressional energy of polyurethane material was 0.046 N·m^{-1}, with a coefficient of variation of 13.5%. Its compressional resilience was 81.02%, with a coefficient of variation of 11.13%. The coefficient of friction (MIU) and surface roughness (SMD) describe the material's friction properties and surface contour. The measured value of the coefficient of friction was 1.064, while the value of surface roughness was 0.221 μm.

The tensile energy of the six polyurethane materials was in the range of 5.3–5.8 N·m^{-1} (Figure 7). As can be seen, the polyurethane material with two joints in the y-direction (i.e., sample designated as PU-2-y) had the highest value of tensile energy. The values of the tensile energy of the remaining materials were quite similar. The increase in tensile energy for materials with joints, in comparison to the polyurethane material without ultrasonic joints, was 2–8%. Figure 8 presents the results of the tested extension of polyurethane materials. As can be seen from the figure, the extension of materials was in the range of 2.2–2.45%. Among the tested polyurethane materials, the highest extension was again characteristic of the material with two joints in the y-direction (sample PU-2-y). Among the materials with one ultrasonic joint, the material with a joint in the y-direction exhibited higher extension (sample PU-1-y). This brought us to the conclusion that ultrasonic joints in the y-direction increased the overall extension of the designed polyurethane material. According to the obtained results, there were small differences between the values of bending stiffness of ultrasonically joint polyurethane materials (Figure 9). This can be explained by the change in thickness of polyurethane material in the area of the joint. Another reason for such behavior is a change in the structure of the material at the joint point, which prevented the material from slipping. The results of the shear stiffness

(Figure 10) show that the polyurethane material with an ultrasonic joint had lower shear stiffness in comparison to the material without a joint. The reason was the rigidity of the ultrasonic joint, which prevented the shear of polyurethane material.

3.2. Measured Thermal Properties of PICs

Figures 11–14 show the results of the measured heat resistance of designed PICs at different pressures and temperatures.

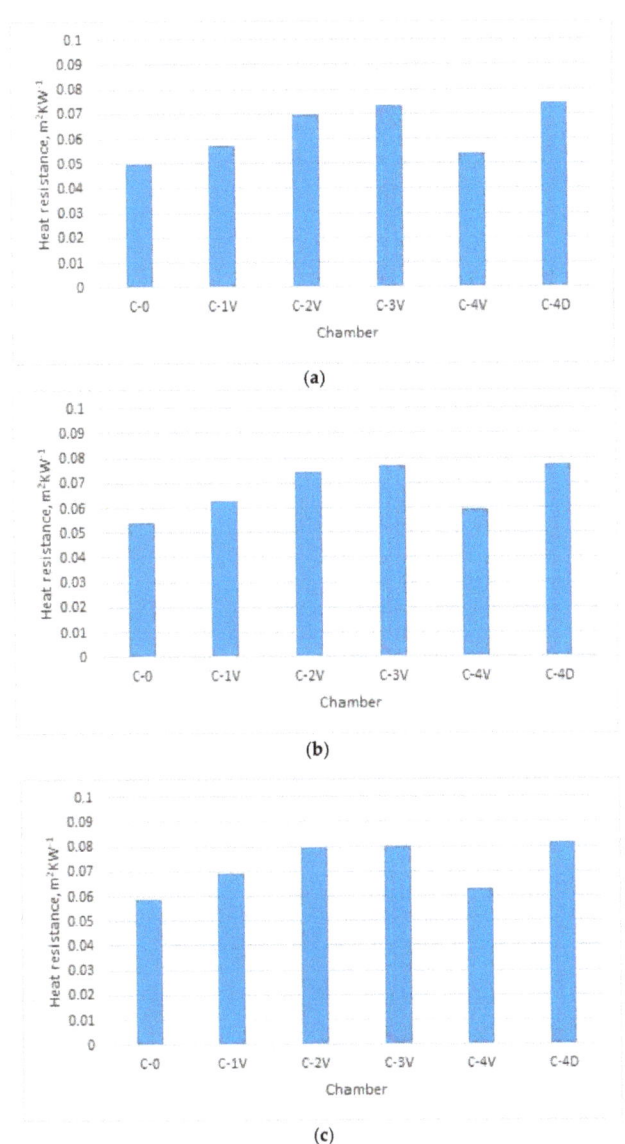

Figure 11. Heat resistance of designed noninflated PICs at different ambient temperatures (T): (a) T = 20 °C; (b) T = 15 °C; (c) T = 10 °C.

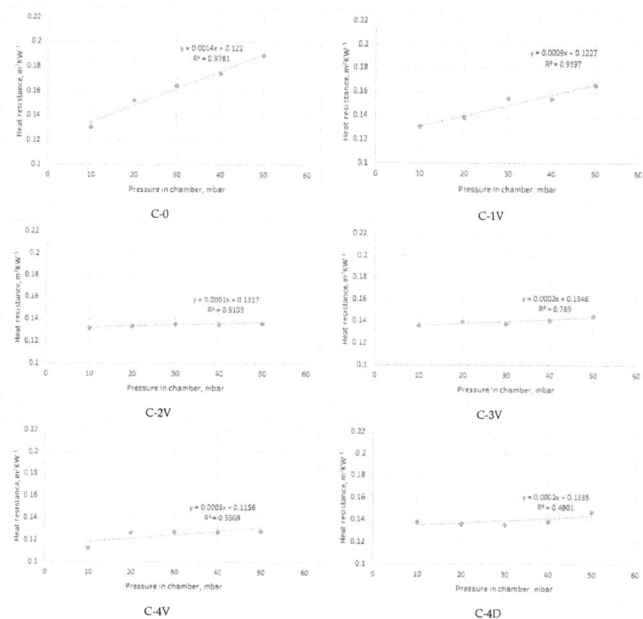

Figure 12. Dependence of heat resistance and air pressure of inflated PICs s at a constant ambient temperature of 20 °C.

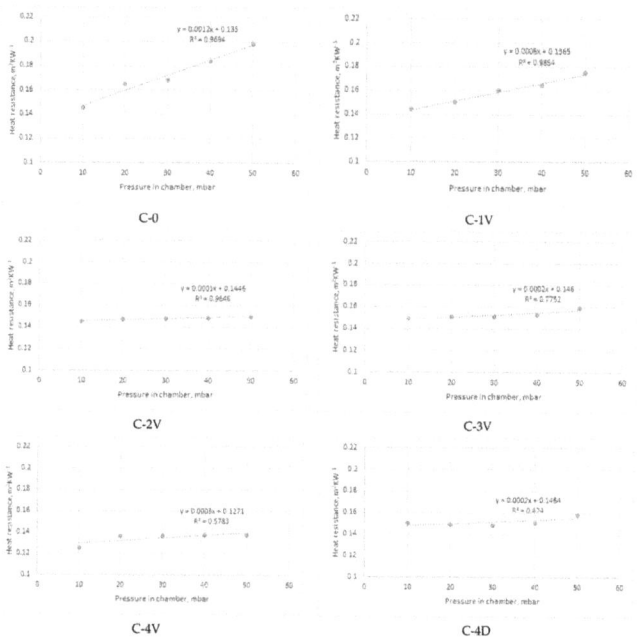

Figure 13. Dependence of heat resistance and air pressure of inflated PICs at a constant ambient temperature of 15 °C.

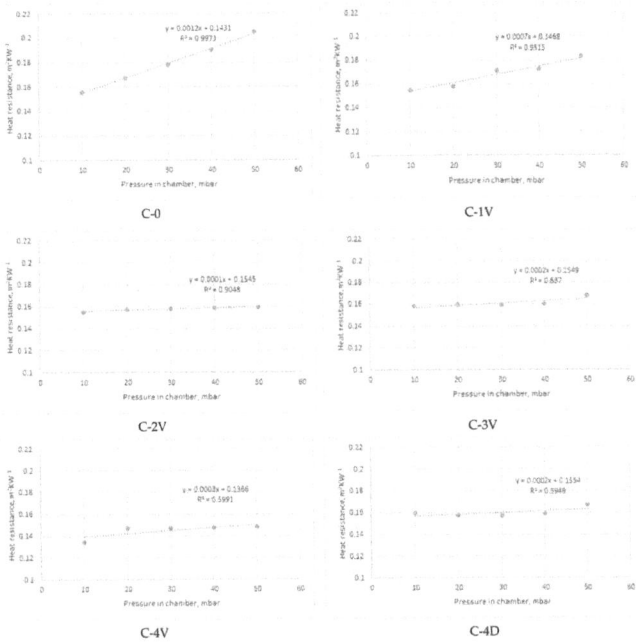

Figure 14. Dependence of heat resistance and air pressure of inflated PICs at a constant ambient temperature of 10 °C.

In this study, a hot plate was used to test the heat resistance of six designed PICs. The PICs were tested under different pressures (0, 10, 20, 30, 40, and 50 mbar) and at different temperatures established in the climate chamber (20 °C, 15 °C, and 10 °C). Upon blowing air into the PIC, the pressure in the PIC increased. This means that there was more air in the PIC at a pressure of 50 mbar than at, for example, a pressure of 30 mbar. The amount of air in the PIC depended on its design. The measured heat resistance of designed PICs was from 0.0496 to 0.0815 $m^2 \cdot K \cdot W^{-1}$ when air was not blown into the chambers (i.e., for nonactivated PICs), and from 0.1309 to 0.2047 $m^2 \cdot K \cdot W^{-1}$ when air was blown into the chambers (i.e., for activated PICs).

The heat resistance of nonactivated PICs (i.e., at 0 bar of air pressure) was in the range of 0.049 to 0.082 $m^2 \cdot K \cdot W^{-1}$. The influence of the design can be well seen in Figure 11. Specifically, under all three ambient temperatures (i.e., T = 10 °C, T = 15 °C, and T = 20 °C), the simplest designed PIC (i.e., the chamber without joints, designated as C-0), provided the lowest resistance to heat transfer. The R_{ct} values were 0.049, 0.054, and 0.058 $m^2 \cdot K \cdot W^{-1}$, respectively. This PIC was followed by the PIC with four vertical joints (designated as C-4V). The highest resistance to heat transfer was achieved by PICs with four diagonal joints and three vertical joints (chambers C-4D and C-3V).

It is interesting to note the changes in the heat resistance of PICs after air was blown into the chambers. Accordingly, the results shown in Figure 12 indicate that the heat resistance of the PIC without an ultrasonic joint (chamber C-0), which in the noninflated state had the lowest value, changed significantly. Already at a pressure of 20 mbar, the heat resistance of PIC C-0 became highest among the tested PICs (0.1531 $m^2 \cdot K \cdot W^{-1}$). The reason for this significant change in the heat resistance of PIC C-0 was the different construction of the PIC. Specifically, the mentioned PIC was the only one among the six tested PICs that did not have an ultrasonic joint. In this PIC, even at a pressure of 20 mbar, a larger amount of air was blown due to the joints that, in other PICs, prevented the chamber from inflating. In this way, a smaller amount of air could be inflated into the chambers.

The PIC with four vertical joints (designated as C-4V) had the lowest heat resistance among all designed PICs. This refers to all investigated pressures and all three ambient temperatures. The reason for such performance can be found in the PIC design. More specifically, four vertical joints disabled the inflation of a higher amount of air for defined pressures.

Measurements of heat resistance were also performed at lower ambient temperatures, i.e., at 15 °C and 10 °C. Due to technical limitations, the temperature of 10 °C was the lowest temperature that could be established and kept constant during the entire measurement cycle in the existing climatic chamber. With these changes in ambient temperature, the heat resistance of all samples increased. Concerning the heat resistance of the PICs at a temperature of 20 °C, the heat resistance of PICs at 15 °C increased to 11%, and that at 10 °C increased to 19%.

Considering the results obtained, an intelligent garment with active thermal protection can benefit from the use of TEIC with four diagonal joints (C-4D). Satisfactory options are also the polyurethane chambers with two ultrasonic joints at a distance of 80 mm (C-2V) and with four ultrasonic joints at a distance of 60 mm (C-4V). All named PICs at a pressure of 50 mbar had a lower heat resistance than the other designed PICs. As a result, the mentioned TEIC would retain a smaller amount of heat next to the body in colder conditions, which would not fulfil its basic purpose. A PIC without an ultrasonic joint (designated as C-0) would not be suitable for practical application. Although it had an extremely high heat resistance when compared to the other tested PICs, it is very unsuitable for use because it inflated significantly, even at low pressures. This would negatively affect wear comfort and limit body movements. The remaining PICs had relatively similar values of heat resistance. Among them, the single-ultrasound PIC (C-1V) also inflated significantly and would affect both wear comfort and body movement.

Figure 12, Figure 13 and Figure 14 present the average values of measured heat resistance. The measuring device (sweating guarded hotplate) calculated the average value on the basis of 30 measurements obtained in short intervals. Since the device is highly precise and was located in the climatic chamber with stable conditions (relative humidity, air temperature, air velocity), the coefficients of variation of measured heat resistance were rather low. Specifically, the coefficients of variation (CV) of heat resistance of PICs measured in different environmental temperatures were as follows:

- for T = 20 °C: CV = 0.12–1.12%;
- for T = 15 °C: CV = 0.15–1.13%;
- for T = 10 °C: CV = 0.12–1.56%.

Figures 12–14 give the linear regression equation and coefficients of determination describing the proportion of variance in the heat resistance that is predic from the pressure in the chamber. As can be seen from the figures, high values of the coefficient were characteristic for PICs C-0, C-1V, and C-2V at all three ambient temperatures (20 °C, 15 °C, and 10 °C). According to the results, the remaining PICs exhibited moderate to weak correlation (the weakest one was for PIC C-4D at all temperatures).

In a previous study, the researchers showed that the ratio of thermal insulation of maximum activated TEIC was almost three times higher than that of nonactivated TEIC [17]. To compare results of the heat resistance of designed PICs, the ratios of maximum activated and nonactivated PICs were calculated and are presented in Table 3.

Table 3 shows that the observed ratio was the highest for the PIC without a joint (C-0). The range was from 3.47 (at 10 °C) to 3.82 (at 20 °C). This outcome was to be expected. The largest amount of air was blown into this PIC because it did not have a joint that would prevent the PIC form expanding. The ratio for the PIC with one joint (C-1V) was 2.65–2.90. Even with this PIC, a larger amount of air was blown. For other PICs, the ratio was about 2, indicating that the amount of blown air was approximately the same.

Table 3. The ratios of maximum activated and nonactivated PICs.

PIC ID	Ambient Temperature		
	20 °C	15 °C	10 °C
C-0	3.82	3.67	3.47
C-1V	2.90	2.78	2.65
C-2V	1.92	2.01	1.99
C-3V	1.97	2.00	2.09
C-4V	2.06	2.03	2.06
C-4D	1.99	2.04	2.05

4. Conclusions

The experiment described in this paper was aimed at investigating the mechanical properties of polyurethane materials used to design polyurethane inflated TEIC, which serves as an adaptive thermal layer in intelligent clothing. Testing of their efficiency considering the thermal protection was also carried out. Throughout the experiment, the influence of the direction and number of ultrasonic joints on the mechanical properties of polyurethane material was determined, especially in terms of the increase in compressional resilience and tensile energy. From the presented results, it can be concluded that the design and construction of PICs are extremely important for the preservation of heat inside the garment. Their optimal choice can greatly affect the wear comfort, especially in a cold environment. The results indicate that a PIC with four diagonal ultrasonic joints is optimal for use within thermally adaptive clothing. In contrast, a PIC without joints is not recommended for practical application, as it would decrease the level of wear comfort and limit body movement. The results also show that the ratios of maximum activated and nonactivated PICs were in the range of 1.92 to 3.82.

It is expected that the outcomes of this study will be used when designing specific sport garments that will enhance the performance of top athletes in cold environments.

Author Contributions: Conceptualization, D.R. and G.Č.; methodology, D.R. and G.Č.; formal analysis, G.Č. and I.S.Č.; investigation, G.Č., D.R. and I.S.Č.; resources, S.F.R.; data curation, I.S.Č.; writing—G.Č. and I.S.Č. writing—review and editing G.Č., I.S.Č., S.F.R., D.R.; visualization G.Č., I.S.Č.; project administration, G.Č., S.F.R.; funding acquisition, D.R., I.S.Č. All authors have read and agreed to the published version of the manuscript.

Funding: This work has been supported in part by the Croatian Science Foundation through the projects IP-2018-01-6363 "Development and thermal properties of intelligent clothing (ThermIC)" and IP-2020-02-5041 "Textile materials for enhanced comfort in sports (TEMPO)" as well as by the University of Zagreb trough research grant TP12/20 "Design of functional sportswear with implication of mathematical models and algorithms".

Institutional Review Board Statement: Not applicable.

Informed Consent Statement: Not applicable.

Data Availability Statement: Data available in a publicly accessible repository.

Conflicts of Interest: The authors declare no conflict of interest.

References

1. Tunakova, V.; Tunak, M.; Tesinova, P.; Seidlova, M.; Prochazka, J. Fashion clothing with electromagnetic radiation protection: Aesthetic properties and performance. *Text. Res. J.* **2020**, *90*, 2504–2521. [CrossRef]
2. Li, R.; Yang, J.; Xiang, C.; Song, G. Assessment of thermal comfort of nanosilver-treated functional sportswear fabrics using a dynamic thermal model with human/clothing/environmental factors. *Text. Res. J.* **2018**, *88*, 413–425. [CrossRef]
3. Wardiningsih, W.; Troynikov, O. Treated knitted fabric for hip protective pads for elderly women. Part II. Performance relevant to thermal comfort. *Text. Res. J.* **2019**, *89*, 5006–5013. [CrossRef]
4. Atasağun, H.G.; Okur, A.; Psikuta, A.; Rossi, R.M.; Annaheim, S. The effect of garment combinations on thermal comfort of office clothing. *Text. Res. J.* **2019**, *89*, 4425–4437. [CrossRef]

5. Erdumlu, N.; Saricam, C. Investigating the effect of some fabric parameters on the thermal comfort properties of flat knitted acrylic fabrics for winter wear. *Text. Res. J.* **2017**, *87*, 1349–1359. [CrossRef]
6. Uren, N.; Okur, A. Analysis and improvement of tactile comfort and low-stress mechanical properties of denim fabrics. *Text. Res. J.* **2019**, *89*, 4842–4857. [CrossRef]
7. Salopek Cubric, I.; Skenderi, Z. Evaluating Thermophysiological Comfort Using the Principles of Sensory Analysis. *Coll. Antropol.* **2013**, *37*, 57–64.
8. Balci Kilic, G.; Okur, A. Effect of yarn characteristics on surface properties of knitted fabrics. *Text. Res. J.* **2019**, *89*, 2476–2489. [CrossRef]
9. Salopek Čubrić, I.; Čubrić, G.; Perry, P. Assessment of Knitted Fabric Smoothness and Softness Based on Paired Comparison. *Fibers Polym.* **2019**, *20*, 656–667. [CrossRef]
10. Fan, J.; Tsang, H.W.K. Effect of Clothing Thermal Properties on the Thermal Comfort Sensation during Active Sports. *Text. Res. J.* **2008**, *78*, 111–118.
11. Wang, S.X.; Li, Y.; Tokura, H.; Hu, J.Y.; Han, Y.X.; Kwok, Y.L.; Au, R.W. Effect of Moisture Management on Functional Performance of Cold Protective Clothing. *Text. Res. J.* **2007**, *77*, 968–980. [CrossRef]
12. Kofler, P.; Nussbichler, M.; Veider, V.; Khanna, I.; Heinrich, D.; Bottoni, G.; Hasler, M.; Caven, B.; Bechtold, T.; Burtscher, M.; et al. Effects of two different battings (sheep wool versus polyester microfiber) in an outdoor jacket on the heat and moisture management and comfort sensation in the cold. *Text. Res. J.* **2016**, *86*, 191–201. [CrossRef]
13. Angelova, R.A.; Reiners, P.; Georgieva, E.; Kyosev, Y. The effect of the transfer abilities of single layers on the heat and mass transport through multilayered outerwear clothing for cold protection. *Text. Res. J.* **2018**, *88*, 1125–1137. [CrossRef]
14. Salopek Čubrić, I.; Skenderi, Z. Effect of Finishing Treatments on Heat Resistance of One- and Two- Layered Fabrics. *Fibers Polym.* **2014**, *15*, 1635–1640. [CrossRef]
15. He, J.; Li, J.; Kim, E. Assessment of the heat and moisture transfer in a multilayer protective fabric system under various ambient conditions. *Text. Res. J.* **2015**, *85*, 227–237. [CrossRef]
16. Firšt Rogale, S.; Rogale, D.; Nikolić, G. Intelligent clothing: First and second generation clothing with adaptive thermal insulation properties. *Text. Res. J.* **2018**, *88*, 2214–2233. [CrossRef]
17. Rogale, D.; Firšt Rogale, S.; Majstorović, G.; Čubrić, G. Thermal properties of thermal insulation chambers. *Text. Res. J.* **2020**. [CrossRef]
18. Potočić Matković, V.M.; Salopek Čubrić, I.; Skenderi, Z. Thermal resistance of polyurethane-coated knitted fabrics before and after weathering. *Text. Res. J.* **2014**, *84*, 2015–2025. [CrossRef]
19. Firšt Rogale, S.; Rogale, D.; Nikolić, G.; Dragčević, Z.; Bartoš, M. *Controllable Ribbed Thermoinsulative Chamber of Continually Adjustable Thickness and Its Application*; EP2254430; European Patent Office (EPO): Munich, Germany, 2010.
20. Čubrić, G.; Geršak, J.; Rogale, D.; Nikolić, G. Determination of mechanical properties of ultrasonically bonded polyurethane foils. *Tekstil* **2009**, *58*, 485–492.
21. Salopek, I.; Skenderi, Z.; Gersak, J. Investigation of knitted fabric dimensional characteristics. *Tekstil* **2007**, *56*, 391–398.
22. Mijović, B.; Skenderi, Z.; Salopek, I. Comparison of Subjective and Objective Measurement of Sweat Transfer Rate. *Coll. Antropol.* **2009**, *33*, 509–514. [PubMed]
23. *Textiles—Physiological effects—Measurement of Thermal and Water-Vapour Resistance under Steady-State Conditions (Sweating Guarded-Hotplate Test)*; ISO 11092:2014; International Organization for Standardization: Geneva, Switzerland, 2014.

Article

Study on the Compression Effect of Clothing on the Physiological Response of the Athlete

Marianna Halász [1,*], Jelka Geršak [2], Péter Bakonyi [3], Gabriella Oroszlány [1], András Koleszár [1] and Orsolya Nagyné Szabó [1]

[1] Institute for Industrial Product Design, Sándor Rejtő Faculty of Light Industry and Environmental Protection Engineering, Óbuda University, Doberdó u. 6, H-1034 Budapest, Hungary; oroszlany.gabriella@uni-obuda.hu (G.O.); koleszar.andras@uni-obuda.hu (A.K.); szabo.orsolya@uni-obuda.hu (O.N.S.)

[2] Research and Innovation Centre for Design and Clothing Science, Faculty of Mechanical Engineering, University of Maribor, Smetanova ulica 17, SI-2000 Maribor, Slovenia; jelka.gersak@um.si

[3] Department of Polymer Engineering, Faculty of Mechanical Engineering, Budapest University of Technology and Economics, Műegyetem rkp. 3, H-1111 Budapest, Hungary; bakonyi@pt.bme.hu

* Correspondence: halasz.marianna@uni-obuda.hu

Citation: Halász, M.; Geršak, J.; Bakonyi, P.; Oroszlány, G.; Koleszár, A.; Nagyné Szabó, O. Study on the Compression Effect of Clothing on the Physiological Response of the Athlete. *Materials* 2022, 15, 169. https://doi.org/10.3390/ma15010169

Academic Editor: Philippe Boisse

Received: 22 November 2021
Accepted: 24 December 2021
Published: 27 December 2021

Publisher's Note: MDPI stays neutral with regard to jurisdictional claims in published maps and institutional affiliations.

Copyright: © 2021 by the authors. Licensee MDPI, Basel, Switzerland. This article is an open access article distributed under the terms and conditions of the Creative Commons Attribution (CC BY) license (https://creativecommons.org/licenses/by/4.0/).

Abstract: The study aimed to analyze whether the high compression of unique, tight-fitting sportswear influences the clothing physiology comfort of the athlete. Three specific sportswear with different compression were tested on four subjects while they were running on a treadmill with increasing intensity. The compression effect of the sportswear on the body of the test persons, the temperature distribution of the subjects, and the intensity of their perspiration during running were determined. The results indicate that the compression effect exerted by the garments significantly influences the clothing physiology comfort of the athlete; a higher compression load leads to more intense sweating and higher skin temperature.

Keywords: clothing physiology; tight-fitting sportswear; running test on a treadmill; thermal comfort; skin temperature; perspiration

1. Introduction

Companies develop advanced high-tech, often high-compression sportswear for professional athletes [1–6]. Researches show that such sportswear can increase the performance of the athlete [7–18]. As a healthy lifestyle involves regular exercise, similar special garments are available for amateur and grassroots sports.

Sportswear has undergone enormous improvement since ancient times. In the ancient Olympic Games, men competed without clothing because they believed the best performance could be achieved this way. Until the 18th century, the manufacturing cost of clothes was so high that only the nobility and wealthy people could afford more clothing, especially for sports. During the Industrial Revolution, however, the textile industry started to develop at a tremendous rate, and so not only rich people could afford a garment or garments intended for sport. Until the 19th century, doing sports was mainly a privilege of men, but this did not bring about considerable changes in sportswear. Women started to play sports in higher numbers in the second half of the 19th century. The demand for women's sportswear started the development of sportswear, continuing intensively to this day [19–22].

The development of sportswear is directly related to the requirements of the sport, the nature and duration of the activities, and the requirements for adequate thermo-physiology comfort of the athlete [23–30]. In addition, scientists are examining the possibility of increasing performance through the compression that the garment exerts on the athlete's body [7–18]. In recent years, fibers for the manufacture of premium sportswear have been extensively researched and considerably improved. With the advancement of technology and the development of high-performance materials, sportswear now has greatly improved properties. Special emphasis is given to high aerodynamic and absorption properties, air and water permeability, strength, and adhesion. Much has also been performed in the field of design [31]. Sportswear is designed not to restrict the athlete's activity but provide them physical support while exercising [32].

These sportswear fabrics are usually composed of polyester or polyamide fibers and elastomer fibers. The light and strong polyester or polyamide fibers provide the necessary strength and clothing physiology parameters. The elastomer fibers, capable of large, completely elastic deformation, ensure that the garment completely fits the athlete's body during exercise and provides compression on it. The knitted fabric structure also facilitates elastic deformation due to the interconnecting loops [1–3].

Compression sportswear is a tight-fitting, compressive form of clothing utilizing the material's elasticity. Professional athletes wear compression suits to improve their athletic performance and speedy recovery from injury. Compression garments are used in high-performance sports such as running, skiing, swimming, and cycling [33]. It has been reported that compression garments improve the perception of muscle damage and increase performance in endurance tests [34]. Compression pants help improve athletic performance and reduce injuries by reducing muscle oscillations [35]. The use of compression stockings minimizes the risk of injury from the overall impact of physical exertion [34].

This case study aims to analyze whether the compression of the tight-fitting sportswear affects the clothing physiology comfort of the athlete and their motion parameters. In our literature search, we did not find any source that investigated the relationship between the compression of sportswear and the clothing physiology of an athlete. Therefore, we believe that the idea that this is worth exploring is novel.

2. Materials and Methods

We tested three tight-fitting sportswear with different compression. In the study, four subjects wore these tight-fitting sports garments while running with increasing intensity on a treadmill.

2.1. The Test Persons

We included four girls (TP1, TP2, TP3, and TP4) of similar age and body type in the study. The age, body mass, and main body measurements of the test subjects are detailed in Table 1.

Table 1. The age, body mass, and main body measurements of the test persons.

Designation	Test Persons			
	TP1	TP2	TP3	TP4
Age (years)	23	23	18	24
Body mass (kg)	59.32	66.88	65.25	70.60
Body height (cm)	166	169	171	171
Chest girth (cm)	96	95	92	94
Waist girth (cm)	73	70	71	77
Hip girth (cm)	97	104	99	111
Thigh girth (cm)	55	58	57	64
Lower leg/calf/girth (cm)	35.5	39	38.5	40.5
Inside leg length (cm)	78	76.5	77	79
Dress size upper	S	S	S	M
Dress size pants	S	M	M	M

The fitness levels of the girls are comparable. Currently, they do sports as a hobby, but as children and teenagers, they were active athletes. They exercise regularly every week, so the running intensity and interval in the tests matched their fitness level.

2.2. Materials

For our purpose, three tight-fitting sportswear were used: one ready-made sportswear made of polyester/elastane knitted fabric and two made-to-measure sportswear sets made of polyamide/elastane knitted fabric. Table 2 gives their designation and material composition.

Table 2. The examined sportswear.

Designation	Description	Composition of the Weft-Knitted Fabrics			
		Long-Sleeved T-Shirt		Pants	
Ready-made Sportswear	Ready-made sportswear	KF1	80% Polyester 20% Elastane	KF2	74% Polyester 26% Elastane
Made-to-measure Sportswear 1	Made-to-measure sportswear with 1% body size reduced	KF3	74% Polyamide 26% Elastane	KF3	74% Polyamide 26% Elastane
Made-to-measure Sportswear 2	Made-to-measure sportswear with 5% body size reduced	KF3	74% Polyamide 26% Elastane	KF3	74% Polyamide 26% Elastane

All three materials are a single weft-knitted fabric and contain a high percentage (20–26%) of elastane fibers. The structure of the knitted fabrics used is shown in Figure 1. The knitted structure with integrated elastane yarn ensures large-scale, multiaxial elastic deformation for the fabrics. In addition, the polyester (PES) and polyamide (PA) fibers provide strength and abrasion resistance and quickly wick sweat away from the body and allow it to evaporate, as they can retain very little moisture.

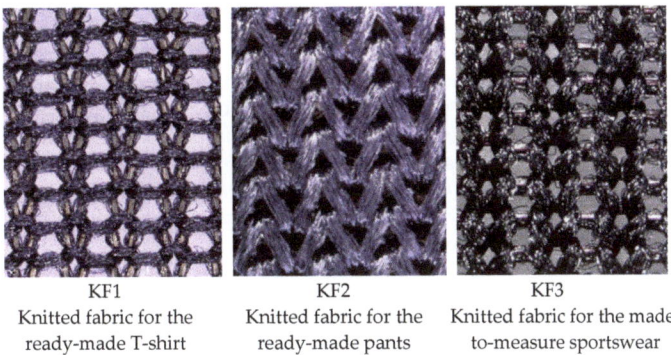

KF1	KF2	KF3
Knitted fabric for the ready-made T-shirt	Knitted fabric for the ready-made pants	Knitted fabric for the made-to-measure sportswear

Figure 1. The structure of the knitted fabrics used (microscopic images at 100× magnification).

Although the composition of the studied knitted fabrics is different, they can be used for our research purpose. This is because, although the properties of PA and PES fibers are not the same, the pressure exerted by sportswear made from these knitted fabrics is determined by the elastane fibers, which are present in a high proportion (20–26%) in the knitted fabrics in addition to PA and PES fibers.

The Ready-made Sportswear was bought in sizes S and M, corresponding to the size of the test persons and contains a long-sleeved T-shirt made from knitted fabric KF1 and pants made from knitted fabric KF2. The two made-to-measure sports suits (a long-sleeved T-shirt and pants) were constructed and made by us according to the cut of Ready-made Sportswear from the knitted fabric, KF3. They differ only in ease allowance (Table 2).

The Made-to-measure Sportswear 1 and 2 are made based on the test persons' body measurements. To ensure comparability, we constructed the patterns of the Made-to-measure Sportswear 1 and 2 to be similar to the bought Ready-made Sportswear. To achieve a compression effect of the sportswear, we have constructed the pattern for the tailored sportswear with negative ease allowance. This negative ease allowance was 1% in the case of the Made-to-measure Sportswear 1 and 5% for the Made-to-measure Sportswear 2, reducing the pattern size by 1% and 5% to under body size, respectively.

Table 3 summarizes the most important properties of the knitted fabrics used. The tests were performed in the Research and Innovation Centre for Design and Clothing Science at Faculty of Mechanical Engineering, University of Maribor, by standard conditions, according to ISO 139 (a temperature of 20 °C and a relative humidity of 65%). The following instruments were used: a Hildebrand thickness gauge, an Akustron Air-Permeability Measuring Instrument, an HC103/01 Moisture Analyzer, a KES-F7 Thermo Labo II for thermal properties, and a Zwick 005 universal testing machine for tensile tests (Zwick Roell GmbH, Ulm, Germany). The static immersion water absorption of the knitted fabrics used was determined according to the ASTM D 583-63 standard.

Table 3. Properties of the knitted fabrics of the examined sportswear.

Code of Knitted Fabric	Mass (g/m^2)	Thickness (mm)	Course Density (Piece/10 mm)	Wale Density (Piece/10 mm)	Air Permeability (L/m^2·s)	Moisture Content (%)	Immersion Water Absorption (%)	Dry Heat Flow (W)	Thermal Resistance R_{ct} (m^2·K/W)	Force at 5% Elongation * (N/m)
KF1	194.1 ± 1.5	0.397 ± 0.005	30	22	388.6 ± 7.2	3.15 ± 0.27	69.0 ± 2.38	2.07 ± 0.09	0.0719	7.44 ± 0.14
KF2	279.1 ± 5.3	0.497 ± 0.012	25	22	427.2 ± 25.9	1.35 ± 0.02	45.6 ± 0.85	2.01 ± 0.04	0.0636	10.36 ± 0.39
KF3	206.2 ± 0.3	0.356 ± 0.003	28	24	133.0 ± 5.3	2.93 ± 0.15	82.2 ± 1.64	2.13 ± 0.12	0.0721	12.35 ± 0.21

* In course direction.

2.3. Test Protocol

The four selected healthy girls participated in the experiment, where we studied the compression effect of clothing on the physiological response of the test subjects. They were selected to be approximately the same age (their average age is 21.0 ± 2.2 years), roughly the same size (body mass 63.3 ± 2.1 kg, height 170.5 ± 2.5 cm), and a similar level of fitness. This way, despite the relatively low number of test persons, we can draw sufficiently well-founded conclusions about the given population.

The wear trial tests were performed under the following ambient conditions: an ambient temperature of 24 °C and relative humidity of 40%.

The test persons wore the previously presented sportswear and ran on a treadmill at increasing intensity. The tests were carried out in the Motion Laboratory of the Department of Mechatronics, Optics and Mechanical Engineering Informatics, and the Biomechanical Research Centre of the Faculty of Mechanical Engineering of the Budapest University of Technology and Economics. The Motion Laboratory also allowed the biomechanical examination of the motion characteristics of the test persons. These results are presented in a separate paper [36].

The test subject performed the same physical activities in each test; a 15-min run, in which running speed was increased every 3 min in the following order:

- 3 min: Load I (running speed of 4 km/h);
- 3 min: Load II (running speed of 7 km/h);
- 3 min: Load III (running speed of 8 km/h);
- 3 min: Load IV (running speed of 10 km/h);
- 3 min: Load V (running speed of 11 km/h).

We performed the tests every morning at the same time. The test girls ran on the treadmill wearing another sports suit once each day. Figure 2 shows one of the test subjects in the Ready-made Sportswear while running on the treadmill, with the markers necessary for the movement tests.

Figure 2. A test person on the treadmill in the Ready-made Sportswear, with the markers for motion tests.

According to the study protocol, we determined the compression effect of the suit on the test subjects' bodies, the temperature distribution of the subjects, and the intensity of their sweating during running. For this purpose, the following measurements were carried out for each test subject while they were wearing their sportswear:

- The compression effect of sportswear on the body (measuring the pressure exerted by the sportswear on the body of the test person);
- Body mass loss due to sweat;
- Thermal imaging during running.

2.3.1. Determination of Pressure Distribution

We measured the compression of sportswear on the test person's body with the PicoPress tester at 17 points on the body (Figure 3).

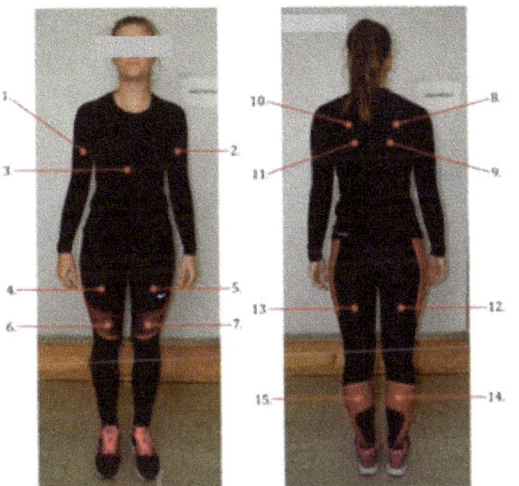 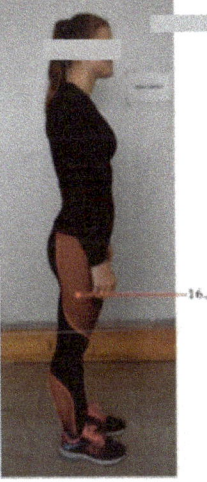

1. Right upper arm
2. Left upper arm
3. Ventral middle
4. Right thigh front
5. Left thigh front
6. Right knee
7. Left knee
8. Right scapula 1.
9. Right scapula 2.
10. Left scapula 1
11. Left scapula 2.
12. Right tight back
13. Left thigh back
14. Right thigh side
15. Left thigh side
16. Right shank back
17. Left shank back

Figure 3. Pressure measurement points.

Compression was measured on all test persons before the subjects started running. The measurement results are given as mean values of pressure on the upper body (Figure 3, pressure measurement points: 1–3, 8–11) and on the lower body (Figure 3, pressure measurement points: 4–7, 12–17).

2.3.2. Determination of Excreted and Absorbed Sweat

The amount of evaporated sweat was determined based on body mass loss, as the difference in the test subject's body mass without any clothing, before and after the study. The amount of sweat absorbed in clothing was determined based on the mass of clothing that a test subject was supposed to wear during the test. Test subjects were measured upon completing the study: first dressed, and then the clothing, piece by piece.

The mass of the clothing was measured with an AA Labor MT -XY 6000 (Midwest Microwave Solutions, Hiawatha, IA, USA) balance, which has an accuracy of 0.1 g, while body mass was measured with a Sartorius IS300IGG (Minebea Intec, Hamburg, Germany) balance with an accuracy of 2 g.

2.3.3. Thermal Imaging Analysis

From the clothing physiology point of view, the critical parameter is the thermal load on the subject during running. For this purpose, the body surface temperature was recorded with a FLIR A325sc thermal imaging camera (FLIR Systems, Wilsonville, OR, USA) (Figure 4).

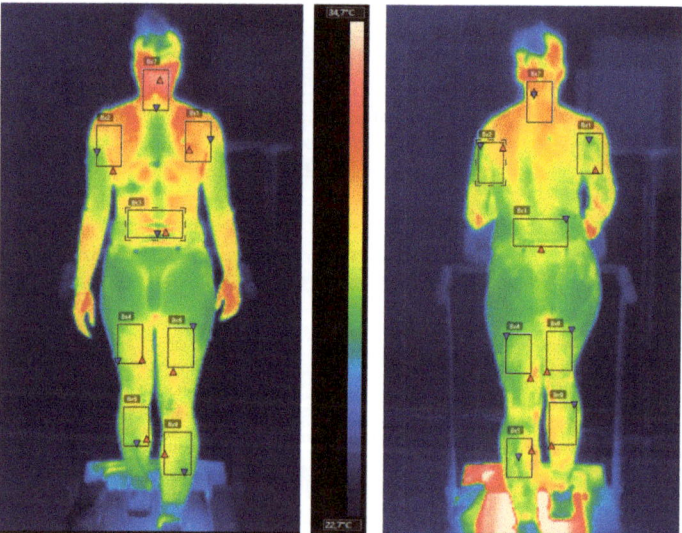

Figure 4. Thermal images at 0.0 min (**left**) and 15.0 min (**right**) with the marked reference areas.

For analysis, we selected eight areas on the body and determined their temperature values based on the thermal images when running speed was changed.

The selected areas were: neck-nape, right and left upper arm, back, right and left thigh, right and left calf. The times of evaluation were: 0 min, 3.0 min, 6.0 min, 9.0 min, 12.0 min, and 15.0 min.

Figure 4 shows the thermal image with the marked selected areas of one test person at 0.0 min and 15.0 min. Except for the neck, all selected areas are covered by sportswear.

3. Results and Discussion

Table 4 contains the average pressure values of each sportswear on each test person. (The detailed pressure values for each test person and sportswear at each measurement point are provided in the Appendix A, Table A1). The average pressure values show that the compression of Made-to-measure Sportswear 2 is higher for each test person than in the case of Made-to-measure Sportswear 1 and that the compression of garments constructed with the same amount of reduction in body size is about the same in the case of each test person. In addition, the Ready-made Sportswear has the highest compression, except for Test Person 2, in whose case the compression of the Ready-made Sportswear is between that of the two other, made-to-measure sports suits. It can also be observed that the standard deviation of the measured compression values for the Ready-made Sportswear is greater than that of the two made-to-measure sports suits. This is probably because the Ready-made Sportswear was not made for the size of the test persons; the persons had to choose from the sizes available. For this reason, the compression of the Ready-made Sportswear shows a higher difference between the test persons than that of made-to-measure sportswear.

Table 4. Average pressure values for each test person and sports suit.

Examined Sportswear	Body Part	Average Pressure (mmHg)			
		TP 1	TP 2	TP 3	TP 4
Ready-made Sportswear	upper	03.43 ± 1.72	2.57 ± 0.79	3.14 ± 1.21	03.72 ± 2.14
	lower	12.08 ± 6.13	7.92 ± 4.76	9.75 ± 5.67	10.75 ± 4.86
Made-to-measure Sportswear 1	upper	1.72 ± 1.70	1.86 ± 0.69	2.00 ± 1.00	2.00 ± 1.41
	lower	6.17 ± 3.27	5.83 ± 2.12	6.25 ± 2.60	6.83 ± 2.52
Made-to-measure Sportswear 2	upper	3.43 ± 1.99	3.57 ± 0.98	3.14 ± 1.21	3.43 ± 1.90
	lower	7.25 ± 2.63	8.25 ± 2.99	8.25 ± 2.86	7.67 ± 2.84

We analyzed the amount of sweat the test persons produced. Figure 5 shows the difference in body mass without clothing, that is, the mass loss of the test persons. Figure 6 shows the amount of sweat absorbed by the sportswear for each test person and suit.

On average, the test persons lost 100–200 g from their body mass during running (Figure 5). Some of it remained in the sportswear as sweat, and the rest evaporated, cooling the body.

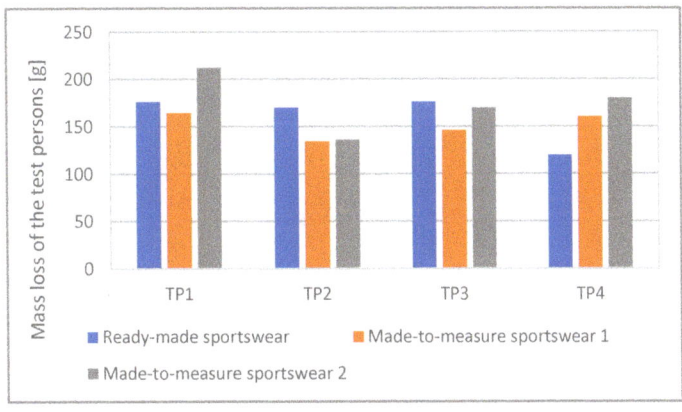

Figure 5. The mass loss of the test persons in the different compression sports suits.

The moisture absorbed by Made-to-measure Sportswear 1 and 2 can be compared well because they were made of the same fabric and the only difference between them is in compression (Figure 6). The higher compression Made-to-measure Sportswear 2 absorbed far more moisture than the lower compression Made-to-measure Sportswear 1 in the case of three test persons (TP2, TP3, and TP4), while in the case of Test Person 1, the two values were nearly identical. The higher compression Made-to-measure Sportswear 2 caused a higher reduction in body mass than the lower compression suit in the case of three test persons (TP1, TP3, and TP4), while with Test Person 2, both sports suits resulted in nearly identical body mass loss (Figure 5). These results indicate that higher compression generated more sweating.

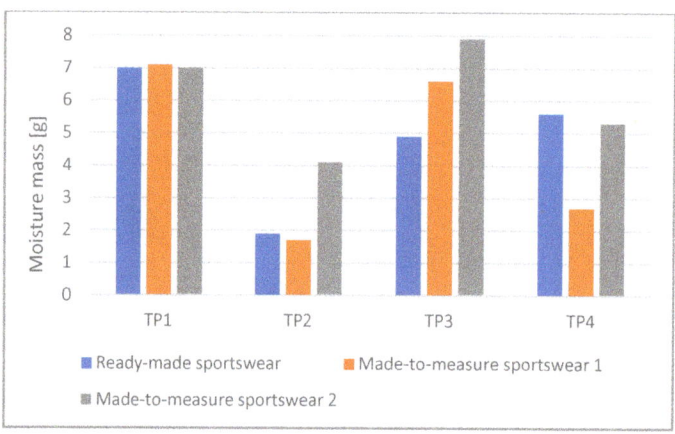

Figure 6. The amount of moisture remaining in the sportswear for each test person and suit.

Figure 7a–d shows the average body surface temperatures as a function of time for each test person and sports suit. As time passes, the average temperature decreases, due to the more intensive sweating, as the evaporating sweat and the moving air caused by running take heat away. The test persons stopped after 15 min.

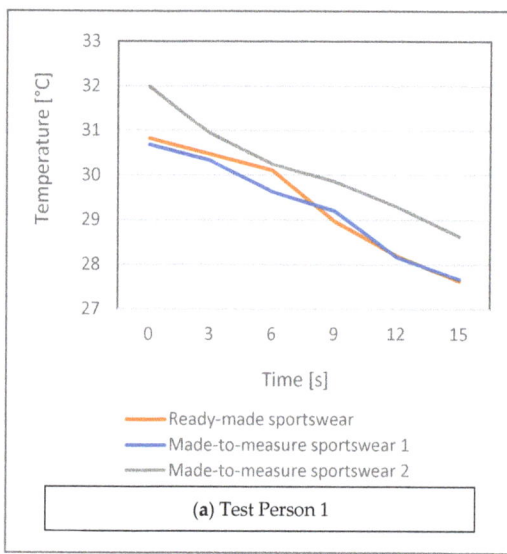

(**a**) Test Person 1

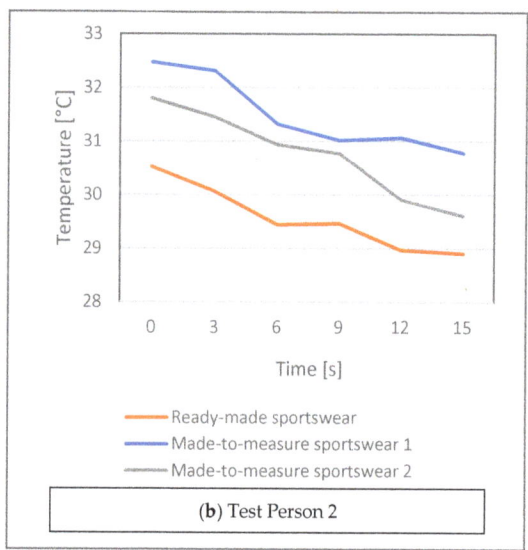

(**b**) Test Person 2

Figure 7. *Cont.*

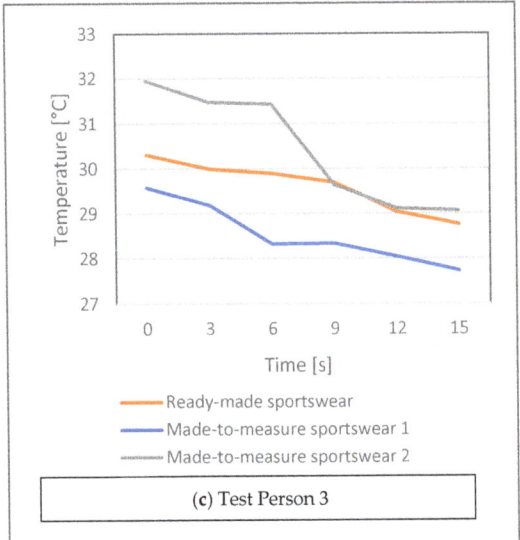
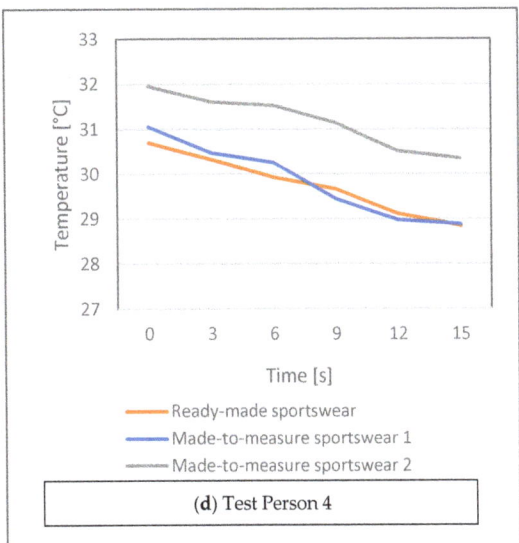

Figure 7. Average body surface temperature as a function of characteristic time points during running for each test person and sports suit.

The results obtained for Made-to-measure Sportswear 1 and 2 can be compared well in the case of thermal imaging as well since they were made from the same fabric; the only difference between them is in compression.

For better comparison, Figure 8 shows the average body surface temperature calculated for the whole running period. While wearing the higher compression Made-to-measure Sportswear 2, three test persons (TP1, TP3, and TP4) had a higher body surface temperature during running than in the case of the lower compression Made-to-measure Sportswear 1. In the case of Test Person 2, the opposite was true. These results indicate that higher compression generated higher body surface temperature.

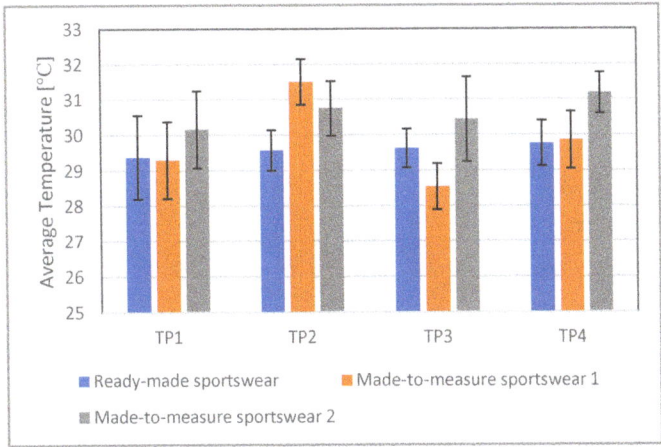

Figure 8. Average body surface temperature calculated for the whole period of running.

4. Conclusions

Our goal was to analyze whether the compression of sportswear on the athlete's body influences the clothing physiology comfort of the athlete. We included four test persons and used three different types of compression sportswear. We measured the compression of sportswear on the test persons, their body surface temperature, and the intensity of their sweating during running at increasing speed.

Our results indicate that the test persons sweated more in the higher compression Made-to-measure Sportswear 2. Their body surface temperature was higher than in the lower compression Made-to-measure Sportswear 1. We plan to conduct more extensive tests with more test persons to reveal more accurate relationships. Furthermore, we wish to supplement our research with measurements with zero compression sportswear.

However, our initial results strongly indicate that in high-tech sports clothing, the compression applied to increase performance has a considerable effect on the clothing physiology comfort of the athlete. This fact cannot be ignored when sports clothing is designed and fitted.

Author Contributions: Conceptualization, M.H. and J.G.; methodology, M.H., J.G. and P.B.; software, P.B.; validation, G.O., A.K. and O.N.S.; formal analysis, G.O., A.K. and O.N.S.; investigation, M.H. and J.G.; resources, M.H. and J.G.; data curation, G.O., A.K. and O.N.S.; writing—original draft preparation, M.H.; writing—review and editing, J.G.; visualization, G.O., A.K., P.B. and O.N.S.; project administration, M.H. and J.G.; funding acquisition, M.H., and J.G. All authors have read and agreed to the published version of the manuscript.

Funding: This research was funded by the Hungarian National Research, Development and Innovation Office through the Project "Relationship between Athletes' Motion and Clothing Physiology" (Grant Agreement No. TÉT_16-1-2016-0068) and the Slovenian Research Agency through the Slovenian–Hungarian Bilateral Scientific Cooperation (Grant Agreement No. BI-HU/17-16-006) and Research Program P2-0123.

Institutional Review Board Statement: Not Applicable.

Informed Consent Statement: Written informed consent was obtained from all subjects involved in the study to the research and publish this paper.

Data Availability Statement: Detailed data can be provided by the corresponding author on request.

Acknowledgments: We would like to thank Emese Antal, the leader of Daquini Activewear, for providing the unique sportswear fabric for the made-to-measure sports suits and our student Anna Gulyás for her active participation in the research.

Conflicts of Interest: The authors declare no conflict of interest.

Appendix A

Table A1. Detailed pressure values for each test person and sportswear at each measurement point.

Person		TP1				TP2		
Number of Points	Name of Points	Pressure (mmHg)			Name of Points	Pressure (mmHg)		
		Ready-made	Made-to-Measure 1	Made-to-Measure 2		Ready-made	Made-to-Measure 1	Made-to-Measure 2
1.	Right upper arm	3	2	2	Right upper arm	2	2	3
2.	Left upper arm	4	2	3	Left upper arm	2	2	4
3.	Ventral middle	1	0	1	Ventral middle	2	1	2
4.	Right thigh front	7	5	7	Right thigh front	4	5	7
5.	Left thigh front	6	5	7	Left thigh front	4	4	6
6.	Right knee	7	5	7	Right knee	5	6	6
7.	Left knee	8	6	5	Left knee	5	5	7
8.	Right Scapula 1	2	1	3	Right Scapula 1	3	2	3

Table A1. Cont.

Person		TP1			TP2			
		Pressure (mmHg)				Pressure (mmHg)		
Number of Points	Name of Points	Ready-made	Made-to-Measure 1	Made-to-Measure 2	Name of Points	Ready-made	Made-to-Measure 1	Made-to-Measure 2
9.	Right Scapula 2	5	2	5	Right Scapula 2	3	3	4
10.	Left Scapula 1	3	0	3	Left Scapula 1	2	1	4
11.	Left Scapula 2	6	5	7	Left Scapula 2	4	2	5
12.	Right tight back	14	11	6	Right tight back	5	5	7
13.	Left thigh back	16	11	6	Left thigh back	6	4	8
14.	Right thigh side	6	5	4	Right thigh side	5	4	5
15.	Left thigh side	5	5	4	Left thigh side	4	3	5
16.	Right shank back	18	5	9	Right shank back	16	9	13
17.	Left shank back	17	4	9	Left shank back	14	8	13
18.	Right waist	20	13	10	Right waist	14	8	12
19.	Left waist	21	11	13	Left waist	13	9	10
	Average [mmHg]:	8.89	5.16	5.84		5.95	4.37	6.53
	Stand. dev. [mmHg]:	6.51	3.85	3.02		4.59	2.61	3.34
Person		TP3			TP4			
		Pressure (mmHg)				Pressure (mmHg)		
Number of Points	Name of Points	Ready-made	Made-to-Measure 1	Made-to-Measure 2	Name of Points	Ready-made	Made-to-Measure 1	Made-to-Measure 2
1.	Right upper arm	4	2	2	Right upper arm	3	1	3
2.	Left upper arm	4	2	4	Left upper arm	3	2	7
3.	Ventral middle	2	2	2	Ventral middle	1	1	1
4.	Right thigh front	5	5	6	Right thigh front	7	5	5
5.	Left thigh front	6	6	6	Left thigh front	7	5	5
6.	Right knee	7	5	8	Right knee	11	5	6
7.	Left knee	7	5	9	Left knee	8	5	6
8.	Right Scapula 1	3	2	2	Right Scapula 1	6	2	2
9.	Right Scapula 2	2	1	5	Right Scapula 2	4	1	4
10.	Left Scapula 1	2	1	3	Left Scapula 1	2	2	3
11.	Left Scapula 2	5	4	4	Left Scapula 2	7	5	4
12.	Right tight back	7	7	6	Right tight back	7	6	6
13.	Left thigh back	10	9	7	Left thigh back	8	9	5
14.	Right thigh side	4	2	5	Right thigh side	7	4	4
15.	Left thigh side	3	2	5	Left thigh side	7	4	3
16.	Right shank back	18	10	11	Right shank back	14	10	10
17.	Left shank back	17	8	12	Left shank back	14	10	13
18.	Right waist	15	7	13	Right waist	19	10	8
19.	Left waist	18	9	11	Left waist	20	9	9
	Average [mmHg]:	7.32	4.68	6.37		8.16	5.05	5.47
	Stand. dev. [mmHg]:	5.56	2.98	3.45		5.30	3.21	2.95

References

1. Shishoo, R. *Textiles in Sports*; The Textile Institute, Woodhead Publishing: Cambridge, UK, 2005.
2. Shishoo, R. *Textiles for Sportswear*; The Textile Institute, Woodhead Publishing: Cambridge, UK, 2015.
3. Hayes, S.G.; Venkatraman, P. *Materials and Technology for Sportswear and Performance Apparel*; CRC Press, Taylor & Francis Group: Boca Raton, FL, USA, 2016.

4. Rödel, H.; Schenk, A.; Herzberg, C.; Krzywinski, S. Links between design, pattern development and fabric behaviours for clothes and technical textiles. *Int. J. Cloth. Sci. Technol.* **2001**, *13*, 217–227. [CrossRef]
5. Venkatraman, P. Compression garments in sportswear: Case studies to explore the effect of body type, tactile sensation and seam position in garments. In *Proceedings of the Indo-Czech International Conference on the Advancements in Speciality Textiles and their Applications in Material Engineering and Medical Sciences, Coimbatore, India, 29–30 April 2014*; Kumaraguru College of Technology: Coimbatore, India, 2014; pp. 1–4. [CrossRef]
6. Jariyapunya, N.; Musilová, B.; Geršak, J.; Baheti, S. The influence of stretch fabric mechanical properties on clothing pressure. *Fibres Text.* **2017**, *24*, 43–48.
7. Kraemer, W.J.; Bush, J.A.; Newton, R.U.; Duncan, N.D.; Volek, J.S.; Denegar, C.R.; Canavan, P.; Johnston, J.; Putukian, M.; Sebastianelli, W.J. Influence of a Compression Garment on Repetitive Power Output Production Before and After Different Types of Muscle Fatigue. *Res. Sports Med.* **1998**, *8*, 163–184. [CrossRef]
8. Doan, B.K.; Kwon, Y.-H.; Newton, R.U.; Shim, J.; Popper, J.E.; Rogers, R.; Bolt, L.; Robertson, M.; Kraemer, W.J. Evaluation of a Lower-Body Compression Garment. *J. Sports Sci.* **2003**, *21*, 601–610. [CrossRef]
9. Ali, A.; Caine, M.P.; Snow, B.G. Graduated Compression Stockings: Physiological and Perceptual Responses During and After Exercise. *J. Sports Sci.* **2007**, *25*, 413–419. [CrossRef]
10. Duffild, R.; Portus, M. Comparison of Three Types of Full-Body Compression Garments on Throwing and Repeat Sprint Performance in Cricket Players. *Br. J. Sport Med.* **2007**, *41*, 409–414. [CrossRef]
11. Tanaka, S.; Midorikawa, T.; Tokura, H. Effects of pressure exerted on the skin by elastic cord on the core temperature, body weight loss and salivary secretion rate at 35 °C. *Eur. J. Appl. Physiol.* **2006**, *96*, 471–476. [CrossRef]
12. Jin, Z.-M.; Yan, Y.-X.; Luo, X.-J.; Tao, J.-W. A Study on the Dynamic Pressure Comfort of Tight Seamless Sportswear. *J. Fiber Bioeng. Inform.* **2008**, *1*, 217–224. [CrossRef]
13. MacRae, B.A.; Cotter, J.D.; Laing, R.M. Compression garments and exercise: Garment considerations, physiology and performance. *Sports Med.* **2011**, *41*, 815–843. [CrossRef]
14. Ashayeri, E. An Investigation into Pressure Delivery by Sport Compression Garments and Their Physiological Comfort Properties. Master's Thesis, RMIT University, Melbourne, Australia, 2012. Available online: https://researchbank.rmit.edu.au/eserv/rmit:161394/Ashayeri.pdf (accessed on 6 July 2018).
15. Senthilkumar, M.; Kumar, L.A.; Anbumani, N. Design and Development of a Pressure Sensing Device for Analysing the Pressure Comfort of Elastic Garments. *Fibres Text. East. Eur.* **2012**, *20*, 64–69. Available online: http://www.fibtex.lodz.pl/article645.html (accessed on 20 November 2020).
16. Beliard, S.; Chauveau, M.; Moscatiello, T.; Cros, F.; Ecarnot, F.; Becker, F. Compression Garments and Exercise: No Influence of Pressure Applied. *J. Sports Sci. Med.* **2015**, *14*, 75–83. Available online: https://www.ncbi.nlm.nih.gov/pmc/articles/PMC4306786/ (accessed on 20 November 2020).
17. Umar, J.; Hussain, T.; Ali, Z.; Maqsood, M. Prediction Modeling of Compression Properties of a Knitted Sportswear Fabric Using Response Surface Method. *Int. J. Mater. Metall. Eng.* **2016**, *10*, 2019–2027. [CrossRef]
18. Xiong, Y.; Tao, X. Compression Garments for Medical Therapy and Sports. *Polymers* **2018**, *10*, 663. [CrossRef]
19. Cunnington, P.; Mansfield, A. *English Costume for Sports and Outdoor Recreation—From the Sixteenth to the Nineteenth Centuries*; Adam & Charles Black: London, UK, 1969.
20. Tyler, M. *The History of the Olympics*; Marshall Cavendish: London, UK, 1975.
21. Riordan, J.; Krüger, A. *European Cultures in Sport*; Intellect: Bristol, UK, 2003.
22. Vadhera, N. Historical sketch of women's participation in sports: An overview. *Int. J. Yogic Hum. Mov. Sports Sci.* **2018**, *3*, 417–422. Available online: http://www.theyogicjournal.com/pdf/2018/vol3issue2/PartG/3-2-70-822.pdf (accessed on 20 November 2020).
23. Fanger, P.O. *Thermal Comfort*; Danish Technical Press: Copenhagen, Denmark, 1970.
24. Mecheels, J. *Körper—Klima—Kleidung*; Schiele & Schön: Berlin, Germany, 1998.
25. Tochihara, Y.; Ohnaka, T. *Environmental Ergonomics—The Ergonomics of Human Comfort, Health, and Performance in the Thermal Environment*; Elsevier: Amsterdam, The Netherlands, 2005.
26. Magyar, Z. Possibilities of Application of Thermal Manikin in Thermal Comfort Tests. Ph.D. Thesis, Szent István University, Gödöllő, Hungary, 2011.
27. Liu, Y.; Hu, H. Compression property and air permeability of weft-knitted spacer fabrics. *J. Text. Inst.* **2011**, *102*, 366–372. [CrossRef]
28. Ramesh, B.V.; Ramakrishnan, G.; Subramanian, V.S.; Kantha, L. Analysis of Fabrics Structure on the Character of Wicking. *J. Eng. Fibers Fabr.* **2012**, *7*, 28–33. [CrossRef]
29. Nagyné Szabó, O. Wear Comfort Improvement of Medical Aids Used for Spine Deformity Treatment. Ph.D. Thesis, University of West Hungary, Sopron, Hungary, 2014. Available online: http://doktori.nyme.hu/484/3/nagyne_szabo_orsolya_angoltezis.pdf (accessed on 20 November 2020).
30. Fangueiro, R.; Filgueiras, A.; Soutinho, F.; Meidi, X. Wicking Behavior and Drying Capability of Functional Knitted Fabrics. *Text. Res. J.* **2010**, *80*, 1522–1530. [CrossRef]
31. Chowdhury, P.; Samanta, K.K.; Basak, S. Recent Development in Textile for Sportswear Application. *Int. J. Eng. Res. Technol.* **2014**, *3*, 1905–1910.

32. Stojanović, S.; Geršak, J. Textile materials intended for sportswear. *Tekstil* **2019**, *68*, 72–88. Available online: https://hrcak.srce.hr/250867 (accessed on 20 November 2020).
33. Engel, F.; Stockinger, C.; Wall, A.; Sperlich, B. Effects of Compression Garments on Performance and Recovery in Endurance Athletes. In *Compression Garments in Sports: Athletic Performance and Recovery*, 1st ed.; Engel, F., Sperlich, B., Eds.; Springer: Cham, Switzerland, 2016; pp. 33–61. [CrossRef]
34. Pérez-Soriano, P.; García-Roig, Á.; Sanchis-Sanchis, R.; Aparicio, I. Influence of compression sportswear on recovery and performance: A systematic review. *J. Ind. Text.* **2019**, *48*, 1505–1524. [CrossRef]
35. Lovell, D.I.; Mason, D.G.; Delphinus, E.M.; McLellan, C.P. Do Compression Garments Enhance the Active Recovery Process after High-Intensity Running? *J. Strength Cond. Res.* **2011**, *25*, 3264–3268. [CrossRef]
36. Pálya, Z.; Kiss, R.M. Biomechanical analysis of the effect of compression sportswear on running. *Mater. Today-Proc.* **2020**, *32*, 133–138. [CrossRef]

Article

Investigation of Flammability of Protective Clothing System for Firefighters

Anica Hursa Šajatović [1,*], Sandra Flinčec Grgac [1,*] and Daniela Zavec [2]

1. Faculty of Textile Technology, University of Zagreb, 10 000 Zagreb, Croatia
2. TITERA Innovative Technologies, 2212 Šentilj, Slovenia; daniela@titera.tech
* Correspondence: anica.hursa@ttf.unizg.hr (A.H.Š.); sflincec@ttf.unizg.hr (S.F.G.)

Abstract: The main characteristic of clothing for protection against heat and flame is the protection of users from external influences and danger in the conditions of elevated temperatures and exposure to flame, fire, smoke, and water. The paper presents research on the clothing system for protection against heat and flame using a fire manikin and systematically analyses the damage caused after testing. As part of the damage analysis, the existence of microdamage and impurities on the clothing system was determined using a USB Dino-Lite microscope. In addition, the intensities and composition of gaseous decomposition products during the thermogravimetric analysis of samples were investigated. The results of the research using a fire manikin showed that the user of the examined clothing system would not have sustained injuries dangerous to health and life, which confirmed the protective properties. The results of the TG-FTIR indicate that the decomposition of the fabric sample of the modacrylic–cotton fiber mixture takes place in three stages, and the identified gaseous degradation products were H_2O, CO_2, and CO.

Keywords: protective clothing for firefighters; flame manikin; fire resistant materials; flammability of materials; thermal decomposition

1. Introduction

The most important garment function is to protect the human body from basic outer influences such as wind, rain, sunlight, dust, and mechanical influences. Protective clothing is clothing that protects individuals who are exposed to life threatening or hazardous environments during work [1–3]. Today, firefighting is one of the most dangerous occupations because they perform their tasks under specific working conditions that arise in accidental situations, creating and spreading fire. This is why their clothing should protect against extreme conditions such as high temperature and flame, rain and water, cold, mechanical action, aggressive and reactive chemicals, chemicals that are hazardous to health, etc. [4]. Attention should be paid to the usual, most commonly used body positions (standing, squatting or sitting) and extreme movements that are used when wearing clothing and to perform a job [5–8]. According to the above-mentioned requirements, functional protective clothing against heat and flame should be developed in cooperation with designers, engineers, and firefighters. The process of designing protective clothing includes the entire design activity for the development of new products with high technological content from the initial idea and first project concept to the feasibility analysis, considers new materials and researchers during the design, prototyping, and manufacturing. In addition, in smart and intelligent protective clothing, the integration of technology with textiles creates enormous possibilities. The combination of technology and fashion can be realized only in a multidisciplinary work where engineers, fashion designers, and scientists will work together to adapt to their environment and create a balance between design requirements, function, performance, ergonomics, protection, and comfort [9–11].

So far, new types of fire resistance fibers, high performance fiber of special properties, and microporous materials have been developed, which provide for the long life and easy care of protective clothing as well as, at the same time, providing the user with an adequate level of protection and safety [12]. Special procedures of evaluating the characteristics of protective materials, tests with a thermal manikin and hot plates for the determination of fabric and clothing characteristics under special conditions have been developed. Scientists from different fields (textile technology, physiology, ergonomics, functional design, etc.) are continually working on research such as the development and producing of protective clothing and materials, research of their durability during use testing by volunteers in simulated environmental conditions, research on moisture transfer through clothing and thermal stress, and how to design and implement the appropriate test methods [13–18].

For clothing in firefighting, fire resistant materials should be used. This could be some fabrics or knits made from aramid fibers such as Nomex®, modacrylic fibers, cotton fibers, or other textile materials with a flame retardant finishing. It is often used to make protective clothing against heat, and flame materials can be made of a blend of the aforementioned fibers with the aim of achieving better comfort, which includes the transport of moisture and heat. This is extremely important for achieving comfort when using clothing systems in various activities that include extreme conditions [19]. Recently, different techniques for the characterization of fire resistant textile materials have been conducted [20]. High thermal protection of firefighter clothing systems can be achieved by wearing multilayer or thick textile materials and it is well-known that the performance of each layer of a firefighters' protective clothing has a significant influence on the level of protection provided. When a clothing system is exposed to high temperature and direct flame or fire, it is still uncertain how destructive different exposures are and how long a piece of firefighting protective clothing can continue to protect to an acceptable level.

In this paper, a protective clothing system made of carefully selected materials was examined. The underwear is made of a wool/modacrylic blend in a ratio of 70%:30%, which by synergy contributes to high protection and comfort. It is known that the inner core of wool fiber has high moisture-absorbing might and can receive twice the amount of moisture of its mass while remaining dry. This property allows wool to adsorb sweat generated during activity or under the influence of extreme conditions. Precisely because of the internal moisture, wool is naturally resistant to flame, while modacrylic, due to its chemical composition, is characterized by high stability to heat, and the flame has the property of self-extinguishing. The overalls were made of a mixture of modacrylic and cotton in a ratio of 55%:45% in order to achieve high resistance to heat and flame resistance and thermophysical comfort [21–23].

Consequently, in this paper, firefighting overalls made from modacrylic/cotton fabric on a fire manikin was tested as well as the mechanical characteristics, and an analysis of damage after fire exposure and the thermal properties of the fabric was conducted.

2. Materials and Methods

The paper investigated the flammability properties of the clothing system for protection against heat and flame, and analyzed the damage caused to the overalls intended for firefighters to extinguish forest fires after testing by using a fire manikin. A clothing system consisting of underwear composed from 70% wool fiber and 30% modacrylic fiber (Figure 1) and overalls for protection against heat and flame intended for extinguishing forest fires (Figure 2), made of 55% modacrylic fiber and 45% cotton fiber, were tested on the fire manikin.

Figure 1. Tested underwear on the fire manikin.

Figure 2. Tested overalls on the fire manikin.

The following testing instruments and systems were used to conduct the experimental part of the work.

The fire manikin (Figure 3) was equipped with 128 temperature sensors placed on its surface (Figure 4). The explosive fire simulation system consisted of 12 gas burners located around the fire manikin. Prior to each test, calibration was performed where the naked manikin was exposed to explosive fire for 3 to 4 s. Therefore, burners must be placed appropriately to always provide heat flux values of about 80 kW/m^2. The data provided by the sensor was collected and displayed using the Labview software solution, and the entire system was controlled by a Mitsubishi Programmable Logic Controller (PLC) unit. Flashfire was achieved by burning the main burners from 2 to 10 s, depending on the duration of the test and the clothing system being tested. Turning off the burner extinguished the fire, and it waits for 120 s until the end of the test when the fan was switched on for faster ventilation of the test room [24].

Figure 3. Fire manikin Žiga (Institut Jožef Štefan, Ljubljana, Slovenia).

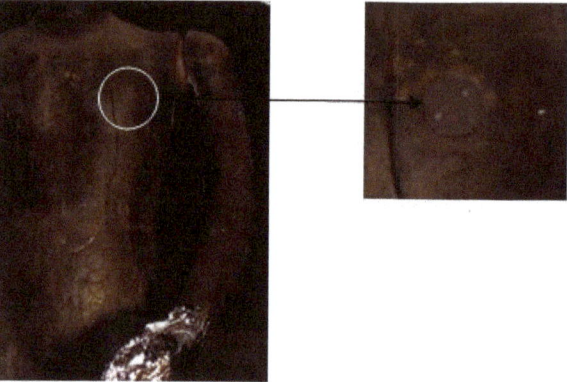

Figure 4. Temperature sensor on the fire manikin.

The clothing system was tested on a fire manikin in accordance with an international standard describing the test method (ISO/DIS13506, 2002). During the test, the clothing system was exposed to open fire for 4 s. Using 128 thermoelements (Figure 4) distributed over the entire surface of the fire manikin ('skin'), the temperature rises on the 'skin' at the time of flame action can be measured. Measurements were recorded every 0.5 s in each area where the thermoelement is located. Based on the temperature data, the heat flux was calculated, which was compared to a human skin model to determine whether burns have occurred. Data were collected for 120 s including the first contact with the flame [20]. After the activity of heat and flame, parts of the overalls were analyzed using a dynamometer, a Dino-Lite USB microscope and a thermal gravimetric device, in order to determine the change in the structure and characteristics of the material. The research was conducted with the aim of determining microdamage and qualitative determination of the present impurities and gases.

Testing of tensile properties on a dynamometer

Determination of breaking force and elongation of the fabric in the direction of the warp and weft was carried out by the strip method according to the HR EN ISO 13934-1:2013 standard using a dynamometer Tensolab 3000, tt. Mesdan, Italy. The dynamometer works at a constant stretching speed of 100 mm/min, and the distance between the clamps according to the norm is 200 mm [25].

Dino-Lite USB microscope

Technical characteristics of Dino-Lite USB microscope are:

- Magnification: 10–90×, 10–50×/200×, fixed 500×
- Resolution: 1.3 megapixels USB:2.0
- Options: calibration, measurement, photography, video recording
- LED and UV lighting, polarizing filter, diffuse lighting
- Outer materials: composite or aluminum alloy [26].

Thermal gravimetry (TG) measurements: A Pyris 1 TGA, PerkinElmer thermogravimetric device measured the loss of sample mass in percent as a function of temperature (and time) during linear or stepwise heating in the temperature range (50 °C–950 °C) and in a certain atmosphere (nitrogen, air, oxygen). The results of TG analysis are presented in the form of curves, and with this method, it is possible to determine the point of degradation or decomposition of the sample. If the device is connected via TG-IR interface (PerkinElmer TL 8000) to an FTIR spectrometer (PerkinElmer, Spectrum 100, Waltham, MA, USA), it is possible to analyze gaseous organic products generated by heating the sample. The sample was heated in a vessel so that the temperature rose evenly at the set rate, while the change in mass was recorded on the balance. Based on the change in mass, it is possible to determine the percentage ratios of the components [25].

Thermogravimetric (TG) experiments were carried out using a Perkin Elmer Pyris 1 TGA thermogravimetric analyzer. Samples were stacked in an open platinum sample pan and the experiment was conducted in an air atmosphere. All samples for TGA were measured from 30 °C to 800 °C at the heating rate of 30 °C/min with a continuous airflow at a rate of 20 mL/min. Samples were studied by the coupled TG-IR technique to better understand the decomposition process of different FR-fabrics. Nitrogen, which does not exhibit IR-absorption, was used as the purge gas, thus the end-products of the decomposition were pyrolyzed rather than the oxidative degraded products. A Thermal Analysis Gas Station (TAGS) equipped with a detector was used for the FTIR analysis. The transfer line, high-temperature flow cell, and TG interface were held at 280 °C for the duration of the run to prevent gas condensation. The evolved gases were transferred through the FTIR flow cell by a peristaltic pump with a flow rate of 60 milliliters per minute.

Microscale Combustion Calorimetry Measurement

The microscale combustion calorimetry (MCC) measurement was performed with a Govmark MCC-2, according to ASTM D7309-2007 (Method A). The fabric samples were first ground in a high-energy vibrating mill RETSCH®-MM 400 at a frequency of 25 Hz for 10 min to form homogeneous powders. Sample weight was in the range of 5 to 6 mg. The sample's thermal degradation products in the nitrogen gas stream were mixed with a 20 cm^3/min stream of oxygen prior to entering a 900 °C combustion furnace. Each sample was run in three replications and the data presented here are all the averages of the three measurements. The MCC provides the peak heat release rate (PHRR) and heat release capacity (HRC) data of the polymeric sample based on the oxygen consumption, the heating rate of the sample, flow rate, and sample weight [27]. This is extremely important for the characterization of materials intended for the manufacture of protective fire uniforms.

3. Results

By testing with a fire manikin and an explosive fire simulator, data on the degree of burns in the case of using a clothing system (underwear and overalls) that was exposed to an explosive flame for a period of 4 s were obtained (Figure 5).

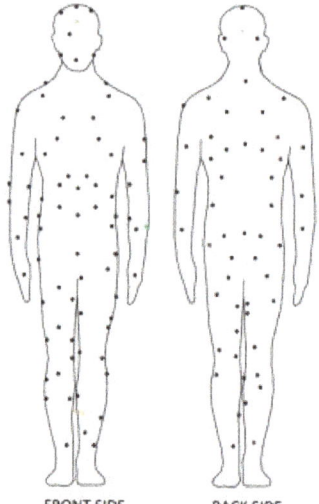
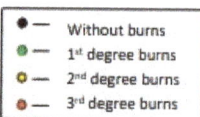

burning time: 4s
time to 1st degree burns: 17s
1st degree burns area: 1%
time to 2nd degree burns: 0s
2nd degree burns area: 2%
time to 3rd degree burns: 6s
3rd degree burns area: 3%

FRONT SIDE   BACK SIDE

Figure 5. Computer display of the results obtained by the fire manikin test.

The clothing system consisting of firefighting underwear and a single-layer overalls intended for firefighters to extinguish open fires after being exposed to explosive fire for 4 s was found to have minor damage (Figure 6). Visual assessment of the damage to the clothing system was performed after a period of 116 s after the flame was extinguished in the chamber in which the fire manikin test was performed. Based on the observations, it was established that there was no damage to the underwear (Figure 6b). There was minor damage to the overalls on the tops of the pockets, and on the sleeves and part of the trousers below the knee (Figure 6a). The visual assessment revealed that there was a shrinkage of material on certain parts of the overalls (shoulder area), and a partial change in color on the material of the overalls (dark brown and black) caused by fire (visible on the folds on the overalls, which appeared during the test, Figure 6).

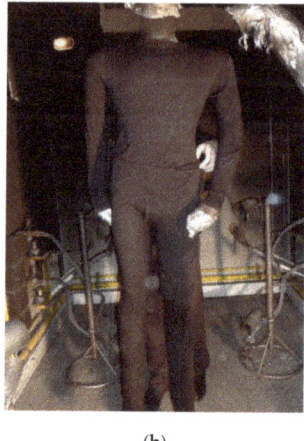

(a)   (b)

Figure 6. Parts of the clothing system after exposure to heat and flame: (**a**) overalls and (**b**) underwear.

Damage to the clothing system confirmed the results obtained by the computer display of test results on the fire manikin and microscopic analyses, which did not show damage to the structure of the material. The results of the fire manikin (Figure 5) showed that after 17 s, there was the first appearance of first-degree burns in the amount of 1% of the total area of human skin in the left forearm. Second-degree burns occurred on the head but were not considered because the test was performed without head protection (without the use of a helmet and/or undercap). Four sensors indicating third-degree burns were identified as invalid because during the calibration of the fire manikin itself, before the test, it was found that the sensors were out of order. Results from fire manikin and damage to the clothing system showed that the user, who would be exposed to fire and heat, would not have injuries dangerous to health and life, and thus shows that the clothing system met the expected properties of use/protection. If the user used such a clothing system when extinguishing a fire and was exposed to direct fire for 4 s, they would survive without major health problems, but should immediately move away from further heat and flame exposure, which is very difficult in real conditions for extinguishing fires indoors. Since this clothing system is intended to extinguish forest fires and low vegetation fires in nature, it was assumed that the firefighter will not come into direct contact with the indoor fire for more than 4 s, and it can be concluded that the tested clothing system provides sufficient protection. When extinguishing forest fires, firefighters may be exposed to fire and flames, losing the protective properties of the outer layer of the clothing system (overalls), so the user should be careful and must understand the basic properties of clothing worn to protect and prevent possible injuries because extinguishing forest fires can take more than 24 h. According to the presented results, the testing method of the fire manikin is suitable for investigated clothing system.

For the purposes of the research, an analysis of the basic material from which the overalls were made and an analysis of the material after exposure to flame during the test on the fire manikin were made. The basic material was made from 55% modacrylic fiber, 45% cotton fiber, with a built-in antistatic grid, mass per unit area 295 g/m^2, woven in canvas. The breaking force in the warp direction was 1049 N and 808 N in the weft direction. After exposing the overalls to explosive fire during the test on the fire manikin and visual assessment of the damage, sampling of parts of the overalls in the direction of the warp and weft was performed. The obtained results are shown in Table 1.

Table 1. Results of the tensile properties of basic materials (before heat exposure) and of the part samples of the overalls after exposure to explosive fire on a fire manikin.

	Direction	Breaking Force [N]	Breaking Elongation [%]	Breaking Time [s]
Basic material	Warp	1049	22.65	27.2
	Weft	808	15.96	19.2
Samples of overalls material after exposure to explosive fire on a fire manikin				
Underside of trousers	Warp	180	13.742	16.5
	Weft	94	5.890	7.1
Back part of overalls	Warp	681	34.366	41.5
	Weft	383	19.244	23.2
Topside of trousers	Warp	182	9.960	12.0
	Weft	92	5.210	6.5

In accordance with the visual assessment of the damage to the overalls after exposure to explosive fire, the results shown in Table 1 confirm that the greatest damage and weakening occurred in the area of the trousers of the overalls. The above research proves that if firefighters wear the tested overalls during firefighting and are exposed to direct flames, they should wear long-legged underwear to avoid skin damage and burns. After testing the whole clothing system using a fire manikin, damage was seen only on the overalls, and no damage was seen on the underwear. Due to the above, all further research was conducted only for the firefighting overalls.

Analysis of the existence of microdamage and the determination of impurities was conducted with a Dino-Lite digital USB microscope. Figure 7 shows the appearance of an undamaged sample of material from which the overalls were made in two different magnifications with a Dino-Lite microscope (magnification ×60, Figure 7a, and magnification ×184, Figure 7b). An analysis of individual parts of the overalls that were tested on a fire manikin are presented below.

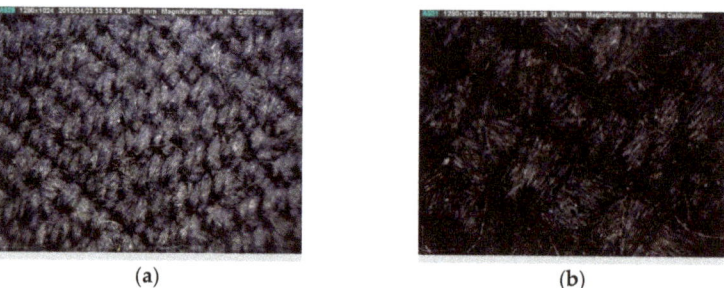

Figure 7. Undamaged sample: (a) magnification ×60 and (b) magnification ×184.

On the cover of the right pocket where the damage was visible, microscopic analysis showed that the material was slightly charred, but there was no damage to the fabric structure itself (Figure 8). It could also be seen that the pocket was functionally shaped and that it expanded and opened only on the lower right side. The pocket cover was also functionally shaped because it was 20 mm larger than the pocket. When making the pocket, non-combustible Velcro tape was used (Figure 8) because there was no damage to the tape or shrinking of the pocket cover after exposure to fire.

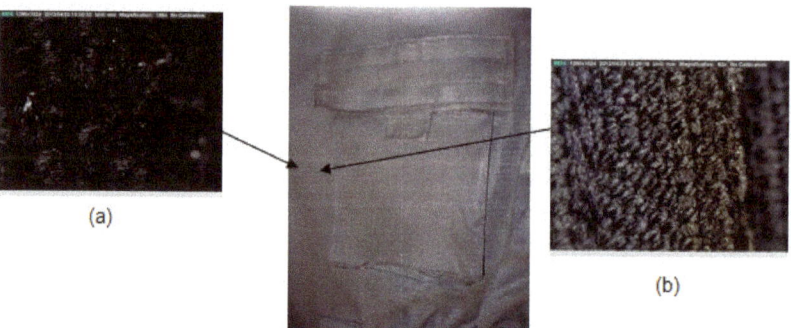

Figure 8. Damage in the area of the upper pocket of the overalls made with a Dino-Lite microscope in two different magnification: (a) 184×; (b) 60×.

On the sleeve, it could be seen that the material was damaged, but microscopic analysis showed that there was no change in the structure of the fabric (i.e., it did not melt, but the fibers under the influence of high temperature were charred and stiffened). Because of this, the material cracked after cooling when removing the suit from the fire manikin (Figure 9). Therefore, after the action of an explosive fire with such damage to the overalls, the user must not continue to extinguish the fire, because in the case of further contact with heat or fire, the overalls will not provide adequate protection.

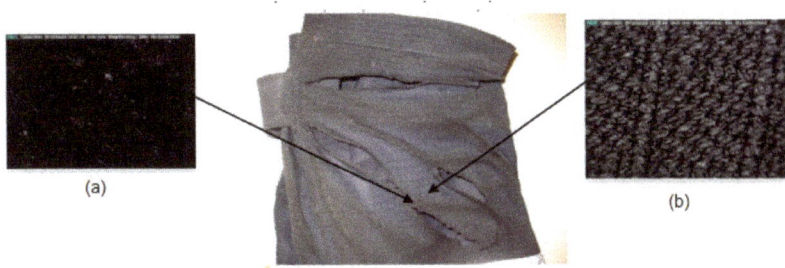

Figure 9. Damage to the sleeve of the overalls made with a Dino-Lite microscope in two different magnification: (**a**) 184×; (**b**) 60×.

Damage to the left leg of the overalls (lower leg area) also occurred during the removal of the overalls from the manikin due to the solidification and charring of the fibers during the exposure of heat and flame (Figure 10). Due to the solidification of the material and its cracking, during further movement of the user, the material would fall off, and such overalls would not provide adequate protection in further use.

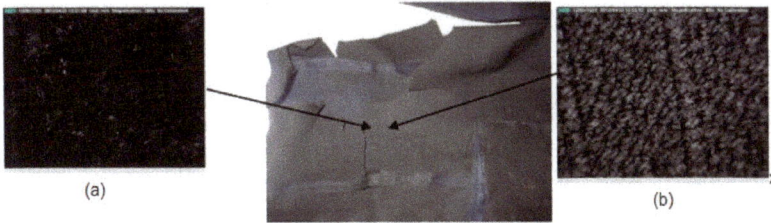

Figure 10. Damage to the left leg of the overalls (lower leg area) made with a Dino-Lite microscope in two different magnification: (**a**) 184×; (**b**) 60×.

TG analysis with monitoring of gaseous decomposition products of the sample

Thermogravimetric analysis (TGA) is the most commonly used method to assess the thermal stability of materials. Using a Pyris and TGA thermogravimetric device, (PerkinElmer, USA), the change (loss) of the sample mass as a function of temperature was monitored. Testing using thermogravimetric analysis shows what happens to the material during contact with heat, and what the decomposition products are.

Figure 11 shows the TG and dTG curves of the thermal decomposition of the basic material from which the overalls was made. From the dTG curve, it can be seen that the degradation of the sample took place in two stages. At a temperature of 290 °C, the first stage of dynamic decomposition was recorded, with a loss of sample weight of 37.150% per minute, and no large quantities of gaseous products were recorded during the decomposition. The second stage of dynamic decomposition began at a temperature of 495 °C, and the peak of the second stage of dynamic decomposition was visible at the temperature of 614 °C (Figure 11), at which a larger amount of gaseous CO_2 and CO products was detected (Figure 12). The thermal stability of the sample was clearly visible in the residue after thermogravimetric analysis of 5.974% at 850 °C.

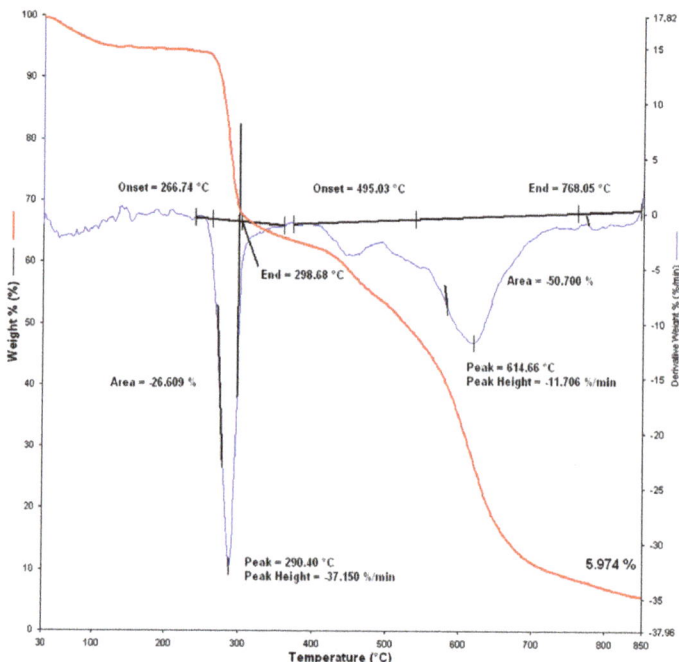

Figure 11. Thermal gravimetric analysis of a sample of the basic material from which the overalls were made (Sample 1) (TG and dTG curve).

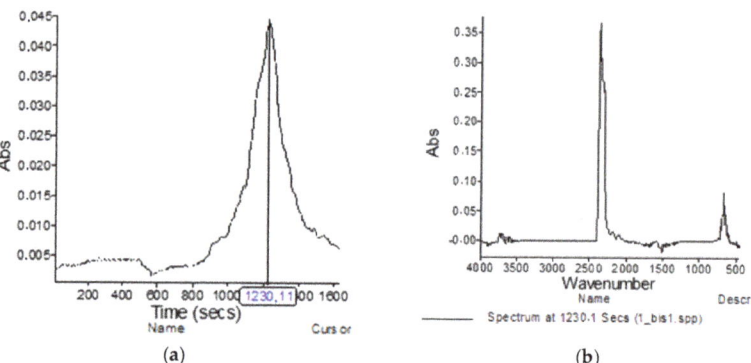

Figure 12. TG-IR analysis of a sample of the basic material from which the overalls were made: (a) absorption spectrum of the highest concentration of gaseous decomposition products; (b) measured gases on IR at temperature 615 °C.

According to the obtained TGA curve (Figure 13), it can be seen that the rapid thermal decomposition started at a temperature of 248.09 °C. The thermogravimetric curve of the modacrylic/cotton fiber sample indicates that during non-isothermal decomposition, the sample decomposes in three stages. This was evident from the dynamic degradation of the sample showing the dTG curve. The first stage started at a temperature of 248 °C. If a person without protection is in contact with such a high temperature, they would obtain fourth degree burns, which would lead to irreversible destruction of subcutaneous tissue. According to the results of testing on the Žiga fire manikin, it is evident that a clothing system consisting of fire-resistant underwear and one-layer overalls provides protection, and that the user would suffer first-degree burns, which manifest as redness and mild swelling, without any other skin damage. The maximum rate of dynamic degradation in the first stage was recorded at a temperature of 271.87 °C with a mass loss of 5.583% per minute. Completion of the first stage of decomposition was at a temperature of 291.40 °C. The second stage of dynamic degradation began at a temperature of 316.36 °C, and the maximum dynamic degradation was recorded at a temperature of 345.90 °C. In the second stage of decomposition, the rate of mass loss increased to 11.096%/min. The third stage of dynamic decomposition began at a temperature of 489.75 °C, and at a temperature of 535.00 °C, the maximum dynamic decomposition was recorded with a weight loss of 10.732%/min. The residue after thermogravimetric analysis for the specified sample was 3.253%, which means that the sample in the specified temperature range lost 96.7% of its mass.

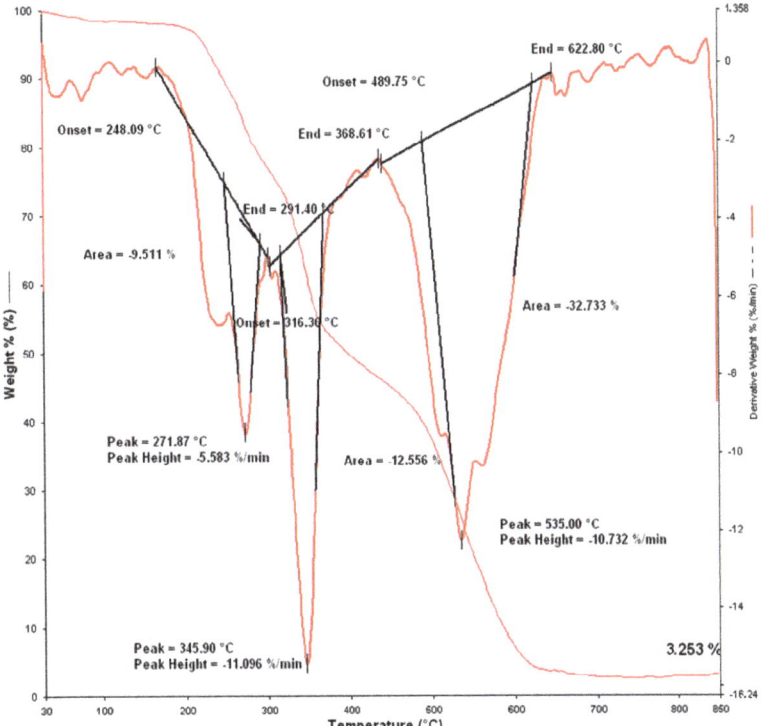

Figure 13. Thermal gravimetric analysis of the material sample of the overalls after exposure to an explosive fire on a fire manikin (Sample 2): TG curve, dTG curve.

Figure 14 shows the gaseous degradation products of the modacrylic/cotton fiber sample (Sample 2). The first intense absorption peak was detected at 509.13 °C (Figure 14d). Evaporative decomposition products were identified as CO_2 (characteristic highest points at 2359 and 2322 cm^{-1}), CO (characteristic highest points at 2179 and 2110 cm^{-1}), and water (characteristic highest points at 3500 to 4000 and 1550 to 1566 cm^{-1}). The second intense absorption peak was measured at 525.85 °C, and the third intense absorption peak was measured at 557.82 °C. The same decomposition gases were detected in all of them.

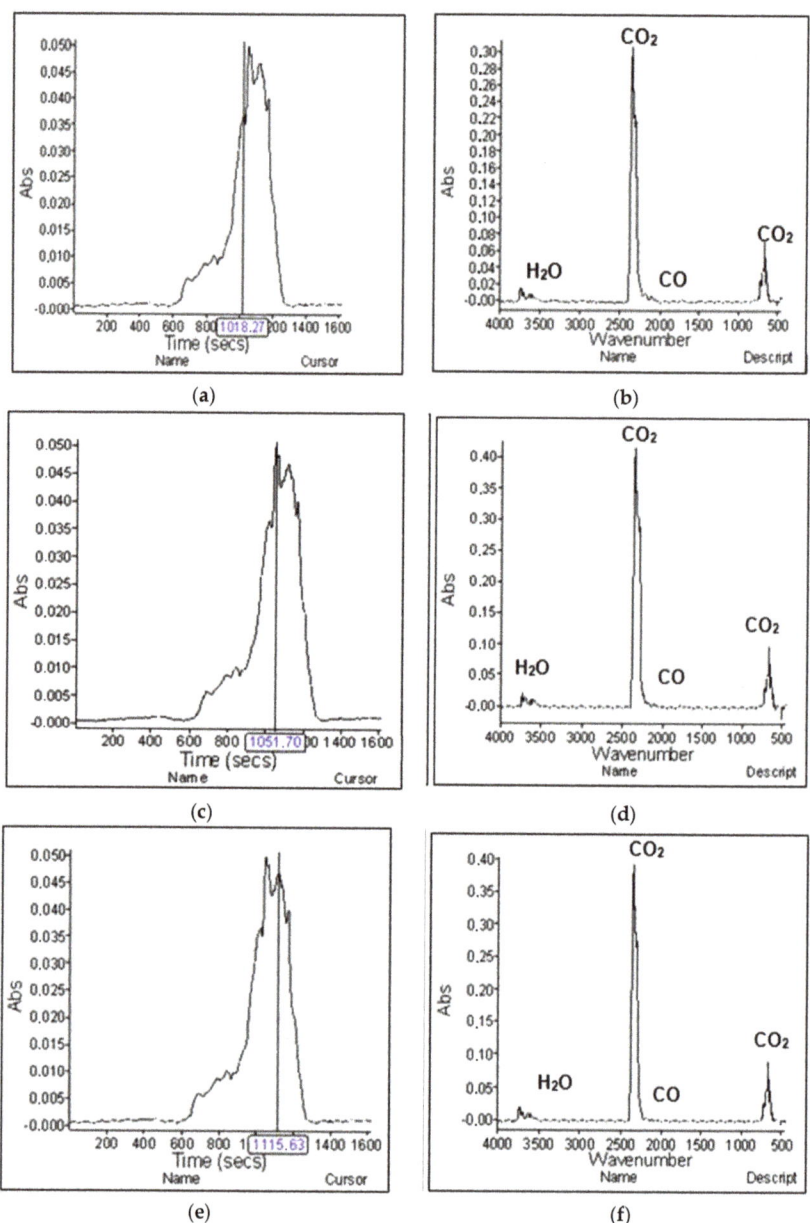

Figure 14. TG-IR analysis of sample. (**a**) The first intense absorption spectrum; (**b**) measured gases on IR at a temperature of 509.13 °C: (**c**) second absorption spectrum; (**d**) measured gases on IR at a temperature of 525.85 °C; (**e**) third absorption spectrum; (**f**) measured gases on IR at a temperature of 557.82 °C.

In order to gain a better insight into the behavior of the sample under the action of heat and flame, research was conducted on a microscale combustion calorimeter (MCC). The obtained results (Table 2) indicate that Sample 1 was thermally more stable than Sample 2. It can be seen that the recorded specific heat release capacity, which for Sample 1 was

34.67 J(g·K)$^{-1}$ and Sample 2 was 45.33 J(g·K)$^{-1}$, and the maximum specific heat release was 26.13 W·g^{-1} in Sample 1 and 37.68 W·g^{-1} in Sample 2. The lower the value of the maximum specific heat release (Qmax), the more stable the sample is to the action of heat and flame. The sample previously damaged by heat and flame during the test on the fire manikin showed slightly less stability compared to the same material before the test. It is assumed that the reason for this is the weakened modacrylic component, which has the role of preventing the development of heat and complete combustion of the cellulosic component. Re-exposure of previously thermally weakened material (Sample 2) to heat and flame resulted in complete combustion of the cellulosic component of the material and partially modacrylic component, which was clearly visible through the amount of charred residue relative to the undamaged one-layer overalls (Sample 1) [19,28].

Table 2. Results of the measurement parameters of the samples of parts of the overalls on a microcalorimeter for combustion.

Measurement Parameters	Sample 1 (Before Exposure to Fire)			Sample 2 (After Exposure to Fire)		
	$\bar{x}$	s	CV, %	$\bar{x}$	s	CV, %
Heat release capacity, ηc J·g^{-1}·K^{-1}	34.67	0.577	0.016654	45.33	1.528	0.033695
Maximum specific heat release, Qmax, W·g^{-1}	26.13	0.220	0.008404	37.68	4.894	0.129864
Heat release temperature, Tmax, °C	386.3	2.326	0.006021	440.03	3.550	0.008067
Specific heat release, hc, kJ·g^{-1}	6.2	0.1	0.016129	6.23	0.153	0.024505
Yield of pyrolysis residue, Yp, g·g^{-1}	0.45	0.012	0.026012	0.33	0.008	0.025397
Specific heat of combustion of fuel gases, hc, gas, kJ·g^{-1}	11.34	0.295	0.026009	9.35	0.144	0.015381
Average release capacity, ηc, J·g^{-1}·K^{-1}	41.33	0.577	0.013968	51.67	1.528	0.029565
Average release temperature, Tmax, K	296.3	5.246	0.017705	331.33	1.401	0.004228

4. Conclusions

Based on theory and conducted research, except for the important factor that is the material selection, great attention should be paid to the functional design of clothing for protection against heat and flame. The tested clothing system provides a high degree of protection, which was in part due to the adequate use of fire-resistant underwear. The results of testing using the fire manikin showed that when using such a clothing system, the user would not suffer tissue damage, which could pose a danger to health or life, if the users are immediately removed from the fire and heat. If the user continues to use a clothing system that has been exposed to explosive fire, the clothing system would not provide adequate protection due to the damage. Without the use of fire-resistant underwear, the user would obtain burns in more areas than with the fire-resistant underwear, especially if they used ordinary cotton underwear, which absorbs moisture faster and the user would "be cooked" when exposed to high temperatures. If the user was exposed to prolonged fire, the damage to the overalls and the injuries sustained would be greater. In practice, when extinguishing fires in forests and low vegetation, such interventions last more than two days under high temperatures, so the studied clothing system would not provide adequate protection if users are exposed to explosive fire.

Research with the help of a fire manikin and an explosive fire simulator (i.e., simulations of the dangers to which firefighters are exposed), greatly helps us to predict the degree of burns of users of the clothing systems, and the possibility of survival. Such research is already of great importance in the first phase of the design of clothing systems, which should provide protection against heat and flame. It is very important that all of the used materials are fireproof, so that the failure does not further endanger the health or life of users. In the design and production processes of protective clothing systems, the cooperation of experts in the field of materials and fibers, designers, constructors, and technologists as well as the end users and manufacturers of protective clothing is necessary.

From the obtained TG-IR and MCC results, it can be concluded that the thermal properties of the protective firefighting clothing system decreased after the exposure to thermal manikin testing (Sample 2) compared to the starting material (Sample 1). However, both tested materials showed relatively good stability against the heat, but based on all of the measured indicators, it requires further development to improve the overall thermal resistance properties.

Author Contributions: Conceptualization, A.H.Š., S.F.G. and D.Z.; Methodology, A.H.Š., S.F.G. and D.Z.; Formal analysis, A.H.Š., S.F.G. and D.Z.; Investigation, A.H.Š., S.F.G. and D.Z.; Resources, A.H.Š., S.F.G. and D.Z.; Writing—original draft preparation, A.H.Š. and S.F.G.; Writing—review and editing, A.H.Š. and S.F.G.; Visualization, A.H.Š. and S.F.G.; Supervision, A.H.Š. and S.F.G. All authors have read and agreed to the published version of the manuscript.

Funding: This work was supported in part by the short-term research support of the University of Zagreb in 2020 and 2021, TP2/20, Processing and maintenance of textiles and leather with environmentally friendly agents and TP16/21, Properties during uses and maintenance of high-performance protective materials and clothing.

Institutional Review Board Statement: Not applicable.

Informed Consent Statement: Not applicable.

Data Availability Statement: Not applicable.

Acknowledgments: A paper recommended by the 14th Scientific and Professional Symposium Textile Science and Economy, The University of Zagreb Faculty of Textile Technology.

Conflicts of Interest: The authors declare no conflict of interest.

References

1. Šajatović, A.H.; Pavlinić, D.Z.; Dragčević, Z. Vatrogasni odjevni sustavi za zaštitu od topline i plamena. *Tekstil* **2013**, *62*, 160–173.
2. Kirin, S.; Dragčević, Z.; Rogale, D. Stanje i normizacija zaštitne odjeće. *Tekstil* **2002**, *51*, 230–237.
3. Horvat, J.; Regent, A. *Osobna Zaštitna Oprema*, 1st ed.; Veleučilište u Rijeci: Rijeka, Croatia, 2009; pp. 32–55.
4. Mikučionienė, D.; Milašiūtė, L.; Baltušnikaitė, J.; Milašius, R. Influence of Plain Knits Structure on Flammability and Air Permeability. *Fibres Text. East. Eur.* **2012**, *20*, 66–69.
5. Rossi, R. Clothing for protection against heat and flames. In *Protective Clothing Managing Thermal Stress*; Series in Textiles; Woodhead Publishing: Cambridge, UK, 2014; pp. 70–89.
6. Coca, A.; Williams, W.J.; Roberge, R.J.; Powell, J.B. Effects of fire fighter protective ensembles on mobility and performance. *Appl. Ergon.* **2010**, *41*, 636–641. [CrossRef] [PubMed]
7. Bogović, S.; Šajatović, A.H. Construction of protective clothing. In *Young Scientist in the Protective Textiles Research*; Faculty of Textile Technology, University of Zagreb: Zagreb, Croatia, 2011; pp. 309–331.
8. Ng, R.; Liu, W.; Yu, W. Dynamic Stress Distribution on Garment. In Proceedings of the 3rd International Textile, Clothing & Design Conference, Dubrovnik, Croatia, 8–11 October 2006; pp. 465–470.
9. Bailey, D.R.F. The Role of the Fashion Designer with Regard the Emergence of Smart Textiles and Wearable Technology. Available online: http://www.gzespace.com (accessed on 20 June 2021).
10. Dammaco, G.; Turco, E.; Glogar, M.I. Design of Functional Protective Clothing. In *Functional Protective Textiles*; Faculty of Textile Technology, University of Zagreb: Zagreb, Croatia, 2012; pp. 37–70.
11. Bullinger, H.J. *Ergonomie, Produkt und Arbeitsplatzgestaltung*, 1st ed.; B.G.Teubner: Stuttgart, Germany, 1994; pp. 5–18.
12. Long, Z.F.; Zhang, W.; Minzhi, C. Investigation of Material Combinations for Fire-fighter's Protective Clothing on Radiant Protective and Heat-Moisture Transfer Performance. *Fibres Text. East. Eur.* **2007**, *15*, 72–75.
13. Trovi, D.A.; Hadjisophocleous, G.V. Research in Protective Clothing for Firefighters: State of the Art and Future Directions. *Fire Technol.* **1999**, *35*, 111–130. [CrossRef]
14. Juričič, D.; Musizza, B.; Gašperin, M.; Mekjavič, I. System for evaluation of fire protective garments. In Proceedings of the 4th International Textile, Clothing & Design Conference—Magic World of Textiles, Dubrovnik, Croatia, 5–8 October 2008; pp. 787–792.
15. Nayak, R.; Houshyar, S.; Padhye, R. Recent trends and future scope in the protection and comfort of fire-fighters' personal protective clothing. *Fire Sci. Rev.* **2014**, *3*, 1–19. [CrossRef]
16. Shaid, A.; Wang, L.; Padhye, R. Textiles for Firefighting Protective Clothing. In *Firefighters' Clothing and Equipment: Performance, Protection, and Comfort*; CRC Press: Boca Raton, FL, USA, 2018; pp. 1–30.
17. Havenith, G.; Heus, R. Ergonomics of Protective Clothing. In Proceedings of the 1st European Conference on Protective Clothing, Stockholm, Sweden, 7–10 May 2000; pp. 26–29.

18. Park, H.; Park, J.; Lin, S.H.; Boorady, L.M. Assessment of Firefighters' needs for personal protective equipment. *Fash. Text.* **2014**, *1*, 1–13. [CrossRef]
19. Kim, H.A.; Kim, S.J. Flame-Retardant and Wear Comfort Properties of Modacrylic/FR-Rayon/Anti-static PET Blend Yarns and Their Woven Fabrics for Clothing. *Fibers Polym.* **2018**, *19*, 1869–1879. [CrossRef]
20. Grgac, S.F.; Bischof, S.; Pušić, T.; Petrinić, I.; Luxbacher, T. Analytical Assessment of the Thermal Decomposition of Cotton-Modacryl Knitted Fabrics. *Fibres Text. East. Eur.* **2017**, *25*, 59–67. [CrossRef]
21. Reeves, W.A.; Summers, T.A. Fire Resistant Cotton Blends Without Fire Retardant Finish. *J. Coat. Fabr.* **1982**, *12*, 92–104. [CrossRef]
22. Khan, J. Shielding Effect to the Flammable Fibres Offered by Inherently Flame Retardant Fibres. Available online: https://hb.diva-portal.org/smash/get/diva2:1371774/FULLTEXT01.pdf (accessed on 8 March 2022).
23. Cardamone, M. Flame resistant wool and wool blends. In *Handbook of Fire Resistant Textiles*; Kilinc, F.S., Ed.; Woodhead Publishing: Cambridge, UK, 2013; pp. 245–271.
24. Pavlinić, D.Z.; House, J.R.; Mekjavić, I.B. Protupožarni odjevni sustavi i njihovo vrednovanje. *Sigurnost* **2010**, *52*, 251–262.
25. Equipment Catalogue. University of Zagreb. Faculty of Textile Technology. Available online: http://www.ttf.unizg.hr/doktorski/KATALOG_OPREME.pdf (accessed on 2 February 2020).
26. Dino—Lite. Available online: http://www.dino-lite.hr/am413t-2/ (accessed on 9 February 2018).
27. Cheng, X.; Yang, C.Q. Flame retardant finishing of cotton fleece fabric: Part, V. Phosphorus-containing maleic acid oligomers. *Fire Mater.* **2009**, *33*, 365–375. [CrossRef]
28. Yang, C.Q.; He, Q. Textile heat release properties measured by microscale combustion calorimetry: Experimental repeatability. *Fire Mater.* **2012**, *36*, 127–137. [CrossRef]

Article

Impact of Washing Parameters on Thermal Characteristics and Appearance of Proban®—Flame Retardant Material

Tea Kaurin, Tanja Pušić, Tihana Dekanić and Sandra Flinčec Grgac *

Faculty of Textile Technology, University of Zagreb, 10000 Zagreb, Croatia
* Correspondence: sflincec@ttf.unizg.hr

Abstract: Proban® is a multiphase treatment of cotton fabrics based on the formation of pre-condensates using the flame retardant (FR) agent tetrakis (hydroxymethyl) phosphonium salts (THPx). The assessment of the durability of a product demands a preliminary understanding of how relevant it is to extend its lifetime. It is therefore important to minimize the risk of agents impacting: (1) the protection level, (2) shape and dimensions, and (3) additional comfort characteristics of the fabric. This research focused on the impact of washing conditions on the durability of FR properties and appearance of Proban® cotton fabrics, which was systematically arranged through the variation in the chemistry distribution in the Sinner's circle. The chemical share was varied in laboratory conditions as a simulation of industrial washing based on component dosing, where the temperature, time and mechanical agitation were constant. The washing of cotton fabrics was performed through 10 cycles in four baths containing high alkali components, medium alkali components, high alkali reference detergent and water. The environmental acceptability of washing procedures through effluent analysis was assessed by physico–chemical and organic indicators. The limited oxygen index (LOI), calorimetric parameters (micro combustion calorimetry), thermal stability and evolved gases during thermal decomposition (thermogravimetric analyzer (TGA) coupled with an infrared spectrometer (TG–IR), surface examination (FE-SEM), spectral characteristics and pH of the aqueous extract of the fabrics before and after 10 washing cycles were selected for proof of durability. The medium alkali bath was confirmed as a washing concept for Proban® cotton fabric through the preservation of FR properties examined through LOI, TGA, TG–IR and MCC parameters and appearance color and low level of fibrillation.

Keywords: cotton fabric; Proban®; environmentally friendly washing procedure; flame retardant durability; MCC; appearance

Citation: Kaurin, T.; Pušić, T.; Dekanić, T.; Flinčec Grgac, S. Impact of Washing Parameters on Thermal Characteristics and Appearance of Proban®—Flame Retardant Material. *Materials* 2022, 15, 5373. https://doi.org/10.3390/ma15155373

Academic Editors: De-Yi Wang and Tao Tang

Received: 8 May 2022
Accepted: 27 July 2022
Published: 4 August 2022

Publisher's Note: MDPI stays neutral with regard to jurisdictional claims in published maps and institutional affiliations.

Copyright: © 2022 by the authors. Licensee MDPI, Basel, Switzerland. This article is an open access article distributed under the terms and conditions of the Creative Commons Attribution (CC BY) license (https://creativecommons.org/licenses/by/4.0/).

1. Introduction

Extreme operating conditions of exposure to the effects of fire, fuel explosions or heat radiation require the use of appropriate protective clothing systems, covering gloves, hoods, upper layers of clothing and appropriate footwear [1]. Particular attention should be paid to the selection of materials that contribute to the comfort and safety of clothes. For this reason, such clothing must, in addition to the protection requirement, meet a number of other requirements, by which it is classified as comfortable and safe to human health and the environment [2,3].

In order to protect humans in open flame fires and extreme heats, the fabric is significant in the preservation and protection of the human body [4].

The composition and structure of the material, its technical characteristics, degree of pre-treatment, finishing and external environment conditions (weather, heat, air flow, oxygen concentration) influence the material's flame retardancy [5]. Cotton is often used because of its high comfort characteristics, but it is easily flammable and generates burns [4]. Various treatments attempt to quench the flames, reduce the combustible gases and strengthen the generation of non-combustible gaseous degradation products in order to achieve flame retardancy. Most cellulose materials are non-durable or semi-durable and as such have limited lifetime [1]. The design of more durable products is the key strategy to preserve material properties and reduce the waste amount [6].

Proban® treatment, introduced by Albright & Wilson from England, whose protective rights and licenses were later transferred to the company Rhodia, is based on the formation of pre-condensates when using a FR agent based on tetrakis (hydroxymethyl) phosphonium salts (THPx) in multiphase treatments. The first is pre-condensation of tetrakis-hydroxymethyl chloride (THPC) or sulphate (THPS) with urea. Prior to application onto cotton, the pH value of the pre-condensate solution is adjusted to 6.0, thus preventing the settling of insoluble products due to passing a higher content of phosphonium salt to the form of phosphine [1]. If the pH rises above pH 6.0, the pre-condensate reactivity decreases due to the formation of phosphine oxide.

Polymerization in the Proban® process is based on the application of gaseous ammonia at room temperature. During the application and thermocondensation process, resins with six reactive P-methylol groups are formed that react with each other [1,7]. The central phosphorous atom in the P-methylol groups may be in one of three states (phosphonium, phosphine and phosphine oxide).

After impregnation, the fabrics are dried and treated with gaseous ammonia in a specially designed chamber where crosslinking takes place. In this phase, phosphorus is still in a lower (III) oxidation state (phosphine oxide). The fabric is treated with an aqueous solution of hydrogen peroxide for oxidation. The key factor is that the final product does not have hydrolyzed bonds on phosphorus, which is completely in a stable phosphine oxide structure. The efficiency of FR cellulose materials depends on process parameters in the domain of technological and use durability or lifetime, so Proban® should be considered a material whose FR properties can be weakened after several washing cycles [4–6]. The assessment of the durability of a product can be based on two aspects, technical and usage (e.g., duration of use, supplier and consumer requirement), as well as on other environmental and economic aspects (e.g., life cycle, cost of products) [4,8,9].

In many cases, some parameters of the Sinner's circle [10], e.g., high washing temperatures, excessive amount of detergents and other treatments to remove soils, stains and odor can impair the properties of personal protective clothing. Washing is a complex process involving physical and chemical parameters: temperature, chemistry (types, composition and amount), time (length of a process) and applied mechanical agitation [11,12]. Parameters depend on each other, so a reduction in one factor has to be compensated by an increase in the others.

Water as a washing medium is the fifth parameter of the Sinner's circle. Detergents contain bleach, fragrances, softeners and other additives that may leave residues on the surface of textiles [13]. Changes in surface properties and pore volume are the main causes of the change in material properties. It is therefore important to minimize the risk of agents impacting the protection level, shape, dimensions and comfort characteristics [14].

The present research focuses on the chemistry as a Sinner's circle parameter to verify the concept of a lower alkalinity washing bath compared to conventional, standard detergent and water, through the assessment of functional properties and appearance, and all guided by the idea of product research according to product life cycle assessment criteria. The limited oxygen index (LOI), calorimetric parameters (MCC), thermal gravimetric analysis (TGA), coupled thermal gravimetry–Fourier transform infrared technique (TG–FTIR), surface examination (SEM), spectral characteristics and pH of an aqueous extract of fabrics before and after washing were selected as criteria for proof of the concept. Future research will investigate the environmental acceptability of washing procedures through effluent analysis to gain complete knowledge of environmental impacts.

2. Materials and Methods

2.1. Material

Research was conducted on a FR functional cotton fabric specified as a Proban® described in Table 1.

Table 1. Characteristics of the cotton fabric.

Surface Mass (g/m^2)		347	
Density: warp/weft (threads/cm)		22/34	
LOI (%)		32	
Spectral parameters	L*	C*	h
face side	31.26	40.67	282.37
back side	32.88	41.01	281.23
Digital image, magnification 54×			
SEM image, magnification 100×			

L* – lightness, C* - chroma.

2.2. Washing Procedures

The textile material was subjected to 10 washing cycles under different conditions specified in Table 2.

Table 2. Description and sample labels.

Proban® Fabric	Description	Washing Bath	Label
Unwashed	Pristine	-	UNW
10 cycles washed	Component dosing in washing process 46	*Det A; *Det B; *A, *B; *NA	46_10x
	Component dosing in washing process 47	Det A; Det B; A, B; NA	47_10x
	Washing process 48	Water (17 ppm)	48_10x
	Washing process 49	*Det ECE-1	49_10x

* Det A: detergent containing fatty alcohol ethoxylate (50–75%), glycol ethers (<20%), amphoteric surfactant (<5%) and propanol (1–5%), A: alkali containing sodium hydroxide (25–35%) and polycarboxylates (<5%), Det B: detergent containing fatty alcohol ethoxylate (<20%), non-ionic surfactants (15–30%), sodium hydroxide (15–20%), sequestering agent (<5%), phosphonates, polycarboxylates (<5%), fluorescent whitening agent, B: disinfecting agent containing hydrogen peroxide (10–20%), peracetic acid (10–20%), acetic acid (25–35%), Det ECE-1: standard detergent according to ISO 105-C06, containing linear alkyl benzene sul-phonate (8%), ethoxylated tallow alcohol, 14 EO (2.9%), sodium soap (3.5%), sodium tripoly-phosphate (43.7%), sodium silicate (7.5%), magnesium silicate (1.9%), carboxymethyl cellulose (1.2%), TAED (0.2%), sodium sulphate (21.2%), water (9.9%), NA: neutralizing agent based on formic acid, p.a.

In the washing process, the fabric labelled as UNW was treated in Wascator FOM71 CLS, (Electrolux, Stockholm, Sweden) for 10 cycles. The component dosage system was used in two variations (46_10x and 47_10x), differing in alkalinity, where 46_10x represented higher alkalinity compared to 47_10x as lower alkalinity. The washing process (49_10x) continued with standard ECE-1 detergent and water (49_10x). All washing protocols are technically specified in Table 3.

Table 3. Specification of laboratory washing protocols 46, 47, 48 and 49.

Stage	Protocol			
	46	47	48	49
Prewash	Det A: 2.94 g/kg	Det A: 2.94 g/kg	-	-
	60 °C, 5 min			

Table 3. Cont.

Stage	Protocol			
	46	47	48	49
1st wash	A: 8 mL/kg Det A: 4.90 g/kg Det B: 2.50 g/kg	A: 2.0 mL/kg Det A: 4.90 g/kg Det B: 2.50 g/kg	-	ECE-1: 5 g/kg
	80 °C, 12 min			
2nd wash	A: 5.30 mL/kg Det A: 2.94 g/kg Det B: 2.0 g/kg	Det A: 2.94 g/kg Det B: 2.0 g/kg	-	ECE-1: 2 g/kg
	80 °C, 1 min			
3rd wash	B: 2.8 mL/kg	B: 2.8 mL/kg	-	-
	80 °C, 20 min			
Cooldown	45 °C, 1 min			
Drain	1 min			
Spin	1 min			
1st rinsing	5 min			
Drain	1 min			
Spin	1 min			
2nd rinsing	3 min			
Drain	1 min			
Spin	1 min			
Neutralization	NA: 3.42 mL/kg	NA: 3.0 mL/kg	-	-
	4 min			
Drain	1 min			
Spin	7 min			

The discharge from the laboratory washing machine after protocols specified in Table 3 were collected and analyzed by standard methods.

2.3. Methods

Different methods were selected for the characterization of the washing baths and effluents as well as fabrics before and after the washing protocols specified in Table 3.

The pH measurement of the washing baths (46, 47, 49), which was taken directly from the machine without fabrics was determined using the pH meter Mettrel.

Physico–chemical characteristics of collected washing effluents (46, 47, 48 and 49) were monitored by determining pH value, conductivity (κ), turbidity (T), total dissolved solids (TDS), total suspended solids (TSS), chemical oxygen demand (COD) using standard analytical methods.

The characterization of fabrics before and after 10 cycles of the washing protocol was based on:

1. residual substances as pH of the aqueous extract [ISO 3071: 2020].
2. limiting oxygen index (LOI).
3. microscale combustion calorimetry (MCC).
4. thermal stability and evolved gases during thermal decomposition (thermogravimetric analyzer (TGA) coupled with an infrared spectrometer (TG–IR).
5. microscopy, SEM and digital.
6. spectral characteristics.

The limiting oxygen index (LOI) technique was used for assessing the ease of ignition as an important parameter for the characterization of FR textile materials. The LOI values of the materials were determined according to ASTM D2863-10 and presented the maximum percentage of oxygen [O_2] in an oxygen-nitrogen gas mixture [O_2] + [N_2] that will sustain burning a standard sample for a certain time. The LOI values were calculated according to Equation (1).

$$LOI = \frac{[O_2]}{[O_2]+[N_2]} \times 100 \; [\%] \qquad (1)$$

The microscale combustion calorimeter (MCC) was used for the thermal characterization of FR samples before and after washing. The measurement was performed using an MCC-2 micro-scale combustion calorimeter (Govmark, Farmingdale, NY, USA) according to ASTM D7309-2007. The sample of 5 mg was heated to a specified temperature using a linear heating rate of 1 °C/s in a stream of nitrogen, with a flow rate of 80 cm^3/min. The thermal degradation products were mixed with a 20 cm^3/min stream of oxygen. Each sample was run in triplicates, and the presented MCC parameters are the averages of the three measurements.

The thermogravimetric analysis (TGA) was carried out using a Pyris 1 TGA (Perkin Elmer, Waltham, MA, USA). The sample weight was adjusted to 5–6 mg and the experiment was conducted in air atmosphere. All samples for the TGA were analyzed in the temperature range from 30 °C to 850 °C in a continuous air or nitrogen flow. The temperature was increased at a rate of 30 °C/min.

The TGA was connected to the Fourier transform infrared spectrometer (FT–IR, PerkinElmer, Spectrum 100S, Shelton, CT, USA) with a TG–IR interface. Evolution profiles of different compounds were tracked by FT–IR. The combination of TGA with FT–IR allowed for the analysis of the nature of the gaseous products formed in TGA and their online monitoring. Nitrogen, which does not exhibit IR-absorption, was used as the purge gas. Air was used as the reaction gas, so the end products of the decomposition were pyrolysis rather than oxidative degraded products. The FT–IR spectrum were acquired throughout the run at a temporal resolution of 4 s and a spectral resolution of 4 cm^{-1}. A thermal analysis gas station (TAGS), equipped with a detector, was used for the FT–IR analysis. The transfer line, high-temperature flow cell, and TG interface were held at 280 °C for the duration of the run to prevent sample condensation. The evolved gases were transferred through the FT–IR flow cell by a peristaltic pump with a flow rate of 60 cm^3/min.

The field emission scanning electron microscope, FE-SEM Mira LMU, Tescan, Brno, Czech Republic, was used for the examination of samples at 20 kV. A digital microscope was also applied for surface examination.

The spectral characteristics were observed as the difference between the washed samples compared to the unwashed sample through: the total difference in color (dE), difference in lightness (dL*), difference in chromaticity (dC*) and difference in hue (dH*) as the average values of four measurements conducted on the remission spectrophotometer DC 3890(Datacolor, Dietlikon, Switzerland).

3. Results and Discussion

Analyses of the properties of the fabric after 10 cycles of the washing process in relation to the pristine fabric were carried out systematically as described in the experimental section. Proban® fabrics were washed over 10 cycles under laboratory conditions according to different protocols: the component wash as a simulation of the industrial process (46_10x) specified as a high alkali and less alkali component wash (47_10x).

Protocols 48 and 49 underwent the same phases but differed in the chemical share in the Sinner's circle of the washing process. Process 49 was carried out with a standard detergent and process 48 in water only. It is known that through the interaction with textiles water has an impact on the textiles' properties [15]. The measurements of the pH washing bath protocols 46 and 47 were conducted without fabrics, Table 4.

Table 4. pH of washing baths.

	46	47	48	49
pre-wash	8.14	8.14	-	-
1st wash	12.28	10.98	-	9.8
2nd wash	12.28	10.98	-	8.6
3rd wash	11.02	7.37	-	-

The results in Table 4 show differences in the pH of the washing baths affected by the dosage of NaOH in the programs 46 and 47. The composition of standard detergent ECE-1 (program 49) reflects an alkali bath.

Results of washing effluents analysis through physico–chemical and organic indicators are presented in Table 5.

Table 5. Characteristics of washing effluents.

	46	47	48	49
pH	2.82	2.89	5.37	5.42
κ (μS/cm)	790	661	52.1	80
T (NTU)	28.9	30.2	36.8	26.7
TS (mg/L)	1029	936	109	121
TDS (mg/L)	1178	1076	18	60
TSS (mg/L)	66	76	96	55
COD (mg/L)	>1500.00	1318.00	190.33	129.67

The values of physico–chemical characteristics of the collected effluents presented in Table 5 show the differences between particular programs. If pH and conductivity are considered, the differences are conditioned by the share of chemistry in the washing process. Component washing (programs 46 and 47) with the neutralization phase caused acidity of effluents (pH lower than 3.0). Washing programs 48 and 49, involving rinsing in water, resulted with higher pH (higher than 5.0). The small difference is obtained in the turbidity values, all effluents possess about 30 NTU. Negligible difference was obtained with total suspended matter (TSS) caused by fibrils released in all of the tested washing processes. The biggest difference is manifested in the load of effluent with chemical substances (COD), where the component washing programs according to the industrial concept (46 and 47) have ten times more values compared to washing with standard detergent and water. The characterization of fabric properties was carried out to verify whether a lower alkaline bath has better potential to preserve the functional properties and appearance of fabrics for a longer life cycle. As such, the residual alkali of fabrics was analyzed by measuring the pH of the aqueous extract, as shown in Table 6.

Table 6. pH of fabrics' aqueous extract.

Fabric	pH
UNW	7.44
46_10x	7.66
47_10x	7.33
48_10x	8.11
49_10x	7.75

The pH values of the aqueous extract of untreated and washed fabrics indicate some differences: the pH value of samples washed in 46, 47 and 49 is slightly above 7.0. Fabrics washed in programs 46_10x (higher alkalinity) and 47_10x (lower alkalinity) were neutralized by an acid, so the impact of higher alkalinity on the pH of the aqueous extract is not so prominent.

An unexpectedly high pH value of the aqueous extract 48_10x can be attributed to the nature of substances released from a Proban®. The swelling ability of cotton in water (48) and the influence of the process parameters of the Sinner's circle, expressed through mechanical agitation and temperature on the cellulose polymer structure, led to the increased alkalinity.

The limiting oxygen index (LOI) technique provides a quantitative measure for the determination of reduced flammability of the material, Table 7.

Table 7. LOI of fabrics in warp and weft directions.

Sample	LOI (%)
UNW warp/weft	32
46_10x warp/weft	32
47_10x warp/weft	32
48_10x warp/weft	32
49_10x warp/weft	32

LOI values for all samples after 10 wash cycle regardless of the treatment conditions (46, 47, 48 and 49) retain the same value as the pristine Proban® sample (32%). Such a high-retained value classifies all samples as self-extinguishing materials.

The MCC analysis results of the samples before and after washing are specified in Table 8 as: heat release capacity (ηc), maximum specific heat release rate (Q_{max}), specific heat release (hc), specific heat of flammable gases (hc, g), temperature at maximum specified heat release rate (T_{max}) and combustion residue.

Table 8. Parameters of MCC analysis unwashed and 10-times-washed Proban® fabric.

Parameters	UNW	46_10x	47_10x	48_10x	49_10x
$\eta c\ (J(g \cdot K)^{-1})$	61.0	65.7	71.7	62.7	69.3
$Q_{max}\ (Wg^{-1})$	62.5	66.7	72.9	64.0	71.0
hc (kJg^{-1})	2.0	2.1	2.4	1.9	2.1
hc, gas (kJg^{-1})	2.0	2.2	3.8	3.1	3.4
Tmax (°C)	314.1	313.0	314.5	311.0	312.8
Residue (%)	40.81	37.40	35.63	39.83	38.78

The resulting values of the released heat capacity (ηc) presented in Table 6 show that the untreated Proban® sample (UNW) has the lowest value, as expected. When considering the sample after 10 washing cycles in water (48_10x), it is obvious that the slight increase in the heat release capacity (ηc) compared to the pristine sample correlated with the resulting LOI value. The heat release capacity (ηc) of the samples 46_10x has the lowest value and the lowest specific heat release, hc, gas = 2.2 kJ/g compared to the other samples. An important parameter for monitoring the thermal stability of the samples is the pyrolysis residue. The sample washed in water (48_10x) had the largest residue—39.83%. Slightly less residue was found in the sample washed with a standard detergent containing phosphates (49_10x)—38.78%, while the smallest residue was found in the sample 47_10x—35.63%. The highest specific heat value of fuel gases was recorded for this last sample, probably due to the combustion of organic sample components.

Figures 1–10 show the TG and dTG curves of all samples and the absorption spectrum of gaseous decomposition products formed during thermogravimetric analyses. All the tested samples showed similar properties during thermal decomposition in an air stream. From the first derivations of thermogravimetric curves (dTG) (Figures 1, 3, 5, 7 and 9) it can be seen that the dynamic degradation in all samples takes place in one step with the maximum speed of degradation at a temperature of about 340 °C, but differs in the onset and end temperatures of degradation and the residue at 850 °C. In order to determine the influence of 10 washing cycles on the thermal stability of the Proban® sample, the thermal properties of the initial sample (UNW) were also examined. The dynamic thermal decomposition of the Proban® sample (UNW) started at a temperature of 321.1 °C and ended at 353.79 °C. From the dTG curve it is evident that the degradation of the sample took place in one step with a mass loss of 41.456%. At the end of the dynamic decomposition, a higher intensity profile FT–IR spectrum of gas evolved was recorded, the composition and quantity of which were detected by IR. The characteristic peaks that appear in the range of 3500–4000 cm^{-1} and 1550–1566 cm^{-1} belong to the released water vapour. CO and CO_2 peaks are visible in the area of 2179 and 2110 cm^{-1}. Peaks belonging to the released CO_2 were also detected in the range of 2359 and 2322 cm^{-1}, while the occurrence of peaks at wave numbers 2951 and 1184 cm^{-1} indicates the release of RCHO [16]. From the absorption curves of the gaseous products of Figure 2b,d, it is apparent that a slightly higher amount of CO_2 was recorded at the second intensity profile FT–IR spectrum of gas evolved (Figure 2c) at 643 °C, which is according to the TG curve at the remaining 30% of the mass. After that the gaseous products are insignificant or absent, which indicates the effect of the treatment aimed at suppressing the development of gaseous products during thermal decomposition. This is very important for monitoring the behavior of the sample in the open flame because it indicates the formation of charred residue and the sample becomes fireproof, which is confirmed by the high LOI value of 32 and low values of specific heat of gas decomposition products of 2.0 KJg^{-1} obtained by the MCC measurements. The best heat resistance results could be found in the sample washed with ECE standard detergent containing 30% phosphate (49_10x), which is visible from the residue at 850 °C, which is 14.932% (Figure 9). From the obtained absorption intensities of gaseous decomposition products (Figure 4a,b, Figure 6a,b and Figure 8a,b) it is obvious that in the samples 46_10x, 47_10x, 48_10x the first intensity profile FT–IR spectrum of gas evolved degree occurs at temperatures from 354 °C to 356 °C, while in the sample 49_10x it is at a slightly lower temperature of 345 °C, indicating an earlier onset of thermal decomposition, which is characteristic of more thermally stable materials. In all samples, gaseous products of CO_2 and CO were recorded, but also the appearance of all the detected gases as in the UNW sample. The second degree intensity profile FT–IR spectrum of gases evolved are visible in Figures 4, 6 and 8, under the c. and d. ranges from 650 to 680 °C. The lowest temperature of the second absorption maximum of gaseous decomposition products was also recorded in the sample 49_10x and was 650 °C (Figure 10c), while the composition of gaseous products shown in Figure 10d indicates the presence of CO_2 and CO with an absorption intensity of CO_2 of 0.1. The comparison of the unwashed sample (UNW) and sample 49_10x clearly indicates that during the maintenance process there was an increase in thermal stability, proving a positive effect of the detergent. However, such applications are limited due to the known problems of phosphate impacting the ecosystem. The high amount of residue after the TG analysis in sample 49_10x was due to residual substances derived from detergent components that were not used to remove stains, but were deposited on the cotton fabric. Very similar behavior in terms of thermal stability and incombustibility to the sample 49_10x is shown by sample 47_10x, with the remark that the adsorption intensity of the decomposition gases recorded at the second maximum (indicating the presence of CO_2—Figure 6c,d) is lower compared to the sample 49_10x and is 0.06, which suggests the possible application of such an environmentally friendly washing procedure in the maintenance of Proban® samples.

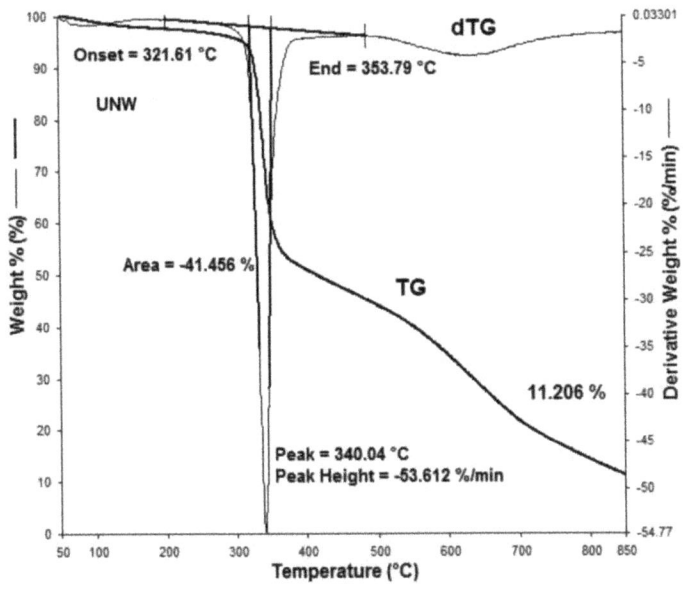

Figure 1. Thermogravimetric analysis of the untreated UNW sample: TG curve, dTG curve.

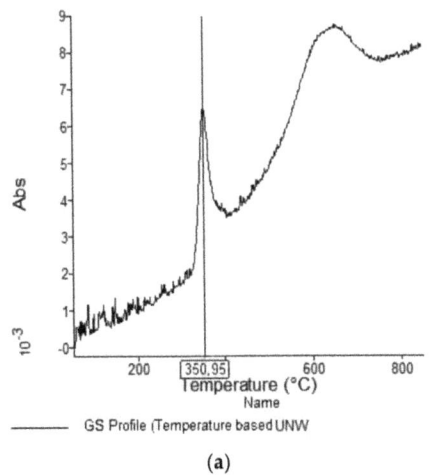

(a)

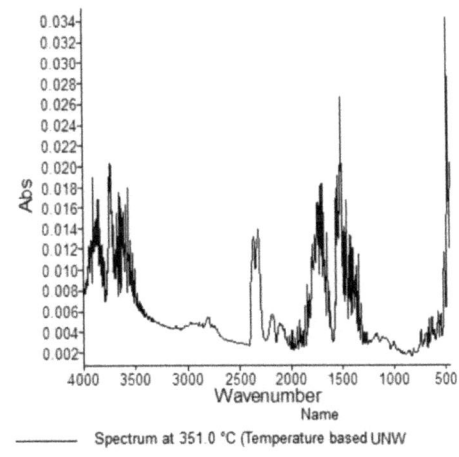

(b)

Figure 2. *Cont.*

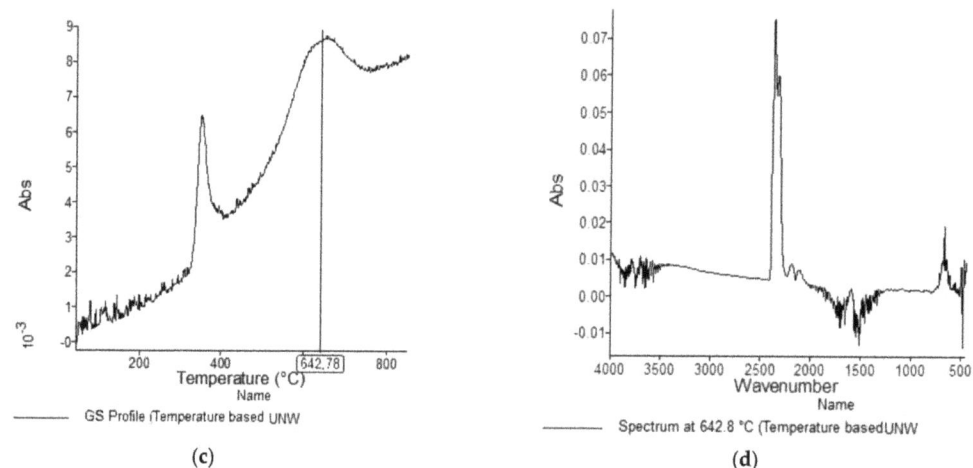

Figure 2. TG–IR analysis of cellulosic material treated with Proban® method during thermooxidative decomposition: (**a**) and (**c**) intensity profile FT–IR spectrum of gas evolved, (**b**) FT–IR spectrum of gas evolved at 351.0 °C, (**d**) FT–IR spectrum of gas evolved at 642.8 °C.

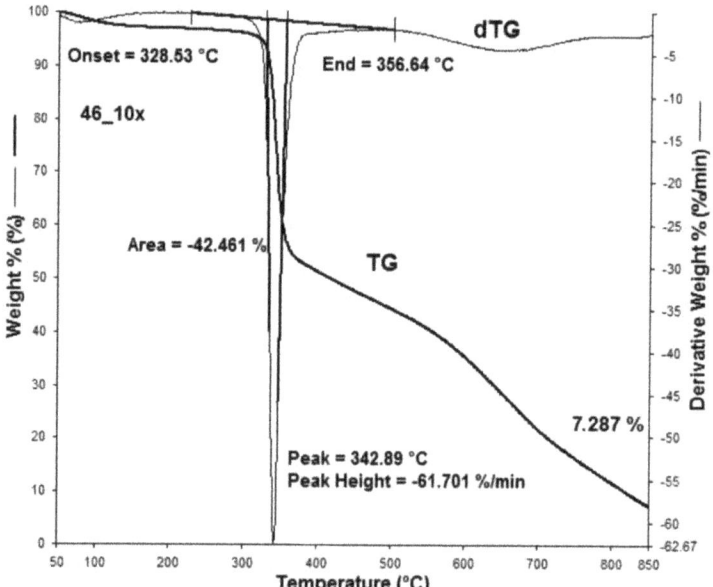

Figure 3. Thermogravimetric analysis of 46_10x sample: TG curve, dTG curve.

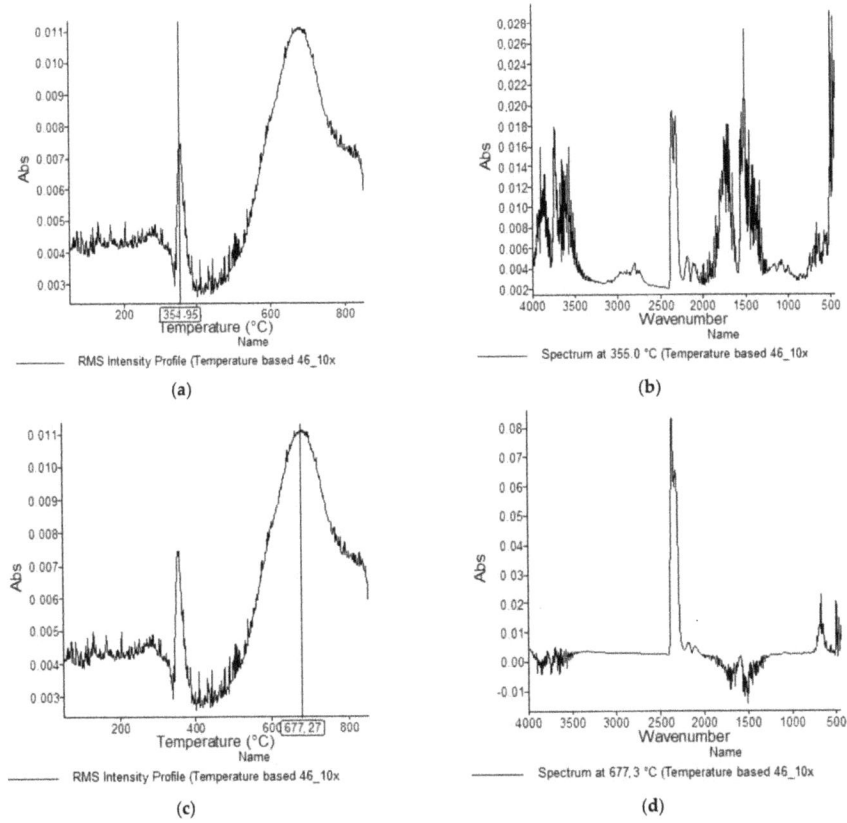

Figure 4. TG–IR analysis of 46_10x during thermo–oxidative decomposition: (**a**) and (**c**) intensity profile FT–IR spectrum of gas evolved, (**b**) FT–IR spectrum of gas evolved at 355.0 °C, (**d**) FT–IR spectrum of gas evolved at 677.3 °C.

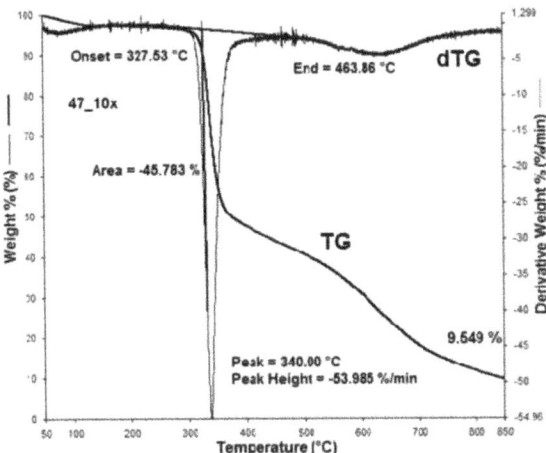

Figure 5. Thermogravimetric analysis of 47_10x sample: TG curve, dTG curve.

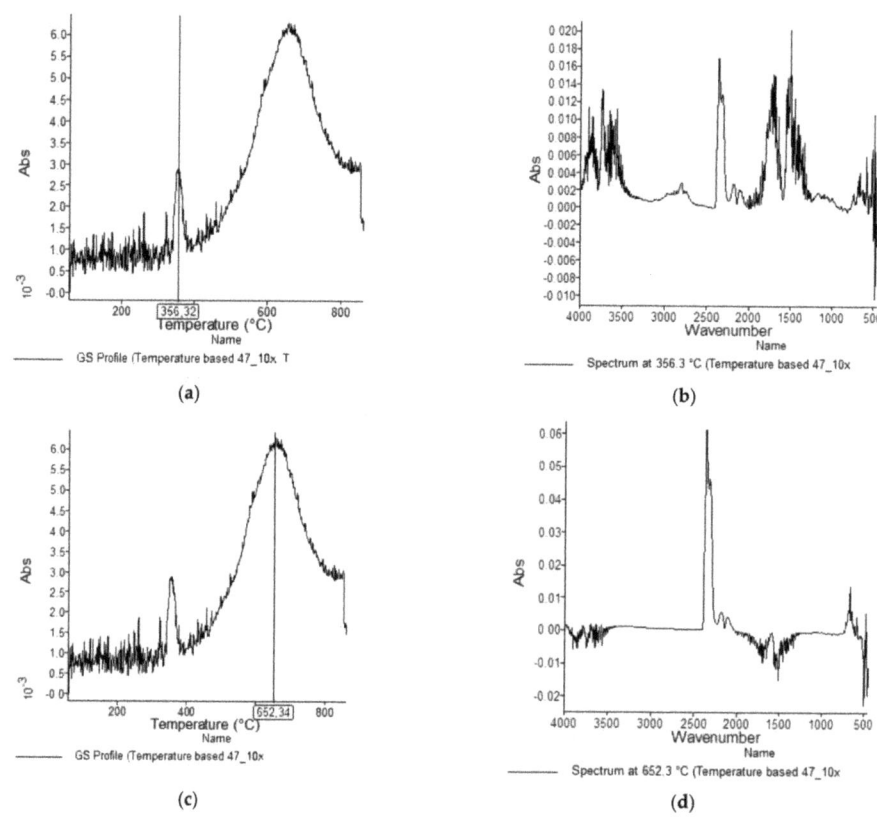

Figure 6. TG–IR analysis of 47_10x during thermo-oxidative decomposition: (**a**) and (**c**) intensity profile FT–IR spectrum of gas evolved, (**b**) FT–IR spectrum of gas evolved at 356.3 °C, (**d**) FT–IR spectrum of gas evolved at 652.3 °C.

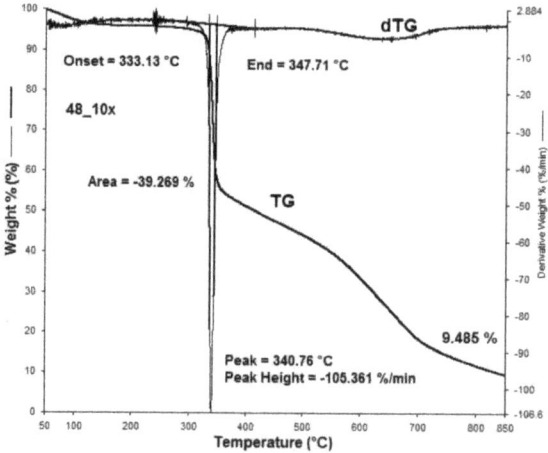

Figure 7. Thermogravimetric analysis of 48_10x sample: TG curve, dTG curve.

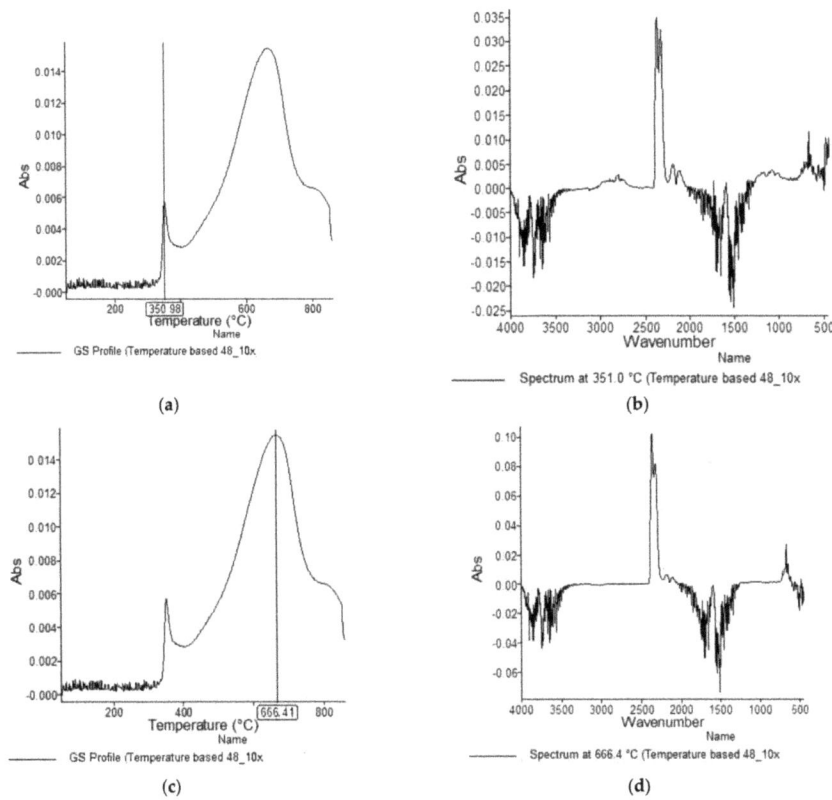

Figure 8. TG–IR analysis of 48_10x during thermo-oxidative decomposition: (**a**) and (**c**) intensity profile FT–IR spectrum of gas evolved, (**b**) FT–IR spectrum of gas evolved at 351.0 °C, (**d**) FT–IR spectrum of gas evolved at 666.4 °C.

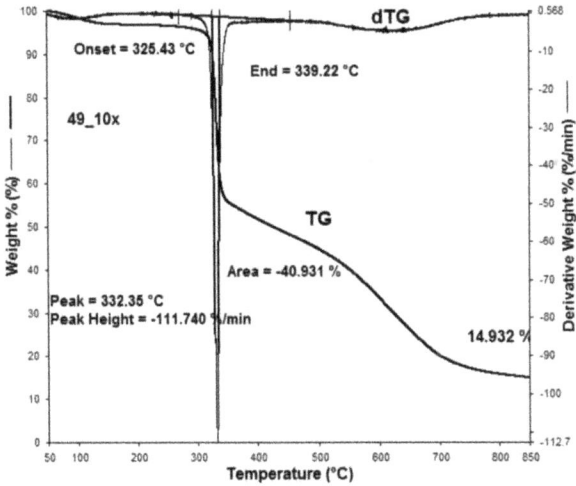

Figure 9. Thermogravimetric analysis of 49_10x sample: TG curve, dTG curve.

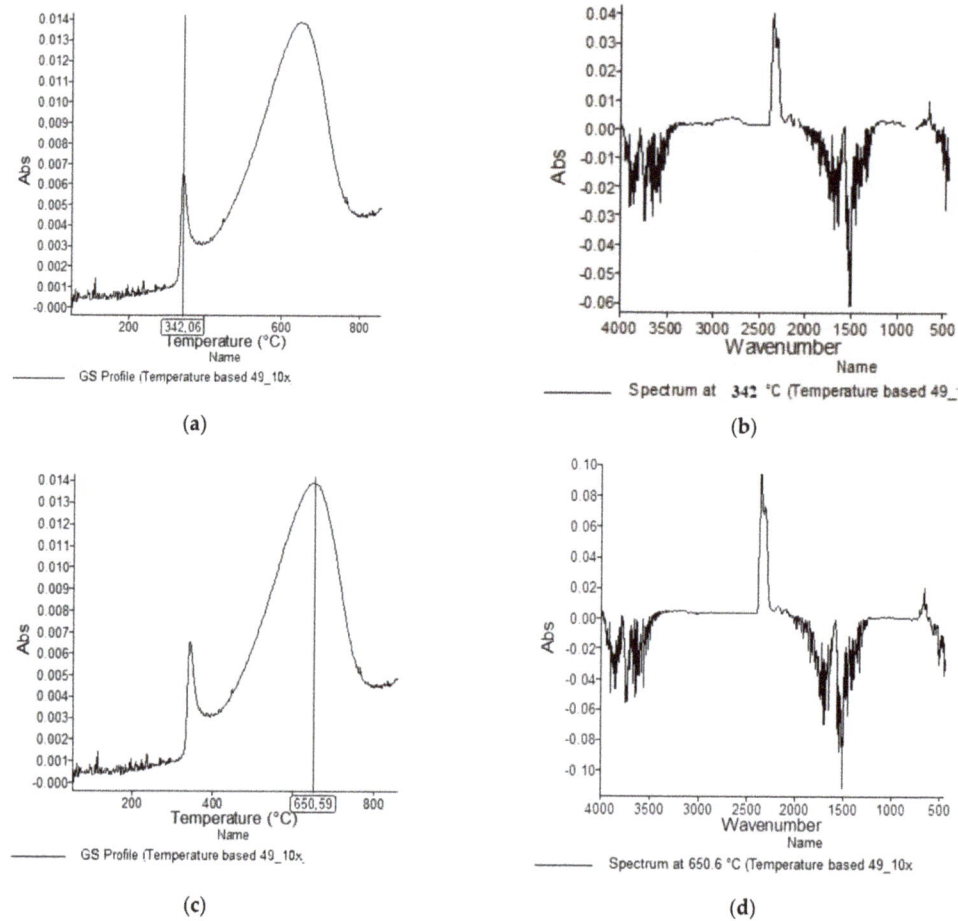

Figure 10. TG–IR analysis of 49_10x during thermo–oxidative decomposition: (**a**) and (**c**) intensity profile FT–IR spectrum of gas evolved, (**b**) FT–IR spectrum of gas evolved at 344.9 °C, (**d**) FT–IR spectrum of gas evolved at 650.6 °C.

The appearance of the fabrics before and after washing was examined by spectral parameters and microscopy, Tables 9 and 10.

Table 9. Differences in spectral characteristics of washed samples compared to the pristine sample.

Sample	Side	Mark (ISO)	Spectral Values			
			dE	dL*	dC*	dH*
46_10x	Face	2-3	4.265	4.176	0.820	−0.248
	Back	3	4.024	3.812	1.213	0.137
47_10x	Face	3-4	3.144	2.464	1.876	0.395
	Back	4-5	3.126	0.801	2.685	1.363
48_10x	Face	2-3	4.781	4.732	0.452	−0.344
	Back	3	3.763	3.298	1.747	0.443
49_10x	Face	2-3	5.577	4.99	−1.801	−1.702
	Back	2-3	4.672	4.355	−1.095	−1.286

dE—difference in color, dL*—difference in lightness, dC*—difference in chromaticity, dH*—difference in hue.

Table 10. Digital images of samples magnified 54 and 230× and SEM image magnified 500×.

	54×	230×	500×
UNW			
46_10x			
47_10x			
48_10x			
49_10x			

According to the results in Table 7, the spectral characteristics of blue cotton fabrics changed through 10 washing cycles, and the intensity of changes depended on the composition of the washing bath. The decrease in alkalinity (47_10x) proved to be the most favourable factor for the preservation of blue color characteristics.

The surface images of the washed samples, Table 8, indicate a specific fibrillation that can be caused by the characteristics of cotton cellulose and the parameters of the Sinner's circle in the washing process. As the mechanical agitation and temperature were constant in the programs 46, 47, 48 and 49, the significant change in appearance specified for the fabric washed in the program 46 may be caused by the stronger swelling ability of cotton cellulose in alkali conditions of this washing bath.

4. Conclusions

The present research focused on the impact of washing conditions on the effluent composition and durability of FR properties and appearance of Proban® cotton fabrics and was systematically arranged through the variation in the chemistry distribution in the Sinner's circle. The chemical share was varied in laboratory conditions as a simulation of industrial washing based on component dosing, where the temperature, time and mechanical agitation were constant. The washing of cotton fabrics was performed through 10 cycles in four baths containing high alkali components, medium alkali components, high alkali reference detergent and water.

The results of effluent analysis showed negligible difference in the characteristics of high and medium alkali process. More optimal characteristics possess effluents from washing with standard detergent and water.

LOI as a parameter is not relevant for precisely determining mutual differences in FR properties, but is a valuable method for the rapid characterization of samples with respect to flame resistance. Proof of the medium alkali bath as a washing concept for Proban® cotton fabric was confirmed through the preservation of FR properties examined through TGA and MCC parameters and appearance based on preserved color and a low fibrillation level. The proposed concept of medium alkali washing can be considered environmentally friendly since it shows the lowest intensity of the gases evolved and economically viable compared to other washing processes in this research, which is in line with LCA product monitoring guidelines.

Finally, the concept of washing of Proban® cotton fabrics at lower pH meets the requirements of functionality and durability, and its acceptability from an ecological point of view requires wastewater treatment using combined methods.

Author Contributions: Conceptualization, T.P. and S.F.G.; methodology, S.F.G. and T.K.; formal analysis, T.P., T.K. and S.F.G.; investigation, T.P., T.K., T.D. and S.F.G.; data curation, T.P. and S.F.G.; writing—original draft preparation, T.P. and S.F.G.; writing—review and editing, T.P. and S.F.G. All authors have read and agreed to the published version of the manuscript.

Funding: This research was funded by the European Union from the European Regional Development Fund under the project KK.01.2.1.02.0064 Development of multifunctional non-flammable fabric for dual use.

Institutional Review Board Statement: Not applicable.

Informed Consent Statement: Not applicable.

Data Availability Statement: Data is available in a publicly accessible repository.

Acknowledgments: The research was performed on equipment purchased by K.K.01.1.1.02.0024 project "Modernization of Textile Science Research Centre Infrastructure" (MI-TSRC). A paper recommended by the 14th Scientific and Professional Symposium Textile Science and Economy, The University of Zagreb Faculty of Textile Technology.

Conflicts of Interest: The authors declare no conflict of interest.

References

1. Yang, C.Q.; Weil, E.D. Functional Protective Textiles. In *Flame Retardant Textiles*; Bischof, S., Ed.; University of Zagreb: Zagreb, Croatia, 2012; pp. 213–230.
2. Pavlinić, D.Z.; House, J.R.; Mekjavić, I.B. Protupožarni odjevni sustavi i njihovo vrednovanje. *Sigurnost* **2010**, *52*, 251–261.
3. Chen, W.B.; Wan, Y.Y.; Que, F.; Ding, X.M. The Durability of Flame Retardant and Thermal Protective Cotton Fabrics during Domestic Laundering. *Adv. Mater. Res.* **2012**, *441*, 255–260. [CrossRef]
4. House, J.R.; Squire, J.D. Effectiveness of Proban® flame retardant in used clothing. *Int. J. Cloth. Sci. Technol.* **2004**, *16*, 361–367. [CrossRef]
5. Pušić, T.; Kaurin, T. Zaštitna odjeća Proban® kvalitete—Mogućnosti i rizici. *Tekstil* **2017**, *66*, 113–120.
6. Alfieri, F.; Cordella, M.; Stamminger, R.; Alexander, B. *Durability Assessment of Products: Analysis and Testing of Washing Machines*; Office of the European Union: Luxembourg, 2018; p. 2760.
7. Carr, C.M. *Chemistry of the Textile Industry*; Blackie Academic & Professional: London, UK, 1995.
8. Alfieri, F.; Cordella, M.; Sanfelix, J.; Dodd, N. An approach to the assessment of durability of energy-related products. *Procedia CIRP* **2018**, *69*, 878–881. [CrossRef]
9. Zhang, D.; Williams, B.L.; Liu, J.; Hou, S.; Smith, A.T.; Nam, S.; Nasir, Z.; Patel, H.; Partyka, A.; Becher, M.A.; et al. An environmentally-friendly sandwich-like structured nanocoating system for wash durable, flame retardant, and hydrophobic cotton fabrics. *Cellulose* **2021**, *28*, 10277–10289. [CrossRef]
10. Sinner, H. *Über das Waschen mit Haushaltswaschmaschinen: In welchem Umfang erleichtern Haushaltwaschmaschinen und -geräte das Wäschehaben im Haushalt*; Haus & Heim Verlag: Hamburg, Germany, 1960; p. 8.
11. Boyano, A.; Cordella, M.; Espinosa, N.; Villanueva, A.; Graulich, K.; Rüdenauer, I.; Alborzi, F.; Hook, I.; Stamminger, R. *Ecodesign and Energy Label for Household Washing Machines and Household Washer-Dryers*; Publications Office of the European Union: Luxembourg, 2017.

12. Bao, W.; Gong, R.; Ding, X.; Xue, Y.; Li, P.; Fan, W. Optimizing a laundering program for textiles in a front-loading washing machine and saving energy. *J. Clean. Prod.* **2017**, *148*, 415–421. [CrossRef]
13. Li, C.; Wang, L.; Yuan, M.; Xu, H.; Dong, J. A New Route for Indirect Mineralization of Carbon Dioxide–Sodium Oxalate as a Detergent Builder. *Sci. Rep.* **2019**, *9*, 12852. [CrossRef] [PubMed]
14. Nayak, R.; Ratnapandian, S. *Care and Maintenance of Textile Products Including Apparel and Protective Clothing*; Textile Institute Professional Publications: London, UK, 2018.
15. Smulders, E. *Laundry Detergents*; Wiley-VCH: Weinheim, Germany, 2002.
16. Flinčec Grgac, S.; Bischof, S.; Pušić, T.; Petrinić, I.; Luxbacher, T. Analytical Assessment of the Thermal Decomposition of Cotton-Modacryl Knitted Fabrics. *Fibres Text. East. Eur.* **2017**, *25*, 59–67. [CrossRef]

Article

The Impact of Elongation on Change in Electrical Resistance of Electrically Conductive Yarns Woven into Fabric

Željko Knezić [1], Željko Penava [1], Diana Šimić Penava [2] and Dubravko Rogale [3,*]

[1] Department of Textile Design and Management, Faculty of Textile Technology, University of Zagreb, 10000 Zagreb, Croatia; zeljko.knezic@ttf.unizg.hr (Ž.K.); zpenava@ttf.unizg.hr (Ž.P.)
[2] Department of Engineering Mechanics, Faculty of Civil Engineering, University of Zagreb, 10000 Zagreb, Croatia; dianas@grad.unizg.hr
[3] Department of Clothing Technology, Faculty of Textile Technology, University of Zagreb, 10000 Zagreb, Croatia
* Correspondence: dubravko.rogale@ttf.unizg.hr

Abstract: Electrically conductive yarns (ECYs) are gaining increasing applications in woven textile materials, especially in woven sensors suitable for incorporation into clothing. In this paper, the effect of the yarn count of ECYs woven into fabric on values of electrical resistance is analyzed. We also observe how the direction of action of elongation force, considering the position of the woven ECY, effects the change in the electrical resistance of the electrically conductive fabric. The measurements were performed on nine different samples of fabric in a plain weave, into which were woven ECYs with three different yarn counts and three different directions. Relationship curves between values of elongation forces and elongation to break, as well as relationship curves between values of electrical resistance of fabrics with ECYs and elongation, were experimentally obtained. An analytical mathematical model was also established, and analysis was conducted, which determined the models of function of connection between force and elongation, and between electrical resistance and elongation. The connection between the measurement results and the mathematical model was confirmed. The connection between the mathematical model and the experimental results enables the design of ECY properties in woven materials, especially textile force and elongation sensors.

Keywords: woven fabric; plain weave; electrically conductive yarn; resistance; tensile force; elongation

1. Introduction

Electrically conductive textiles represent a new generation of textile fibers, yarns, fabrics, and knits, with a wide range of uses [1]. Mattila therefore expands the field of conventional textiles, regarding their integrative complexity and their degree of integration. Conventional textiles retain low complexity and a low degree of integration. So-called functional textiles retain low complexity but take on a high degree of integration. Wearable electronics are attached to but not integrated into textile structures. When electronics and other devices are integrated directly into fibers and textiles, it is called textronics and fibertronics, and represents the highest degree of integration and complexity [2]. Therefore, electrically conductive textiles play an increasingly important role in the production of intelligent clothing, surveillance and protective clothing, energy-producing clothing, clothing for medical diagnostics, for heating, etc., and are used in the production of uniforms for the army, police, firefighters, and other special services. Gao et al. [3] note that highly stretchable sensors for wearable biomedical applications have attracted substantial attention since their appearance in the early twenty-first century, due to their unique characteristics, such as low modulus, light weight, high flexibility, and stretchability.

According to the working environment, Cheng et al. [4] divided flexible force sensors into wearable and implantable sensors. Wearable sensors are still limited by their complex structure, difficult material handling, and dependence on external power supply.

Nascimento et al. [5] consider the importance of sensors for the monitoring and protection of human health, as well as methods applied for physical and motor rehabilitation, which they predict will be linked to machine learning, artificial intelligence, and the Internet of things (IoT) in the near future.

Liu et al. [6] made an overview of the development of indicators and sensors for health monitoring, and divided them into five groups: body motion, skin temperature, heart rate/ECG/pulse, metabolism, and respiration. Measurement parameters for the respiration indicator are strain, pressure, and humidity for detection of cardiac arrest, apnea, and emotional control. In particular, wearable sensors installed on the chest/abdomen can be used to detect respiration.

Ehrmann et al. [7] grouped the possible applications of ECYs for electromagnetic shielding, textile pressure sensors, antielectrostatic textiles, and electronic circuits in plain textile materials, as well as their most common use in elongation sensors.

Such textile materials consist of electrically conductive fibers and yarn and thread, irrespective of whether such electrically conductive yarn is only partially incorporated or the whole flat textile product is made of it. In ECY, in addition to the electrical conductivity of the fibers, the total conductivity will also depend on the average length of the fiber within the thread, the formation of permanent joints, and the electrical resistance present at the contact areas with adjacent fibers [8]. It is known that twisting yarn compresses the fibers as well as shearing, joining, and arranging them into a more compact structure. It is to be expected that this creates additional and alternative electrically conductive contacts that directly affect the overall electrical conductivity of the yarn [9]. Accordingly, it can be assumed that similar or identical behavior of ECY used for textile sensors increases the number of shares or twists required to improve the linearity such that the sensitivity could achieve the minimum requirements [10].

Some studies have described the design and development of a flexible sensor for measuring textile deformations based on a conductive polymer composite [11].

In electrically conductive textiles, the length-associated electrical resistance and the contact resistance contribute the most to the overall resistance [12]. The length-related resistance depends only on the length of the electrically conductive yarn when the tensile force acts in parallel with the electrically conductive yarn, while the contact resistance depends on the contact surface under the tensile force [13–15]. Ryu et al. provided a theoretical analysis showing the effect of changes in the deformation of electrically conductive yarn on electrical resistance [16]. From theoretical analysis and experimental research, it was determined that the contact resistance of folded yarns in the fabric is a key factor affecting the sensitivity of the sensor embedded in the fabric [14]. The application of sensors in the conductive fabric is based on changes in electrical resistance that respond to stimuli such as deformation, temperature, and humidity [17]. Several papers have described experimental studies of the relationship between load/elongation and electrical resistance of electrically conductive fabrics [18,19].

Roh [20], Tangsirinaruenart and Stylios [21], and Ruppert-Stroescu and Balasubramanian [22] successfully fastened ECY to textile materials by sewing or embroidery using different types of machine sewing stitches, obtaining similar results of changes in resistance with elongation as in the case of weaving and knitting. Much of the research on textile strain sensors is based on the use of metal wires or electrically conductive yarns from well-known manufacturers. Eom et al. [23] show that it is possible to treat textiles with their own chemical substances and then sew them to the textile materials with flat and zigzag sewing stitches.

Ivšić et al. [24] show that with the embroidery technique it is possible to make wearable antennas on clothes with the help of conductive threads for high frequencies. Garnier et al. show that antennas on clothing at a short-wave frequency can also be obtained using the embroidery technique [25]. Chang et al. [26] dedicated themselves to the development of stretchable microwave antenna systems, considering that wireless functionality is essential for the implementation of wearable systems. They replaced the textile substrate with elastomeric substrates, and obtained better results due to changes in the dimensions of the antenna elements.

It is evident that ECYs have a good potential for future application. Thus, Yang et al. see the use of conductive yarns as flexible actuators for soft robotics [27], while Kan and Lam foresee a future trend in wearable electronics in the textile industry [28].

Yao and Zhu [29] state that the development of wearable multifunctional sensors assumes the extent to which woven or textile materials are replaced by printed, stretchable conductors made of silver nanowires with excellent properties.

Ha et al. [30] have researched wearable and flexible sensors based on engineered functional nano/micro-materials, with unique sensing capabilities for detection of the physical and electrophysiological vital signs of humans. One easy way to make stretchable conductors is the use of conductive nanomaterials—such as graphene, conducting polymers, and liquid metals—in combination with an intrinsically stretchable elastomeric polymer matrix.

However, some authors have already discovered some defects of the electrically conductive filaments, so this will need to be considered in future research. Stavrakis et al. indicate the peeling of silver-plated layers [31] when too much electric current passes through the yarn. Another problem is cyclic loading, at which unwanted hysteresis occurs, such that the data are not repeatable after a certain number of load cycles; this is indicated by McKnight et al. [32]. A similar problem was noted by Wang et al., who examined intrinsically stretchable and conductive textiles by a scalable process for elastic wearable electronics [33]. The durability of the conductive threads used for integration of electronics into smart clothing, over several washing cycles of clothing, was examined by Briedis et al., who found that the values of the parameters of the ECYs decrease with more washing cycles [34].

The research presented in this paper is focused on determining the impact of elongation on the change in electrical resistance of the ECYs woven into fabric, and later on the development of motion sensors that change electrical resistance due to elongation.

ECY was used in a traditionally woven ribbon with variable electrical resistance when voltage is changed. Textile ribbon with ECY could be used in the garment as a motion sensor to monitor the breathing of people with apnea, and this innovation was recognized and awarded at international exhibitions by two special awards, five gold medals, and one silver medal. These awards show the innovation of the traditionally woven ribbon, and the applicability of electrically conductive yarn. This innovation was used in the breathing simulator, i.e. chest movement simulator.

The aim of this paper is to investigate how the change in the elongation of the samples under the action of tensile force affects the change in the electrical resistance of ECY woven into fabrics.

2. Materials and Methods

2.1. Measuring the Electrical Resistance

The ohmic resistance of a conductor (or of a resistor) can be measured by several instruments and by various methods, e.g., by measuring current and voltage with ammeters and voltmeters; by comparing the known with unknown resistance; or by using ordinary and digital ohmmeters, as well as different measuring bridges. For this research, the so-called comparative method of resistance measurement is used.

According to this method, a resistor of known resistance R_N connects to the voltage of the source U in series (Figure 1), or with a resistor of unknown resistance R_X. The value of the unknown resistance is obtained from the value of the known resistance and the ratio of voltage drops—i.e., currents—on both resistors [35].

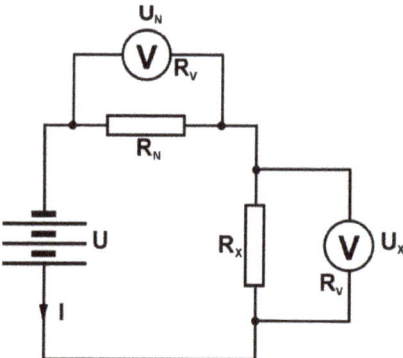

Figure 1. Scheme of the series measuring circuit.

In the case of a series connection with resistors, the voltages are measured with an internal resistance voltmeter R_V. With this connection, the voltage drops across the resistors R_X and R_N are:

$$U_X = I \cdot \frac{R_X \cdot R_V}{R_X + R_V}; \; U_N = I \cdot \frac{R_N \cdot R_V}{R_N + R_V} \quad (1)$$

If the resistances R_X and R_N are significantly less than R_V, the impact of R_V can be neglected, from which follows the expression (2):

$$R_X = R_N \cdot \frac{U_X}{U_N} \quad (2)$$

A series connection is used to measure low-value resistance. To measure smaller resistances (of the order of magnitude of several hundred ohms), a series measurement connection (Figure 1) is used, which was also used in this research.

2.2. Force–Elongation Diagram for the Fabric

The functional relationship between tensile force and elongation cannot be determined theoretically, but only by experimentally testing samples made of a specific material. Fabrics are a special type of anisotropic, inhomogeneous material [36,37]. In the biaxial structure of weaving, two main directions are defined: longitudinal (warp), and transverse (weft). Experiments determine the relationship between force and elongation in the form of diagram. Mechanical characteristics are investigated in the elastic range [38,39]. Under the action of tensile force F, elongation ε of the sample occurs. Figure 2 shows the force–elongation curve of the fabric.

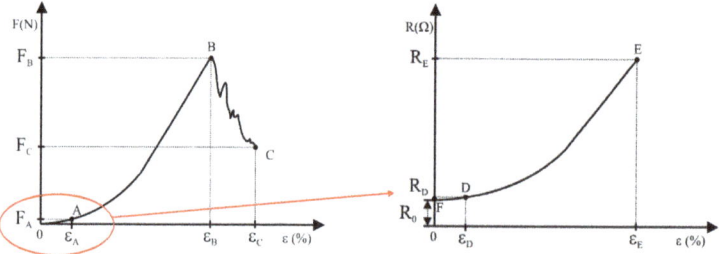

Figure 2. Characteristic diagrams of the tensile force–elongation (F-ε) curve of the woven fabric and the electrical resistance–elongation (R-ε) curve of the conductive fabric yarn.

The yarns in the fabric are arranged in a wavy shape. Due to the corrugation of the yarns, the fabric behavior in the elongation at the warp and the weft will consist of a range that mainly occurs due to the correction of the corrugation of the yarns and a range that results from the elongation of the yarns themselves. The final elongation at the break of the fabric will depend on the size of the corrugation, and on the possibility of this corrugation being straightened, which is determined by the pliability of the other yarn system.

The curve in the tensile force–elongation diagram (Figure 2) consists of two parts [40,41]: The first part 0A is a linear region representing the elastic area of the fabric, in which the yarns move into the fabric embroidery and the corrugation of the yarns in the fabric is corrected. Up to point A, the linear dependence of tensile force and fabric elongation applies. Thus, Hooke's law of anisotropic material behavior can be applied to this region. The second part AB is nonlinear. As the force increases, elastoplastic deformation occurs. In this part, in addition to the movement of the yarns in the fabric, the yarns are unloaded in their elastic range, and the fabric sample is elongated due to the elongation of the sample itself. With a further increase in force, the fabric takes on plastic deformations. Individual yarns of the sample break, and then the sample breaks completely. Point B is the maximum elongation force.

In further considerations, the behavior of the fabric in the elastic range—i.e., up to point A—is observed. The elastic response of fabrics to tensile force is modelled using the idea that tensile force F is expressed in the region 0A as a linear function of elongation ε:

$$F = k \cdot \varepsilon \tag{3}$$

where k is the elasticity coefficient of the material.

The polynomial function is applied for modelling the elastoplastic deformation, expressed as:

$$F = Q(\varepsilon) = a_n \cdot \varepsilon^n + \cdots + a_1 \cdot \varepsilon + a_0 \tag{4}$$

where $a_0, a_1, \ldots$, and a_n are known coefficients.

The least squares method is used for fitting part 0A and determining the coefficients k, $a_0, a_1, \ldots$, and a_n. Determining point A between the linear and nonlinear parts (Figure 2) represents the most sensitive part of the problem.

2.3. Mathematical Model of Force and Elongation Dependence

A mathematical model of the dependence of the force F and the elongation ε is set up, which can be expressed as:

$$F(\varepsilon) = \begin{cases} k \cdot \varepsilon, & 0 \leq \varepsilon \leq \varepsilon_A \\ Q(\varepsilon), & \varepsilon_A \leq \varepsilon \leq \varepsilon_B \end{cases} \tag{5}$$

At transition point A, the conditions of equality must be met as follows:

$$k \cdot \varepsilon_A = Q(\varepsilon_A) \tag{6}$$

The curve F-ε (Figure 2) must be continuous—that is, the transition at point A must be smooth. Thus, the equality condition (Equation (6)) and the condition of matching the differentials of Equation (7) at point A must be satisfied:

$$k = Q'(\varepsilon_A) \tag{7}$$

The polynomial Q of the lowest degree that would meet both the above conditions of Equations (6) and (7) must have four coefficients, or the degree of the polynomial Q must be n = 3, expressed as:

$$Q(\varepsilon) = a_3 \cdot \varepsilon^3 + a_2 \cdot \varepsilon^2 + a_1 \cdot \varepsilon + a_0 \tag{8}$$

The parameters of the function F (ε) shown in Equation (5) are estimated by the least squares method. Estimates of regression parameters are determined such that Equation (9) holds:

$$\sum_{i=1}^{N}(F(\varepsilon_i) - F_i)^2 \to \min \qquad (9)$$

where ε_i and F_i are experimentally obtained values of elongation and associated force, respectively, while F is the theoretical value of force.

2.4. Electrical Resistance–Elongation Diagram for the Fabric, and Mathematical Model

When a tensile force acts on a fabric sample, the sample is stretched in the direction of the force, and its lateral constriction—which is perpendicular to the direction of the force. This results in a change in the electrical resistance and conductivity of the conductive yarns woven into the fabric. Based on the experimentally obtained values, a connection can be obtained between the electrical resistance of the conductive yarns and the corresponding relative elongation, as shown by the R-ε diagram in Figure 2.

The initial resistance of the conductive yarn is denoted by R_0. The curve in the electrical resistance–elongation diagram (Figure 2) consists of a linear part AD and a nonlinear part located on the curve between points D and E. Point E represents the maximum value of electrical resistance R_E at elongation ε_E. In the initial range of the diagram up to point D the electrical resistance is represented as a linear function of extension:

$$R = p \cdot \varepsilon + s \qquad (10)$$

where p is the coefficient of the slope of the line and s is section of the line.

A polynomial function can be applied to model the nonlinear part of the curve, expressed as:

$$R = H(\varepsilon) = b_n \cdot \varepsilon^n + \cdots + b_1 \cdot \varepsilon + b_0 \qquad (11)$$

where $b_0, b_1, \ldots,$ and b_n are known coefficients.

Determination of the coefficients $p, b_0, b_1, \ldots,$ and b_n is solved by the least squares method. A mathematical model of the dependence of the electrical resistance R and the extension ε is set, which can be written as:

$$R(\varepsilon) = \begin{cases} p \cdot \varepsilon + s, & 0 \leq \varepsilon \leq \varepsilon_D \\ H(\varepsilon), & \varepsilon_D \leq \varepsilon \leq \varepsilon_E \end{cases} \qquad (12)$$

To determine the transition point D, the condition of equality must be met:

$$p \cdot \varepsilon_D + s = H(\varepsilon_D) \qquad (13)$$

The curve R-ε (Figure 2) must be continuous—i.e., the transition at point D must be smooth—and, therefore, the equality condition of Equation (13) must be satisfied, and the differentials of the functions at these points must coincide:

$$p = H'(\varepsilon_D) \qquad (14)$$

The function that meets the conditions of Equations (13) and (14) is a polynomial B, which must be third-degree (n = 3):

$$H(\varepsilon) = b_3 \cdot \varepsilon^3 + b_2 \cdot \varepsilon^2 + b_1 \cdot \varepsilon + b_0 \qquad (15)$$

The parameters of the function R (ε) shown in Equation (12) can be estimated by the least squares method. Estimates of regression parameters can be determined such that Equation (16) holds:

$$\sum_{i=1}^{N}(R(\varepsilon_i) - R_i)^2 \to \min \qquad (16)$$

where ε_i and R_i are experimentally obtained values of elongation and associated resistance, respectively, while R is the theoretical value of resistance.

3. Experimental Part

3.1. Samples of the Conductive Yarns

The conductive yarns used for this paper were purchased from Shieldex Trading Inc. (Bremen, Germany). The ECYs are silver-coated products that have antibacterial properties and are thermally and electrically conductive [42].

In the experimental part of the paper, three different ECYs are used (designations X, Y, and Z). All three ECYs are made of high-strength polyamide (PA 6.6), and are coated with 99% pure silver. The yarn with designation X is single and has 17 filaments; its yarn count after coating with silver is 142 dtex. The yarn with designation Y is double and has 34 filaments; its yarn count after silver plating is 295 dtex. The yarn with designation Z is double and has 72 filaments; its yarn count after silver plating is 604 dtex. The most important properties of these ECYs are shown in Table 1.

Table 1. Characteristics parameters of the ECYs.

Code Name	X	Y	Z
Material	Polyamide 6.6 filament	Polyamide 6.6 filament	Polyamide 6.6 filament
Metal-plated	99% Pure silver	99% Pure silver	99% Pure silver
Coating	Yes	Yes	Yes
Filaments	17	17	36
Ply	1	2	2
Yarn count, raw (dtex)	117f17	117f17	235f36
Yarn count, silverized (dtex)	142	295	604
Resistivity	<500 Ω/m	<300 Ω/m	80 Ω/m

3.2. Fabrics Samples

Fabric samples were made in a particularly careful manner in a real industrial process on an air-jet loom, which was computer-controlled, with the required weaving parameters. In this way, the electrically conductive threads were woven into the textile material during the weaving process, in the same manner as they would be in the case of real production of the electrically conductive textiles, with controlled and repeatable parameters.

The experiments were performed on fabric samples a with structurally identical plain weave from a mixture of cotton (50%) and polyamide (50%) yarns, a warp yarn count of Tt = 33 × 2 tex, and a weft yarn count of Tt = 50 tex. Table 2 shows the data for the underlying fabric.

Table 2. Test results for the basic fabric parameters.

	Warp Direction			Weft Direction				
Fabric Structure	Yarn Fibers	Yarn Count (tex)	Density (cm^{-1})	Yarn Fibers	Yarn Count (tex)	Density (cm^{-1})	Weight (g/m^2)	Fabric Thickness (mm)
Plain weave	50%Cotton/ 50%PA	33 × 2	33	50%Cotton/ 50%PA	50	25	208.2	0.463

The yarn count of the yarn was determined via the gravimetric method, according to ISO 2060:1994. Fabric density was tested according to ISO 7211-2:1984 [43]. The thickness of the fabric was determined according to ISO 5084:1996 [44]. Prior to the testing, the fabrics with woven ECYs were conditioned for 24 h under normal conditions: 65 ± 2% humidity, at a temperature of 20 °C ± 2 °C. ECYs were woven into the fabric in the direction of the weft and warp yarns in the plain weave. ECYs are shown in red, and warp yarns in blue.

For the purposes of this experiment, three fabrics with different ECY yarn counts of (X, Y, and Z) were designed and woven. The samples of fabrics with ECYs were cut in three different directions: the direction of the weft (90°), where the length of the electrically conductive yarn was equal to the length of the sample, and amounted to $l_0 = 20$ cm, (Figure 3a); the direction of the warp (0°), where the length of the conductive yarn was equal to the width of the sample, and amounted to $c_0 = 5$ cm (Figure 3b); and at an angle of 45° to the weft, where the length of the conductive yarn $d_0 = c_0 \cdot \sqrt{2} = 7.05$ cm (Figure 3c). The direction of action of the tensile force during the performance of the experiment was always the same. For each specified cutting direction of the electrically conductive fabric sample, and for each yarn count, five measurements were performed.

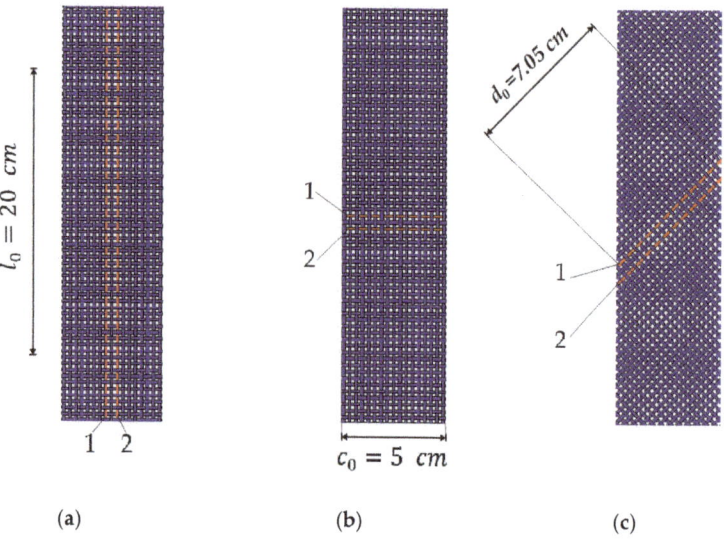

Figure 3. Fabric samples with woven ECYs: (a) sample cut in the weft direction (0°); (b) sample cut in the warp direction (90°); (c) sample cut at a 45° angle.

In each cut fabric sample, there are two ECYs denoted with the numbers 1 and 2. Both ECYs in the sample were of equal yarn counts; the electrical resistance of yarn 1 is denoted by R_1, and that of yarn 2 is denoted by R_2.

3.3. Method and Manner of Measuring the Electrical Resistance of Electrically Conductive Yarns

For the purposes of this research, the following measuring equipment was used: a device for measuring the breaking force of the Textechno Statimat ME+ sample tensile strength tester, (Textechno H. Stein GmbH & Co. KG, Moenchengladbach, Nordrhein-Westfalen, Germany) which is fully automated, microprocessor-controlled, and works on the principle of constant strain rate; a measurement assembly (Faculty of Textile Technology, Zagreb, Croatia), purpose-designed via the resistance comparative method, an optoelectronic start signal adjustment assembly, an analog-to-digital converter (NI USB-6212), (National Instruments, Austin, TX, USA) and PCs (HP, Houston, TX, USA) to control the measurement process and collect the measured data. In addition to measurement, the PCs were used for the development of appropriate measurement and analytical software (LabVIEW 2019, NI, Austin, TX, USA), as well as for conducting a complete analysis of measured data, and for the presentation of results.

To measure and collect the research results, a measuring system was designed and manufactured, as shown in concept in Figure 4, and realized in Figure 5.

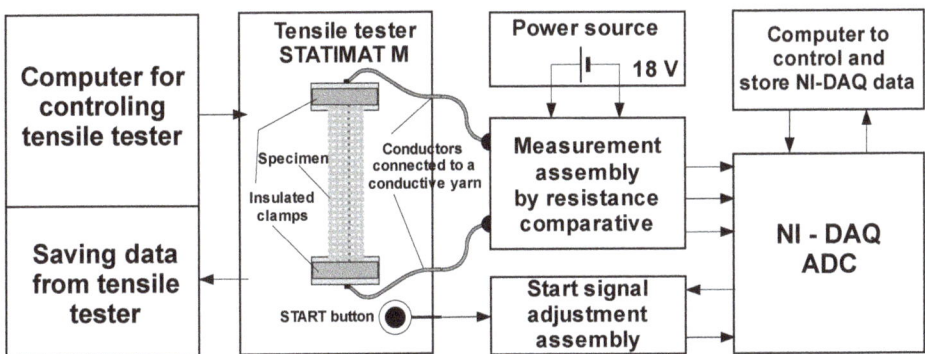

Figure 4. System for measuring changes in the electrical resistance of woven conductive yarns.

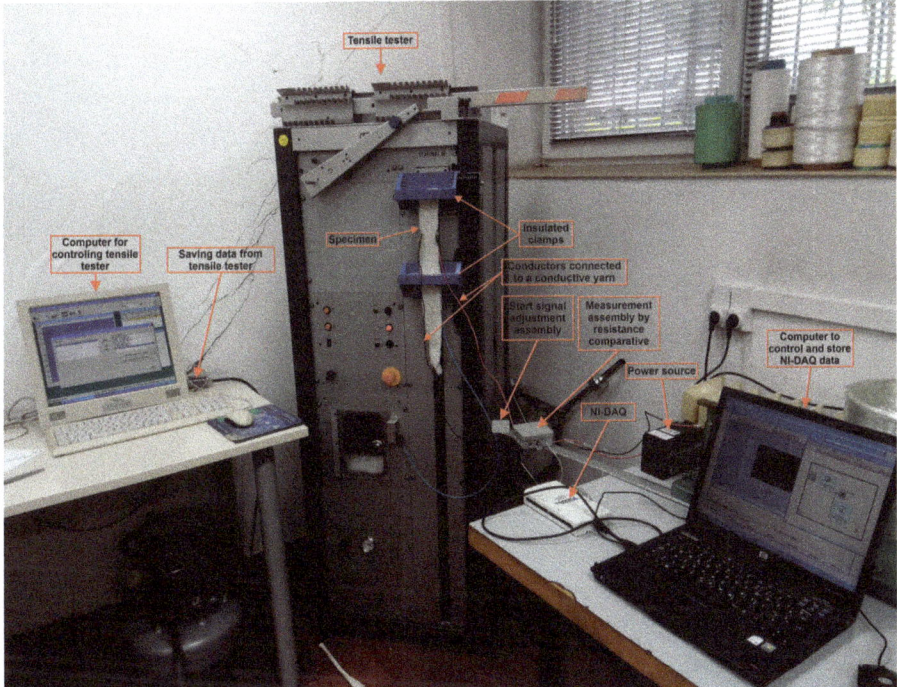

Figure 5. Photo of the measuring system.

In addition to the preparation and processing of samples, measurement and control software was used for the precise monitoring of changes in the resistance of the electrically conductive thread for every 0.02 mm of elongation. For this purpose, the values of the magnitude of the stress from the resistance-measuring assembly were read by a comparative method, and a change in the resistance value of the conductive filament was processed for each elongation value. The data were read at a speed of 50 measurements per second and stored for further processing.

The sample of standard dimensions was fixed to the galvanically insulated clamps of the tensile tester at a distance of l_0 = 200 mm, and copper conductors for measuring voltage change—i.e., resistance—were connected to the ECYs by the process of fixing copper clips (crimping). The samples were subjected to uniaxial tensile loading at a tensile speed of v = 100 mm/min until a break was reached. The tensile properties of all samples were tested according to ISO 13934-1:2008 with a tensile tester.

The START button of the tensile tester is simultaneously connected to the ADC via the start signal adjustment assembly. This allows us, by pushing the START button, to simultaneously start measuring and storing the force at the break and the elongation data from the tensile tester, as well as the voltage signals from the sample, to calculate the resistance change.

4. Results and Discussion

Under the action of tensile force F, the corresponding longitudinal deformation is measured, i.e., the elongation ε of the sample and the electrical resistance of the conductive yarns R. Mean values of the results of measuring the action of tensile force F and associated elongation ε, and mean values of electric resistance R for the samples cut at 0°, 45°, and 90° angles, are shown in the Figures 6–8.

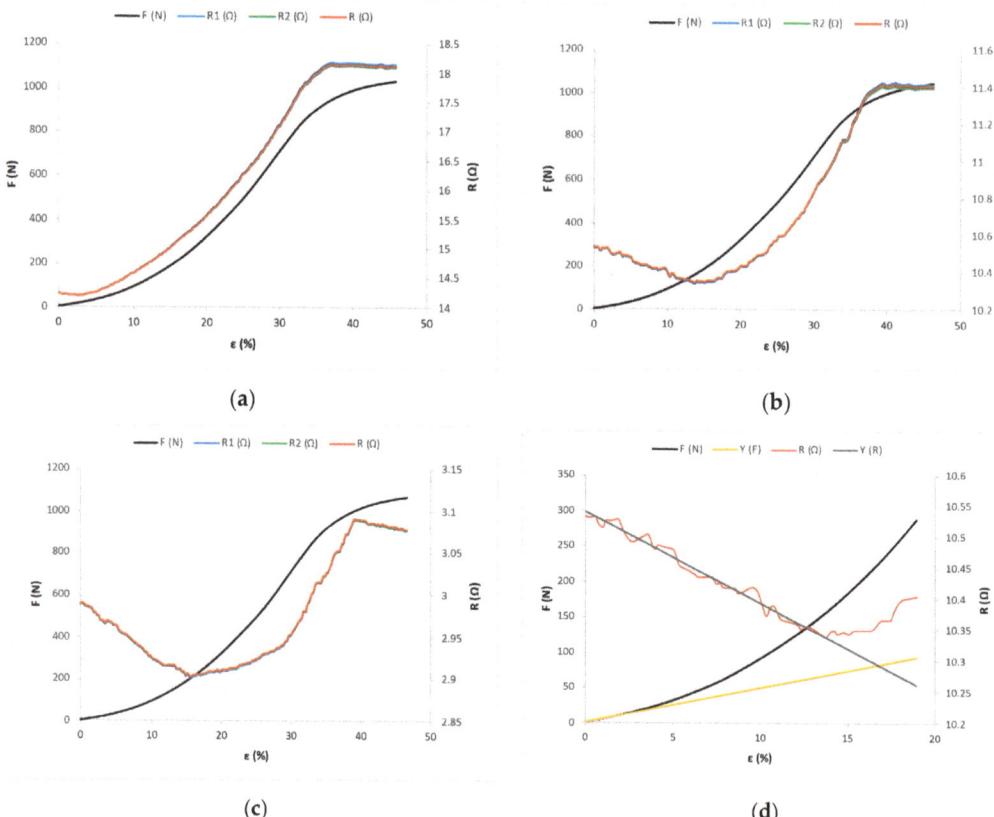

Figure 6. Force–elongation (F-ε) and electrical resistance–elongation (R-ε) diagrams for fabric samples cut in the warp direction: (**a**) for sample X-0; (**b**) for sample Y-0; and (**c**) for sample Z-0. (**d**) Experimental and mathematical models of F-ε and R-ε curves for sample Y-0.

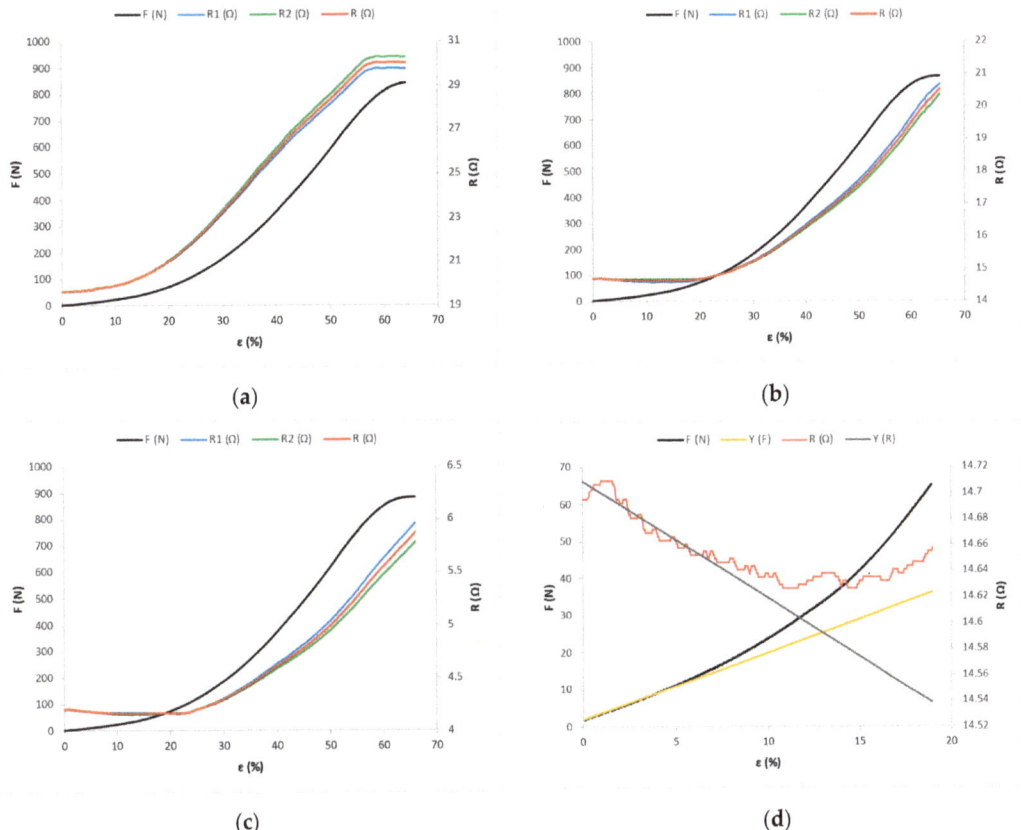

Figure 7. Force–elongation (F-ε) and electrical resistance–elongation (R-ε) diagrams for fabric samples cut at a 45° angle: (**a**) for sample X-45; (**b**) for sample Y-45; and (**c**) for sample Z-45. (**d**) Experimental and mathematical models of F-ε and R-ε curves for sample Y-45.

X-0 is the designation of a fabric sample with a yarn count (X) cut at a 0° angle; X-45 is the designation of a fabric sample with a yarn count (X) cut at a 45° angle; X-90 is the designation of a fabric sample with a yarn count (X) cut at a 90° angle.

Y-0 is the designation of a fabric sample with a yarn count (Y) cut at a 0° angle; Y-45 is the designation of a fabric sample with a yarn count (Y) cut at a 45° angle; Y-90 is the designation of a fabric sample with a yarn count (Y) cut at a 90° angle.

Z-0 is the designation of a fabric sample with a yarn count (Z) cut at a 0° angle; Z-45 is the designation of a fabric sample with a yarn count (Z) cut at a 45° angle; Z-90 is the designation of a fabric sample with a yarn count (Z) cut at a 90° angle.

R_1 denotes the mean value of the measured electrical resistance of the electrically conductive yarn 1; R_2 denotes the mean value of the measured electrical resistance of the electrically conductive yarn 2 (Figure 3). The mean value of the electrical resistance of conductive yarns 1 and 2 is denoted by $R = (R1 + R2)/2$.

Figure 6a shows the distributions of tensile force and electrical resistance as functions of elongation for the fabric sample X-0, Figure 6b for the fabric sample Y-0, and Figure 6c for the fabric sample Z-0.

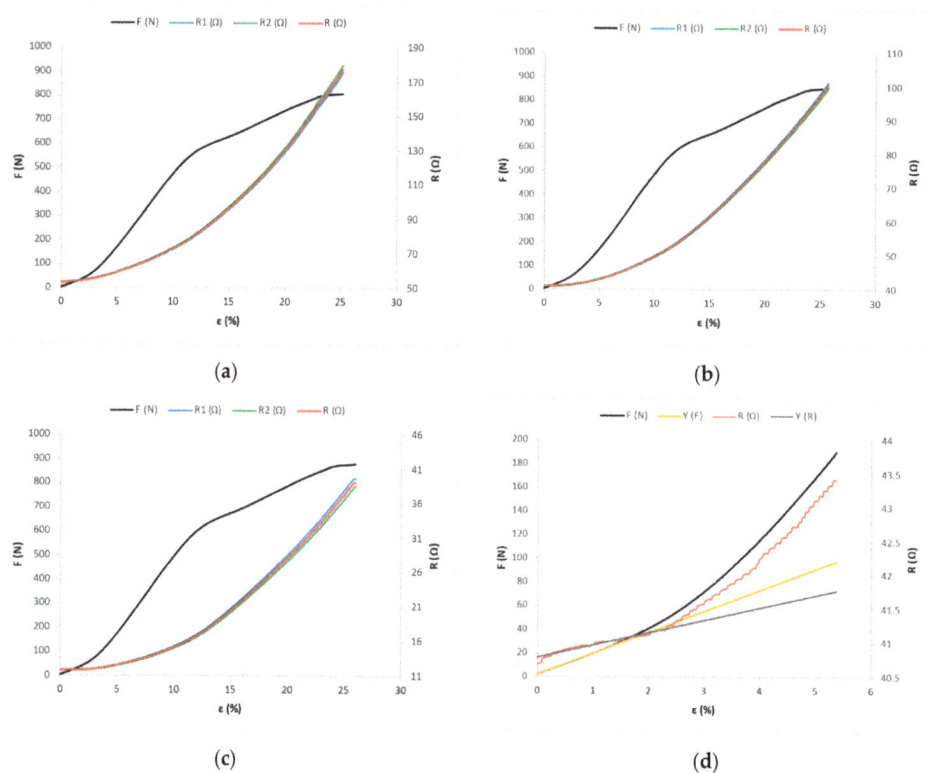

Figure 8. Force–elongation (F-ε) and electrical resistance–elongation (R-ε) diagrams for fabric samples cut in the weft direction: (**a**) for sample X-90; (**b**) for sample Y-90; and (**c**) for sample Z-90. (**d**) Experimental and mathematical models of F-ε and R-ε curves for sample Y-90.

In the linear area, the slope of the line is important for the analysis—not its section on the ordinate—and, therefore, in further considerations, the section on the ordinate will not be shown or considered (Figure 6d, Figure 7d, and Figure 8d). The experimental force–elongation curves (F-ε) and electrical resistance–elongation curves (R-ε) are presented, as well as the corresponding analytical models in the first zone (linear) predicting the force–elongation curve and the electrical resistance–elongation curve for the fabric samples: Y-0 (Figure 6d); Y-45 (Figure 7d); and Y-90 (Figure 8d). Figure 6d shows the representation only for sample Y-0, and the representations for X-0 and Z-0 are very similar and are made in a similar way, so they are not shown here. The same reasoning applies to the representations in Figures 7d and 8d.

Figure 7a shows the tensile force and electrical resistance distributions as functions of the X-45 fabric sample, Figure 7b shows the Y-45 fabric sample, and Figure 7c shows the Z-45 fabric sample.

Figure 8a shows the tensile force and electrical resistance distributions as functions of the X-90 fabric sample, Figure 8b shows the Y-90 fabric sample, and Figure 8c shows the Z-90 fabric sample.

The mean values for the experimentally obtained maximum force F_B and the corresponding elongation ε_B, along with the coefficients of variation CV, are shown in Table S1.

The maximum force F_B has the highest value for sample Z-0 with the highest yarn count cut in the warp direction—amounting to F_B = 1068.8 N—and the lowest value for sample X-90 with the lowest yarn count, cut in the weft direction, amounting to F_B = 809.1 N. The maximum elongation ε_B has the highest value for sample Z-45—amounting to ε_B = 65.98%—and the lowest value for sample X-90, amounting to $\varepsilon_B = 25.22\%$. For samples cut in the same direction, the value of maximum force and elongation increases with the increase in the yarn count of the electrically conductive yarn.

The mean values of the measured initial electrical resistance of the electrically conductive yarn 1 shown in Table S2 are denoted by R_{01}. The mean values of the initial electrical resistance of the electrically conductive yarn 2 are denoted by R_{02}, and the mean value of the initial resistance of the conductive yarns 1 and 2 is $R_0 = (R_{01} + R_{02})/2$. The mean values of the measured maximum electrical resistance at point E of the electrically conductive yarn 1 are denoted by R_{E1}. The mean values of the measured maximum electrical resistance of the electrically conductive yarn 2 are denoted by R_{E2}, and the mean value of the maximum electrical resistance of the electrically conductive yarns 1 and 2 is $R_E = (R_{E1} + R_{E2})/2$.

In samples cut in the same direction, the mean values of the initial electrical resistance and the maximum electrical resistance decrease with the increase in the yarn count of the electrically conductive yarn. The maximum electrical resistance R_E has the highest value for sample X-90 with the lowest yarn count—which is cut in the weft direction and amounts to $R_E = 177.69\ \Omega$—and has the lowest value for sample Z-0 with the highest yarn count, which is cut in the warp direction and amounts to $R_E = 3.09\ \Omega$. The initial electrical resistance has the highest value for sample X-90— amounting to $R_0 = 53.40\ \Omega$—and the lowest value for sample Z-0, amounting to $R_0 = 2.99\ \Omega$.

In further considerations, the change in tensile force and electrical resistance, along with the behavior of the fabric with ECYs, will be observed in the linear region alone. The values of point A at which the force–elongation curve passes from the elastic range—which is linear—to the nonlinear range are shown by the coordinates (ε_A, F_A) given in Table S3 for different samples of fabrics with ECYs. Point A is common to the experimental curve and the mathematical model; it is obtained from Equations (6) and (7).

For the corresponding mathematical models, the corresponding coefficients of the linear function were calculated, i.e., the lines k (range OA) and the coefficients a_0, a_1, a_2, and a_3 of the cubic parabola (nonlinear range). The values of these coefficients were calculated using Equations (4), (5) and (8), and are shown in Table S3. The correlation coefficients r between the experimental curve and the mathematical model were calculated, and are shown in Table S3. The correlation coefficients show very high congruence of the experimental curves and the mathematical model in the linear and nonlinear ranges. The equation of lines and of the third-order polynomial describe very well the curve of the ratio between the experimentally obtained values of the forces and the corresponding elongations.

For samples cut in the same direction, the values of force and elongation at point A increase with the increase in the yarn count of the conductive yarn. Furthermore, in the linear part of the diagram, the slope of the line increases with these samples. For fabric samples cut in the weft direction (X-90, Y-90, and Z-90) the slopes of the lines in the linear range are the highest, and for samples cut at an angle of 45° the slopes of the lines are the lowest. Therefore, the samples X-90, Y-90, and Z-90 have the highest values of force and the lowest values of elongation at point A.

The values of point D at which the electrical resistance–elongation curve passes from an elastic range that is linear to a nonlinear range are shown by the coordinates (ε_D, F_D) given in Table S4 for different samples of fabric with ECYs. Point D is common to the experimental curve and the mathematical model, and is obtained from Equations (13) and (14).

For the corresponding mathematical models, the associated coefficients of the linear function—i.e., the lines p (range OD) and the coefficients b_0, b_1, b_2, and b_3 of the cubic parabola (nonlinear range)—were calculated. The values of these coefficients were calculated using Equations (11), (12) and (15), and are shown in Table S4. The associated correlation coefficients r between the experimental curve and the mathematical model were also calculated, and are given in Table S4. The correlation coefficients show good congruence of the experimental curves and the mathematical model in the linear and nonlinear ranges. The equation of lines and the third-order polynomial describe well the curve of the ratio of the experimentally measured values of electrical resistance and the corresponding elongations.

For samples cut in the same direction, the electrical resistance values decrease and the elongations at point D increase with the increase in the yarn count of the conductive yarn. For fabric samples cut in the warp direction (X-0, Y-0, and Z-0) and at a 45° angle (X-45, Y-45, and Z-45), in the OD range, the line direction coefficient p has a negative value (Table S4), so the values of electrical resistance decrease from the initial value of resistance R_0 to point D, where they have the value R_D (Figures 6a–c and 7a–c). For these samples, in the linear part of the force–elongation dependence diagram, the direction coefficient of line k has a positive value (Table S3, Figures 6a–c and 7a–c). Thus, in the linear part, the values of tensile forces and elongation increase, and at the same time the values of electrical resistance decrease (Tables S3 and S4). It can be freely assumed that in the structure of the electrically conductive yarn there is an increase in the number of parallel contact resistances between the filaments, with a simultaneous decrease in series resistances on the surface, which causes an overall decrease in electrical resistance at the tested length of the electrically conductive yarn.

For fabric samples cut in the weft direction (X-90, Y-90, and Z-90), in the OD range, the direction coefficient of line p has a positive value (Table S4)—i.e., the electrical resistance values increase from the initial resistance value R_0 towards point D, where they have the value R_D (Figure 8a–c). For these samples, in the linear part of the force–elongation diagram, the direction coefficient of line k also has a positive value (Table S3), and the values of the tensile forces and elongation increase, as do the values of electrical resistance (Tables S3 and S4). In these samples, there is a direct action of force and elongation on the conductive yarns, which are consequently elongated, reducing their waviness, which increases their electrical resistance. Due to the action of force and tension, there is an increased number of surface series resistances with simultaneous interruptions of contacts in the structure (between filaments), which eliminates parallel joints; thus, there is a noticeable increase in total electrical resistance on the tested length of woven electrically conductive yarn.

Figure 9 shows the correlation between the slope of the line k, showing the dependence of force and elongation, and the slope of the line p, showing the dependence of electrical resistance and elongation. High correlation coefficients r were obtained, and the highest were for samples X-90, Y-90, and Z-90, amounting to r = 0.9933.

If the values of electrical resistance at point D (R_D) are compared (Table S4) for samples of the same yarn count that are cut in different directions (X-0, X-45, and X-90), (Y-0, Y-45, and Y-90), and (Z-0, Z-45, and Z-90), it can be concluded that the values of electrical resistance are the lowest when the samples are cut in the warp direction, and the highest for the samples cut in the weft direction (14.19 Ω, 19.53 Ω, and 54.87 Ω), (10.34 Ω, 14.64 Ω, and 41.20 Ω), and (2.92 Ω, 4.17 Ω, and 11.81 Ω), respectively. Samples cut in the direction of the warp (X-0, Y-0, and Z-0) have a 5-cm length of electrically conductive yarn, which is located at the sample perpendicular to the direction of the tensile force. Under the action of force, the sample is elongated in the direction of the force, and in the transverse direction (perpendicular to the direction of force action) there is a lateral narrowing of the sample and an increase in the corrugation of the electrically conductive yarn. In doing so, its cross-section increases on parts of the yarn due to compression, which has the effect of reducing the electrical resistance. Here we should also assume surface interruptions of series-connected elements, with a simultaneous significant increase in the number of parallel joints in the yarn structure, so it is logical that the total resistance at the measured yarn length decreases.

Samples cut at a 45° angle (X-45, Y-45, and Z-45) have an electrically conductive yarn length of 7.05 cm. In these samples, the length of the electrically conductive yarn is greater than in the sample cut in the direction of the warp, so this assumes a higher value of the initial electrical resistance R_0.

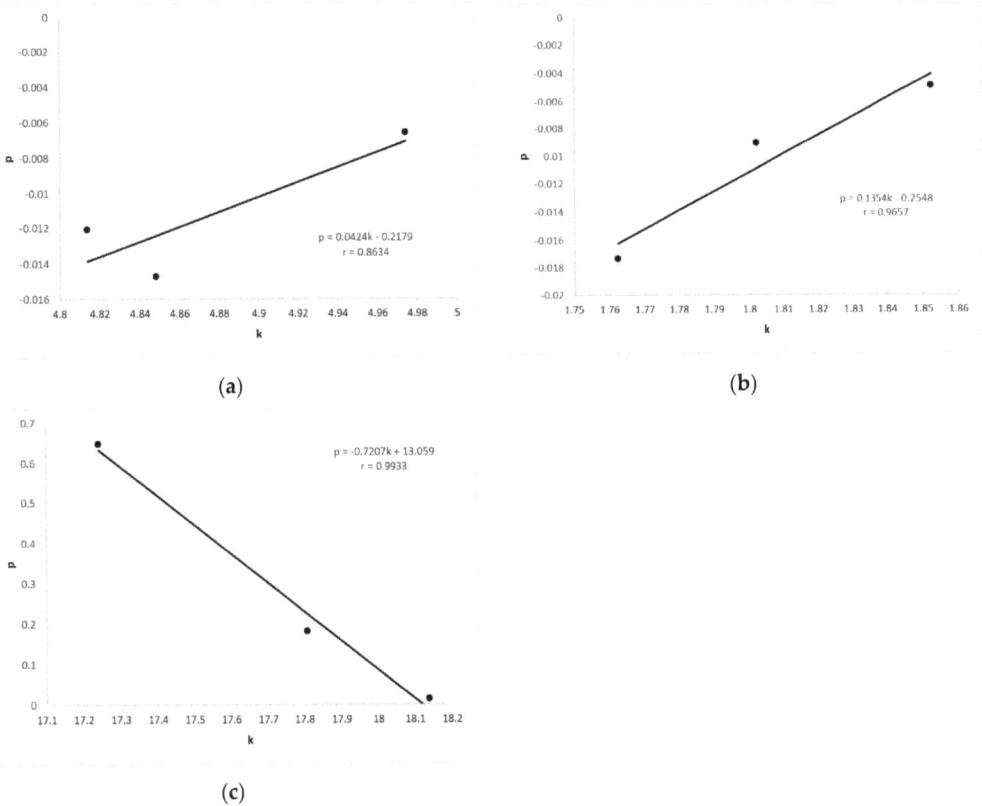

Figure 9. Correlation between the slope of the line k of force and the slope of the line p of electric resistance for samples: (**a**) X-0, Y-0, and Z-0; (**b**) X-45, Y-45, and Z-45; and (**c**) X-90, Y-90, and Z-90.

The samples cut in the weft direction (X-90, Y-90, and Z-90) have an electrically conductive yarn length of 20 cm, which is placed in the sample parallel to the direction of action of the tensile force. Under the action of force, the sample and the length of the electrically conductive yarn extend in the direction of the force, which leads to a large increase in electrical resistance. As the length of the electrically conductive yarn is proportional to the value of the electrical resistance, it has been proven that the value of electrical resistance is affected by yarn count, but also by the change in the lengths of the electrically conductive yarn in the sample in relation to the direction of action of the tensile force.

5. Conclusions

The results of the experiment conducted in this paper indicate that the electrical resistance of the woven ECY increases or decreases depending on the strength of the force as well as the direction of force (i.e., elongation of the fabric). This effect can be characterized from a negative or a positive aspect. If fabrics with woven ECY are used as energy conductors, then in the event of elongation the resistance changes, so part of the conducted energy is converted into joule heat, and the supply voltage also changes—which is unfavorable. Likewise, if a fabric with woven ECY is used as a conductor for the transmission of measurement signals, then as a result of the elongation of the material, a measurement error may occur due to a change in resistance.

The positive side of the change in the resistance of woven yarns may be for the production of textile sensors for registering changes in the state of textile materials by measuring force, i.e., elongation.

Therefore, the examination of the impact of elongation on the change in the electrical resistance of the ECYs woven into the fabric is very significant. The aim of this paper is to examine how the change in the elongation of the samples under the action of tensile force affects the change in the electrical resistance of the ECYs woven into fabrics.

The impact of the action of tensile force in different directions on the fabric with woven ECY was examined. Elongation of the fabric results in a change in the electrical resistance of the woven electrically conductive yarn. Based on the experimental results, we obtained the ratio curves of the magnitudes of tensile forces and associated elongations, as well as the curves of the ratios of electrical resistance and elongation values, for directions when the force acts in the warp direction, in the weft direction, and at an angle of 45°.

A mathematical analysis was performed by which the appropriate analytical model was connected to the experimental curves of force–elongation and electrical resistance–elongation. The analytical model describing the curves of force–elongation and electrical resistance–elongation has two ranges: linear and nonlinear.

The associated correlation coefficients r between the experimental curve and the mathematical model were calculated, and showed a very high correspondence of the experimental curve and the mathematical model in the linear and nonlinear ranges. The correlation coefficient r is between 0.9945 and 0.9999 for the force–elongation curve, and between 0.9813 and 0.9999 for the electrical resistance–elongation curve.

Research into the impact of the action of force on the change in the electrical resistance of an electrically conductive yarn in a fabric was conducted in a linear elastic range. The following conclusions can be drawn from the conducted research and the analysis of the results:

For samples cut in the same direction, the values of tensile force and elongation increase, while the values of electrical resistance decrease, with the increase in the yarn count of the electrically conductive yarn.

The value of electrical resistance is impacted by the yarn count, but also by the length of the electrically conductive yarn in the sample in relation to the direction of action of the tensile force. As the length of the electrically conductive yarn increases and its cross-section decreases, the electrical resistance increases. It can be freely assumed that the change in the total resistance is also impacted by changes in series-parallel elements in the structure and surface of the electrically conductive threads when they are exposed to the action of elongation force.

This research provides information and conclusions about the influence of tensile force on changes in the electrical resistance of conductive threads woven into fabrics. The obtained results will be used for future research aimed at developing a stretchable and sensitive strain sensor for smart textiles and clothing.

Based on the conducted measurements and experimentation for the purpose of incorporating woven sensors into e-textiles—smart and intelligent clothing—the textile ribbon with ECYs was made using the traditional hand-weaving technique. Such sensors are integrated into smart clothing for apnea patients and infant clothing for the purpose of monitoring breathing as textile-embedded interconnects [32], recording the change in resistance, which is converted into voltage signals, leading to the input of electronic circuits for breathing detection, i.e., chest movements. The realized prototype of this type of clothing has been awarded a number of prizes at world innovation exhibitions, which is an indicator of the innovativeness of the approach [45]. The presented sensor had the simple task of detecting respiratory arrest, where the accuracy of the sensor and its hysteresis were not of primary importance, as with most of the sensors presented in the introductory part of this paper. The main advantages of the presented sensor are its reliability, simplicity of construction, suitability for conventional textile production processes, and pleasant textile feel.

Woven textile sensors with ECYs have proven to be excellent for incorporation into clothing, given that they are predominantly textile materials that are comfortable interacting with the skin of the human body.

Supplementary Materials: The following are available online at https://www.mdpi.com/article/10.3390/ma14123390/s1, Table S1: The mean values of maximum force FB, the corresponding elongation εB, and the coefficients of variation CV; Table S2: The mean values of initial R0 and maximum electrical resistance RE, and respective variation coefficients; Table S3: The coordinates of point A, coefficients k, a0, a1, a2, and a3 of the curves of the mathematical model, and correlation coefficient r; Table S4: The coordinates of point D, coefficients p, b0, b1, b2, and b3 of the curves of the mathematical model, and correlation coefficient r.

Author Contributions: Conceptualization, Ž.K., Ž.P., and D.R.; investigation, Ž.K., Ž.P., D.Š.P., and D.R.; methodology, Ž.K. and Ž.P.; experimental research, Ž.K. and Ž.P.; formal analysis, D.R. and D.Š.P.; writing—original draft preparation, D.Š.P. and Ž.P.; writing—review and editing, Ž.K. and D.R.; software, Ž.P.; supervision, Ž.K. and D.R.; project administration, Ž.K.; funding acquisition, D.R. All authors have read and agreed to the published version of the manuscript.

Funding: This work has been supported in part by the Croatian Science Foundation through the project IP-2018-01-6363 "Development and thermal properties of intelligent clothing (ThermIC)", as well as by the University of Zagreb through research grant TP11/20 "Development of measuring equipment for determining temperature gradients in multilayer clothing thermal insulation".

Institutional Review Board Statement: Not applicable.

Informed Consent Statement: Not applicable.

Data Availability Statement: Not applicable.

Conflicts of Interest: The authors declare no conflict of interest.

References

1. Tao, X.S. *Smart Fibres, Fabrics and Clothing*; Woodhead Publishing Limited: Cambridge, UK; CRC Press LLC: Cambridge, UK, 2001; pp. 1–6, 35–57, 124–149, 247–253.
2. Mattila, H. Yarn to Fabric: Intelligent Textiles. In *Textiles and Fashion—Materials, Design and Technology*; Sinclair, R., Ed.; Woodhead Publishing Limited: Cambridge, UK; CRC Press LLC: Cambridge, UK, 2014. [CrossRef]
3. Gao, Q.; Jinjie, Z.; Zhenwen, X.; Olatunji, O.; Jinyong, Z.; Lei, W.; Hui, L. Highly stretchable sensors for wearable biomedical applications. *J. Mater. Sci.* **2019**, *54*, 7. [CrossRef]
4. Cheng, M.; Zhu, G.; Zhang, F.; Tang, W.I.; Jianping, S.; Yang, J.; Zhu, L. A review of flexible force sensors for human health monitoring. *J. Adv. Res.* **2020**, *26*, 53–68. [CrossRef]
5. Nascimento, L.M.S.d.; Bonfati, L.V.; Freitas, M.L.B.; Mendes Junior, J.J.A.; Siqueira, H.V.; Stevan, S.L., Jr. Sensors and Systems for Physical Rehabilitation and Health Monitoring—A Review. *Sensors* **2020**, *20*, 4063. [CrossRef] [PubMed]
6. Liu, Y.; Wang, H.; Zhao, W.; Zhang, M.; Qin, H.; Xie, Y. Flexible, Stretchable Sensors for Wearable Health Monitoring: Sensing Mechanisms, Materials, Fabrication Strategies and Features. *Sensors* **2018**, *18*, 645. [CrossRef] [PubMed]
7. Ehrmann, A.; Heimlich, F.; Brücken, A.; Weber, M.; Haug, R. Suitability of knitted fabrics as elongation sensors subject to structure, stitch dimension and elongation direction. *Text. Res. J.* **2014**, *84*, 2006–2012. [CrossRef]
8. Chawla, S.; Naraghi, M.; Davoudi, A. Effect of twist and porosity on the electrical conductivity of carbon nanofiber yarns. *Nanotechnology* **2013**, *24*, 255708. [CrossRef]
9. Maity, S.; Chatterjee, A. Polypyrrole Based Electro-Conductive Cotton Yarn. *Int. J. Text. Sci.* **2014**, *4*, 1–4. [CrossRef]
10. Liu, T.; Xu, Y.H.; Chen, H.; Zou, F.Y. Different number of shares and twists effect to the sensing performance of conductive yarns contains coated carbon fibers. *Adv. Mat. Res.* **2011**, *148*, 803–807. [CrossRef]
11. Cochrane, C.; Koncar, V.; Lewandowski, M.; Dufour, C. Design and development of a flexible strain sensor for textile structure based on a conductive polymer composite. *Sensors* **2007**, *7*, 473–492. [CrossRef]
12. Li, L.; Song, L.; Feng, D. Electromechanical analysis of length-related resistance and contact resistance of conductive knitted fabrics. *Text. Res. J.* **2012**, *82*, 2062–2070. [CrossRef]
13. Kim, B.; Koncar, V.; Dufour, C. Polyaniline-coated PET conductive yarns: Study of electrical, mechanical, and electro-mechanical properties. *J. Appl. Polym. Sci.* **2006**, *101*, 1252–1256. [CrossRef]
14. Zhang, H.; Tao, X.M.; Wang, S.Y.; Yu, T. Electro-mechanical properties of knitted fabric made from conductive multifilament yarn under unidirectional extension. *Text. Res. J.* **2005**, *75*, 598–606. [CrossRef]
15. Yang, B.; Tao, X.M. Effect of tensile condition and textile structure on resistance response of stainless steel fibers fabric. *Rare Metal Mat. Eng.* **2008**, *37*, 227–231.

16. Ryu, J.W.; Hong, J.K.; Kim, H.J.; Jee, Y.J.; Kwon, S.Y.; Yoon, N.S. Effect of strain change of electrically conductive yarn on electric resistance and its theoretical analysis. *Sen'i Gakkaishi.* **2010**, *66*, 209–214. [CrossRef]
17. Kincal, D.; Kumar, A.; Child, A.D.; Reynolds, J.R. Conductivity Switching in Polypyrrole-Coated Textile Fabrics as Gas Sensors. *Synth. Met.* **1998**, *92*, 53–56. [CrossRef]
18. Oh, K.W.; Park, H.J.; Kim, S.H. Stretchable Conductive Fabric for Electrotherapy. *J. Appl. Polym. Sci.* **2003**, *88*, 1225–1229. [CrossRef]
19. Xue, P.; Tao, X.M.; Wang, S.; Yu, T. Electro-mechanical behavior and mechanistic analysis of fiber coated with electrically conductive polymer. *Text. Res. J.* **2004**, *74*, 924–937. [CrossRef]
20. Roh, J.-S. *Conductive Yarn Embroidered Circuits for System on Textiles*; IntechOpen: London, UK, 2018; pp. 161–174. [CrossRef]
21. Tangsirinaruenart, O.; Stylios, G. A Novel Textile Stitch-Based Strain Sensor for Wearable End Users. *Materials* **2019**, *12*, 1469. [CrossRef]
22. Ruppert-Stroescu, M.; Balasubramanian, M. Effects of stitch classes on the electrical properties of conductive threads. *Text. Res. J.* **2017**, *88*, 004051751772511. [CrossRef]
23. Eom, J.; Jaisutti, R.; Lee, H.; Lee, W.; Sang Heo, J.; Lee, J.Y.; Park, S.K.; Kim, J.H. Highly Sensitive Textile Strain Sensors and Wireless User-Interface Devices using All-Polymeric Conducting Fibers. *ACS Appl. Mater. Interfaces* **2017**, *9*, 10190–10197. [CrossRef]
24. Ivšić, B.; Galoić, A.; Bonefačić, D. Durability of Conductive Yarn Used for Manufacturing Textile Antennas and Microstrip Lines. In *Proceedings of the IEEE International Symposium on Antennas and Propagation, Fajardo, Puerto Rico, USA, 26 June–1 July 2016*; IEEE Publisher: Piscataway, NJ, USA, 2016; pp. 1757–1758.
25. Garnier, B.; Mariage, P.; Rault, F.; Cochrane, C.; Končar, V. Electronic-components less fully textile multiple resonant combiners for body-centric near field communication. *Sci. Rep.* **2021**, *11*, 2159. [CrossRef]
26. Chang, T.; Tanabe, Y.; Wojcik, C.C.; Barksdale, A.C.; Doshay, S.; Dong, Z.; Liu, H.; Zhang, M.; Chen, Y.; Su, Y.; et al. A General Strategy for Stretchable Microwave Antenna Systems using Serpentine Mesh Layouts. *Adv. Funct. Mater.* **2017**, 1703059. [CrossRef]
27. Yang, Y.; Wu, Y.; Li, C.; Yang, X.; Chen, W. Flexible Actuators for Soft Robotics. *Adv. Intell. Syst.* **2019**, *2*, 1900077. [CrossRef]
28. Kan, C.-W.; Lam, Y.-L. Future Trend in Wearable Electronics in the Textile Industry. *Appl. Sci.* **2021**, *11*, 3914. [CrossRef]
29. Yao, S.; Zhu, Y. Wearable multifunctional sensors using printed stretchable conductors made of silver nanowires. *Nanoscale* **2014**, *6*, 2345–2352. [CrossRef] [PubMed]
30. Ha, M.; Lim, S.; Ko, H. Wearable and flexible sensors for user-interactive health monitoring devices. *J. Mater. Chem. B* **2018**, *6*, 4043–4064. [CrossRef]
31. Stavrakis, A.K.; Simić, M.; Stojanović, G.M. Electrical Characterization of Conductive Threads for Textile Electronics. *Electronics* **2021**, *10*, 967. [CrossRef]
32. McKnight, M.; Agcayazi, M.T.; Ghosh, T.; Bozkurt, A. Fiber-Based Sensors. In *Wearable Technology in Medicine and Health Care*; Tong, T., Ed.; Elsevier: Amsterdam, The Netherlands, 2018. [CrossRef]
33. Wang, C.; Zhang, M.; Xia, K.; Gong, X.; Wang, H.; Yin, Z.; Guan, B.; Zhang, Y. Intrinsically Stretchable and Conductive Textile by a Scalable Process for Elastic Wearable Electronics. *ACS Appl. Mater. Interfaces* **2017**, *9*, 13331–13338. [CrossRef]
34. Briedis, U.; Vališevskis, A.; Ziemele, I.; Abele, I. Study of Durability of Conductive Threads Used for Integration of Electronics into Smart Clothing. *Key Eng. Mater.* **2019**, 320–325. [CrossRef]
35. Malarić, R. *Instrumentation and Measurement in Electrical Engineering*; Brown Walker Press: Boca Raton, FL, USA, 2011.
36. Hu, J. *Structure and mechanics of woven fabrics*; Woodhead Publishing: Boca Raton, FL, USA, 2004; pp. 91–122.
37. Kovar, R.; Gupta, B.S. Study of the Anisotropic Nature of the Rupture Properties of a Plain Woven Fabric. *Text. Res. J.* **2009**, *79*, 506–516. [CrossRef]
38. Zouari, R. Experimental and numerical analyses of fabric off-axes tensile test. *J. Text. Inst.* **2010**, *101*, 8–68. [CrossRef]
39. Penava, Ž.; Šimić Penava, D.; Knezić, Ž. Determination of the Elastic Constants of a Plain Woven Fabrics by Tensile Test in Various Directions. *Fibres Text. East. Eur.* **2014**, *22*, 57–63.
40. Penava, Ž.; Šimić Penava, D.; Lozo, M. Experimental and analytical analyses of the knitted fabric off-axes tensile test. *Text. Res. J.* **2021**, *91*, 62–72. [CrossRef]
41. Bassett, R.J.; Postle, R.; Pan, N. Experiment Methods for Measuring Fabric Mechanical Properties: A Review and Analysis. *Text. Res. J.* **1999**, *69*, 866–875. [CrossRef]
42. Shieldex Trading, Inc. "Yarns/Threads," Shieldex Trading, Inc. 1 January 2018. [Online]. Available online: https://www.shieldextrading.net/products/yarns-threads/ (accessed on 21 May 2019).
43. ISO 7211-2:1984. *Textiles—Woven fabrics—Construction—Methods of analysis—Part 2: Determination of Number of Threads per Unit Length (Last Reviewed and Confirmed)*; International Organization for Standardization: Geneva, Switzerland, 2017.
44. ISO 5084:1996. *Textiles—Determination of Thickness of Textiles and Textile Products, 2019 (Last Reviewed and Confirmed)*; International Organization for Standardization: Geneva, Switzerland, 2019.
45. Intelligent Clothing for Patients with Apnea and Snoring. Available online: https://moodle.srce.hr/eportfolio/view/view.php?id=134843 (accessed on 23 May 2021).

Article

Development of 3D Models of Knits from Multi-Filament Ultra-Strong Yarns for Theoretical Modelling of Air Permeability

Tetiana Ielina [1], Liudmyla Halavska [1], Daiva Mikucioniene [2], Rimvydas Milasius [2,*], Svitlana Bobrova [1] and Oksana Dmytryk [1]

[1] Department of Textile Technology and Design, Kyiv National University of Technologies and Design, 01001 Kyiv, Ukraine; yelina.tv@knutd.edu.ua (T.I.); galavska.ly@knutd.edu.ua (L.H.); bobrova.sy@knutd.edu.ua (S.B.); dmytryk.om@knutd.edu.ua (O.D.)
[2] Department of Production Engineering, Kaunas University of Technology, 44249 Kaunas, Lithuania; daiva.mikucioniene@ktu.lt
* Correspondence: rimvydas.milasius@ktu.lt

Abstract: The work is devoted to the study of the geometric parameters of a knitted loop. It has been found that the optimal model is a loop model detailed at the yarn level, which considers the change in the cross-sectional shape and sets the properties of the porous material in accordance with the internal porosity of the yarn. A mathematical description of the coordinates of the characteristic points of the loop and an algorithm for calculating the coordinates of the control vertices of the second order spline, which determine the configuration of the yarn axes in the loop, are presented in this work. To create 3D models, Autodesk AutoCAD software and Structura 3D software, developed in the AutoLisp programming language, were used. The simulation of the air flow process was carried out in the Autodesk CFD Simulation environment. For the experimental investigation, plane knits from 44 tex × 3 linear density ultra-high molecular weight polyethylene yarns were produced, and their air permeability was tested according to Standard DSTU ISO 9237:2003. The results obtained during the laboratory experiment and simulation differed by less than 5%.

Keywords: knit; plain stitch; multi-filament yarn; 3D model; stretching; air permeability

Citation: Ielina, T.; Halavska, L.; Mikucioniene, D.; Milasius, R.; Bobrova, S.; Dmytryk, O. Development of 3D Models of Knits from Multi-Filament Ultra-Strong Yarns for Theoretical Modelling of Air Permeability. *Materials* **2021**, *14*, 3489. https://doi.org/10.3390/ma14133489

Academic Editor: Dubravko Rogale

Received: 6 May 2021
Accepted: 16 June 2021
Published: 23 June 2021

Publisher's Note: MDPI stays neutral with regard to jurisdictional claims in published maps and institutional affiliations.

Copyright: © 2021 by the authors. Licensee MDPI, Basel, Switzerland. This article is an open access article distributed under the terms and conditions of the Creative Commons Attribution (CC BY) license (https://creativecommons.org/licenses/by/4.0/).

1. Introduction

Current approaches to the modelling of the physical properties and mechanical behaviour of knitted structures are based on the use of specialized software. To a greater or lesser extent, theoretical models consider the internal hierarchy of textile structures and provide an opportunity to choose the optimal level of modelling. They are usually based on the use of the principle of homogenization and consist of generalizations of the properties of the structural parts of the lower level in their transition to the upper one. The methods of homogenization have received the greatest development in relation to predictions of the properties of textile reinforced composite materials [1–3]. Studies have also shown the application of various theoretical approaches to the homogenization of the properties of textile fabrics and products. The multiscale hierarchical model for the determination of the properties of woven fabrics was proposed by Carvelli and Poggi [4]. Multiscale modelling systems that support the interaction between fine-scale and coarse-scale models make it possible to assess material properties based on the fine-scale details. Talebi et al. developed an open-source framework for multiscale modelling and the simulation of cracks in solids [5]. In order to form a theoretical background of the software, three categories of multiscale modelling techniques were investigated: hierarchical, semi-concurrent, and concurrent. However, a knitted structure cannot be considered as a continuum, but rather as a rod construction and requires other homogenization algorithms. The results of computer simulation depend on the level of geometric modelling. A representation of the geometry of the yarn (knitted into a stitch) and its structural parts, such as single-strand yarns, filaments or even fibres, is a rather difficult task. In addition, micro level modelling is associated with an increase in the computational time. When deciding on the level of detail of the model, it is important to assess the relevance of the complexity of the model geometry. On the other hand, attempts to simplify the geometry of the model and the transition to the macro level can give adequate results only if there are reliable, proven algorithms that consider their physical properties and structural parameters.

An assessment of the air permeability of textile fabrics is important in the context of ensuring the necessary functionality of filter materials, sportswear, outerwear and underwear, medical and rehabilitation products, as well as other groups of textile products. The air permeability of knitted structures depends on many factors, the most important of which are the type of raw material, the parameters of yarns, the knitting pattern, the technical parameters of the knitting process, the operating conditions, the processing methods, etc. Cotton and other natural fibres are often used for knitted clothes because of the high air and water vapor permeability. At the same time, it is strongly recommended to combine natural and synthetic fibers; this allows for better operational characteristics [6–9]. Wilbik-Hałgas et al. used a computer image analysis for an assessment of the surface porosity of knitted fabrics. According to the results of the study, air permeability is a function of the structural characteristics of knitted fabrics, such as surface porosity and thickness [10]. In their research [11], Muraliene and Mikucioniene investigated the changes in the air permeability of knitted samples with an elastomeric in-lay yarn insertion and without it in a stretched state. Špelić et al. studied the effects of walking on the thermal properties of clothing and subjective comfort [12]. Thermo-mechanical technological processes that textile materials undergo in the manufacturing process may affect their geometrical and physical properties [13]. The air permeability of textile materials used for clothing greatly influences a person's sense of comfort. There are theoretical models of the air permeability of knitted fabrics presented in the literature [14–17]. Ogulata and Mavruz [16] presented the relationship between the porosity of knitwear and its air permeability and proposed a method for its theoretical determination. The results of the use of this technique in the computer modelling of air permeability are presented in [17]. Air permeability is an important characteristic and is unique to materials which exhibit porosity. Usually, there are two types of pores in the structure of textile fabrics: inter-structural (inter-yarns) spaces and air channels in yarns. Xiao et al. [18] considered the pores of textile fabrics as tubes with variable cross sections and, considering the curvature of the pore walls, presented expressions describing the air flow. Another expression connecting the drop of air pressure with the dynamic viscosity of air, the air flow rate, as well as the length and hydraulic diameter of the pore, was given in the work of Kulichenko [19]. In the presence of through-structural pores, the air permeability of the textile material is mostly determined by the through-structural porosity. Moreover, the size and configuration of the pores are also important. Thus, for the same value of the through-structural porosity, fabric made from the thin yarns will have a lower air permeability than a fabric made from the thick yarns [19]. With an increase in the size of through-structural pores, the significance of the fibrous composition as a factor in the air permeability of the textile fabric decreases. However, the hairiness and roughness of the yarn creates aerodynamic friction on the surface of the yarn and determines an increase in the resistance to air flow. Therefore, the hairiness and roughness of the yarn surface reduces the flow rate and, accordingly, reduces the air permeability. Moreover, the orientation and method of fixing the fibers and individual sections of the yarn in the fabric structure are also important [14].

Due to an increased interest in the 3D modeling of knitted structures, many papers have been dedicated to the development of a geometric description of a knitted stitch. Different approaches have been suggested by Kyosev et al. [20], Trujevcev [21], Kaldor et al. [22], Wadekar et al. [22], and others. Topological models for weft knitted structure descriptions are proposed in [23–25]. Puszkarz and Krucinska [26] investigated the air permeability of two-layer knitted fabrics experimentally and by means of computer modeling. The authors considered the compliance of the results. A comparison of the experimental results and those obtained during the simulation of the process of air flow through models of the knitted structure, detailed at the level of single yarns and at the level of filaments, confirmed that the data obtained by using models, detailed at the level of filaments, were closer to the experimental results. The study of knitted fabric thermal insulation [27], realized with mono-fiber and multi-fiber 3D models, showed that both measurements and simulations yielded comparable results.

Later scientific publications have enhanced the theoretical approaches to the development of materials-by-design. Ultra-high molecular weight polyethylene (UHMWPE) threads can be used for the production of protective clothes. An optimization of the design process of such items can be reached by studying their physical and mechanical behavior and improving the means of creating their three-dimensional models. It is necessary to consider the possibility of modeling deformations in knitwear samples yielding to changing physical properties. In this study we focus on the peculiarities of unidirectional stretching and an air permeability simulation of plain knits made from UHMWPE threads.

2. Materials and Methods

Computational fluid dynamics (CFD) methods are based on the continuity equation, which in the case of a liquid or gas is reduced to the Navier-Stokes' equations and the equations of conservation. Modern software systems are equipped with a mathematical apparatus to solve a system of differential equations that describe the task and consider the geometry of the three-dimensional model and the user-specified boundary conditions. In order to simplify the work with porous materials, popular CFD programs provide the ability for the permeability, or the porosity of materials to be considered by setting one of the available indicators, depending on the available set of initial data. For example, the Autodesk CFD Simulation software [28] used for modelling a porous object provides the option to choose such a type of material as the so-called distributed resistance to flow. Such materials can be characterized by an additional pressure drop after flowing through the material. A description of the characteristics in the settings of the material can be achieved by one of the possible ways:

i. through the pressure drop coefficient, where the overpressure gradient can be determined using the Expression (1):

$$\frac{\partial p}{\partial X_i} = K_i \frac{\rho V_i^2}{2} \quad (1)$$

where i is the direction of the global coordinate. Index K may be found by determining the pressure drop against flow.

ii. through the friction index [14]. In such case, the gradient of additional pressure can be expressed by the Equation (2):

$$\frac{\partial p}{\partial X_i} = \frac{f}{D_h} \frac{\rho V_i^2}{2} \quad (2)$$

where f is the coefficient of friction; D_h is the hydraulic diameter in mm.

iii. through the Darcy ratio (3):

$$\frac{\partial p}{\partial X_i} = C \mu V i \quad (3)$$

where C is the filling factor which is the inverse value to the permeability; μ is the viscosity of the flow [14].

In [19], a dynamic model of the isothermal filtration of low humidity air through multilayer porous structures was proposed. The dependence of the conductivity on the rate of air filtration in packets of barrier clothing as quasi-linear generalizations of Darcy's law was established. According to this model, the air velocity in the pore V (m/s) can be calculated using the Expression (4):

$$V = \frac{\Delta P}{80\mu} \cdot \left(\frac{d_h^2}{L}\right) \tag{4}$$

where ΔP is the differential pressure with the laminar flow of liquid or gas; μ is the dynamic viscosity of liquid or gas in Pa·s; d_h is the hydraulic diameter of the pore in mm; L is the thickness of the porous material in mm.

The hydraulic diameter of the pore is equal to:

$$d_h = 4 \cdot \left(\frac{S}{P}\right) \tag{5}$$

where S is the area of the pore in mm^2; P is perimeter of the pore in mm.

For experimental investigations, 3 samples were knitted on the flat weft-knitting machine of gauge 8E from DOYENTRONTEX with 44 tex ×3 linear density ultra-high molecular weight polyethylene yarns (UHMWPE) in the single jersey pattern. To stretch knitted specimens in a uniaxial direction along wales and courses, the tensile-testing machine Kao Tiech KTO-7010AZ (Beijing TIME High Technology Ltd., Beijing, China) was used. Knitted samples of 200 mm × 50 mm size were prepared and a size of 100 mm was marked for fixation in the clamps of the tensile tester. On each sample the control points were marked as shown in Figure 1.

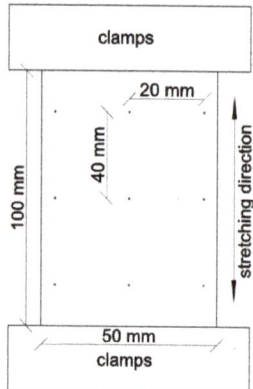

Figure 1. Schematic of the stretching of a knitted sample.

Macrophotographs of the knitted structure were made before and after the stretching forces were applied by means of the digital microscope Microsafe ShinyVision MM-2288-5X-S . (Taiwan) After the stretching distance between the control points changed, however, the number of stitches did not. The course and wale spacing was assessed by dividing the average distance between the control points by the numbers of stitches in the corresponding direction. Structural parameters were determined for the knitted fabrics in a relaxed stable state (Figure 2a) and after maximal unidirectional stretching along wales (Figure 2b) and courses (Figure 2c). Images of these structures are presented in Figure 2.

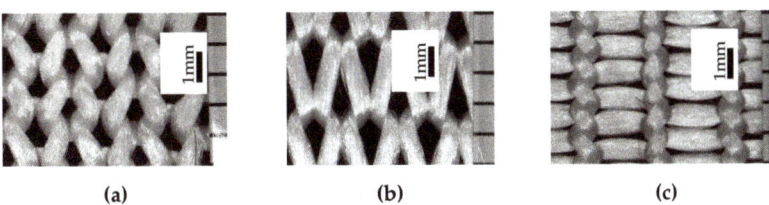

Figure 2. Images of knitted samples: (**a**) in a relaxed state; (**b**) after being stretched along wales; (**c**) after being stretched along courses.

The experimentally determined structural parameters of the knitted specimens before stretching and after maximal stretching in wale and course directions are presented in Table 1.

Table 1. Structural parameters.

Stretching Direction	Wale Spacing w, mm	Course Spacing c, mm	Yarn Diameter D, mm	Fabric Thickness M, mm	Loop Length l, mm
Before stretching	1.92 ± 0.1	1.85 ± 0.1		0.85 ± 0.05	
Stretched along wales	1.58 ± 0.1	2.78 ± 0.15	0.7 ± 0.05	0.75 ± 0.05	7.95 ± 0.4
Stretched along courses	3.46 ± 0.2	0.85 ± 0.05		0.82 ± 0.05	

Air permeability is characterized by the amount of air (dm^3) that passes through 1 m^2 of textile fabric in 1 second at a certain pressure difference on both sides of the fabric. The air permeability of knitted fabrics was determined in accordance with DSTU ISO 9237: 2003. Ten experimental tests were performed for each sample variant. All measurements were carried out in a standard atmosphere according to Standard EN ISO 139:2005 (20 °C ± 2 °C temperature and 65% ± 4% humidity).

3. Results and Discussions

3.1. Modelling of the Knitted Loop

The geometrical description of a yarn bent into a knitted loop includes the option to describe the axial line of the yarn and the pattern of the changes in the shape and size of the yarn along its axial line. In this research, the model described in [29,30] was used for the determination of the axial line geometry, as the yarn used for both the experimental knitted fabrics and the theoretical modelling was multifilament and consisted of 420 filaments, each of which had a diameter of 0.025 mm. To construct the volume of the yarn, a monofilament and multifilament model was considered as an option for use. The shape and size of the sections (the boundaries of the dense bundle of filaments) changes along the axial line of the yarn. Microlevel models are built in AutoCAD using Structura-3D software in addition. The selected method of (a) the arrangement of the cross-sections of the individual filaments in the structure of the yarn and (b) the alternation of circular and elliptical sections of the yarn along its axial line is shown in Figure 3.

Figure 3. (**a**) Principal arrangement of the cross-sections of individual filaments in the structure of the yarn and (**b**) in the yarn bent into the loop.

A visualization of a structural fragment of the knitted sample, built considering the geometry of the individual filaments in the yarn, is presented in Figure 4. However, the disadvantages of this model are a significant mutual penetration (cross-section) of the volumes that represent individual filaments in the model, and high demand for RAM.

Figure 4. Visualization of the knitted fabric structure with detailing at the level of individual filaments, built in the Autodesk AutoCAD software environment.

A comparison of the models of different detailing levels with a photographic image of the corresponding knitted sample (made by using digital microscope Microsafe ShinyVision MM-2288-5X-S (Taiwan)) is shown in Figure 5.

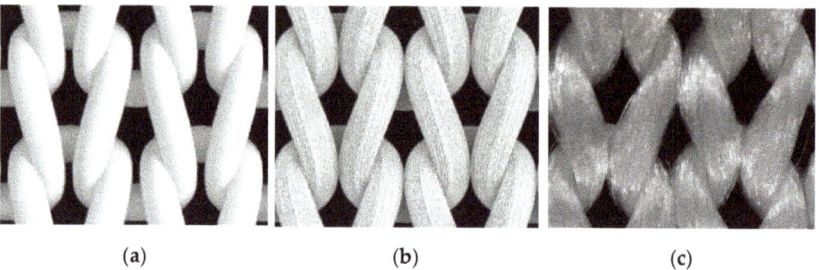

(a) (b) (c)

Figure 5. (a) Model of a knitted sample in a free state detailed at the level of yarn; (b) individual filaments; (c) image of the knitted sample in a free state.

As has been mentioned, the presence of sufficiently large through-structural (inter-yarn) pores in the case of a dense arrangement of filaments in the structure of the yarn means that air passes only through these pores, since the distance between the individual filaments does not exceed the thickness of the boundary layer. In addition, according to Bernoulli's law, when passing through narrow sections, the air flow velocity increases, while in the zone of increased velocity, the pressure decreases, which leads to the suction of an additional volume of air adjacent to these flow tubes. To illustrate this phenomenon, we developed in the Autodesk AutoCAD environment the model of a straight filament fragment detailed at the level of filaments (Figure 5a) and the monofilament model of a yarn fragment with an equivalent volume (Figure 5b). These models were placed in tube models of identical geometric parameters and imported into the Autodesk Simulation CFD environment, where the same boundary conditions were set at a pressure drop of 49 Pa. Graphical results of the analysis are presented in Figures 6 and 7.

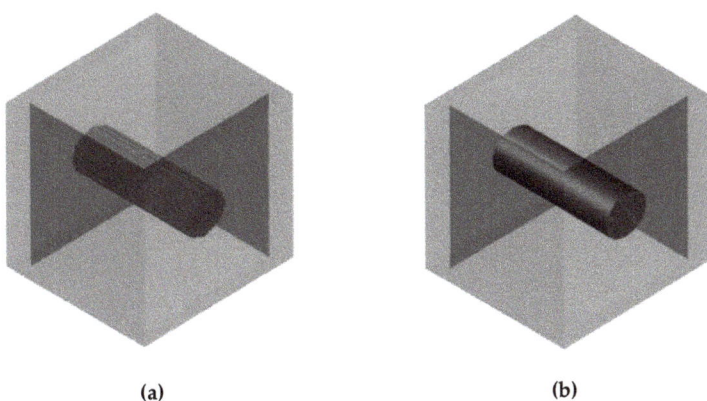

Figure 6. Models of a straight yarn segment made with filaments (**a**) and without filaments (**b**).

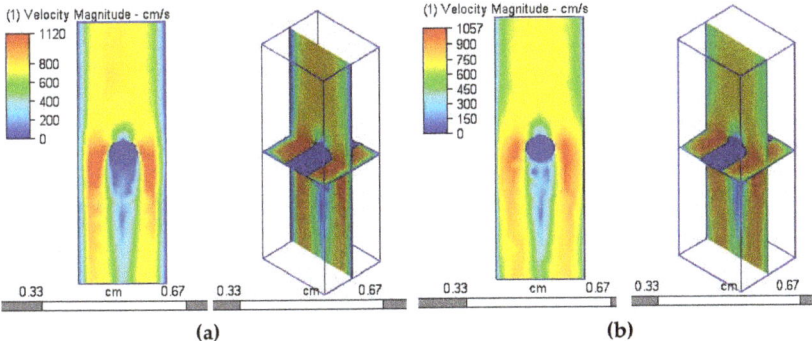

Figure 7. Graphical results of the analysis of the air flow through a tube with a linear yarn fragment made as multi-fiber (**a**) and mono-fiber (**b**).

In Figure 7, where the zones of the highest air flow velocity are shown in red and the lowest in blue, the program algorithms considered the set of filaments (bundle) impenetrable (Figure 7a) and the air bent around the yarn fragment, as it is shown in Figure 7b.

The obtained data from the calculations show that the filtration rate is not higher, but lower than in the case of an impermeable and smooth monofilament yarn model. This can be explained by the greater surface roughness of the multifilament model because the longitudinal protrusions on the surface form ribbing, with a height equal to the radius of an individual filament (in our case, 0.0125 mm).

In order to validate the geometric model of a stretched plain knitted structure, both laboratory and simulation methods of air permeability assessment were completed. To simulate the process of air flow through the structure of knitted fabrics, models with geometry detailed at the yarn level, with an adjustment of the flow resistance coefficient, were selected. As the initial data for the algorithm of selected modeling [29], the following parameters were provided: wale spacing w, course spacing c, thickness of the knitted fabric M, yarn diameter D, and the tangent angle at the interlacing point γ, as well as the angle of inclination of the loop legs α (as shown in Figure 9b). These initial data can be obtained experimentally or calculated according to well-known methods, for example, those presented in [21]. For the model of the knitted structure, maximally stretched along wales, $w = w_{max}$ and $c = c_{min}$, while for the model of knitted structure, maximally stretched along courses, $w = w_{min}$ and $c = c_{max}$. The effective radius R_0 of a multifilament yarn [31] can be determined by the formula (6):

$$R_0 = \frac{d}{2}\sqrt{\frac{N}{\varphi'}} \qquad (6)$$

where d is the diameter of the elementary filament in mm; N is the number of filaments in a yarn; φ is the coefficient of the packing density.

In the case of the ultimate packing density, $\varphi = 0.92$. Then, for our multifilament yarn R_0:

$$R_0 = \frac{0.025}{2}\sqrt{\frac{420}{0.92}} = 0.267 \qquad (7)$$

It can be assumed that the diameter of the yarn in the compressed state is $D_0 = 0.534$ mm.

It is known that in the process of stretching the knitted single jersey fabric along the wales, the wale spacing w takes the value w_{min} when the yarn is redistributed and pulled from the loops' heads and feet into the loop legs. In this case $w_{min} = 4 \cdot D_0$, id est, the minimum wale spacing is equal to four minimum yarn diameters (because the yarn is in a compressed state). If we accept the assumption of a circular cross-section of the yarn, then the value of the minimum wale spacing, considering Expression (6), will be equal to 2.14 mm, which significantly exceeds the experimentally obtained value for the stretched knitted fabric, which is 1.58 mm (see in Table 1). However, if the bundle of filaments has an elliptical cross-section, and the position of the major and minor axes of the ellipse changes along the axial line of the yarn as shown in Figure 8b, the ratio of the values of the major and minor semiaxes of the ellipse can be selected, at which the area of the ellipse is equal to the area of the circle of the radius determined by Equation (6), and their projection onto the plane of the fabric is equal to the width of the projection of the yarn onto the plane of the fabric, determined experimentally, as shown in Figure 8a.

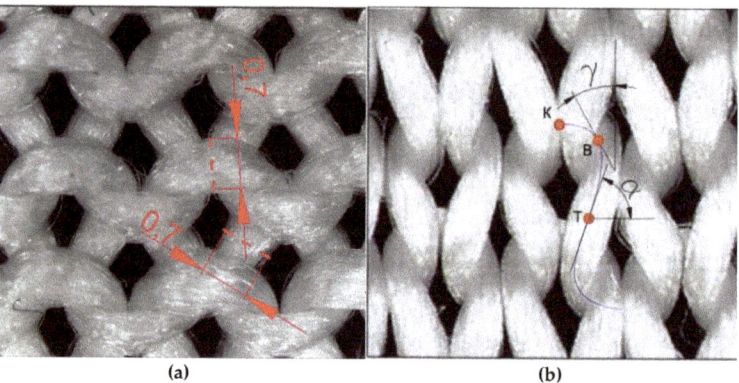

Figure 8. Layout of the elliptical section (a); and characteristic points (b).

As it is presented in [30], when simulating in a 3D environment, a space curve representing the axial line of the yarn in the model must pass through the characteristic points that lie directly on the axial line of the yarn and are located so that in 3D space their position can be determined using traditional ideas about the shape of the loop. In Figure 8a, the schematical view of the area of the interlacing of the yarns in the loop of the single jersey knitted fabric is presented. Characteristic points K, B, T, which belong to the axial line of the yarn, and control points P_0, P_1, P_2, P_3, which lie at the intersection of tangents to the axial line drawn at points K, B and T, are shown accordingly in Figure 9a,b. To describe the axes of the ellipse, the following notation is used: P_{max} is the large, and P_{min} is the minor axis of the ellipse. The axial line equation is built on the basis of the coordinates of the control points, and those, in turn, can be calculated based on traditional ideas about the geometry of the knitted loop, taking into account certain assumptions.

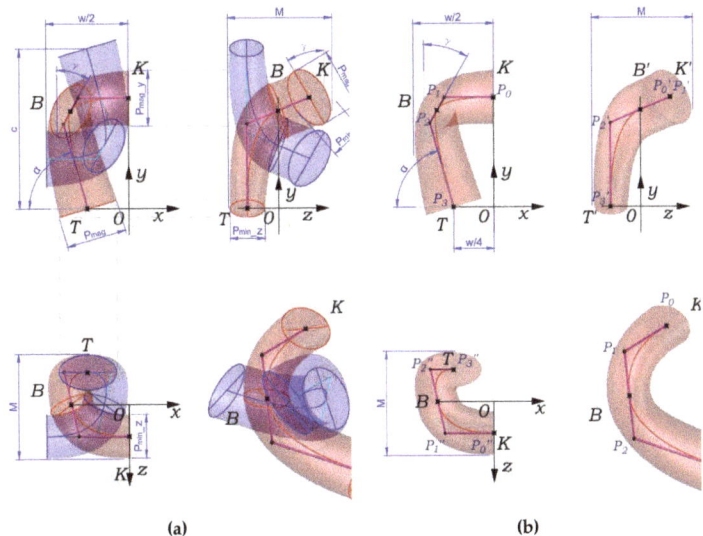

Figure 9. Characteristic points K, B, T (**a**) and the control points P0, P1, P2, P3 (**b**) of the spline curve, which describes a quarter of the axial line of the yarn in the knitted loop of single jersey structure.

In the coordinate system, situated as shown in Figure 10, the coordinates of points $K(X_k, Y_k, Z_k)$, $A(X_a, Y_a, Z_a)$, $B(X_b, Y_b, Z_b)$, $C(X_c, Y_c, Z_c)$, $T(X_t, Y_t, Z_t)$ can be calculated taking into account the accepted assumptions about the symmetric shape of the loop and the known characteristics of the structure: w, c, P_{max}, P_{min}, M. Mathematical expressions for determining the coordinates of the characteristic points of the loop in the selected coordinate system are presented in Table 2.

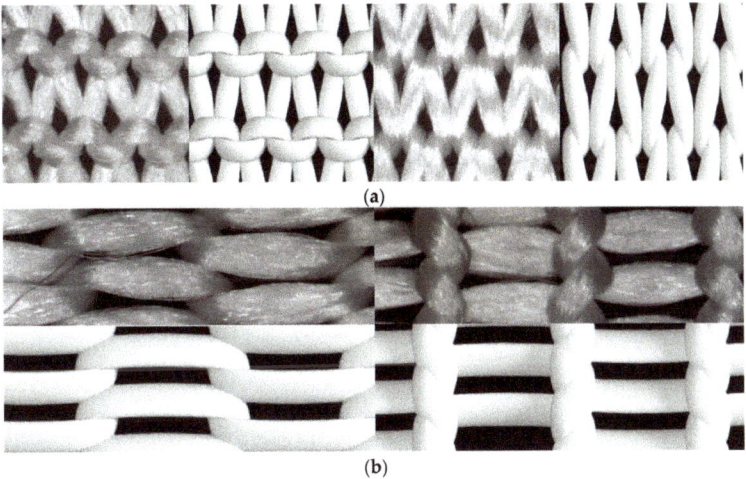

Figure 10. Meso-model and image of the single jersey knitted fabric made from high molecular weight polyethylene yarns at maximum stretching along wales (**a**) and along courses (**b**).

Table 2. Mathematical expressions for determining the coordinates of characteristic points in coordinate system of the loop.

Point	Abscissa	Ordinate	Applicata
K	$X_k = 0$	$Y_k = \frac{c + P_{mag}}{2}$	$Z_k = \frac{M - P_{min-z}}{2}$
B	$X_b = \frac{w}{4} + \frac{P_{min}}{2} \cos \gamma$	$Y_b = \frac{c}{2} + \frac{D_y}{2} \sin \gamma$	$Z_b = 0$
T	$X_t = \frac{A}{4}$	$Y_t = 0$	$Z_t = -\frac{M - P_{min}}{2}$

The direction of the tangents to the points K, B and T in the projection onto the XOY plane can be determined as follows. The tangent at point K (segment P_0P_1) is parallel to the OX axis and the XOY plane (plane of the fabric). In the projection onto the YOZ plane, the segment P_0P_1 is expressed to a point that coincides with the point K. The tangent at point T (segment P_2P_3) is located at an angle $(180 - \alpha)$ to the X-axis. It is also parallel to the XOY plane but located on the opposite side of the XOY than the segment P_0P_1. This is clearly seen in the projection onto the XOZ plane (Figure 10b). The segment P_1P_2 intersects with the plane of the fabric (XOY) at point B. Point P_1 belongs to the tangent line drawn through the point K, and point P_2 belongs to the tangent line drawn through the point T. In accordance with the previously accepted assumptions, the projection of the tangent at point B onto the plane of the fabric is located at an angle γ to the OY axis.

The presented algorithm of the calculation allows the coordinates of the characteristic points of the loop and the direction of the tangents at these points to be determined. It allows the creation of a B-spline, which reproduces the axial line of the yarn in 3D modelling systems of the knitted structure. Since in systems of finite element modelling mechanical properties are assigned to individual design objects, in meso-models used for the modelling of deformations, it is advisable to use the cross-section of the yarn in a compressed state in all sections of the loop.

The three-dimensional model and the images of the sample of single jersey knitted fabric made of high molecular weight polyethylene yarns at maximum stretching along the wales are presented in Figure 10a, and along courses in Figure 10b. The yarn cross-section of the model is shown as an ellipse.

3.2. Determination of Air Permeability and Its Simulation

In order to simulate a laboratory experiment in the Autodesk Simulation CFD software environment for a three-dimensional model, the boundary conditions were set as following: a pressure drop corresponded to 5 mm water column height or 49 Pa; the environmental temperature was set at 20 °C.

The internal porosity of the yarn was set at 8% for all specimen models. The results of air permeability obtained experimentally and during simulation using the Autodesk CFD Simulation software are presented in Table 3.

Table 3. Air permeability of the single jersey knitted fabric.

Sample No	Experimental Value of Air Permeability, dm^3/(m^{2}s)	Simulated Value of Air Permeability, dm^3/(m^{2}s)	Discrepancy of the Simulated Value, %
1	1617	1687	0.04
2	2353	2301	−0.02
3	1646	1645	0.00

The images obtained during the analysis of the simulation results are presented in Figure 11. The images present the distribution of zones with different air velocities. In accordance with the theory laid down in the algorithms of the program, the highest air velocity was in the narrowest places (shown in red in the diagrams). The flow rate decreased at the pore walls, which also corresponded to the positions of hydro and aerodynamics, which explains the decrease in the flow rate when approaching the tube (pore) walls due to the viscosity of the liquid or gas and friction forces.

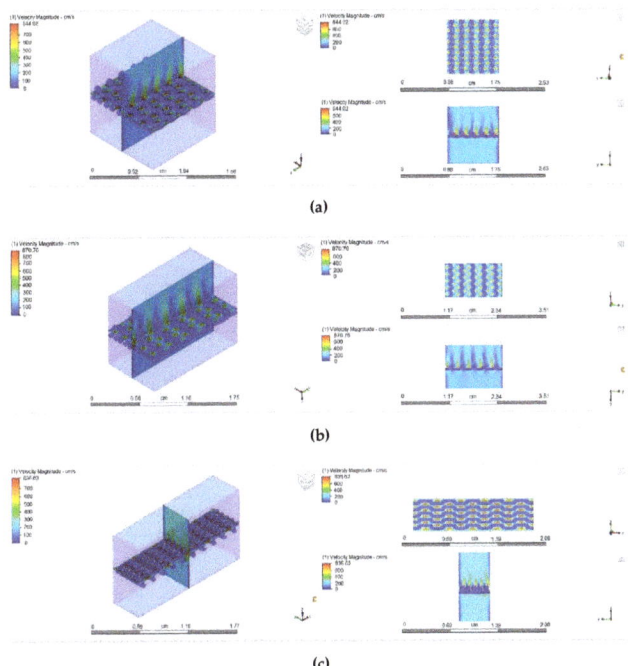

Figure 11. Illustration of the process of air passing through the knitted fabric under the condition of a pressure drop of 49 Pa: (**a**) model of the knitted fabric in a free state, (**b**) model of the knitted fabric in the state of uniaxial stretching along the wales, (**c**) model of the knitted fabric in the state of uniaxial stretching along the courses.

A comparison of the air permeability results obtained experimentally and after virtual simulation confirmed good accuracy. However, it should be noted that in real textile materials, the air flow can change the geometry of the holes depending on the pressure drop and the strength of the fastening of individual sections of the yarn in the knitted fabric. So far, it is difficult to take this into account in the available geometric models of knitted structures and the physical models of processes. Further research in this direction will improve the accuracy of the results obtained during the simulation.

4. Conclusions

Current trends in the use of digital tools for the modelling of physical processes occurring in knitted structures, particularly air permeability, were examined in this research. The features of constructing the model of the loop knitted in single jersey pattern, detailed at the level of yarns and individual filaments, were considered in this work. It was proposed that the air channels formed between the individual filaments in the yarn with special characteristics of porous materials were considered. The experimental results of the air permeability of single jersey knitted fabric from 44 tex ×3 linear density ultra-high molecular weight polyethylene yarns confirmed good accuracy of the model implemented using the Autodesk CFD Simulation software. The results obtained allow the proposed yarn-level geometric model to be used in assessments of the air permeability of plain knitted fabrics from synthetic hydrophobic yarns. Further development of the model could be performed by studying the hydraulic diameter changes of the pores under air or liquid flow in order to enhance its applicability for hydrophilic yarns. Also, experimental verification of the presented mathematical model will be done during the further investigations.

Author Contributions: Conceptualization, T.I., R.M., D.M. and L.H.; methodology, T.I., D.M. and L.H.; formal analysis, D.M., R.M. and L.H.; investigation, T.I., L.H. and O.D.; resources, L.H., S.B. and O.D.; data curation, D.M., T.I. and S.B.; writing—original draft preparation, T.I., D.M. and L.H.; writing—review and editing, R.M.; visualization, D.M. and T.I.; supervision, R.M. and L.H.; project administration, R.M. and L.H. All authors have read and agreed to the published version of the manuscript.

Funding: This research received no external funding.

Institutional Review Board Statement: Not applicable.

Informed Consent Statement: Not applicable.

Acknowledgments: This work was supported by the Lithuanian–Ukraine cooperation project "Knitted Materials for Personal Protective Equipment Against Mechanical and Flame Damages (PERPROKNIT)" by the Research Council of Lithuania and the Ministry of Education and Science of Ukraine.

Conflicts of Interest: The authors declare no conflict of interest.

References

1. Bakhvalov, N.S.; Panasenko, G.P. *Homogenization: Averaging Processes in Periodic Media: Mathematical Problems in the Mechanics of Composite Materials*; Springer: New York, NY, USA, 1989. Available online: https://doi.org/10.1007/978-94-009-2247-1 (accessed on 1 June 2021).
2. Lomov, S.; Dufort, V.; Luca, P.D.; Verpoest, I. Meso-macro integration of modeling of stiffness of textile composites. In Proceedings of the 28th International Conference of SAMPE Europe, Porte de Versailles Expo., Paris, France, 2–4 April 2007; pp. 4003–4008.
3. Loginov, A.; Grishanov, S.; Harwood, R. Modelling the load-extension behavior of plain-knitted fabric. Part I: A unit-cell approach towards knitted-fabric mechanics. *J. Text. Inst.* **2002**, *93*, 218–238. [CrossRef]
4. Carvelli, V.; Poggi, C. A homogenization procedure for the numerical analysis of woven fabric composites. *Composites Part A* **2001**, *32*, 1425–1432. [CrossRef]
5. Talebi, H.; Silani, M.; Bordos, S.P.A.; Kerfriden, P.; Rabczuk, T. A computational library for multiscale modeling of material failure. *Comput. Mech.* **2014**, *53*, 1047–1071. [CrossRef]
6. Khankhadjaeva, N.R. Role of Cotton Fiber in Knitting Industry. In *Cotton Science and Processing Technology*; Springer: Singapore, 2020. Available online: https://doi.org/10.1007/978-981-15-9169-3_11 (accessed on 1 June 2021).
7. Mikučionienė, D.; Arbataitis, E. Comparative Analysis of the Influence of Bamboo and other Cellulose Fibres on Selected Structural Parameters and Physical Properties of Knitted Fabrics. *Fibres Text. East. Eur.* **2013**, *21*, 76–80.
8. Bivainyte, A.; Mikucioniene, D. Influence of Shrinkage on Air and Water Vapour Permeability of Double-Layered Weft Knitted Fabrics. *Mater. Sci.* **2012**, *18*, 271–274. [CrossRef]
9. Hyun, A.K.; Seung, J.K. Physical properties and wear comfort of bio-fiber-embedded yarns and their knitted fabrics according to yarn structures. *Autex Res. J.* **2019**, *19*, 279–287. [CrossRef]
10. Wilbik-Halgas, B.; Danych, R.; Wiecek, B.; Kowalski, K. Air and water vapour permeability in double-layered knitted fabrics with different raw materials. *Fibres Text. East. Eur.* **2006**, *14*, 77–80.
11. Muraliene, L.; Mikucioniene, D. Influence of structure and stretch on air permeability of compression knits. *Int. J. Clothing Sci. Technol.* **2020**, *32*, 825–835. [CrossRef]
12. Špelić, I.; Rogale, D.; Mihelić Bogdanić, A. The Study on Effects of Walking on the Thermal Properties of Clothing and Subjective Comfort. *Autex Res. J.* **2020**, *20*, 228–243. [CrossRef]
13. Kalendraite, B.; Krisciunaite, J.; Mikucioniene, D. Influence of sublimation process on air permeability and water absorption dynamics. *Int. J. Clothing Sci. Technol.* Available online: https://www.emerald.com/insight/content/doi/10.1108/IJCST-04-2020-0050/full/html (accessed on 1 June 2021). [CrossRef]
14. Kukin, G.N.; Solovjov, A.N. Textile Materials Science. Chapter 3, Legkaya Industriya, Moscow Moskow. 1967, p. 226. (In Russian). Available online: https://scholar.google.com/scholar_lookup?title=Textile%20Materials%20Science&publication_year=1992&author (accessed on 1 June 2021).
15. Kulichenko, A.V. Theoretical and Experimental Models for Prediction of Air –Permeability of Textile. In Proceedings of the 4th International Textile, Clothing & Design Conference, Dubrovnik, Croatia, 5–8 October 2008; pp. 799–802.
16. Oğulata, R. Tugrul; Mavruz, S. Investigation of porosity and air permeability values of plain knitted fabrics. *Fibres Text. East. Eur.* **2010**, *5*, 71–75.
17. Mezarciöz, S.; Mezarcioz, S.; Oğulata, R. Tugrul. Prediction of air permeability of knitted fabrics by means of computational fluid dynamics. *Tekstil ve Konfeksiyon* **2014**, *24*, 202–211.

18. Xiao, X.; Zeng, X.; Long, A.; Lin, H.; Clifford, M.; Saldaeva, E. An Analytical model for through-thickness permeability of woven fabric. *Textile Res. J.* **2012**, *82*, 492–501. [CrossRef]
19. Kulichenko, A.V. Theoretical Analysis, Calculation, and Prediction of the Air Permeability of Textiles. *Fibre Chemistry.* **2005**, *37*, 371–380. [CrossRef]
20. Kyosev, Y.; Angelova, Y.; Kovar, R. 3D modeling of plain weft knitted structures of compressible yarn. *Res J Text Appar.* **2005**, *9*, 88–97. [CrossRef]
21. Trujevcev, A.V. The Stitch Model by of Dalidovich Loop in Context of Modern Theoretic S tudies. (In Russian). 2002. Available online: https://ttp.ivgpu.com/wp-content/uploads/2021/04/268_29.pdf (accessed on 1 June 2021).
22. Kaldor, J.M.; James, D.L.; Marschner, S. Simulating knitted cloth at the yarn level. *ACM T Graphic* **2008**, *27*, 65.
23. Wadekar, P.; Goel, P.; Amanatides, C.; Dion, G.; Kamien, R.D.; Breen, D.E. Geometric modeling of knitted fabrics using helicoid scaffolds. *J. Eng. Fibers Fabr.* **2020**, *15*, 1–15. [CrossRef]
24. Ielina, T.V.; Halavska, L.Y. Development of Algorithm of Data Analysis of Topological Model of Filament in the Knitted Fabric Structure. 2013. (In Ukrainian). Available online: http://journals.khnu.km.ua/vestnik/pdf/tech/2013_3/14eli.pdf (accessed on 1 June 2021).
25. Kapllani, L.; Amanatides, C.E.; Dion, G.; Shapiro, V.; Breen, D. TopoKnit: A Process-Oriented Representation for Modeling the Topology of Yarns in Weft-Knitted Textiles. Available online: https://arxiv.org/abs/2101.04560 (accessed on 1 June 2021).
26. Puszkarz, A.; Krucinska, I. Modelling of air permeability of knitted fabric using the computational fluid dynamics. *AUTEX Res. J.* **2018**, *18*, 364–376. [CrossRef]
27. Puszkarz, A.; Krucinska, I. The study of knitted fabric thermal insulation using thermography and finite volume method. *Text. Res. J.* **2017**, *87*, 643–656. [CrossRef]
28. Fluid Flow Definitions. Autodesk Knowledge Network. Available online: https://knowledge.autodesk.com/support/cfd/learn-explore/caas/CloudHelp/cloudhelp/2014/ENU/SimCFD/files/GUID-4DAF0D6D-F1F4-4E90-A9C8-5CACB85E79BE-htm.html (accessed on 1 June 2021).
29. Bobrova, S.; Ielina, T.; Beskin, N.; Bezsmertna, V.; Halavska, L. The use of 3D geometric models in special purpose knitwear design and predicting of its properties. *Vlákna a textil (Fibres Text.)* **2018**, *25*, 19–26.
30. Ausheva, N.; Halavska, L.; Yelina, T. Geometric representation features of textile yarn in the 3D modeling systems. In Proceedings of the Scientific Conference "Unitech-13", Gabrovo, Bulgaria, 22–23 November 2013; pp. 199–202.
31. Cherous, D.A.; Shilko, S.V.; Charkovski, A.V. Prognozirovaniye effektivnykh mekhanicheskikh kharakteristik trikotazha. *Fizicheskaya mekhanika* **2008**, *11*, 107–114. (In Russian)

Article

Impact of the Elastane Percentage on the Elastic Properties of Knitted Fabrics under Cyclic Loading

Tea Jovanović, Željko Penava * and Zlatko Vrljičak

Department of Textile Design and Management, Faculty of Textile Technology, University of Zagreb, 10000 Zagreb, Croatia
* Correspondence: zpenava@ttf.unizg.hr or zeljko.penava@ttf.unizg.hr; Tel.: +385-1-3712-576

Abstract: Elastic knitted fabrics find numerous applications in the industry for compression stockings, sports and leisure wear, swimwear, ballet wear, etc. During its use, knitwear is subjected to dynamic loading due to body movements. The loading and unloading of the knitted fabric affect the size of the elastic region in which unrecovered deformation completely disappears. This paper deals with the influence of the elastane percentage in the knitted fabric on the elastic properties of the knitted fabric under dynamic loading. For this experiment, three types of yarn were used in different combinations: polyamide (PA), wrapped elastane yarn and bare elastane. The mentioned yarns were used to knit three different groups of plated weft-knitted fabrics (two yarns in a knitted fabric row): without elastane, knitted fabric with a percentage of wrapped elastane, and knitted fabric with a percentage of bare elastane. The percentage of elastane ranged between 0% and 43%. First, standard uniaxial tensile tests were performed on knitted fabric samples until breakage under static load. The force–elongation diagrams obtained are used to determine the elastic limit up to which Hook's law applies. All knitted fabrics were cyclically tested to the elastic limit. From the obtained loading and unloading curves, unrecovered deformation (unrecovered elongation), elastic elongation and hysteresis index were determined and calculated. The results showed that the percentage of elastane significantly affects the size of the elastic region of the knitted fabric and has no effect on the hysteresis index. Therefore, it is necessary to optimize the elastane percentage for different knitted fabric designs to achieve the best dynamic recovery of the knitted fabric and to design a more stretchable knitted garment that fits the body as well as possible.

Keywords: knitted fabric; cycle load; elastic limit; elastane; hysteresis curve

Citation: Jovanović, T.; Penava, Ž.; Vrljičak, Z. Impact of the Elastane Percentage on the Elastic Properties of Knitted Fabrics under Cyclic Loading. *Materials* **2022**, *15*, 6512. https://doi.org/10.3390/ma15196512

Academic Editor: Hubert Rahier

Received: 4 July 2022
Accepted: 15 September 2022
Published: 20 September 2022

Publisher's Note: MDPI stays neutral with regard to jurisdictional claims in published maps and institutional affiliations.

Copyright: © 2022 by the authors. Licensee MDPI, Basel, Switzerland. This article is an open access article distributed under the terms and conditions of the Creative Commons Attribution (CC BY) license (https://creativecommons.org/licenses/by/4.0/).

1. Introduction

The main function of many elastic knitted fabrics is to provide stability and wearing comfort, and not to restrict body movement. For example, knitted fabrics for sports and leisure wear should be comfortable, less sweat-absorbent, breathable, lightweight, durable and suitable for all weather conditions and body movements. Such requirements are affected by the climatic changes to which the knitted fabric is exposed, the number of wears, washes and drying processes. All of the above affect their deformation, change in structure and properties, on the basis of which the results of the force drop and hysteresis index are obtained. Knitted fabrics are textile fabrics that have a much higher elongation in the course direction, the so-called transverse direction, than in the wale direction, the so-called longitudinal direction [1]. Increasingly, highly elastic knitted fabrics are used that are stretchable, comfortable and fit easily on the body. They are usually made with two yarns interlooped into a horizontal row of the knitted fabric. The first yarn is a ground yarn, mainly made of natural fibers, and the second yarn is elastane [2]. For most fabrics, it is very difficult to achieve a stretch of 10–50% and avoid deformations due to stress recovery. Therefore, at the beginning of the use of elastic textiles, especially in sportswear, only knitted fabrics were used, and the next innovation was the addition of elastomeric fibers [3].

Elastomeric fibers have been present for about 70 years. With the improvement of production processes, they are used in a wide and diverse field of textiles [4]. Elastane fibers are also known under trade names such as Lycra, Spandex and Dorlastan. They are like rubber and are very stretchy. Elastane fibers contain a polyurethane bond, which means that the structure of elastic fibers is complex [5]. Elastane is used in all areas where unrecovered elasticity is required, e.g., compression stockings, elastic trousers, sportswear and swimwear, corsets and other knitwear. Elastic garments have a wide range of applications in the field of tight sportswear [6]. Multifilament yarns made of polyamide (PA) or polyester (PES) are often used instead of cotton to produce leisure and sportswear. In a sense, all fibers, except the cheapest raw materials, are high-performance fibers. Single cotton yarns have an extension at break of 3 to 8%, while the elastane has an extension at break of 200 to 900%. Force at break and extension at break are two important factors of any yarn. Force at break is the force required to break the yarn. The percentage increase in length at break is referred to as extension at break. The extension of yarn is the percentage elongation of yarn from the initial length. The mentioned percentages of yarn extension to breakage are important because of the compression force of the knitted fabric on the body. By using the harmonized parameters of yarn, density of knitted fabric and knitted structure, the appropriate structure of the knitted fabric which is suitable for a particular construction of the garment and its application is achieved. Different combinations of yarns, knitted structures and knitted fabric densities provide the appropriate elongation values that correspond to the construction of a garment [7–9].

Mukhopadhyay [10] interlooped lycra in combination with cotton and analyzed its effect on the elongation and recovery of the knitted fabric. He found that recovery was better, and that elongation and elasticity were greater in cotton knitted fabrics with a combination of lycra.

Plain single jersey fabrics for making summer T-shirts or undershirts have an elongation at break in the course direction of 150 to 250%, and in the wale direction of 50 to 150%. Double-face knitted fabrics are mostly used in making men's winter underwear and are four times more stretchable in the course direction than in the wale direction. Chain knitted fabrics made on warp knitting machines with elastane yarns are intended for bathing suits and have approximately equal elongation in the course direction and in the wale direction. Each knitted structure basically gives a different structure of the knitted fabric and even tensile properties [11].

One of the methods for producing knitted fabrics with different elongation is to use plated and partially plated knitted structures in various combinations. When making a plated fabric, a course is formed from two yarns. If it is intended to produce a fuller, and thus more solid structure of the knitted fabric, two identical yarns should be used. However, if it is intended to get knitted fabrics of different tensile properties, two yarns of significantly different tensile properties are used, e.g., the ground yarn is cotton or polyamide, and the other, plating yarn, is elastane. Other combinations include knitting with different structures and counts of the ground and elastane yarn. The combination of plated and partially plated structures, different yarns and sinking depths results in a wide range of elastic knitted structures for different applications. This combination is most commonly used in the production of preventive compression stockings, which exert a compression of 10 to 25 hPa on the leg. In other combinations, instead of the plated structure or with the plated structure, a certain tuck-knitted structure with the specified or similar combinations of yarns can be used. Depending on the function of the product and the construction of the machine, a certain combination will be applied [12,13].

Elastic knitwear should follow body movements. The purpose of the knitted fabric is to follow this elongation and to recover after relaxation. The purpose of elastic knitted fabrics is not to hinder the body movement, but to follow it and return to its original shape [14,15]. Knitted fabrics containing elastane provide a high level of comfort and ease of usage because of elastic properties.

The mechanical properties of textile fibres, i.e., the reactions of the fibres to acting forces and deformations, are technically very important properties that ultimately affect the quality of the end product [16]. Kisilak believes that because of the elasticity and high elasticity of the fibres, the deformations of the knitted fabric are only temporary unless the stresses are too high or last too long, resulting in unrecovered or irreversible deformation [17].

In use, the knitwear is subjected to complex and variable dynamic loads in individual loading and unloading cycles that change the shape of the knitwear. Cyclical measurements are used to analyze the behaviour of the knitted fabric in use. Decrease in force and residual deformation of the knitted fabric are two parameters which well describe a change in the knitted fabric structure after cyclic loads [18]. Su and Yang also found in their studies that elastic recovery decreases with an increasing number of cycles [4]. The results show that with an increasing number of repeat cycles, the plasticity of the elastane increases, while the elasticity decreases, and the elastomeric yarn loses its recovery ability.

If the knitted fabric is successively loaded and unloaded, the stress–strain diagram will have the shape of a loop. This phenomenon of the lagging of strain behind stress is called elastic hysteresis. The surface of the hysteresis loop represents the energy spent on unrecovered deformations during a loading cycle. When repeating the cycle, the surface of the loop slowly increases until breakage occurs. The larger the hysteresis area, the greater the energy loss, i.e., the lower the recovery of the knitted fabric. Elastic knitted fabrics have a higher recovery when the energy loss is lower [19]. Liang and Stewart conducted dynamic tests and defined hysteresis as the difference between stretching (loading) and relaxation (unloading) [20]. Different researchers have studied the cyclic loading of textiles and the occurrence of hysteresis [21–24].

Penava conducted the knitted fabric off-axes tensile test and for the resulting force–elongation curves proposed a process of determining the zones corresponding to elastic deformation, elastic–plastic deformation, and plastic deformation [25].

The aim of this work was to investigate the influence of the elastane percentage in the knitted fabric on the size of the elastic deformation region of the knitted fabric and on the hysteresis index under cyclic loading.

2. Theoretical Analysis

Elastic weft-knitted fabrics (jersey) are usually made of two, three or four yarns interlooped in a horizontal row of loops, where all the yarns form loops and only one is made of elastane. In the basic form of the plated knitted fabric, there are two yarns in a row, one of which is made of elastane (Figure 1a). These knitted fabrics are used for the production of preventive compression stockings, which are mainly worn by pregnant women. In the case of a partially plated knitted fabric, e.g., 1 + 1, the elastane yarn is interlooped into every second row of loops next to the ground yarn. This structure is used in the production of casual trousers, especially for women and children. In the case of partially plated elastic knitted fabric 1 + 3, an elastane yarn is interlooped into every fourth row of loops and the knitted fabric is used to produce elastic fine cotton lingerie containing elastane.

One of the most significant properties of the knitted fabric for clothing is its stretch and elasticity, both in the wale direction and the course direction of stitches. Due to their structure, weft-knitted fabrics are several times more stretchier in the direction of the courses than in the direction of the wales. Knitted fabrics are anisotropic materials whose anisotropy is reflected in different behavior during force action in different directions. During stretching, the loops are deformed in the direction of the forces. Stretching in the course direction changes the shape of the loops, while stretching in the wale direction is achieved exclusively by stretching the yarn.

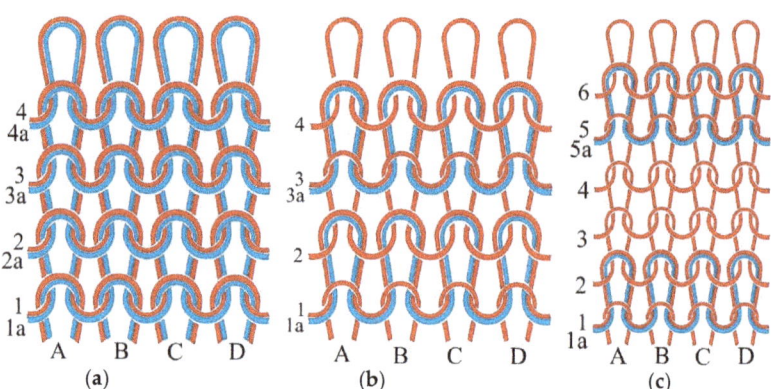

Figure 1. Schematic of plated jersey fabric: (**a**) basic structure; (**b**) partial plating 1 + 1–elastane in every second row of loops; (**c**) partial plating 1 + 3–elastane in every fourth row of loops. 1–6 are basic yarns in course; 1a–4a are plating yarns in course; A–D are wales.

2.1. Load-Elongation Diagram of Knitted Fabric

Two main directions are distinguished in the structure of the knitted fabric: longitudinal direction (wales) and transverse direction (courses). Under the influence of tensile force F, normal stresses occur in the knitted fabric and its elongation. Figure 2 shows the curve of the ratio of tensile force and elongation of the knitted fabric.

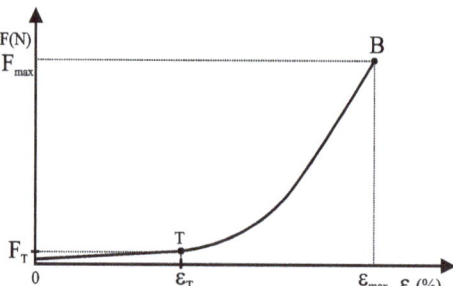

Figure 2. The typical load-elongation (F − ε) diagram of knitted fabric.

The curve in the tensile force-relative elongation diagram (Figure 2) consists of two parts [25,26]. The first part 0T is a linear region representing the elastic part of the knitted fabric. The second part TB is nonlinear and elastoplastic deformation occurs. Point B represents the maximum force (F_{max}) and maximum elongation (ε_{max}) of the knitted fabric. In further investigations, the behaviour of the knitted fabric in the elastic region, i.e., up to the point T, is observed [26]. The determination of the point T (elastic limit) is solved by the method of least squares. Tensile force F in the region 0T is expressed as a linear function of the elongation ε [26].

2.2. Cycle Force–Elongation Response

To observe the behaviour of the knitted fabric in use and to analyse the phenomenon of hysteresis, the knitted fabric should be cyclically loaded [27]. A sample of the knitted fabric is fixed in a tensile tester with two clamps. The upper clamp is fixed, i.e., there is a force-measuring probe on it. The lower clamp is automatically moved down and up by the motor. In this way, the motion is generated for a certain number of cycles, which leads to stresses in the knitted fabric. By moving the lower clamp up and down, the elongation value ε changes, as shown in Figure 3a The elongation values lie in the interval between the minimum elongation value ε_{min} (in our test ε_{min} is zero) and the elongation value ε_T at the elastic limit.

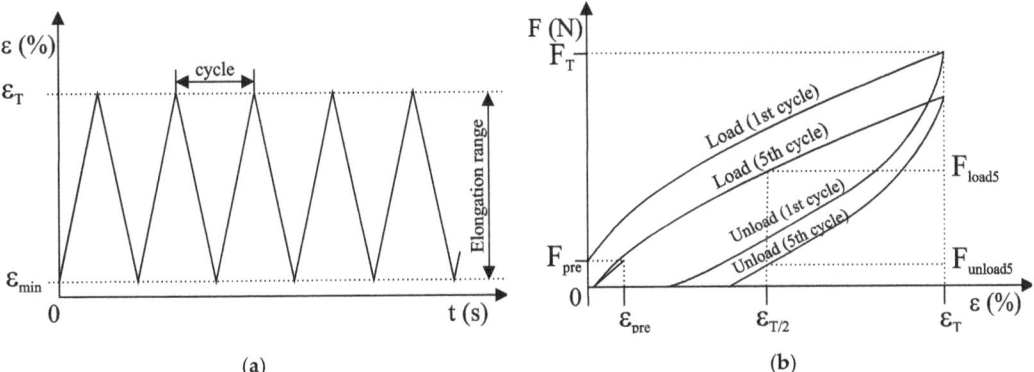

Figure 3. Cyclic testing: (**a**) schematic representation of the change in elongation over time; (**b**) loading–unloading curve.

In dynamic testing, the lower clamp of the tensile tester moves until it reaches the elongation at a predetermined preload. The sample is then loaded–unloaded five times between two defined elongation points ($0 - \varepsilon_T$) and reloaded in the next cycle until it reaches the preload value, Figure 3b.

For the 5th cycle, the elastic elongation is calculated according to Equation (1):

$$\varepsilon_{ee} = \varepsilon_T - \varepsilon_{pre} \tag{1}$$

ε_{ee} is the elastic elongation, ε_{pre} is the elongation under the action of a preload force at the beginning of the sixth cycle. The hysteresis index is analysed at half the given elongation $\varepsilon_T/2$ and presented by Equation (2):

$$HYI = \frac{F_{unload5}}{F_{load5}} \tag{2}$$

HYI is the hysteresis index, $F_{unload5}$ is the force read at half the given loading $\varepsilon_T/2$ in the fifth unloading cycle, F_{load5} is the force read at half the given loading $\varepsilon_T/2$ in the fifth loading cycle.

For the 5th cycle, unrecovered elongation (ε_{ue}) is calculated according to Equation (3):

$$\varepsilon_{ue} = \varepsilon_T - \varepsilon_{ee} \tag{3}$$

3. Experimental Part

In the experimental part of the work, tensile tests were carried out until the sample of a plated jersey fabric broke under static load. In this test, the values of the tensile forces and the corresponding elongations were determined, and the elastic ranges (elastic limit) were also found. For this purpose, the classical methods, and instruments for testing the tensile properties of knitted fabrics, were used. Dynamic tests of knitted fabric samples, cyclic loading and unloading of knitted fabrics were also carried out. Cycle load-elongation diagrams were obtained.

The experiment was carried out by measuring the deformation of the knitted fabric at the static and dynamic loads, both in the course and wale directions. The aim of this experiment was to determine the influence of the elastane percentage on the elastic range of the knitted fabric, the residual deformation and the hysteresis index.

3.1. Materials

3.1.1. Characteristic Parameters of Yarns

Three types of yarn were used for this experiment: polyamide (PA), wrapped elastane yarn and bare elastane. The most important properties of these yarns are listed in Table 1.

Table 1. Characteristic parameters of yarns.

Material Type	Polyamide (PA)	Wrapped Elastane Yarn	Wrapped Elastane Yarn	Wrapped Elastane Yarn	Bare Elastane	Bare Elastane	Bare Elastane
Yarn Count (dtex)	156	156	130	78	156	195	235
Breaking Elongation (%)	32.64	30.24	26.56	22.08	405.28	482.40	513.92
Breaking Force (N)	6.38	1.73	1.28	0.55	1.56	1.60	2.41
Tenacity (cN/tex)	40.87	11.07	9.84	6.99	10.01	8.22	10.26
Work of Rupture (N·mm)	124.90	29.13	19.64	8.27	270.82	361.99	469.38

The tensile strength properties of the yarn were measured using a Statimat M tensile tester (Textechno H. Stein GmbH & Co. KG, Moenchengladbach, Nordrhein-Westfalen, Germany). The measurements were carried out according to ISO 2062: 2009 and method B. The force–elongation diagram for polyamide and wrapped elastane yarns of different counts is shown in Figure 4a, that for bare elastane of different counts in Figure 4b.

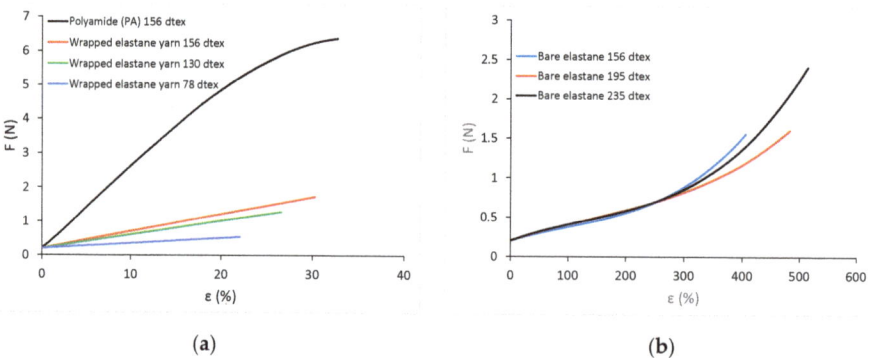

Figure 4. Force–elongation diagrams: (**a**) polyamide yarn and wrapped elastane yarn; (**b**) bare elastane yarn.

The force–elongation relation for polyamide yarn and for bare elastane yarns is non-linear. For the wrapped elastane yarn, the relation between force and elongation is linear. In the case of bare elastane yarns and of wrapped elastane yarns, the values for breaking forces, breaking elongation and work of rupture increase with increasing yarn counts.

3.1.2. Characteristic Parameters of Knitted Fabrics

Three basic groups of knitted fabric samples were made. In all samples, the course was made from two basically different yarns. Also, in all samples, the ground yarn is polyamide (PA) multifilament with a count of 156 dtex f 68, which is very often used in the production of socks. The first sample (NE) is ground and was made from two equal, abovementioned PA yarns. This sample is used as a reference, and the structures and tensile properties of other samples are compared with it. In the production of the second sample group, in besides the PA multifilament ground yarn with a count of 156 dtex, single-wrapped elastane yarns (elastane core and PA sheath with different PA and elastane percentages) in three different yarn counts were used, which are used in the production of preventive or compression stockings. In the first sample of this group (WE17), a knitted row was made with the abovementioned 156 dtex PA yarn and an elastane yarn, also with a count of 156 dtex-(156/33/10). The second sample of this group (WE16) was made with a 156 dtex PA yarn and a slightly finer elastane yarn, i.e., yarn with a count of 130 dtex-(130/33/20). In the third sample of this group (WE14), a knitted row was made with one of the PA yarns mentioned and an even finer elastane yarn-count 78 dtex-(78/13/7). In the production of these four samples, the tensile force of the PA multifilament ground yarn at the entrance of the knitting zone was 2.5 ± 0.5 cN, and the tensile force of the elastane yarns was 8 ± 2 cN.

The characteristic parameters of the analyzed knitted fabrics are listed in Table 2.

Table 2. Test results for plaited knitted fabric parameters.

Fabric Code	Fabric Type	ℓ_{PA} (mm)	ℓ_E (mm)	U_{PA} (%)	U_E (%)	Knitted Fabric Spirality (°)	Thickness (mm)	Mass per Unit Area (g/m^2)
Non-elastane knitted fabric								
NE	Plaited	3.93	0	100	0	0	1.09	200
Knitted fabrics with a percentage of wrapped elastane								
WE14	Plaited	3.76	2.46	86	14	2	1.13	225
WE16	Plaited	3.65	1.56	84	16	3	1.18	260
WE17	Plaited	3.81	1.89	83	17	5	1.23	280
Knitted fabrics with a percentage of bare elastane								
CE32	Plaited	3.86	2.39	68	32	3	1.21	310
CE38	Plaited	3.88	2.65	62	38	1	1.17	320
CE43	Plaited	3.81	2.01	57	43	5	1.27	380

ℓ_{PA}—length of interloping the ground (PA) yarn in one loop (mm); ℓ_E—length of interloping the elastane yarn in one loop (mm); U_{PA}—percentage of PA yarn in the knitted fabric (%); U_E—percentage of the elastane yarn in the knitted fabric (%); NE—non-elastane; WE14—knitted fabric with 14% wrapped elastane; CE32—knitted fabric with 32% bare elastane.

In the third sample group, the abovementioned PA multifilament yarn with a count of 156 dtex as the ground yarn was interloped into one row of the knitted fabric, and bare elastic yarns type 162C (100% elastane) with three different counts were interlooped next to it. The first sample of this group (CE32) with the ground 156 dtex PA yarn in one row also contains an interlooped bare elastane yarn with a count of 156 dtex. In the second sample of this group (CE38), a coarser elastane yarn with a count of 195 dtex was used besides the ground PA yarn, and in the third group (CE43), an even coarser elastane yarn with a count of 235 dtex was used. With these two coarser yarns, i.e., the PA multifilament yarn with a count of 156 dtex and the bare elastane yarn with a count of 235 dtex, practically the highest possible load on knitting needles was achieved for the specified knitting machine gauge. From these three groups of samples or a total of seven basic samples, the structures of knitted fabrics with different elongations were obtained.

The measurement method and the procedure used to examine the thickness of the knitted fabric are specified in the ISO 5084:2003 standard. DIN EN 14971 was used to determine the number of courses and wales of the knitted fabric per unit of length. Standards that were also used to determine the parameters of knitted fabrics:

- ISO 2062:2003: Textiles—Yarn from packages. Determination of single-end breaking force and elongation at break.
- ASTM D8007-15 (2019) was used to determine the wale and course counts of weft-knitted fabrics per unit of length.
- ISO 3801:1977. Determination of mass per unit length and mass per unit area.

3.2. Methods

A single bed circular knitting machine with a gauge of E17, needle bed diameter 95 mm (3¾″) knitted with 200 needles to make tubular elastic knitted fabrics. The individual samples were 500 to 800 mm long and have a uniform structure that is used in the production of prevention and compression stockings. The circumference of the tubular knitted fabric ranged from 150 to 200 mm. This structure of the knitted fabric is suitable for making the part of the stocking that wraps above the ankle and in the lower part of the leg calf, where the circumference is usually from 200 to 300 mm. Prior to testing, all samples were conditioned under standard atmosphere conditions (relative humidity 65 ± 2 %, at a temperature of 20 ± 2 °C). For this test, standard samples measuring 200 mm × 50 mm were cut and clamped in the clamps of the device at a distance of 100 mm. Five measurements were taken for each knitted fabric sample.

The tensile properties of all knitted fabric samples were tested according to ISO 13934-1:2013 using the test strip method on a Statimat M tensile tester (Textechno H. Stein GmbH & Co. KG, Mönchengladbach, North Rhine-Westphalia, Germany). After determining the elastic limits (point T) for all knitted fabric samples, the cyclic loading of the knitted fabric samples was carried out up to a predetermined elastic limit.

Dynamic tests of knitted fabric samples were performed according to DIN 53835-2-Testing of textiles; determination of the elastic behavior of single and plied elastomeric yarns by repeated application of tensile load between constant extension limits.

4. Results and Discussion

Under the effect of the tensile force F, the corresponding longitudinal deformation, i.e., elongation ε, was measured. The mean values of the measurement results of the effect of the tensile force F and the corresponding elongation ε for the non-elastane knitted fabric samples (NE) and with a different percentage of wrapped elastane (WE) are shown in Figure 5a for the wale direction and in Figure 5b for the course direction.

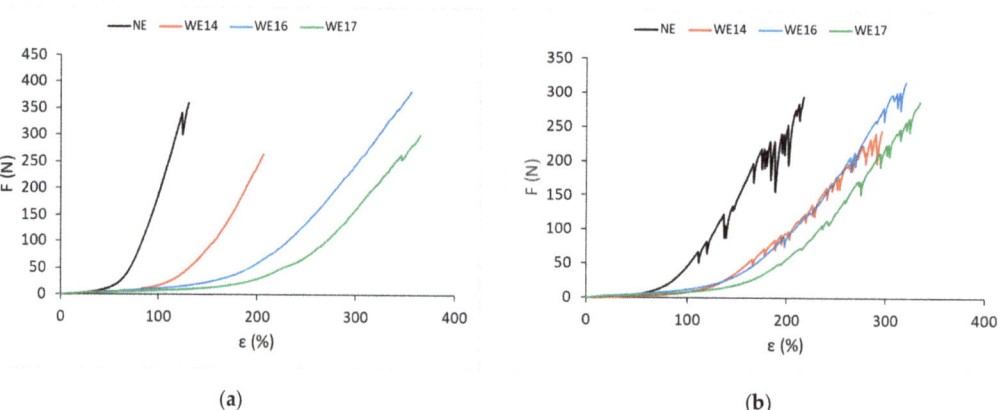

(a)   (b)

Figure 5. Force–elongation diagrams for knitted fabric samples with polyamide yarn and wrapped elastane yarns: (**a**) wales direction; (**b**) courses direction.

The mean values of the measurement results of the effect of the tensile force F and the corresponding elongation ε for the knitted samples with different percentages of bare elastane (CE) are shown in Figure 6a for the wale direction and in Figure 6b for the course direction.

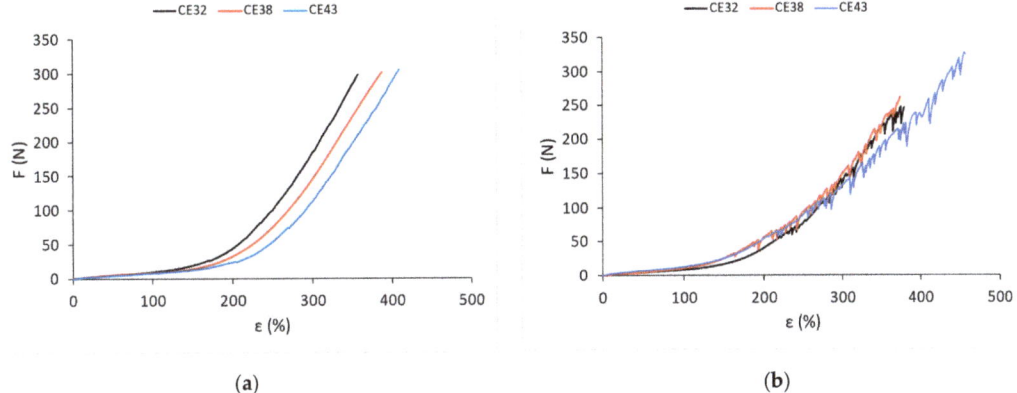

Figure 6. Force–elongation diagrams for knitted fabric samples with bare elastane: (**a**) wales direction; (**b**) courses direction.

The corresponding mean values of breaking force, elongation at break and work of rupture for the non-elastane knitted fabric samples (NE) and with a different percentage of wrapped elastane (WE) are listed in Table 3, and for the knitted fabric samples with bare elastane (CE), are listed in Table 4. The results presented refer to the wale and course directions of the knitted fabrics.

Table 3. Mean values of elongation at break, breaking force, and work of rupture for the non-elastane knitted fabrics and with a percentage of wrapped elastane.

	Wales				Courses			
	NE	WE14	WE16	WE17	NE	WE14	WE16	WE17
Breaking Elongation (%)	130.4	206.6	356.3	364.5	217.3	295.7	320.3	334.4
Breaking Force (N)	359.5	263.6	383.5	302.0	293.8	244.7	316.9	287.8
Work of Rupture (kN·mm)	12.55	12.44	35.33	25.25	19.71	20.62	28.45	24.60

Table 4. Mean values of elongation at break, breaking force, and work of rupture for the knitted fabrics with bare elastane percentage.

	Wales			Courses		
	CE32	CE38	CE43	CE32	CE38	CE43
Breaking Elongation (%)	358.1	387.8	409.3	379.0	373.4	456.6
Breaking Force (N)	297.3	301.4	305.3	247.4	261.8	326.8
Work of Rupture (kN·mm)	27.11	29.94	30.36	26.07	27.49	46.72

The knitted fabrics with a percentage of wrapped elastane have significantly higher values of breaking elongation than the non-elastane knitted fabrics. The elongation at break increases with an increasing percentage of wrapped elastane. Elongation at break values increases in the wale and course directions with increasing elastane percentage.

When increasing the percentage of bare elastane in the knitted fabrics, the values of breaking elongation, breaking force, and work of rupture increase.

When knitted fabric samples are stretched with uniaxial tensile force after the linear region, elastoplastic and plastic (unrecovered) deformations occur that are undesirable during the use of knitted fabrics. Up to the elastic limit (point T), the knitted fabric behaves completely elastic. The values of tensile force and corresponding elongation at the point of elasticity are given in Table 5.

Table 5. Elongation values and corresponding force (ε_T, F_T) at the elastic limit for all knitted fabric samples.

Fabric	Wale Direction		Course Direction	
	ε_T (%)	F_T (N)	ε_T (%)	F_T (N)
NE	21.76	3.70	32.32	2.90
WE14	58.24	6.58	50.80	3.23
WE16	90.22	11.26	64.60	7.29
WE17	105.24	9.20	73.40	6.57
CE32	95.32	15.91	65.40	9.20
CE38	107.03	11.23	77.08	8.22
CE43	119.16	10.17	87.32	7.57

Non-elastane knitted fabric (NE) has the lowest elastic limit, namely, in the wale direction amounting to 21.76% and 32.32% in the course direction. The action of tensile forces higher than 2.9 N on such knitted fabrics already leads to the occurrence of elastoplastic and unrecovered deformations. Such knitted fabrics have a small elastic region. In the case of elastane-wrapped knitted fabrics, the elastic limit increases (elongation values and force increase) in relation to non-elastane knitted fabric (NE), so that elastane-wrapped knitted fabrics can withstand higher loads in the elastic range. When the percentage of wrapped elastane increases, the elastic limit rises from 50.8 to 73.4% in the course direction and from 58.24 to 105.24% in the wale direction. The knitted fabric with the highest percentage of bare elastane (CE 43) has the highest elastic limit at which the value of elongation is 119.16%. The elongation at the T-point of knitted fabric CE43 is six times greater in the wale direction than in non-elastane knitted fabric (NE) and almost three times greater in the course direction. The elastic limit (T) increases with the percentage of elastane in the knitted fabric in both the wale direction and in the course direction, whereby this increase is greater in the wale direction. Thus, different percentages of elastane in the knitted fabric can change its elastic properties.

The statistical concepts correlation and regression, which are used to evaluate the relationship between two variables, are used in this paper [28,29]. Figure 7a shows the relation between the values of the knitted fabric elongation at the elastic limit ε_T and the elastane percentage. This relation is linear both in the wale direction and in the course direction. When the elastane percentage increases, the elongation at the elastic limit also increases linearly. For the wale direction, the correlation coefficient is r = 0.85, and for the course direction, r = 0.88, which is a very high correlation. However, when comparing samples made with elastane yarns wrapped with PA filaments, the scatter of results is greater than when analyzing samples made with bare elastane yarns. These differences are patterned by the structure of the yarn and the different number and yarn count of the filaments that wrap the elastane core of the yarn. By interloping the bare elastane yarn with the PA textured ground yarn, gradually and linearly, without major variation, the elasticity limit of this type of knitted fabric increases.

Figure 7b shows the relation between the values of the tensile force of the knitted fabric at the elastic limit ε_T and the elastane percentage. This relation is linear both in the wale direction and in the course direction. For the wales direction, the correlation coefficient is r = 0.70, and for the courses direction, r = 0.80, which is significantly less than the elongation at the elastic limit. Spreads are more irregular in knitted fabric made with elastane wrapped yarns. It is interesting to note that with an increase in the yarn count of the bare elastane yarn, i.e., the proportion of elastane, the tensile force decreases at the end of the elastic deformation of the knitted fabric. This phenomenon can be explained by the higher density of the knitted fabrics and lower porosity.

Figure 8 shows the diagrams of the results of cyclic loading and unloading of the knitted fabrics up to the elastic limit for all tested samples.

(a) (b)

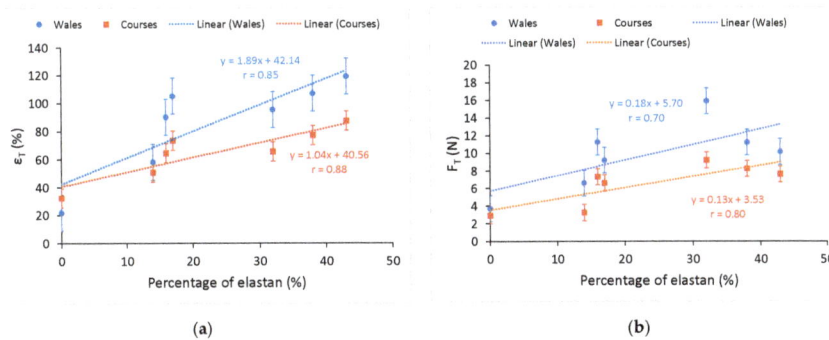

Figure 7. Elastic limit values depending on the percentage of elastane: (**a**) elongation ε_T at point T; (**b**) tensile force F_T at point T.

(a) (b)

(c) (d)

(e) (f)

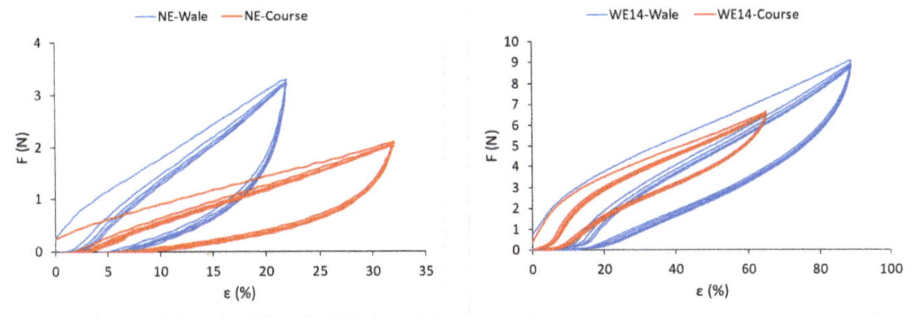

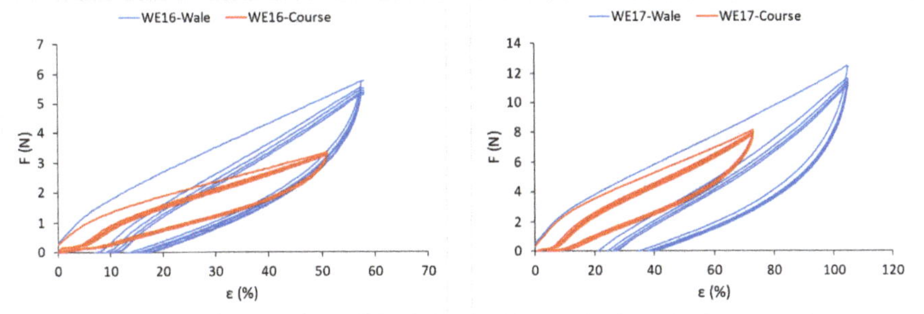

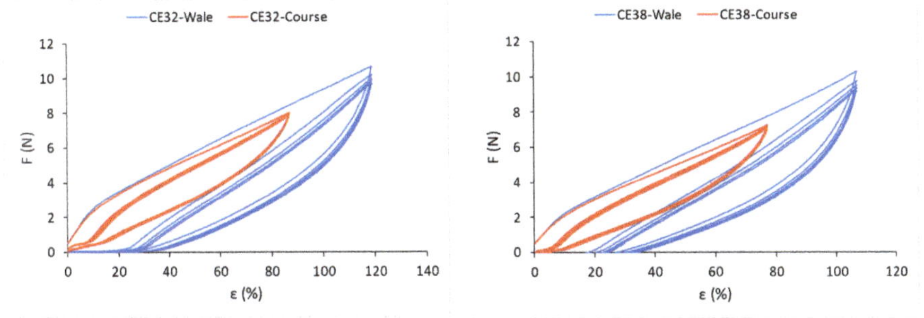

Figure 8. *Cont.*

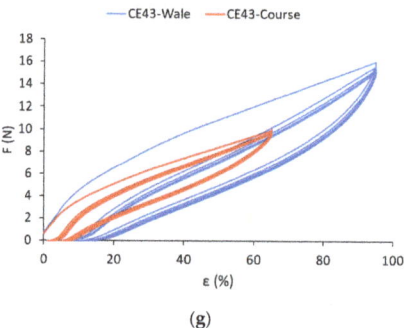

(g)

Figure 8. Diagrams of the results of cyclic loading and unloading of the knitted fabrics up to the elastic limit: (**a**) non-elastane knitted fabric; (**b**) knitted fabric with 14% wrapped elastane; (**c**) knitted fabric with 16% wrapped elastane; (**d**) knitted fabric with 17% wrapped elastane; (**e**) knitted fabric with 32% bare elastane; (**f**) knitted fabric with 38% bare elastane; (**g**) knitted fabric with 43% bare elastane.

The values of unrecovered elongation (ε_{ue}), elastic elongation (ε_{ee}), and hysteresis index (HYI) in the fifth loading–unloading cycle of the knitted fabrics in the wale direction are listed in Table 6, and for the course direction in Table 7. Equations (1)–(3) were used to calculate the values.

Table 6. Unrecovered elongation (ε_{ue}), elastic elongation (ε_{ee}), hysteresis index (HYI) in the 5th cycle in the wale direction.

Fabric	NE	WE14	WE16	WE17	CE32	CE38	CE43
ε_{ue} (%)	4.3	13.8	31.1	17.8	14.8	29.9	33.0
ε_{ee} (%)	17.7	44.2	73.9	71.2	80.2	77.1	86.0
HYI	0.181	0.382	0.385	0.477	0.598	0.411	0.437

Table 7. Unrecovered elongation (ε_{ue}), elastic elongation (ε_{ee}), hysteresis index (HYI) in the 5th cycle in the course direction.

Fabric	NE	WE14	WE16	WE17	CE32	CE38	CE43
ε_{ue} (%)	6.6	5.5	1.1	6.3	6.8	8.8	7.0
ε_{ee} (%)	25.4	45.5	71.9	58.7	58.2	68.2	80.0
HYI	0.204	0.507	0.547	0.653	0.625	0.536	0.565

The relation between the elastane percentage in the knitted fabric and the value of the elastic elongation of the knitted fabric in the fifth cycle of cyclic loading and unloading in the wale direction is shown in Figure 9a, and in the course direction, in Figure 9b.

This relation is linear both in the wale direction and in the course direction. By increasing the elastane percentage, the elastic elongation also increases linearly in the fifth cycle of cyclic loading. For the wale direction, the correlation coefficient is r = 0.85, and for the course direction, r = 0.81, which is a very high correlation. The elastic elongation in the fifth cycle for the knitted fabric with 43% bare elastane (CE43) is 4.86 times greater than the elastic elongation of the non-elastane knitted fabric (NE) in the wale direction and 3.15 times greater in the course direction. An increase in the percentage of elastane (wrapped and unwrapped) in the knitted fabric causes an increase in the elastic deformation of the knitted fabric even under cyclic loading. As in the previous cases, the analysis results of knitted fabrics made with elastane-wrapped yarns have larger and more irregular scatters than knitted fabrics made with bare elastane yarns. The reason is also in the different structures of elastane yarns.

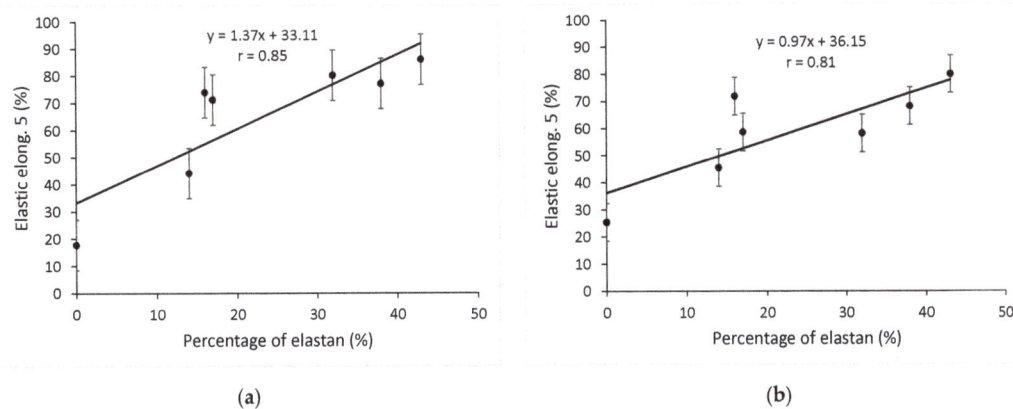

Figure 9. Relation between the percentage of elastane in the knitted fabric and the value of the elastic elongation of the knitted fabric in the fifth cycle of cyclic loading–unloading: (**a**) in the wale direction; (**b**) in the course direction.

Figure 10a shows the relation between the elastane percentage of the knitted fabric and the hysteresis index in the wale direction and Figure 10b in the course direction.

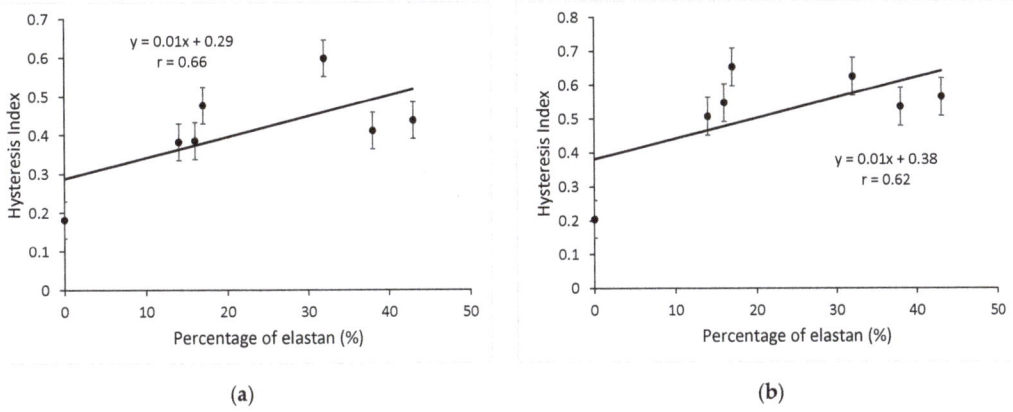

Figure 10. Relation between the percentage of elastane in the knitted fabric and the hysteresis index: (**a**) in the wale direction; (**b**) in the course direction.

The correlation coefficient is low, and it is not possible to talk about the linear dependence of the elastane percentage in the knitted fabric and the hysteresis index. If only elastane knitted fabrics are analyzed (Tables 6 and 7), it can be seen that elastane percentage has no effect on the hysteresis index. However, it is interesting to relate the hysteresis index to the yarn count of the interloped elastane yarns. Knitted fabrics with finer wrapped elastane yarns (WE14) and knitted fabrics with coarser bare elastane yarns (CE43) in the course direction give a similar hysteresis index, but in the wale direction the difference between the hysteresis index is greater.

The designed, manufactured and analyzed samples of elastic knitted fabrics with PA and wrapped elastane yarns can be used in the production of preventive compression stockings that are gladly worn by pregnant women or the elderly. When making the legs of these stockings, different lengths of yarn will be woven into the individual courses of knitted fabrics in order to obtain a structure with a different amount of stretching, and thus compression on a particular part of the leg. Knitted fabrics made with PA and bare elastane yarns can be used in making recreational long pants or T-shirts and underwear for mountaineers. When making cuts for this type of clothing, the amount of elastic stretch must be used in order to obtain clothing that fits comfortably against the body.

5. Conclusions

In static testing of knitted fabrics, an increase in the elastane percentage in the knitted fabric leads to an increase in the elastic limit of the knitted fabric, i.e., the knitted fabric is stretched to higher elongation values in the elastic region.

Cyclic loading affects the tensile properties of knitted fabrics.

From the test results, it can be concluded that if the occurrence of unrecovered deformation of the knitted fabric under cyclic loading is to be prevented, it is recommendable that the predetermined value to which the knitted fabric stretches (upper elongation limit), Figure 3a, is less than or equal to the value of the elongation at the elastic limit ε_T.

To reduce the effect of dynamic load on the knitted fabric, it is definitely recommended to use bare, single- or double-wrapped elastane yarn.

The results of cyclic tests have shown that the elastane percentage significantly affects the size of the elastic region of the knitted fabric and has no influence on the hysteresis index. Therefore, it is necessary to optimize the percentage of elastane for different designs of elastic knitted fabrics in order to achieve the best possible dynamic recovery of the knitted fabric and the fit of the garment to the body.

With an increase in the percentage of elastane from 0 to 38%, the elastic area for knitted fabrics increases from 21.76 to 107.03%, the elastic limit is higher and this enables the knitted fabrics to withstand higher loads in the elastic area and return to its original state after unloading (Table 5).

With less yarn interloping in the course of knitted fabric, the stretching of this kind of knitted fabric is also less, and it is used when making compression stockings around the ankle. A higher stretching is obtained with a longer length of yarn interloping in the course of knitted fabric, whereby a higher stretch of the knitted fabric is also achieved, which is needed on the sole of the leg or below the groin.

When making any compression garments, especially preventive compression stockings, it is necessary to match the fineness of the basic and elastane yarns and their structure. When making stockings with less compression, e.g., up to 42 hPa (32 mmHg), the basic PA yarn and elastane yarn can be of the same or similar fineness, e.g., 78, 110 or 156 dtex. However, when making stockings with higher compression, the elastane yarn is much coarser than the base yarn and has a fineness of, for example, 195, 235 or 300 dtex.

Author Contributions: Conceptualization, Ž.P. and Z.V.; investigation, Ž.P., Z.V. and T.J.; methodology, Ž.P., Z.V. and T.J.; experimental research, Ž.P.; formal analysis, Ž.P.; writing—original draft preparation, Ž.P. and Z.V.; writing—review and editing, Ž.P. and Z.V. All authors have read and agreed to the published version of the manuscript.

Funding: This research received no external funding.

Conflicts of Interest: The authors declare no conflict of interest.

References

1. Chowdhary, U. Stretch and Recovery of Jersey and Interlock Knits. *Int. J. Text. Sci. Eng.* **2018**, *1*, 1–8.
2. Lozo, M.; Penava, Ž.; Lovričević, I.; Vrljičak, Z. The Structure and Compression of Medical Compression Stockings. *Materials* **2022**, *15*, 353. [CrossRef] [PubMed]
3. Bardhan, M.K.; Sule, A.D. Anatomy of Sportswear and Leisurewear: Scope for Spandex Fibres. *Man-Made Text. India* **2001**, *3*, 81–86.
4. Su, C.-I.; Yang, H.-Y. Structure and Elasticity of Fine Elastomeric Yarns. *Text. Res. J.* **2004**, *74*, 1041–1044. [CrossRef]
5. Kielty, C.M.; Sherratt, M.J.; Shuttleworth, C.A. Elastic fibres. *J. Cell Sci.* **2002**, *115*, 2817–2828. [CrossRef]
6. Senthilkumar, M.; Anbumani, N.; Hayavadana, J. Elastane fabrics—A tool for stretch applications in sports. *Indian J. Fibre Text. Res.* **2011**, *36*, 300–307.
7. Cavezzi, A.; Michelini, S. *Phlebolymphoedema: From Diagnosis to Therapy*; Edizioni P. R.: Bologna, Italy, 1998; p. 50.

8. Hearle, J.W.S. *High-Performance Fibres*; Woodhead Publishing Limited with The Textile Institute: Cambridge, UK, 2001; p. 62.
9. Ramelet, A.A.; Monti, M. *Phlebology: The Guide*; Elsevier: Amsterdam, The Netherlands, 1999; pp. 154–196.
10. Mukhopadhyay, A.; Sharma, I.C.; Mohanty, A. Impact of lycra filament on extension and recovery characteristics of cotton knitted fabric. *Indian J. Fibre Text. Res.* **2003**, *28*, 423–430.
11. Vrljičak, Z. *Pletiva*; University of Zagreb Faculty of Textile Technology: Zagreb, Croatia, 2019; p. 52.
12. Kowalski, K.; Karbowski, K.; Klonowska, M.; Ilska, A.; Sujka, W.; Tyczynska, M.; Wlodarczyk, B.; Kowalski, T.M. Influence of a Compression Garment on Average and Local Changes in Unit Pressure. *Fibres Text. East. Eur.* **2017**, *25*, 68–74.
13. Gries, T. *Elastische Textilien*; Deutscher Fachverlag: Frankfurt am Main, Germany, 2005; p. 23.
14. Senthilkumar, M.; Anbumani, N.; de Araujo, M. Elastic Properties of Spandex Plated Cotton Knitted Fabric. *J. Inst. Eng.* **2011**, *92*, 9–13.
15. Eryuruk, S.H.; Kalaoglu, F. Analysis of the performance properties of knitted fabrics containing elastane. *Int. J. Cloth. Sci. Technol.* **2016**, *28*, 463–479. [CrossRef]
16. Nyoni, A.B.; Brook, D. The Effect of Cyclic Loading on the Wicking Performance of Nylon 6.6 Yarns and Woven Fabrics Used for Outdoor Performance Clothing. *Text. Res. J.* **2010**, *80*, 720–725. [CrossRef]
17. Kisilak, D. A New Method of Evaluating Spherical Fabric Deformation. *Text. Res. J.* **1999**, *69*, 908–913. [CrossRef]
18. Jovanović, T.; Penava, Ž.; Vrljičak, Z. Deformation of Elastic Knitted Fabrics Under Cyclic Loading. In Proceedings of the 14th Scientific-Professional Symposium Textile Science & Economy, Zagreb, Croatia, 26 January 2022; pp. 1–10.
19. Mani, S.; Anbumani, N. Dynamic elastic behavior of cotton and cotton/spandex knitted fabrics. *J. Eng. Fibers Fabr.* **2014**, *9*, 93–100. [CrossRef]
20. Liang, A.; Stewart, R.; Bryan-Kinns, N. Analysis of sensitivity, linearity, hysteresis, responsiveness, and fatigue of textile knit stretch sensors. *Sensors* **2019**, *19*, 3618. [CrossRef] [PubMed]
21. Lee, H.S.; Ko, J.H.; Song, K.S.; Choi, K.H. Segmental and chain orientational behavior of spandex fibers. *J. Polym. Sci. Part B Polym. Phys.* **1997**, *35*, 1821–1832. [CrossRef]
22. Messiry, M.E.; Mohamed, A. Analysis of the effect of cyclic loading on cotton-spandex knitted fabric. In Proceedings of the 5th International Conference on Advances in Mechanical, Aeronautical and Production Techniques, Kuala Lumpur, Malaysia, 12–13 March 2016; pp. 1–6.
23. Kawabata, S. *Nonlinear Mechanics of Woven and Knitted Materials*; Textile Structural Composites; Elsevier: Amsterdam, The Netherlands, 1989; pp. 67–116.
24. Kawabata, S.; Inoue, M.; Niwa, M. Theoretical Analysis of the Non-Linear Deformation Properties of a Triaxial Weave under Biaxial Stress Fields. *Compos. Sci. Technol.* **1996**, *5*, 261–271. [CrossRef]
25. Penava, Ž.; Šimić Penava, D.; Lozo, M. Experimental and analytical analyses of the knitted fabric off-axes tensile test. *Text. Res. J.* **2021**, *91*, 62–72. [CrossRef]
26. Knezić, Ž.; Penava, Ž.; Šimić Penava, D.; Rogale, D. The Impact of Elongation on Change in Electrical Resistance of Electrically Conductive Yarns Woven into Fabric. *Materials* **2021**, *14*, 3390. [CrossRef]
27. Matuso, M.; Yamada, T. Hysteresis of Tensile load – Strain Route of Knitted Fabrics under Extension and Recovery Processes Estimated by Streain History. *Text. Res. J.* **2009**, *79*, 275–284. [CrossRef]
28. Simon, F.; Strangfeld, C.; Gussen, L.; Lang, S.; Wölfling, B.-M.; Notz-Lajtkep, H. Prediction model for the analysis of the haptic perception of textiles. *J. Text. Eng. Fash. Technol.* **2021**, *7*, 79–85. [CrossRef]
29. Akoglu, H. User's guide to correlation coefficients. *Turk. J. Emerg. Med.* **2018**, *18*, 91–93. [CrossRef] [PubMed]

Article

Usage Durability and Comfort Properties of Socks Made from Differently Spun Modal and Micro Modal Yarns

Antoneta Tomljenović *, Juro Živičnjak and Ivan Mihaljević

Department of Materials, Fibers and Textile Testing, Faculty of Textile Technology, University of Zagreb, Prilaz baruna Filipovića 28a, 10000 Zagreb, Croatia
* Correspondence: antoneta.tomljenovic@ttf.unizg.hr

Abstract: Socks, being a necessary item of clothing, must be comfortable and maintain their quality throughout their life. Since the applicability of modal fibers and microfibers, as well as yarns produced using unconventional processes, in sock knitting has been insufficiently researched, this paper evaluated three groups of medium sized socks knitted in a plain single jersey pattern produced with the highest percentage of ring, rotor and air-jet spun modal or micro modal yarns of the same linear density in full plating with different textured polyamide 6.6 yarns compared to conventional cotton socks. The sock quality was evaluated through an investigation of the physical properties, wear resistance and dimensional stability, as well as the water vapor absorption, air permeability and thermal resistance using the thermal foot model before and after five repeated washing and drying cycles, according to the proposed methodology. The results showed that the fiber fineness, the structure of the differently spun yarns and the sock plain knits, the polyamide content and the implementation of the pretreatment of the socks had an influence on the obtained results. The socks made from modal and micro modal yarns differed in their properties. Compared to cotton socks, they have better comfort properties, a generally better pilling resistance and, after pretreatment, a comparable abrasion resistance.

Keywords: socks; modal; micro modal; yarn type; durability; comfort; textile testing

1. Introduction

Socks (Latin soccus) are knitted next-to-skin-type garments worn on the feet, often covering the ankle and part of the calf. They usually come in calf and over-calf lengths and are worn inside shoes. They consist of the top part, the part that covers the leg (leg part) and the part that covers the foot (foot part). More precisely, the main parts of socks (Figure 1) are the cuff, leg, heel, sole, toe and foot [1,2].

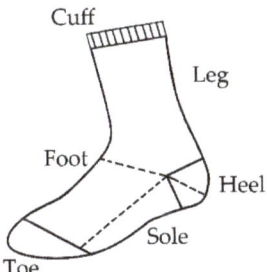

Figure 1. Different parts of socks.

Socks have to fulfill high requirements for their usage durability and comfort. Therefore, it is very important to select the right fibers and yarns for their production [3]. Socks come in a variety of fibers and fiber combinations. Usually, casual socks are made from a high percentage of cotton or wool to provide softness and comfort. They are blended with polyamide and/or elastane to improve the fit, durability and shrink resistance. Occasionally, acrylic, polyester, polypropylene or luxury fibers such as silk, linen, cashmere or mohair are also added to improve the sock properties [4]. Man-made artificial fibers made from cellulose (e.g., viscose, modal and lyocell) offer a silky handle, exceptional contact comfort and better hydrophilicity than cotton and play an important role in the textile market for lingerie items worn in direct contact with the skin [5]. Their applicability in the knitting of socks, especially for modal fibers and microfibers, is still insufficiently researched. The comparison of the modal and cotton fibers properties, relevant to this study, is shown in Table 1.

Table 1. Properties comparison of cotton and modal staple fibers [6–8].

Fiber Type	Cotton	Modal
Elongation at break (%)		
- conditioned	8–10	10–15
- wet	12–14	11–16
Tenacity (cN/tex)		
- conditioned	25–30	36–42
- wet	26–32	27–30
Moisture regain (%)	7.0–9.5	11.5–12.5
Water retention value (%)	42–53	55–70
Micrographs of the fiber cross and longitudinal view (scanning electron microscopy) *		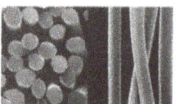

* For cotton and modal fibers images [9].

Many scientific papers have analyzed the improvement of the properties of socks using different types of fibers, e.g., cotton, wool, acrylic or their blends, and either blended or not with polyester, polyamide and/or elastane [1,5,10–15]. Few authors include viscose and bamboo, soybean, flax [16–19] and even reclaimed the cotton fibers [20,21] in the analysis of the sock knits. A group of authors used fibers such as modal, micro modal, viscose, bamboo, soybean and chitosan to produce plain single jersey sock samples to compare their comfort-related properties with those of cotton socks [22,23]. However, it should be noted that polyamide and elastane were not added, in contrast to the commercial production conditions for socks. The results indicated that the fiber type, together with the moisture regain and fabric thickness, affected the comfort properties in the socks knits, with modal, viscose, chitosan and seacell fabrics performing relatively better than the others. Compared to micro modal sock fabrics, the modal sample was found to have a lower thermal resistance, suggesting that microfiber fabrics have a lower thermal conductivity and, thus, better insulation properties.

Different types of yarns are used for socks, the most important of which are the main yarn and the plating yarn. Socks are usually knitted from single spun main yarns using a conventional ring spinning system. Unconventional rotor and air-jet spinning technologies result in yarns with different structures and properties [24]. Numerous comparative studies have been published on the properties of ring, rotor and air-jet spun yarns and the knitted fabrics made from them [24–28]. Air-jet spun yarns exhibit a lower hairiness and fewer irregularities because they have a uniformly distributed layer of wrapped fibers compared to yarns produced with rotor and ring spinning systems. Consequently, fabrics made from air-jet spun yarns are known to have a very smooth handle and are less prone to surface pilling. Ring spun yarns are hairier but more compactly structured and have a higher tenacity than rotor and air-jet spun yarns. Therefore, knitted fabrics made from ring spun yarns have a better abrasion resistance [29]. However, the applicability of unconventionally spun yarns in the knitting of socks has not yet been sufficiently researched. A few published studies have reported the effect of unconventionally spun main yarn on the properties of sock fabrics, such as a study aimed at determining the relative effect of the wool fiber fineness, yarn type (ring two-ply conventional high twist, unconventional low twist and single conventional) and fabric structure on the thermal and moisture transfer properties, cyclic compression and friction properties of sock fabrics [30–32]. It was found that the effect of the yarn type was directly related to the physical properties of the fabric, especially the fabric thickness. The thermal resistance of the fabrics composed of ring two-ply yarns with a low twist was higher [31]. The thermal comfort and physical properties—as well as the air permeability, bursting strength and pilling grade of the socks made from two types of cotton open-end rotor spun yarns, one made from a blend of reclaimed cotton fibers and polyester, and the other made from cotton fibers—were investigated along with the influence of elastane [20,21].

The knitted structures used for socks must have an adequate elasticity and fit for the feet and legs. The very stretchy rib knit structures (e.g., rib 1×1, rib 2×2) are generally used for the top part and rarely for the leg part of casual socks, while the plain knit stich is used in the foot part [1,4]. In this case, due to the advantageous properties contributing to socks, the elasticity in the leg and foot parts is provided using synthetic highly elastic plating yarns. Usually, polyester, polyamide and/or elastane yarns are used [3,20].

It is well known that the air, moisture and heat transfer through the garment are the most important factors in clothing comfort [19]. Socks should provide more comfort than other garments as there is less air circulation in socks and in shoes [2,33]. Air permeability is an important comfort factor for transporting moisture or vapor from the skin. Air permeability is a function of the thickness and surface porosity of the knit [33,34] and ensures the breathability of the socks.

The moisture regain is a parameter that quantifies the hydrophilicity of a textile material and depends on the relative humidity and temperature. The swelling properties are related to the moisture regain; the more hydroscopic a fiber is, the more it swells with water [35]. The absorption capacity of textiles has been reported to be influenced by many factors, including the fiber chemistry and morphology, fabric thickness and structure, yarn type, surface properties of the fibers and fabrics, size and shape of the yarn and fiber interstices and finishing treatments of the fibers, yarns and fabrics [30].

The thermal resistance of textiles is influenced by the fiber conductivity, fabric porosity and fabric structure [16]. The thickness of the fabric and the volume of the trapped air generally determine the thermal resistance of the sock fabric under static conditions, with less effect from the fiber itself. Thinner fabric traps less air and is, therefore, typically less insulating than thicker fabric. The effect of the fibers on the thermal resistance is usually limited due to their packing density in the fabric structure, which contributes to the volume of the trapped air [30,36]. In the case of socks, the extensibility of the knitted fabric at the wearer's foot must also be taken into account [3]. Given the increasing demand for comfortable clothing, numerous studies have been conducted on the thermal resistance of fabrics. Thermal foot manikins, available since the late 1990s, have been used primarily to study the thermal resistance of footwear. Despite the important contribution of the sock to the thermal protection provided by the footwear, there are significantly fewer studies investigating the thermal resistance of sock fabrics or socks alone [30]. The thermal resistance of socks has been measured under steady state conditions using the thermal foot model [3,16,36] and the sock-shaped hot plate [22,23]. The thermal resistance of the sock fabrics was measured by a standardized method according to EN ISO 11092:2014 using a sweating guarded hot plate [2,30,37]. Since this test requires rectangular seamless samples measuring 510×510 mm, the thermal resistance of the sock fabrics was also measured on the smaller knit samples using the Permetest or Alambeta tester (equivalent to ISO 8301:1991) [16,21] and calculated as the quotient of the fabric thickness and thermal conductivity parameters of the sample [17–19]. Several studies have been conducted on the wearing of socks in humans, generally focusing on the perception or measurement of the thermal comfort and foot dryness directly on the wearer's feet [14,38].

Socks should maintain their quality throughout their lifetime. They must meet high wear resistance requirements, especially a higher abrasion resistance and a lower propensity to surface pilling. Abrasion, which is an unavoidable problem, usually occurs on the heel, sole and toes of the socks. The sock rubs in the shoes, slippers or even on the floor. In the first stage of abrasion, the small pill balls get tangled due to the loose fibers that shed from the knitted surface when worn and washed, resulting in an undesirable appearance and unpleasant hassle. Eventually, the fibers that bind the pills to the surface break down and a hole or thinning occurs [4,15]. The abrasion resistance and propensity to surface pilling of sock knits are influenced by many factors, such as the type of fibers and blends, the structure of the yarns, the construction of the knit and the finish [4,17,39]. Various researchers have studied the abrasion resistance and surface pilling tendency [3,4,15,20,29,33] of sock knits. It has been found that the abrasion resistance of socks knitted in a single jersey structure can be increased by a number of measures: the use of yarns with a higher linear density, yarns with a higher twist coefficient and folded yarns, the addition of polyester fibers in blends and the addition of elastic yarns (polyamide or elastane filaments) to the structure [4,15,33]. In recent research, highly functional socks were knitted in the plain single jersey pattern in full plating using three-ply and two-ply twisted yarns composed of various high-performance fibers with PCMs, insect repellent, bio-ceramic, silver and carbon additives [2,37]. The use of spiral auxiliary yarns consisting of various combinations of high-performance para-aramid, elastomeric, polyamide, polyester and conventional cotton fibers has also been investigated for the development of abrasion-resistant socks [13]. When evaluating the sock-to-skin friction, the plain single jersey structures knitted with two cotton yarns were found to be the most suitable for running socks [35]. A review of the literature shows that no studies were found on the wear resistance of socks knitted with three single spun yarns in full plating using the plain knit stitch.

Dimensional stability is one of the basic requirements for socks [11]. The leg and sole parts can undergo major shrinkage after washing, which affects the usability of the socks due to size mismatching problems [5,10]. However, most published work has been concerned with the dimensional stability of knitted fabrics rather than socks [11]. The dimensional properties of single jersey knitted fabrics are mainly influenced by the constituent fibers, yarn properties, knitting machine variables, processing and finishing [40]. The shrinkage of knitted fabrics was studied considering the knit properties and wet treatments. It was found that fabric shrinkage is strongly influenced by the type of yarn and fiber blend [41]. The dimensional differences between knitted jersey fabrics made from open-end rotor and ring spun yarns were also investigated. The fabric made from open-end yarns had a relatively good dimensional stability [42]. Studies concerning the sock dimensional stability are very few. The effects of three different parameters, namely the linear density of yarn, the loop length and the construction, on the dimensional stability of 100% cotton socks during wet processing were compared. It was found that the loop length had a significant inverse effect on the dimensional stability of the socks [5]. Cotton/polyamide 70/30 and cotton/polyamide/elastane (79/20/1) socks were found to shrink significantly after the first wash. The second wash had no significant effect on the dimensional variations, as the socks had already assumed their fully relaxed dimensions [10]. It was found that, as the elastane content increases, the shrinkage of the socks decreases [11], and that the drying temperature and the external force during the drying time also have a great influence on the dimensional change of the socks [43].

The number of European standards for the testing of knitwear and socks are low [3]. Since the properties of the socks change after domestic care, and the pretreatment of the samples in some standardized test methods is only given as an option without an obligation for use, it is necessary to expand the research in the field for a quality assessment of socks. It should be noted that there is no standard or official procedure specifically for determining the dimensional stability of socks, and there are very few studies on the thermal resistance of socks measured on the thermal foot model alone.

The literature review showed that the applicability of modal fibers and microfibers, as well as the unconventionally spun yarns made from them, in knitting socks has not been sufficiently researched. Additionally, the analysis of the usage durability and comfort-related properties of these kind of socks produced under commercial production conditions with the addition of polyamide and elastane plating yarns, especially those knitted with three single spun yarns in full plating, was not founded in the literature. There are no research results on the influence of conventional cotton yarns used in combination, as well as polyamide plating yarns of different properties. Therefore, this work evaluated three groups of socks knitted in single jersey pattern with the highest percentage of single ring, rotor and air-jet spun modal or micro modal yarns of the same linear density in full plating with different textured polyamide 6.6 yarns. The influence of the different yarn combinations in socks produced under the same conditions was investigated in comparison to the conventional sock samples produced from cotton yarns. The evaluation of the sock quality was carried out through an investigation of the basic physical properties, wear resistance and dimensional stability, as well as the water vapor absorption, air permeability and thermal resistance using the thermal foot model before and after five repeated washing and drying cycles, according to the proposed methodology.

In this context, the aim of this work was to determine the applicability of modal and micro modal yarns for the production of socks, to determine the influence of the differently spun main yarn types and the polyamide plating yarns used on the tested sock properties and to confirm the applicability of the proposed methodology for the evaluation of the sock quality, which includes the implementation of a pretreatment.

2. Materials and Methods

2.1. Sock Production and the Yarns Used

Six differently spun yarns with a nominal linear density of 20 tex were selected as the main yarns for the knitting of socks—conventional single ring spun, unconventional single rotor and air-jet spun yarns, all composed of bright modal staple fibers (MD) with a linear density of 1.3 dtex or modal microfibers (MMD) of 1.0 dtex and a length of 38/40 mm.

- The ring spun yarns (Ri) of the modal and micro modal fibers were produced using the following process: the preparation process (opening, blending and mixing), carding process, spinning preparation (drawing, pre-spinning and ring spinning), winding and cleaning. The yarns were spun on a Zinser 351 ring spinning machine (ring diameter: 42 mm, ring type: f2, spindle speed: 16,500 min^{-1}), wound up and cleaned on an Autoconer X5. The number of twists for the modal ring spun yarns was 746 m^{-1} and 734 m^{-1} for the micro modal spun yarns.
- The rotor spun yarns (Ro) from the modal and micro modal fibers were produced using the fiber preparation process (opening and blending), carding, spinning preparation (drawing) and rotor spinning using a Schlafhorst A8 rotor spinning machine with a rotor diameter of 33 mm. The nominal twist number of the rotor yarns calculated from the rotor speed was 750 twists per meter.
- The air-jet spun yarns (Ai) from the modal and micro modal fibers were produced using the preparation process (opening and blending), the carding process, spinning preparation (three drawing passages) and air-jet spinning on a Rieter J20 machine with an inner spindle diameter of 1.2 mm. The air-jet yarn twists were determined according to the high pressure of the air in the rotating vortex at 0.6 MPa [27].

Figure 2 shows the longitudinal view with the characteristic surface structure and twists for the modal and micro modal ring, rotor and air-jet spun yarns used.

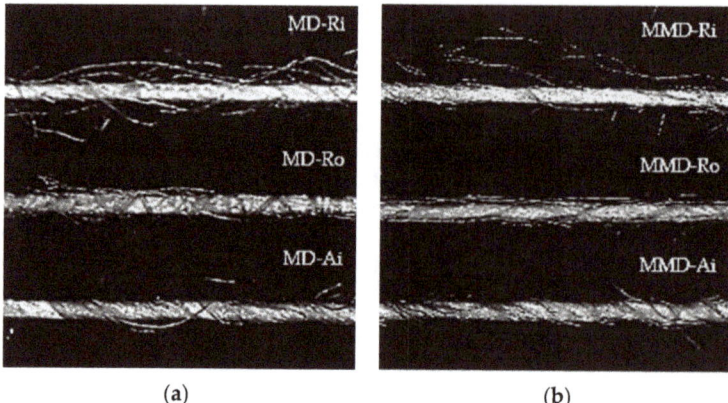

Figure 2. Optical microscopy images of the yarns' longitudinal view (magnification 60×): (**a**) modal ring (MD-Ri), rotor (MD-Ro) and air-jet (MD-Ai) spun yarns; (**b**) micro modal ring (MMD-Ri), rotor (MMD-Ro) and air-jet (MMD-Ai) spun yarns.

Two textured multifilament polyamide 6.6 plating yarns with different properties (PA 6.6 (1); designation: PA 6.6 156 dtex f 42 and PA 6.6 (2); designation: PA 6.6 220 dtex f 68) and one single Lycra with a linear density of 54 tex (usually used in commercial production) were selected as the plating yarns. A polyamide plating yarn was knit into the toe, heel, foot, leg and top part of the socks for reinforcement and support, along with the three single spun yarns in each knit course. An additional single Lycra yarn was knitted into the cuff at the top of the socks only to keep them from falling.

With the aim of determining the influence of the differently spun main yarn types, the selected polyamide plating yarns and the conventional cotton yarn used in combination with the sock properties, three groups (A, B and C) of medium sized calf length socks were designed and produced.

- Sock group A, knitted with three single spun modal or micro modal yarns of the same type (spun by a ring, rotor or air-jet spinning system) in full plating with a polyamide PA 6.6 (1) yarn
- Sock group B, knitted with three single spun modal or micro modal yarns of the same type (spun by a ring, rotor or air-jet spinning system) in full plating with a polyamide PA 6.6 (2) yarn
- Sock group C, in which one of the single spun modal or micro modal yarns was replaced by a coarser cotton ring spun yarn (CO-Ri (2)). This group of socks was, thus, knitted with two single spun modal or micro modal yarns of the same type (spun by a ring, rotor or air-jet spinning system) in combination with a cotton ring spun yarn in full plating with a polyamide PA 6.6 (2) yarn.

The influence of the different yarn combinations in the socks was investigated in comparison to the conventional sock samples (of groups A, B and C) produced under the same conditions using a single cotton ring spun yarn (CO-Ri (1)) with a nominal linear density of 20 tex. The properties of all the yarns used for the knitting of socks are given in Table 2.

A description of the socks produced, including the values of the fiber content in the specific parts of the sock knitted structure, is given in Table 3.

All the sock samples of the same size (EU 42) were knitted using a Lonati automatic sock knitting machine with an E9 gauge and a cylinder diameter of 95 mm (3 3/4″) with 108 needles and two knitting systems (Figure 3). After sewing the toes, the socks were ironed and stabilized on flat leg forms using an industrial Cortese ironing machine under the following conditions: pressing for 4 s and steaming (saturated steam at 3.17 kg/m^3)

with a pressure of 0.7 bar for 6 s at a temperature of 110 °C. The plain single jersey pattern was used in the foot and leg part, and a 1 × 1 rib structure was used in the cuff of the socks. The toe seams were placed high over the toes, the toe and heel parts were smooth, and the square heel shape were made.

Table 2. Properties of the yarns used in the knitting of socks.

Yarn Type	Linear Density, Tex	Breaking Force, cN	Breaking Elongation, %	Tenacity, cN/Tex	Work of Rupture, cNcm
Main yarn *					
MD-Ri	20.0	487 ± 10	10.2 ± 0.2	24.3 ± 0.5	1436 ± 47
MD-Ro	20.0	325 ± 9	7.2 ± 0.2	16.3 ± 0.5	738 ± 32
MD-Ai	20.0	406 ± 10	9.0 ± 0.2	20.3 ± 0.5	1067 ± 42
MMD-Ri	20.0	506 ± 11	9.5 ± 0.2	25.3 ± 0.6	1421 ± 50
MMD-Ro	20.0	344 ± 10	7.3 ± 0.2	17.2 ± 0.5	777 ± 42
MMD-Ai	20.0	365 ± 11	8.2 ± 0.2	18.2 ± 0.5	886 ± 46
CO-Ri (1)	20.0	302 ± 5	3.7 ± 0.1	15.1 ± 0.3	301 ± 10
CO-Ri (2)	25.0	326 ± 8	3.8 ± 0.1	13.3 ± 0.3	333 ± 15
Plating yarn *					
PA 6.6 (1)	15.6	652 ± 8	26.07 ± 0.6	41.8 ± 0.5	4775 ± 180
PA 6.6 (2)	22.0	991 ± 4	28.5 ± 0.2	45.0 ± 0.2	7846 ± 81
Lycra	54.0	551 ± 14	321.0 ± 18	10.2 ± 0.4	2467 ± 324

* MD—modal fibers, MMD—micro modal fibers, CO—cotton fibers, PA 6.6—polyamide fibers; Ri—ring spun yarn, Ro—rotor spun yarn, Ai—air-jet spun yarn.

Table 3. Specification and structure characteristics of the socks produced.

Sock Group	Yarn Type		Fiber Content, %	
			Leg and Foot/ Plain Pattern	Cuff/ Rib Pattern
A	Main yarn *	Modal: Ri, Ro or Ai × 3 Micro modal: Ri, Ro or Ai × 3 Cotton: Ri (1) × 3	79 ± 1	55 ± 1
	Plating yarn	Polyamide 6.6 (1) × 1 Lycra × 1	21 ± 1 -	14 ± 1 31 ± 1
B	Main yarn *	Modal: Ri, Ro or Ai × 3 Micro modal: Ri, Ro or Ai × 3 Cotton: Ri (1) × 3	71 ± 1	52 ± 1
	Plating yarn	Polyamide 6.6 (2) × 1 Lycra × 1	29 ± 1 -	19 ± 1 29 ± 1
C	Main yarn *	Modal: Ri, Ro or Ai × 2 Micro modal: Ri, Ro or Ai × 2 Cotton: Ri (1) × 2	44 ± 1	42 ± 1
		+ Cotton: Ri (2) × 1	28 ± 1	18 ± 1
	Plating yarn	Polyamide 6.6 (2) × 1 Lycra × 1	28 ± 1 -	16 ± 1 24 ± 1

* Ri—ring spun yarn, Ro—rotor spun yarn, Ai—air-jet spun yarn.

2.2. Socks Evaluation Methodology

In this paper, the usage durability of socks was evaluated through an investigation of the wear resistance, dimensional stability and comfort-related properties, as well as the water vapor absorption, air permeability and thermal resistance. In addition, the physical properties of the socks consisting of the density parameters, mass and thickness were measured to define their influence on the mentioned properties.

Figure 3. Automatic sock knitting machine, Lonati, Goal FL 626.

Since the properties of socks change after domestic care, and the pretreatment of the samples is only indicated in some standardized test methods as an option without an obligation for use, all the tested properties were measured before and after one and the five repeated washing and drying cycles of the socks, according to the proposed methodology (Figure 4). The socks were washed according to the procedure 3M of EN ISO 6330:2012 at a temperature of 30 °C with mild agitation during heating, washing and rinsing using a non-phosphate ECE reference detergent without an optical brightener in the Electrolux Wascator FOM71 CLS. After each wash cycle, the socks were line dried in the open air (procedure A).

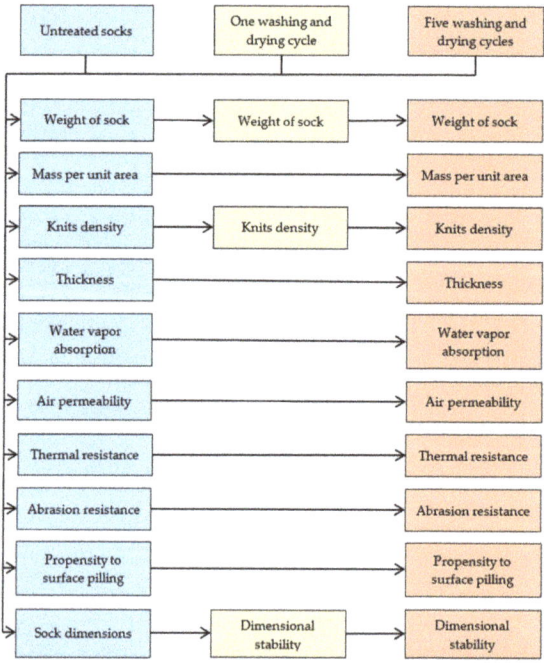

Figure 4. Schematic view of the sock quality evaluation methodology that includes the selected performance properties.

Prior to measurement, the untreated and the pretreated sock samples were conditioned on a flat surface for at least 24 h in a standard atmosphere with a temperature of 20 ± 2 °C and a relative humidity of 65 ± 4%. For all the tests on the sock knits, the socks were cut open and sampled as necessary, as shown in Figure 5.

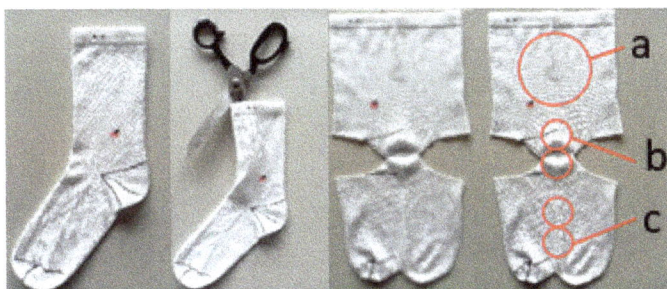

Figure 5. Cutting the socks and the sampling of the test specimens: (**a**) for testing the propensity to surface pilling, mass per unit area and water vapor absorption; (**b**,**c**) for testing the abrasion resistance.

2.2.1. Sock Physical Properties Measurements

The following physical properties of the socks were determined:

- The weight of one sock expressed in grams
- The sock plain knit mass per unit area according to the EN 12127:2003 expressed in g/m^2
- The sock plain knit thickness according to the EN ISO 5084:2003 expressed in millimeters, using the Hess MBV GmbH thickness tester 2000-U
- The sock plain knit density parameters consisting of the number of wales/cm, courses/cm and stitches/cm^2, according to the EN 14971:2008

The number of measurements was five for the mass and density tests and ten for the thickness test, with the mean values given as the result.

2.2.2. Sock Usage Properties Measurements

The wear resistance of the socks was determined by measuring the following properties:

- The sock plain knit abrasion resistance was measured using the Mesdan-Lab Martindale abrasion and pilling tester (Figure 6a) in accordance with the EN 13770:2002, method 1, through the determination of the specimen breakdown, where the plain knit specimens were abraded against the reference wool abrasive fabric with a cyclic planar motion in the form of a Lissajous figure. The SDC Enterprises Limited UK Martindale woven wool abradant with a mass per unit area of 250 g/m^2, as specified in the EN ISO 12947-1:1998+AC:2006, was used for the test. Two circular specimens with a diameter of 38 ± 5 mm were taken from the heel and two from the sole of the socks (Figure 5). During the test, the specimens were stretched over a flattened rubber surface of the holders and loaded with the corresponding weight of 12 kPa. The endpoint was defined as the occurrence of the specimen breakage (the breakage of the thread in the knitted structure, usually resulting in a hole) or significant thinning (wear of the main spun yarns), which was periodically checked. During the inspection, the pills were removed using sharp scissors with curved blades. The number of rubs until the endpoint was reached was recorded.
- The sock plain knit propensity to surface pilling was measured according to the EN ISO 12945-2:2020 using the modified Martindale method (Figure 6a). Three specimens were rubbed according to the Lissajous figure against the same reference wool abradant loaded with the corresponding weight of 415 g. The specimens were cut in a circular shape with a diameter of 140 ± 5 mm and taken from the socks leg part (Figure 5). During the test, the specimens were visually assessed after 125, 500, 1000, 2000,

5000 and 7000 pilling rubs, according to the EN ISO 12945-4:2020, with grades 1 to 5 corresponding to the appropriate pilling degrees. Each specimen was evaluated separately by three experts and the result was expressed as a mean value.

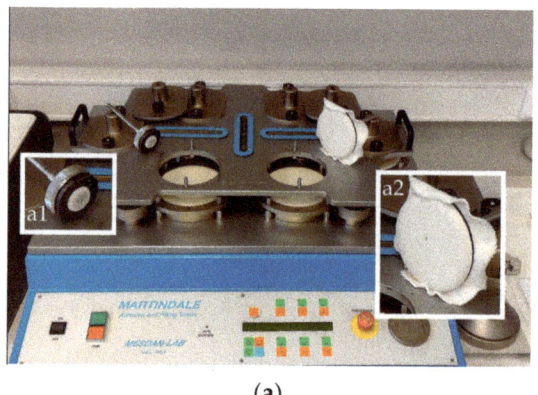 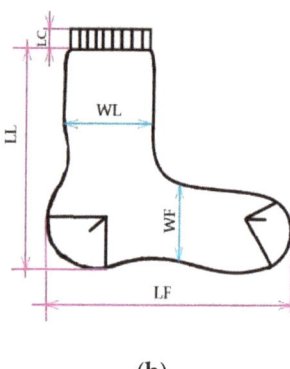

(a) (b)

Figure 6. Evaluation of the usage properties of socks: (**a**) the device for determining the wear resistance of sock knits, Mesdan-Lab Martindale abrasion and pilling tester 2561 E, with (a1) the specimen holder for the abrasion resistance test and (a2) the specimen holder for the pilling test; (**b**) the measurement scheme for the dimensions of the socks: length of cuff, leg and foot part (LC, LL and LF); width of leg and foot part (WL and WF).

The EN ISO 3759:2011 and EN ISO 5077:2008 standards, normally used for ordinary knitted fabrics, was adopted to determine the dimensional variation of the socks after one and five consecutive washing and drying cycles. The length and width dimensions, measured at the top, foot and leg of the untreated and pretreated socks were determined according to the specification shown in Figure 6b. The percentage change in the length and width of the socks was calculated, and for the condition of whether the dimension decreased, the shrinkage was expressed as a minus.

2.2.3. Socks Comfort Properties Measurements

The following comfort-related properties of the socks were studied:

- The water vapor absorption of the sock plain knits was determined according to the ASTM D 2654-89a. The circular specimens of the plain knits with an area of 100 cm^2 were cut from the leg part of the conditioned socks, weighed, then dried in an oven at 105 ± 2 °C for 24 h and reweighed. The difference between the mass of the conditioned and the mass of the oven-dried specimens was calculated as the moisture regain and expressed as a percentage. The mean value of three measurements was given as the result.
- The permeability of the sock plain knits to air was determined according to the EN ISO 9237:1995 using the air permeability tester shown in Figure 7a. The arithmetic mean of the individual air flow readings in a test area of 5 cm^2 and a pressure drop of 100 Pa was noted. The air permeability was calculated according to Equation (1)

$$R = \frac{\overline{qv}}{A} \cdot 167 \qquad (1)$$

where qv is the arithmetic mean of the air flow expressed in dm^3/min, A is the test surface area expressed in cm^2 and 167 is the conversion factor from dm^3/min cm^2 to mm/s.

- The thermal resistance of the socks was defined as the ability of the socks to resist the heat flow through their knitted structure, using the thermal foot manikin system

(Figure 7b). It consisted of the thermal foot (EU size 42), a stainless steel support structure, shock absorbers and a heating subsystem. The heating subsystem, controlled by the personal computer was connected to the thermal foot using highly flexible cables. The thermal foot was divided into 13 silver alloy surface segments that were independently heated to a temperature of 35 ± 0.5 °C. Since the heaters and temperature sensors were installed in each segment, the thermal resistance on each segment or the total resistance could be determined using a special algorithm. The two upper segments were excluded from the measurement, so the total measurement area at the thermal foot was 88.190 mm². The apparatus constant (R_{ct0}) needed to be determined first under the defined environmental conditions of 20 ± 2 °C air temperature, 65 ± 4% relative humidity, an air speed of 1 m/s. The sock to be tested was then placed on thermal foot and the total thermal resistance (R_{ctt}) of the apparatus and the sock was measured. The thermal resistance of the tested sock sample (R_{ct}) was calculated from the difference between R_{ctt} and R_{ct0} according to Equation (2)

$$R_{ct} = R_{ctt} - R_{ct0} \qquad (2)$$

where R_{ct} is the thermal resistance of the tested sock, R_{ctt} is the total thermal resistance of the apparatus and the sock and R_{ct0} is the thermal resistance of the apparatus (thermal foot), all in m² °C/W. As a result, the mean value of the measurements on the three sock samples of the same group was provided.

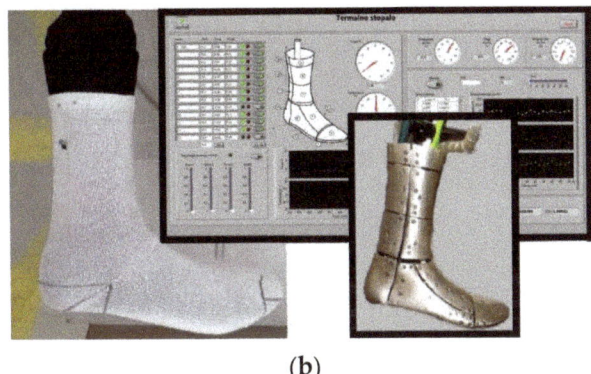

(a) (b)

Figure 7. Evaluation of the comfort-related properties of socks: (**a**) the device for determining the air permeability of the sock knits, Air Tronic Mesdan S.p.A.; (**b**) the thermal foot manikin system.

3. Results and Discussion

The results include the measured physical, usage and comfort properties of the designed socks. The influence of the different spun main yarn types, polyamide plating yarns and conventional cotton yarns in combination with the sock properties was discussed.

3.1. Sock Physical Properties

The results of the measured physical properties of the untreated and pretreated sock samples (determined before and after the five repeated washing and drying cycles) with the corresponding standard deviation are shown in Figures 8–10 and in Table 4. Table 5 contains a comparison of the mean values of the physical properties of the untreated and pretreated socks (made from differently spun modal or micro modal yarn) calculated for groups A, B and C with the corresponding standard deviation.

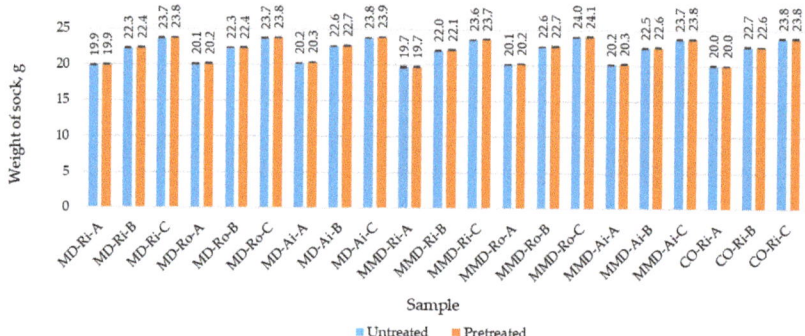

Figure 8. Weight of the untreated and pretreated sock samples determined before and after five repeated washing and drying cycles (where MD—modal fibers, MMD—micro modal fibers, CO—cotton fibers; Ri—ring spun yarn, Ro—rotor spun yarn, Ai—air-jet spun yarn; A, B, C—sock group).

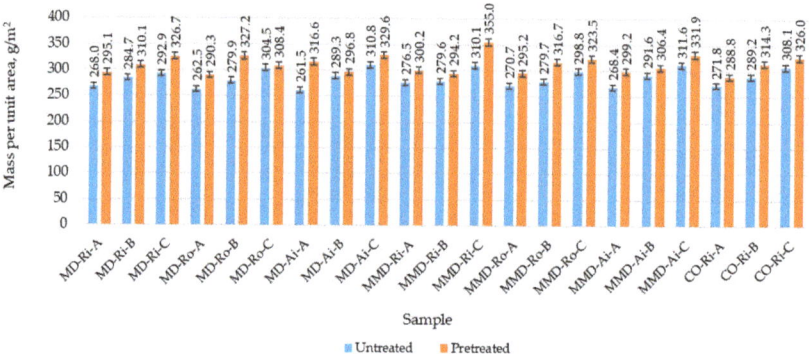

Figure 9. Mass per unit area of the untreated and pretreated sock plain knit samples determined before and after five repeated washing and drying cycles of the socks (where MD—modal fibers, MMD—micro modal fibers, CO—cotton fibers; Ri—ring spun yarn, Ro—rotor spun yarn, Ai—air-jet spun yarn; A, B, C—sock group).

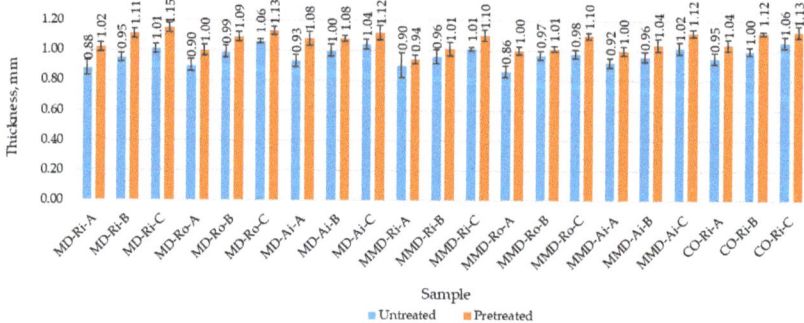

Figure 10. Thickness of the untreated and pretreated sock plain knit samples determined before and after five repeated washing and drying cycles of the socks (where MD—modal fibers, MMD—micro modal fibers, CO—cotton fibers; Ri—ring spun yarn, Ro—rotor spun yarn, Ai—air-jet spun yarn; A, B, C—sock group).

Table 4. The untreated and pretreated sock plain knit samples' density parameters (number of wales per cm, courses per cm and stitches per cm^2) determined before and after five repeated washing and drying cycles of the socks.

Sock Sample *	Wales, cm^{-1}		Courses, cm^{-1}		Stitches, cm^{-2}	
	Untreated	Pretreated	Untreated	Pretreated	Untreated	Pretreated
MD-Ri-A	6	7	7	8	42	56
MD-Ri-B	6	7	7	9	42	63
MD-Ri-C	6	6	7	8	42	48
MD-Ro-A	6	6	8	9	48	54
MD-Ro-B	6	6	8	9	48	54
MD-Ro-C	6	6	7	9	42	54
MD-Ai-A	6	6	7	9	42	54
MD-Ai-B	6	6	7	8	42	48
MD-Ai-C	6	6	7	8	42	48
MMD-Ri-A	6	7	8	8	48	56
MMD-Ri-B	6	7	8	8	48	56
MMD-Ri-C	6	6	8	8	48	48
MMD-Ro-A	6	6	8	9	48	54
MMD-Ro-B	6	7	7	8	42	56
MMD-Ro-C	6	6	7	8	42	48
MMD-Ai-A	6	6	8	9	48	54
MMD-Ai-B	6	6	7	9	42	54
MMD-Ai-C	6	6	7	8	42	48
CO-Ri-A	6	6	8	9	48	54
CO-Ri-B	6	6	8	9	48	54
CO-Ri-C	6	6	8	8	48	48

* MD—modal fibers, MMD—micro modal fibers, CO—cotton fibers; Ri—ring spun yarn, Ro—rotor spun yarn, Ai—air-jet spun yarn; A, B, C—sock group.

Table 5. Mean values of the physical properties calculated for groups A, B and C of the untreated and pretreated socks made from differently spun modal and micro modal yarns.

Sock Group *	Weight of Sock, g		Mass per Unit Area, g/m^2		Thickness, mm	
	Untreated	Pretreated	Untreated	Pretreated	Untreated	Pretreated
MD-A	20.1 ± 0.14	20.1 ± 0.16	264.0 ± 2.86	300.6 ± 11.43	0.90 ± 0.021	1.03 ± 0.034
MD-B	22.4 ± 0.16	22.5 ± 0.17	284.6 ± 3.86	311.4 ± 12.45	0.98 ± 0.022	1.09 ± 0.012
MD-C	23.8 ± 0.04	23.8 ± 0.05	302.7 ± 7.41	321.6 ± 9.37	1.04 ± 0.021	1.13 ± 0.012
MMD-A	20.0 ± 0.22	20.1 ± 0.25	271.9 ± 3.37	298.2 ± 2.16	0.89 ± 0.025	0.98 ± 0.028
MMD-B	22.4 ± 0.26	22.5 ± 0.25	283.6 ± 5.63	305.8 ± 9.21	0.96 ± 0.005	1.02 ± 0.014
MMD-C	23.8 ± 0.17	23.8 ± 0.17	306.8 ± 5.70	336.8 ± 13.33	1.00 ± 0.017	1.11 ± 0.009

* MD—modal fibers, MMD—micro modal fibers, A, B, C—sock group.

As can be seen in Figure 8 and Table 5, all the sock samples of group A made from modal and micro modal yarns had a lower weight than the socks in groups B and C. The values for the untreated socks were 20.0 g, 22.4 g and 23.8 g, respectively, with almost no variation within the same group compared to the cotton socks made with the highest percentage of ring yarns. This indicates a high quality in the sock production and suggests that the coarser polyamide plating yarn PA 6.6 (2) used in sock groups B and C and the cotton ring spun yarn CO-Ri (2) used in group C change the weight of the socks and their structure in such a way that these socks become thicker and heavier.

Therefore, the lowest values of the thickness and mass per unit area of the untreated sock plain knits made from modal and micro modal yarns were determined for the sock samples in group A and the highest values for the sock samples in group C (Figures 9 and 10). According to the results presented in Table 5, despite the minor variations within the sock groups, there were small differences between the mean thickness values calculated for the same group of socks (A, B and C) made from modal and micro modal yarns (Table 5).

According to the previously published results, the number of modal fibers in the yarn cross-section was between 155 and 156, while the number of microfibers in the yarn cross-section was between 200 to 202. The number of twists was also uniform for all the yarn types; the twist coefficient ranged from 3.280 to 3.350 m^{-1} tex$^{0.5}$ [27]. However, the values of the overall unevenness of the ring, rotor and air-jet spun modal yarns (10.21%, 13.95% and 12.33%, respectively) were higher than the same values of the micro modal yarns (9.67%, 12.69% and 12.12%, respectively). The values of the hairiness of the ring, rotor and air-jet spun modal yarns (6.09, 4.34 and 3.71) were also higher than those of the micro modal yarns (5.28, 4.08 and 3.56). Thus, the hairiness and the overall unevenness of the yarn depend mainly on the spinning technique and the fineness of the fiber. This led us to the conclusion that a more uniform structure of micro modal yarns has a major influence on the slightly lower mean values of the determined thickness (Table 5) for the same sock group.

As can be seen in Figure 10 and Table 5, the mass per unit area results showed a greater variability within and between the same sock groups. This can be explained by the differences in the yarn structure as well as in the structure of the plain knit of the socks, which have a major effect on the mass per unit area of the fabric. Despite the fact that all the socks were knitted under the same conditions, the number of courses/cm of the untreated socks was uneven, which consequently led to uneven changes in the number of stitches/cm^2 in the pretreated samples (Table 4).

After one and five repeated washing and drying cycles, as shown in Figure 8, the weight of the socks made from the highest percentage of differently spun modal, micro modal and conventional cotton yarns did not change significantly. It remained almost the same for the cotton socks, while a minimal increase of up to 0.5% was observed for the socks made from MD and MMD yarns. This can be explained by the full relaxation of the socks after the water pretreatment and the presence of the structural differences, owing to which they have a better ability for absorbing moisture, as shown in Table 1.

From the results presented in Table 4 for the density of the sock plain knits, it can be seen that the number of wales per cm, courses per cm and stitches per cm^2 increased mainly from washing, with most of the changes occurring after one and remaining the same after five pretreatment cycles. It is well known that a knitted fabric is brought to a tension-free state, i.e., a state with minimal energy, using a full relaxation treatment that includes a wet treatment with mechanical agitation and drying [20]. Consequently, after pretreatment, when a sock knit structure approaches its minimum energy, the fabric width and length shrinkage occurs. The increase in the density values of the sock plain knits achieved after the five pretreatment cycles resulted from the sock shrinkage, which was higher in the length of the socks.

After the five repeated washing and drying cycles, the mass per unit area and the thickness largely reflected the density parameters of the sock plain knits. Due to the shrinkage potential of the socks, the socks became more voluminous as the relaxation process progressed, resulting in an increase in the mass per unit area and the thickness of up to 16%, with the same traceability of the results.

3.2. Socks Usage Properties

In this study, the abrasion resistance was estimated from the number of abrasion rubs to the significant thinning of one component in all the untreated and pretreated sock plain knits tested, where the spun staple yarns wore out and the base of the polyamide multifilament plating yarns remained, as shown in Figure 11. The polyamide filaments were more difficult to liberate from the fabric structure and were generally considered highly abrasion resistant, while the cotton and modal fibers were moderately abrasion resistant [2,44].

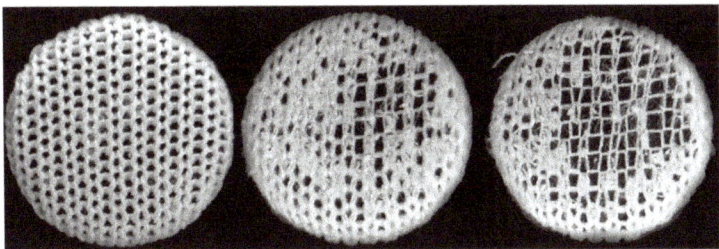

Figure 11. Surface appearance of the untreated heel plain knit sample MD-Ai-A, before, during and at the end of the abrasion resistance test (determined at 0, 8000 and 12,000 abrasion rubs).

The endpoints reached in the plain knits sampled from the heels and soles of the tested socks, are shown in Figure 12a, b. The untreated sock knits made from modal ring spun yarns exhibited better abrasion resistance than those made from rotor and air-jet modal spun yarns, which was particularly noticeable in the sock heel (Figure 12a). Ring spun yarns are hairier but more compactly structured, which means that the fibers are more twisted at their surface than in rotor yarns (Figure 2), which do not promote slight fiber wear. This may also be related to the fact that ring spun yarns have a higher breaking strength, elongation at break, tenacity and work of rupture (Table 2) than air-jet and rotor spun yarns, which means that they can better withstand repeated distortion. In general, for the untreated socks samples in group A produced from modal yarns in full plating with textured polyamide yarn PA 6.6 (1), the specimen breakdown occurred at the highest values of the abrasion rubs recorded. Despite the fact that the polyamide content was increased in sock groups B and C (Table 3), it was found that this had no effect on the improvement of the abrasion resistance. Increasing the fineness of the fibers further changes the properties of sock knits. The use of microfibers in the production of yarns leads to an increase in the number of fibers in the cross-section with a higher cohesion, which—together with the addition of PA 6.6 (2) elastic plating yarn with a higher linear density, breaking force and work at rupture (Table 2)—primarily leads to a better abrasion resistance determined in the untreated sock group B (from ring and rotor spun yarns). In the untreated sock group C, where one of the single spun modal or micro modal yarns was replaced by a cotton ring spun yarn CO-Ri (2), no improvement in the abrasion resistance was observed. When comparing the untreated socks, lower abrasion resistance values were obtained for MD and MMD socks than for conventional cotton socks (especially those with a thicker and heavier structure). The determined abrasion resistance was lowest for the untreated sock plain knits with the highest proportion of micro modal yarns.

From the results shown in Figure 12a,b, it can be seen that the abrasion resistance increased greatly after the five washing and drying cycles for all the sock samples tested. The pretreated socks were found to increase the abrasion resistance by up to 77% for the socks made from modal fibers, 80% for the socks made from micro modal fibers and up to 56% for cotton socks. Overall, the abrasion resistance of the pretreated socks made from modal and, in most cases, micro modal yarns was comparable to that of conventional cotton socks, with significant thinning occurring within the interval of 30,000 to 45,000 abrasion rubs. During the wet pretreatment, the fibers stick to the surface of the knitted fabric, so that the fabric reaches a tighter state and the movement of the fibers within the yarn is limited. The result is a more stable, thicker and voluminous sock structure. Another parameter affecting the abrasion is the stitch density of the sock plain knit (see Section 3.1). The more stitches per unit area in the fabric, the lower the force on each thread. The plain knits with a tighter structure of socks had a higher abrasion resistance. It can be concluded that the structural abrasion of the pretreated sock samples was lower than the surface abrasion, which was due to the shrinkage of the socks and their full relaxation. The structural changes also influenced the higher elasticity of the tested specimens, which were uniformly stretched over a flattened rubber surface of Martindale specimen holders. This was supported by the fact that most of the pretreated sock samples in group B made from a higher percentage of elastic PA yarn (Table 3) showed a significant increase in the measured endpoints (compared to socks in group A).

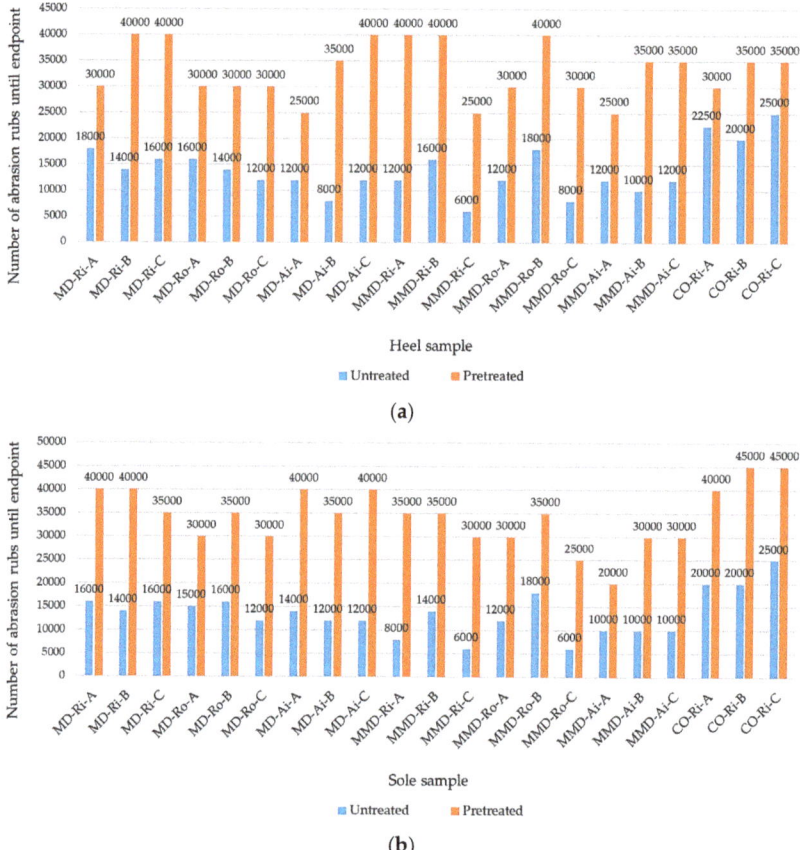

Figure 12. Abrasion resistance of the untreated and pretreated plain knits sampled from (**a**) the heel of the socks; (**b**) the sole of the socks, determined before and after five repeated washing and drying cycles (where MD—modal fibers, MMD—micro modal fibers, CO—cotton fibers; Ri—ring spun yarn, Ro—rotor spun yarn, Ai—air-jet spun yarn; A, B, C—sock group).

According to the EN ISO 12945-4:2020, a pill is defined as the entangling of fibers into balls that protrude from the fabric and are of such a density that light cannot penetrate it and it casts a shadow. This change may occur during washing and/or wearing. In the pretreated sock samples, no pilling was observed on the surface after the five repeated washing and drying cycles. However, during the wear simulation, the propensity to surface pilling increased with the increasing number of abrasion cycles for all the untreated and pretreated sock plain knits, as shown in Table 6.

Table 6. Visually assessed propensity to surface pilling in the untreated and pretreated sock plain knits using grades of pilling.

Sock Sample *	Untreated						Pretreated					
	Number of Pilling Rubs											
	125	500	1000	2000	5000	7000	125	500	1000	2000	5000	7000
MD-Ri-A	4/5	4/5	4/5	4	4	3/4	4/5	4	3	2/3	2	2
MD-Ri-B	4	4	3/4	3/4	3	2/3	4/5	4/5	4	3/4	3	1
MD-Ri-C	3/4	3	2/3	2	1	1	4/5	4	3	2/3	2	1
MD-Ro-A	4/5	4/5	4/5	4	4	3	4/5	4/5	4	3/4	3	2/3
MD-Ro-B	4/5	4	4	3/4	2/3	2/3	4/5	4/5	4	4	3/4	3
MD-Ro-C	4	3/4	3	2/3	2	1	4/5	4/5	4	3/4	3/4	3
MD-Ai-A	5	5	5	5	4/5	4/5	4/5	3/4	3	2/3	2	1
MD-Ai-B	4/5	4/5	4/5	3/4	3	2/3	4/5	3	2/3	2	1	1
MD-Ai-C	4	3/4	3	3	1/2	1	4/5	4/5	3/4	3	2/3	2
MMD-Ri-A	5	5	5	5	4/5	4	4	3	2	2	1/2	1
MMD-Ri-B	5	4/5	4/5	4/5	4	2/3	4/5	3	2/3	2	1/2	1
MMD-Ri-C	3/4	3	2/3	2/3	2	1	4/5	4	3	2/3	2	1
MMD-Ro-A	5	4/5	4/5	4/5	4	3	4/5	4	3/4	3	2/3	2
MMD-Ro-B	5	4/5	4/5	4/5	4	3	4/5	4	3/4	3	2/3	2
MMD-Ro-C	3	2/3	2	1/2	1	1	4/5	4	3	2/3	2	1
MMD-Ai-A	5	5	5	5	4/5	4	4/5	3/4	3	2/3	1/2	1
MMD-Ai-B	5	4/5	4/5	4/5	4	3/4	4/5	3	2/3	2	1	1
MMD-Ai-C	5	4/5	4	3/4	3	2/3	4/5	4	3/4	3	1	1
CO-Ri-A	4/5	4	3/4	3	3	2/3	5	4	3	2/3	1/2	1
CO-Ri-B	4/5	4	4	3/4	3	2/3	4/5	4	3	2/3	2	1/2
CO-Ri-C	4/5	4	3/4	3	3	2	4/5	4	3	3	2	1

* MD—modal fibers, MMD—micro modal fibers, CO—cotton fibers; Ri—ring spun yarn, Ro—rotor spun yarn, Ai—air-jet spun yarn; A, B, C—sock group.

The best rated untreated plain knits were in sock group A, especially those made from air-jet spun modal and micro modal yarns, as they were less hairy, specifically regular and densely structured when compared to ring and rotor spun yarns (Figure 2). These sock plain knits showed only partially formed pills on their surface after 7000 pilling rubs (grade 4 and 4/5). Within sock group A, the pilling tendency of the conventional cotton socks was higher than that of the other tested socks. The results presented in Table 6 show that the coarser polyamide plating yarn PA 6.6 (2) used in the socks in groups B and C, as well as the cotton spun yarn CO -Ri (2) used in the socks in group C, had a negative effect on the tendency of the sock plain knits for surface pilling, resulting in a reduction in the pilling grades.

After 7000 pilling rubs, lower final grades were found for all the tested pretreated sock knit samples, as shown numerically in Table 6 and visually in Table 7 for the MMD-Ai sock samples. The best grades (2, 2/3 and 3), and thus, a lower propensity to surface pilling after the five consecutive domestic washing and drying cycles, were exhibited by the sock knits made from the highest percentage of rotor spun modal and micro modal yarns. These pretreated sock knits exhibited moderate and distinct pilling on their surfaces and, in general, the propensity to surface pilling was less than that of conventional cotton socks. These results of the wear resistance obtained by both tests confirm the justification of the proposed evaluation methodology and the tests performed on the sock samples after simulating domestic care.

Table 7. Surface appearance of the untreated and pretreated sock plain knits made from micro modal air-jet spun yarns (MMD-Ai) at the end of the pilling test.

Sock Sample	MMD-Ai-A	MMD-Ai-B	MMD-Ai-C
Untreated *			
No. Rubs/Pilling Grade	7000/4	7000/3–4	7000/2–3
Pretreated *			
No. Rubs/Pilling Grade	7000/1	5000/1	5000/1

* Optical microscopy images taken with a magnification of 10×.

Socks tend to change their dimensions greatly with repeated washing and drying. In this study, the dimensional changes were observed in the leg and foot length and the leg and foot width of the socks. Major changes were observed after the first pretreatment cycle and remained almost unchanged after five cycles. Therefore, the dimensions measured before and after the five repeated washing and drying cycles in the length and width directions are shown in Figure 13, and the calculated percentage change is shown in Table 8.

From the results shown in Figure 13, it can be seen that the measured dimensions decreased after the five pretreatment cycles for all the sock samples tested, resulting in the shrinkage in both the length and width. The difference in the dimensions between the untreated and pretreated socks, was higher in the lengthwise direction of the socks, as shown in Table 8. The results obtained could be related to the increase in the density values of the pretreated sock plain knits discussed in Section 3.1. The calculated shrinkage of the sock leg length ranged from 4.17% to 13.46%, with the lowest values obtained for the sock samples MD-Ri-A and MMD-Ri-C and the highest for the sock sample MMD-Ai-A. In general, the determined shrinkage was lower and comparable for the leg length of the socks made from the highest percentage of modal and cotton ring yarns (MD-Ri and Co-Ri). The shrinkage of the sock foot length ranged from 4.17% to 18.00% for the MMD-Ai-C and MD-Ri-A sock samples, respectively.

The leg and foot width of the socks measured in several samples remained unchanged after the pretreatment. In most cases, these were the heaviest and thickest socks in group C, as shown in Table 5. Since the elasticity of the sock plain knits depends primarily on the elasticity of the yarn used for the plating, it was found that increasing the percentage of polyamide (in sock groups B and C) mainly reduces the shrinkage of the socks in the width direction. Since the higher values of the transverse elasticity and stable width of the socks provide easier wearing, the present test results point in favor of this.

As the sock construction allows for shrinkage of up to 40% [5], it could be concluded that the dimensional stability of the socks is satisfactory and is strongly influenced by their physical properties and the fiber and yarn blend used for their production.

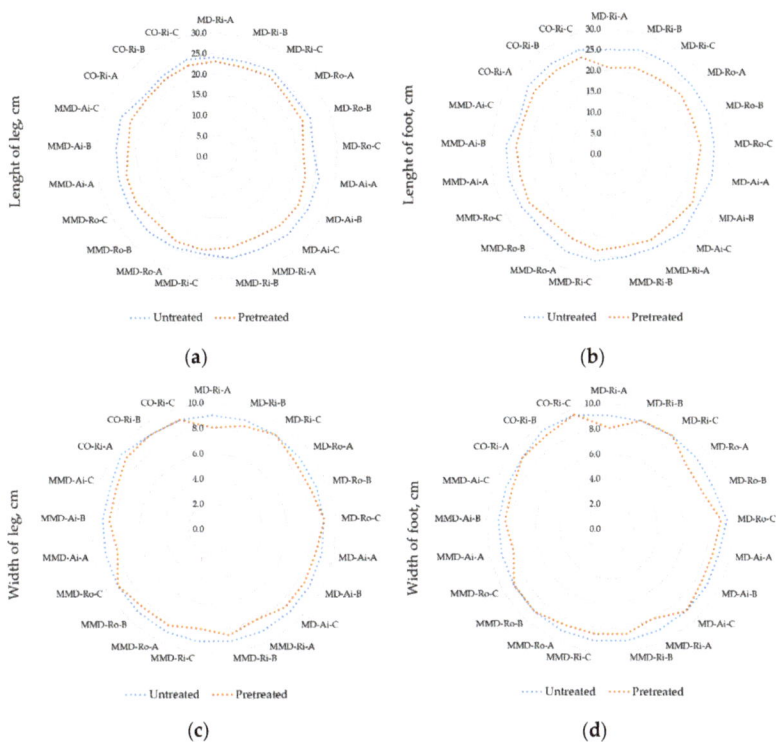

Figure 13. Measured sock dimensions: (**a**) length of the leg part; (**b**) length of the foot part; (**c**) width of the leg part; (**d**) width of the foot part of the untreated and pretreated socks determined before and after five repeated washing and drying cycles (where MD—modal fibers, MMD—micro modal fibers, CO—cotton fibers; Ri—ring spun yarn, Ro—rotor spun yarn, Ai—air-jet spun yarn; A, B, C—sock group).

From the results shown in Figure 14, it can be seen that the determined moisture regain decreased for the thicker and heavier sock plain knits. For the untreated sock samples, the highest values were found in sock group A and the lowest in sock group C. Within the same group of the untreated socks made from the highest percentage of differently spun modal yarns (Table 9), the moisture regain mean value for sock group A was 9.28%, for sock group B was 8.22% and for sock group C was 7.02%. For the socks made from differently spun micro modal yarns, the mean values of the moisture regain were 9.40%, 8.38% and 7.05% for groups A, B and C, respectively, with the lowest values found in the socks made from air-jet spun yarns. In the untreated conventional cotton socks made from ring spun yarns, the water vapor absorption was lower, ranging from 5.91% to 4.68% (Figure 14). Despite the minor variation within the sock groups, the slightly higher mean values for the moisture regain were calculated for the same group of socks (A, B and C) made from micro modal yarns (Table 9). This suggests that a more uniform structure of the micro modal yarns (as discussed in Section 3.1), which has a major influence on the slightly lower mean values of the determined sock knit thickness (Table 5), also influences the higher mean values of the moisture regain determined for the same group of socks (Table 9).

In support of this finding, it was found that, with the increase in the polyamide content in sock groups B and C and the cotton content in group C, the moisture regain values decreased. The differences found between the fiber types were related to the hygroscopicity of the fibers. Modal fibers provided better hydrophilicity than cotton, as shown in Table 1. When measuring the moisture absorption, the fiber type had the greatest influence, as the sock knits made from more hydrophobic polyamide fibers absorbed less moisture than the knits made from the two types of modal, which were hydrophilic. Any difference in the behavior of the moisture properties between the modal and micro modal fibers may be due to the morphology of the fibers and yarns since their affinity for water was the same.

Table 8. Dimensional stability of the socks after five pretreatment cycles.

Sock Sample *	Changes in Dimension, %			
	Leg Length	Foot Length	Leg Width	Foot Width
MD-Ri-A	−4.17	−18.00	−11.112	−11.11
MD-Ri-B	−6.25	−17.31	−5.56	0.00
MD-Ri-C	−6.00	−17.31	0.00	0.00
MD-Ro-A	−6.25	−13.46	−5.56	−11.11
MD-Ro-B	−8.00	−16.98	−5.56	−11.11
MD-Ro-C	−10.41	−13.46	0.00	−5.26
MD-Ai-A	−13.46	−15.38	−5.56	−5.56
MD-Ai-B	−9.61	−6.00	−5.56	−5.56
MD-Ai-C	−11.53	−13.46	−5.56	0.00
MMD-Ri-A	−12.00	−8.00	−11.11	−11.11
MMD-Ri-B	−10.00	−10.00	−5.56	−5.56
MMD-Ri-C	−4.17	−9.61	−11.11	−5.56
MMD-Ro-A	−6.12	−13.46	−5.56	−5.56
MMD-Ro-B	−12.00	−12.00	−5.56	0.00
MMD-Ro-C	−8.00	−7.84	0.00	0.00
MMD-Ai-A	−8.00	−11.53	−11.11	−11.11
MMD-Ai-B	−12.00	−9.61	−5.56	5.56
MMD-Ai-C	−9.80	−4.17	−5.56	−5.56
CO-Ri-A	−4.35	−7.69	−5.56	0.00
CO-Ri-B	−4.25	−7.69	0.00	−5.26
CO-Ri-C	−6.12	−7.69	0.00	0.00

* MD—modal fibers, MMD—micro modal fibers, CO—cotton fibers; Ri—ring spun yarn, Ro—rotor spun yarn, Ai—air-jet spun yarn; A, B, C—sock group.

3.3. Socks Comfort Properties

The results of the measured comfort-related properties of the untreated and pretreated sock samples are shown in Figures 14–16. Table 9 contains a comparison of the mean values of the moisture regain, air permeability and thermal resistance of the untreated and pretreated socks (made from differently spun modal or micro modal yarn) calculated for groups A, B and C with the corresponding standard deviation.

The influence of the structure of the plain knits on the investigated properties was confirmed after the wet pretreatment of the socks, during which the water vapor absorption increased in all the tested sock samples (Figure 14). It was found that, in the pretreated socks, the moisture regain increased up to 27% for the socks made from modal fibers and up to 16.8% for the socks made from micro modal fibers and cotton. The highest increase was found in the sock plain knits in groups B and C (Table 9). This can also be explained by the full relaxation of the socks and the presence of the structural differences from which they provide a better ability for absorbing moisture (as explained in Sections 3.1 and 3.2).

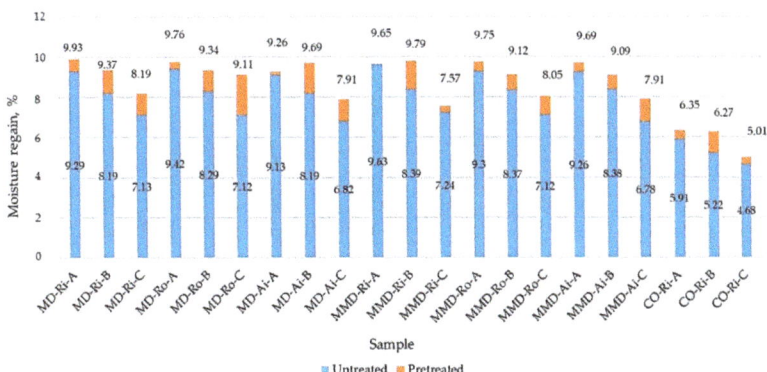

Figure 14. Moisture regain of the untreated and pretreated sock plain knit samples determined before and after five repeated washing and drying cycles of the socks (where MD—modal fibers, MMD—micro modal fibers, CO—cotton fibers; Ri—ring spun yarn, Ro—rotor spun yarn, Ai—air-jet spun yarn; A, B, C—sock group).

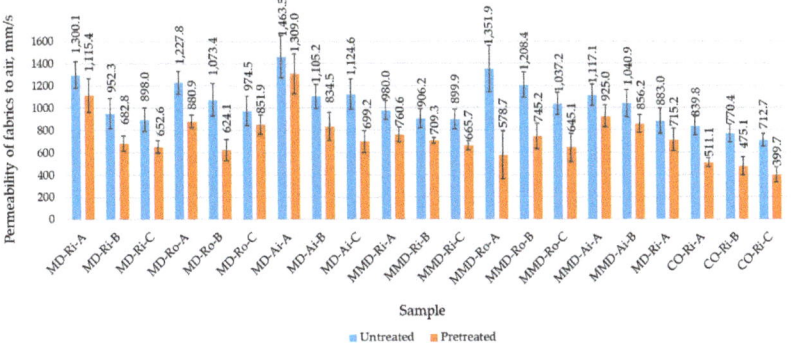

Figure 15. Air permeability of the untreated and pretreated sock plain knit samples determined before and after five repeated washing and drying cycles of the socks (where MD—modal fibers, MMD—micro modal fibers, CO—cotton fibers; Ri—ring spun yarn, Ro—rotor spun yarn, Ai—air-jet spun yarn; A, B, C—sock group).

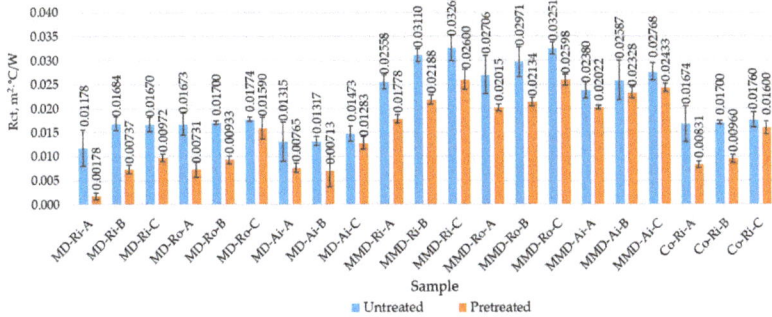

Figure 16. Thermal resistance (Rct) of the untreated and pretreated socks determined before and after five repeated washing and drying cycles (where MD—modal fibers, MMD—micro modal fibers, CO—cotton fibers; Ri—ring spun yarn, Ro—rotor spun yarn, Ai—air-jet spun yarn; A, B, C—sock group).

Table 9. Mean values of the comfort properties calculated for groups A, B and C of the untreated and pretreated socks made from differently spun modal and micro modal yarns.

Sock Group *	Moisture Regain, %		Air Permeability, mm/s		Thermal Resistance, m^2·C/W	
	Untreated	Pretreated	Untreated	Pretreated	Untreated	Pretreated
MD-A	9.28 ± 0.119	9.65 ± 0.284	1330.47 ± 98.59	1101.77 ± 175.04	0.01389 ± 0.00209	0.00558 ± 0.00269
MD-B	8.22 ± 0.047	9.47 ± 0.158	1043.63 ± 65.87	713.80 ± 88.65	0.01567 ± 0.00177	0.00794 ± 0.00099
MD-C	7.02 ± 0.144	8.40 ± 0.513	999.03 ± 94.12	734.57 ± 85.12	0.01639 ± 0.00125	0.01282 ± 0.00252
MMD-A	9.40 ± 0.166	9.70 ± 0.041	1149.67 ± 153.56	754.77 ± 141.44	0.02548 ± 0.00133	0.01938 ± 0.00113
MMD-B	8.38 ± 0.008	9.33 ± 0.323	1051.83 ± 123.61	770.23 ± 62.53	0.02889 ± 0.00221	0.02217 ± 0.00082
MMD-C	7.05 ± 0.195	7.84 ± 0.202	940.03 ± 69.05	675.33 ± 29.42	0.03093 ± 0.00230	0.02544 ± 0.00078

* MD—modal fibers, MMD—micro modal fibers, A, B, C—sock group.

The higher determined values of the water vapor absorption certainly have an effect on the feeling of the higher wearing comfort due to sweating. However, when the fiber swells, the free space in the sock decreases and so does the transport of sweat from the inside to the outside. Therefore, the air permeability is an important comfort factor to transport moisture away from the skin and ensure the breathability of the socks. Since the air permeability also depends on the thickness and surface porosity of the sock knits [33], the results shown in Figure 15 indicate that all the untreated sock samples in group A had the highest values for the air permeability and, thus, the highest porosity and breathability. Increasing the thickness and mass of the sock knits, i.e., the tightness and compactness of their structure, prevented the passage of air and reduced the air permeability in sock groups B and C.

As can be seen in Figure 15 and Table 9, the air permeability results showed a higher variability within the measurements for the sock samples, especially within the same sock group made from differently spun modal and micro modal yarns. This was confirmed by the fact that, in the group of the untreated socks made from modal yarns, the values of the air permeability in sock group A ranged from 1227.8 mm/s to 1463.5 mm/s, in sock group B from 952.3 mm/s to 1208.4 mm/s and in sock group C from 898.0 mm/s to 1124.6 mm/s, with the highest values found for the socks made from air-jet spun yarns. For the untreated socks made from micro modal yarns, the air permeability values in sock group A ranged from 980.0 mm/s to 1351.9 mm/s, in sock group B from 906.2 mm/s to 1105.2 mm/s and in sock group C from 883.0 mm/s to 1037.2 mm/s, with the highest values found for the socks made from rotor spun yarns and the lowest for the socks made from ring spun yarns. In the untreated conventional cotton socks made from ring spun yarns, the permeability of air was lower and in the range of 712.7 mm/s to 839.8 mm/s (Figure 15).

It was found that the socks made from spun yarns with hairy surfaces, such as ring spun yarns, primarily have a lower air permeability than the socks made from rotor and especially air-jet yarns with more uniform surfaces, as shown in Figure 2 and discussed in Section 3.1. This is due to the fact that, as the hairiness of the yarn surface decreases, the protruding fibers become more firmly attached to the yarn and channels (open pores) open up between the neighboring yarn loops in the same area of the sock knit. The untreated socks made from modal yarns are mainly more breathable than the socks made from micro modal yarns, probably due to the different fineness of the fibers and their packing density in the yarns used for knitting. The untreated conventional cotton socks made from ring spun yarns are less breathable compared to MD and MMD socks. This can also be partially attributed to the convolutions in the cotton fibers and their surface properties, as well as the non-smooth cross-sections, as shown in Table 1, that impede the passage of air through the fabric.

After the five repeated washing and drying cycles, the air permeability of all the tested socks plain knits was significantly reduced due to the shrinkage and relaxation of the sock structure (Table 9). It was found that the air permeability of the pretreated socks decreased up to 72% for the socks made from modal yarns, up to 130% for the socks made from micro modal yarns and up to 78% for cotton socks. Reducing the air permeability can be an advantage in colder months. However, since all the socks were elastic due to the addition of plating a polyamide multifilament yarn in each row of knitting, they stretch at the foot of the wearer and the open porosity and breathability of the socks are increased, which can have a positive effect on wearing comfort.

The results of the thermal resistance of the untreated and pretreated socks, determined as the arithmetic mean of the three individual sock measurements with a corresponding standard deviation, are shown in Figure 16.

As presented in Table 5, the mean values of the thickness calculated in groups A, B and C of the untreated socks made from differently spun modal yarns (0.90 mm, 0.98 mm and 1.04 mm, respectively) were minimally higher than the same values of the sock groups made from differently spun micro modal yarns (0.89 mm, 0.96 mm and 1.00 mm).

From the results presented in Table 9, it is clear that the socks in group A had the lowest mean values of the thermal resistance. As the thickness and mass of the socks increased (Table 5), as well as the tightness of their structure, which prevents the passage of heat, the thermal resistance in sock groups B and C increased. It was confirmed that the volume of trapped air determines the ability of the socks to resist the heat flow through their knitted structure. Thinner knits trap less air and are, therefore, less insulating than thicker knits.

The effect of the fiber fineness on the thermal resistance was mostly limited to the packing density in the yarn structure, which contributed to the volume of trapped air. The literature [23] indicated that fabrics made from microfibers have lower thermal conductivity and, thus, better insulation properties. Since micro modal yarns used for the knitting of socks contain a higher number of fibers per unit length than modal yarns, their packing density and surface area are higher. This resulted in higher thermal resistance values for the sock groups with the highest percentage of micro modal yarns compared to those made from modal yarns (Table 9), although the mean thickness values were lower in the MMD sock groups (Table 5).

However, as Figure 16 shows, the thermal resistance varies even within the same sock group. The sock samples made from modal rotor spun yarns (in sock groups A, B and C) had higher thermal resistance values than the socks made from ring and air-jet spun yarns. The finding that the overall unevenness of rotor spun modal and micro modal yarns was higher compared to ring and air-jet spun yarns (as discussed in Section 3.1 and [27]) led to the conclusion that the uneven structure of rotor spun yarns, especially in the outer layer (Figure 2), and the kitted structure produced from them contains a larger volume of air and, therefore, has a major influence on the obtained results. It should also be noted that, in the groups of the untreated socks made from modal and micro modal yarns, the lowest values of thermal resistance were found primarily in the socks made from air-jet spun yarns with the lowest hairiness. This can be related to the previously discussed results of the determined air permeability, as shown in Figure 15. It follows that the sock samples with a higher air permeability and, thus, a higher porosity offer less resistance to the heat flow.

The thermal resistance of conventional cotton socks made from ring spun yarns was higher than that of the socks made from modal ring spun yarns. This could be due to the fact that cotton socks have a slightly thicker structure and, thus, a lower water vapor absorption and air permeability, as shown in Figures 10, 14 and 15.

The foot length of the untreated socks was approx. 20% shorter than the length of the wearer's foot and, after the pretreatment of the socks, it was additionally shortened up to 18%. Due to the observed shrinkage of the pretreated socks, as shown in Table 8, the socks were additionally stretched during the test on the thermal foot (as well as on the wearer's foot), which led to an increase in the porosity of their structure and, thus, to a higher breathability of the socks. This can have a positive effect on the wearing comfort in warmer weather and is reflected in the significantly lower resistance to the passage of heat in all the tested sock samples after the five repeated pretreatment cycles (Figure 16, Table 9).

4. Conclusions

The study described in this paper aimed to determine the applicability of modal and micro modal yarns for the manufacture of socks, to investigate the influence of the differently spun main yarn types and the polyamide plating yarns used on the physical usage and comfort properties of socks and to confirm the proposed methodology for sock evaluation. Three groups of socks were designed—groups A and B with three single spun modal or micro modal yarns of the same type (ring, rotor or air-jet) and linear density, and group C where one of these yarns was replaced by a coarser cotton ring spun yarn. The socks were knitted in a single jersey pattern in full plating with different polyamide 6.6 yarns and compared to conventional cotton socks. From the results presented, the following can be concluded.

The socks samples in group A made from MD and MMD yarns had a lower weight than the socks in groups B and C, with almost no variation within the same group and in comparison with cotton socks. The coarser polyamide plating yarn used in sock groups B and C and the cotton yarn in group C changed their structure, and the socks became thicker and heavier. The more uniform structure of the MMD yarns had a major impact on the slightly lower thickness values obtained in the MMD sock groups. The mass per unit area showed a greater variability within and between the same sock groups due to the differences in the yarn and sock knit structure. After the five repeated washing and drying cycles, the sock weight remained almost the same for cotton socks, while a minimal increase of up to 0.5% was observed for the socks made from MD and MMD yarns. Due to the shrinkage potential, the socks became more voluminous and had higher density, resulting in an increase in the mass per unit area and thickness of up to 16%.

The untreated sock knits made from modal ring spun yarns had a better abrasion resistance than those made from rotor and air-jet modal spun yarns. The untreated MD group A sock samples exhibited the highest abrasion resistance. Although the polyamide content was increased in sock groups B and C, no improvement in the abrasion resistance was observed. The abrasion resistance was lower in the untreated MMD socks, with sock group B (made from ring and rotor spun yarns) achieving better results. The abrasion resistance increased greatly after the five repeated pretreatment cycles for all the sock samples tested—up to 77% for MMD socks, up to 80% for MD socks and up to 56% for cotton socks. Overall, the abrasion resistance of the pretreated MD socks and, in most cases, the MMD socks was comparable to that of conventional cotton socks, with significant thinning occurring in the interval from 30,000 to 45,000 abrasion rubs.

The group A untreated socks exhibited the best pilling resistance, especially those made from air-spun MD and MMD yarns. These untreated socks showed only partially formed pills on the surface after 7000 pilling rubs. Within sock group A, the pilling tendency was higher in the conventional cotton socks. The polyamide plating yarn used in sock groups B and C and the cotton yarn used in the group C socks had a negative influence on the obtained results. Lower final grades were found in all the tested pretreated socks, with the best grades obtained by the MD and MMD socks made from rotor spun yarns.

After the five pretreatment cycles, the shrinkage in the lengthwise direction of the socks was higher in all the tested socks. The determined shrinkage in the leg and foot length was maximum 13.5% and 18.0%, respectively. The determined shrinkage was lower in leg length for the socks made from MD and CO ring yarns. Increasing the polyamide content (in sock groups B and C) primarily reduced the shrinkage of the socks in the width direction.

The moisture regain decreased for the thicker and heavier socks in groups B and C, with the lowest values found for the MD and MMD socks made from air-jet spun yarns. Slightly higher mean values for the moisture regain were calculated for all the MMD sock groups. With the increase in the polyamide content in sock groups B and C and the cotton content in group C, the values for the moisture regain values decreased, since modal fibers have a better hydrophilicity than cotton and PA. After the wet pretreatment, the water vapor absorption increased in all the tested socks—up to 27% in MD socks and up to 16.8%

in MMD and cotton socks, which can also be explained by the full relaxation of the socks and the presence of the structural differences from which they provided a better ability for absorbing moisture.

Increasing the thickness and mass of the sock knits, i.e., the tightness and compactness of their structure, decreased the air permeability in sock groups B and C compared to group A. The results showed a greater variability within the same sock groups, with the highest values found in the untreated MD socks for the socks made from more uniform air-jet spun yarns. The untreated MMD socks had a lower air permeability, with the highest values found for the socks made from rotor yarn and the lowest for the socks made from ring spun yarn with a hairy surface. The untreated MMD socks were more breathable, primarily due to the different fineness of the fibers and their packing density in the yarns used for knitting. The untreated cotton socks were less breathable compared to the MD and MMD socks. The pretreated socks had a significantly lower air permeability—up to 72% for MD socks, up to 130% for MMD socks and up to 78% for cotton socks.

The socks in group A had the lowest mean thermal resistance values. As the thickness and mass of the sock knits increased, the thermal resistance in sock groups B and C increased. The packing density and surface area of MMD yarns are higher, resulting in a higher thermal resistance in the MMD socks, although their average thickness was lower compared to the MD socks. The MD sock samples from rotor spun yarns with a more uneven structure had higher thermal resistance values. The lowest thermal resistance values were found primarily in the socks made from air-jet spun yarns with the lowest hairiness. Due to the observed shrinkage, the pretreated socks were additionally stretched during the test, which was reflected in the significantly lower thermal resistance.

The results obtained in all the tests confirmed the justification of the proposed evaluation methodology and the tests performed on the sock samples after the simulated domestic care.

The applicability of unconventionally spun modal and micro modal yarns in knitting socks was confirmed through the comparison with conventional cotton socks. It is expected that the outcome of this study will be used in the selection of fibers and yarns for the production of socks with specific properties. Future studies will investigate the applicability of the various innovative blends of viscose and lyocell fibers and the yarns made from them using the unconventional spinning processes for sock knitting.

Author Contributions: Conceptualization, A.T.; methodology, A.T., J.Ž. and I.M.; formal analysis, A.T.; investigation, I.M. and J.Ž.; resources, A.T.; data curation, J.Ž and I.M.; writing—original draft preparation, A.T. and J.Ž.; writing—review and editing, A.T., J.Ž. and I.M.; visualization, A.T. and J.Ž.; supervision, A.T.; project administration, A.T. and J.Ž.; funding acquisition, A.T. All authors have read and agreed to the published version of the manuscript.

Funding: This research was funded by the Croatian Science Foundation within the Project IP-2016-06-5278 "Comfort and antimicrobial properties of textiles and footwear", as well as by the University of Zagreb within the research grant TP 20/21 "High functionality textiles and possibilities of objective evaluation".

Institutional Review Board Statement: Not applicable.

Informed Consent Statement: Not applicable.

Data Availability Statement: Data available in a publicly accessible repository.

Acknowledgments: The authors acknowledge Zenun Skenderi and Zlatko Vrljičak, the University of Zagreb Faculty of Textile Technology and Ivan Kraljević, Jadran Hosiery Zagreb for their great technical support.

Conflicts of Interest: The authors declare no conflict of interest.

References

1. Tsujisaka, T.; Azuma, Y.; Matsumoto, Y.-I.; Morooka, H. Comfort pressure of the top part of men's socks. *Text. Res. J.* **2004**, *74*, 598–602. [CrossRef]
2. Stygienė, L.; Varnaitė-Žuravliova, S.; Abraitienė, A.; Sankauskaite, A.; Skurkyte-Papieviene, V.; Krauledas, S.; Mažeika, S. Development, investigation and evaluation of smart multifunctional socks. *J. Ind. Text.* **2020**, *51*, 2330–2353. [CrossRef]
3. Tomljenović, A.; Skenderi, Z.; Kraljević, I.; Živičnjak, J. Durability and comfort assessment of casual male socks. In *Advances in Applied Research on Textile and Materials-IX. CIRATM 2020*; Msahli, S., Debbabi, F., Eds.; Springer Proceedings in Materials; Springer Nature Switzerland AG: Cham, Switzerland, 2022; Volume 17, pp. 210–215.
4. El-Dessouki, H.A. A study on abrasion characteristics and pilling performance of socks. *Int. Des. J.* **2014**, *4*, 229–234.
5. Khan, A.; Ahmad, S.; Amjad, A.; Khan, I.A.; Ibraheem, W. Development of a statistical model for predicting the dimensional stability of socks during wet processing. *J. Textile Sci. Eng.* **2017**, *7*, 1000304.
6. Wulfhorst, B.; Külter, H. Fiber tables according to P.-A. Koch. Cotton. *Chem. Fibers Int.* **1989**, *39*, 12–34.
7. Albrecht, W.; Wulfhorst, B.; Külter, H. Fiber tables according to P.-A. Koch. Regenerated cellulose fibers. *Chem. Fibers Int.* **1991**, *40*, 26–44.
8. Albrecht, W.; Reintjes, M.; Wulfhorst, B. Fiber tables according to P.-A. Koch. Lyocell fibers (Alternative regenerated cellulose fibers). *Chem. Fibers Int.* **1997**, *47*, 298–304.
9. CEN ISO/TR 11827:2016; Textiles-Composition Testing-Identification of Fibres. CEN: Brussels, Belgium, 2016.
10. Abdessalem, S.B.; Abidi, F.; Mokhtar, S.; Elmarzougui, S. Dimensional Stability of Men's Socks. *Res. J. Text. Appar.* **2008**, *12*, 61–69. [CrossRef]
11. Basra, S.A.; Asfand, N.; Azam, Z.; Iftikhar, K.; Irshad, M.A. Analysis of the factors affecting the dimensional stability of socks using full-factorial experimental design method. *J. Eng. Fiber. Fabr.* **2020**, *15*, 1–10. [CrossRef]
12. Soltanzade, Z.; Najar, S.S.; Haghpanahi, M.; Mohajeri-Tehrani, M.R. Effect of socks structure on planar dynamic pressure distribution. *Proc. Inst. Mech. Eng. H J. Eng. Med.* **2016**, *230*, 1043–1050. [CrossRef]
13. Khalid, R.; Jamshaid, H.; Mishra, R.; Ma, P.; Zhu, G. Performance analysis of socks produced by auxetic yarns for protective applications. *J. Ind. Text.* **2022**, *51*, 6838–6863. [CrossRef]
14. West, A.M.; Havenith, G.; Hodder, S. Are running socks beneficial for comfort? The role of the sock and sock fiber type on shoe microclimate and subjective evaluations. *Text. Res. J.* **2021**, *91*, 1698–1712. [CrossRef]
15. Özdıˆl, N.; Marmarali, A.; Oğlakcioğlu, N. The abrasion resistance of socks. *Int. J. Cloth. Sci. Technol.* **2009**, *21*, 56–63. [CrossRef]
16. Mansoor, T.; Hes, L.; Bajzik, V.; Noman, M.T. Novel method on thermal resistance prediction and thermo-physiological comfort of socks in a wet state. *Text. Res. J.* **2020**, *90*, 1987–2006. [CrossRef]
17. Hashan, M.; Hasan, K.M.F.; Khandaker, F.R.; Karmaker, K.C.; Deng, Z.; Zilani, M.J. Functional properties improvement of socks items using different types of yarn. *Int. J. Text. Sci.* **2017**, *6*, 34–42.
18. Čiukas, R.; Abramavičiūtė, J.; Kerpauskas, P. Investigation of the thermal properties of socks knitted from yarns with peculiar properties. Part I. Thermal conductivity coefficient of socks knitted from natural and synthetic textured yarns. *Fibres Text. East. Eur.* **2010**, *18*, 89–93.
19. Čiukas, R.; Abramavičiūtė, J.; Kerpauskas, P. Investigation of the thermal properties of socks knitted from yarns with peculiar properties. Part II: Thermal resistance of socks knitted from natural and stretch yarns. *Fibres Text. East. Eur.* **2011**, *19*, 64–68.
20. Demiroz Gun, A.; Nur Akturk, H.; Sevkan Macit, A.; Alan, G. Dimensional and physical properties of socks made from reclaimed fibre. *J. Text. Inst.* **2014**, *105*, 1108–1117.
21. Gun, A.D.; Alan, G.; Macit, A.S. Thermal properties of socks made from reclaimed fibre. *J. Text. Inst.* **2015**, *107*, 1112–1121. [CrossRef]
22. Cimilli, S.; Nergis, B.U.; Candan, C.; Özdemir, M. A comparative study of some comfort-related properties of socks of different fiber types. *Text. Res. J.* **2010**, *80*, 948–957. [CrossRef]
23. Avcı, H.; Özdemir, M.; İridağ, B.Y.; Duru, C.S.; Candan, C. Comfort properties of socks from seacell fibers. *J. Text. Inst.* **2017**, *109*, 419–425. [CrossRef]
24. Iqbal, S.; Eldeeb, M.; Ahmad, Z.; Mazari, A. Comparative study on viscose yarn and knitted fabric made from open end and rieter airjet spinning system. *Tekst. Konfeksiyon* **2017**, *27*, 234–240.
25. Rameshkumar, C.; Anandkumar, P.; Senthilnathan, P.; Jeevitha, R.; Anbumani, N. Comparitive studies on ring rotor and vortex yarn knitted fabrics. *AUTEX Res. J.* **2008**, *8*, 100–105.
26. Tripathi, L.; Behera, B.K. Comparative studies on ring, compact and vortex yarns and fabrics. *J. Text. Sci. Fash. Technol.* **2020**, *6*, 1–14.
27. Skenderi, Z.; Kopitar, D.; Ercegović Ražić, S.; Iveković, G. Study on physical-mechanical parameters of ring-, rotor- and air-jet-spun modal and micro modal yarns. *Tekstilec* **2019**, *62*, 42–53. [CrossRef]
28. Kim, H.A.; Kim, S.J. Mechanical Properties of Micro Modal Air Vortex Yarns and the Tactile Wear Comfort of Knitted Fabrics. *Fibers Polym.* **2018**, *19*, 211–218. [CrossRef]
29. Özdil, N.; Özçelik Kayseri, G.; Süpüren Mengüç, G. Analysis of Abrasion Characteristic in Textiles. In *Abrasion Resistance of Materials*; Adamiak, M., Ed.; IntechOpen: Rijeka, Croatia, 2012; pp. 119–146.
30. Van Amber, R.R.; Wilson, C.A.; Laing, R.M.; Lowe, B.J.; Niven, B.E. Thermal and moisture transfer properties of sock fabrics differing in fiber type, yarn, and fabric structure. *Text. Res. J.* **2015**, *85*, 1269–1280. [CrossRef]

31. Van Amber, R.R.; Lowe, B.J.; Niven, B.E.; Laing, R.M.; Wilson, C.A.; Collie, S. The effect of fiber type, yarn structure and fabric structure on the frictional characteristics of sock fabrics. *Text. Res. J.* **2015**, *85*, 115–127. [CrossRef]
32. Van Amber, R.R.; Lowe, B.J.; Niven, B.E.; Laing, R.M.; Wilson, C.A. Sock fabrics: Relevance of fiber type, yarn, fabric structure and moisture on cyclic compression. *Text. Res. J.* **2015**, *85*, 26–35. [CrossRef]
33. Mohammad, A.A.; Abdel Megeid, Z.M.; Saleh, S.S.; Abdo, K.M. Studying the performance of men's socks. *Res. J. Text. Appar.* **2012**, *16*, 86–92. [CrossRef]
34. Wang, W.Y.; Hui, K.T.; Kan, C.W.; Maha-In, K.; Pukjaroon, S.; Wanitchottayanont, S.; Mongkholrattanasit, R. An Analysis of Air Permeability of Cotton-Fibre-Based Socks. *Key Eng. Mater.* **2019**, *805*, 76–81. [CrossRef]
35. Baussan, E.; Bueno, M.; Rossi, R.; Derler, S. Analysis of current running sock structures with regard to blister prevention. *Text. Res. J.* **2013**, *83*, 836–848. [CrossRef]
36. Skenderi, Z.; Mihelić-Bogdanić, A.; Mijović, B. Thermophysiological Wear Comfort of Footwear. *J. Leather Footwear* **2017**, *66*, 12–21.
37. Stygienė, L.; Krauledas, S.; Abraitienė, A.; Varnaitė-Žuravliova, S.; Dubinskaitė, K. Thermal comfort and electrostatic properties of socks containing fibers with bio-ceramic, silver and carbon additives. *Materials* **2022**, *15*, 2908. [CrossRef]
38. Bertaux, E.; Derler, S.; Rossi, R.M.; Zeng, X.; Koehl, L.; Ventenat, V. Textile, physiological, and sensorial parameters in sock comfort. *Text. Res. J.* **2010**, *80*, 1803–1810. [CrossRef]
39. Kumpikaitė, E.; Tautkutė-Stankuvienė, I.; Simanavičius, L.; Petraitienė, S. The influence of finishing on the pilling resistance of linen/silk woven fabrics. *Materials* **2021**, *14*, 6787. [CrossRef]
40. Sakthivel, J.C.; Anbumani, N. Dimensional properties of single jersey knitted fabrics made from new and regenerated cellulosic fibers. *J. Text. Appar. Technol. Manag.* **2012**, *7*, 1–10.
41. Onal, L.; Candan, C. Contribution of Fabric Characteristics and Laundering to Shrinkage of Weft Knitted Fabrics. *Text. Res. J.* **2003**, *73*, 187–191. [CrossRef]
42. Mckinney, M.; Broome, E.R. The Effects of Laundering on the Performance of Open-End and Ring-Spun Yarns in Jersey Knit Fabrics. *Text. Res. J.* **1977**, *47*, 155–162. [CrossRef]
43. Hashimoto, Y.; Kim, K.O.; Hashimoto, K.; Takatera, M. Effect of washing and drying conditions on dimensional change in various articles of knitted clothing. *J. Fiber Bioeng. Inform.* **2018**, *11*, 227–240. [CrossRef]
44. Savile, B.P. *Physical Testing of Textiles*, 1st ed.; Woodhead Publishing Limited: Cambridge, UK, 1999; pp. 184–207.

Disclaimer/Publisher's Note: The statements, opinions and data contained in all publications are solely those of the individual author(s) and contributor(s) and not of MDPI and/or the editor(s). MDPI and/or the editor(s) disclaim responsibility for any injury to people or property resulting from any ideas, methods, instructions or products referred to in the content.

Article

Textile Pattern Design in Thermal Vision—A Study on Human Body Camouflage

Catarina Pimenta [1,*], Carla Costa Pereira [2] and Raul Fangueiro [3]

1. Centre for Textile Science and Technology, University of Minho, Campus de Azurém, 4800-058 Guimarães, Portugal
2. Faculty of Architecture, University of Lisbon, Rua Sá Nogueira, 1349-063 Lisbon, Portugal; carlota.morais@gmail.com
3. Department of Mechanical Engineering, University of Minho, Campus de Azurém, 4800-058 Guimarães, Portugal; rfangueiro@dem.uminho.pt
* Correspondence: catarina.rpimenta@hotmail.com

Citation: Pimenta, C.; Pereira, C.C.; Fangueiro, R. Textile Pattern Design in Thermal Vision—A Study on Human Body Camouflage. *Materials* 2021, *14*, 4364. https://doi.org/10.3390/ma14164364

Academic Editors: Dubravko Rogale and John T. Kiwi

Received: 2 June 2021
Accepted: 2 August 2021
Published: 4 August 2021

Publisher's Note: MDPI stays neutral with regard to jurisdictional claims in published maps and institutional affiliations.

Copyright: © 2021 by the authors. Licensee MDPI, Basel, Switzerland. This article is an open access article distributed under the terms and conditions of the Creative Commons Attribution (CC BY) license (https://creativecommons.org/licenses/by/4.0/).

Abstract: This paper reports on a new approach to the creation process in fashion design as a result of the exploitation of thermal camouflage in the conceptualization of clothing. The thermal images' main variation factors were obtained through the analysis of their color behavior in a (diurnal and nocturnal) outdoor beach environment, with the presence and absence of a dressed human body (through the use of a thermal imaging camera), such as the analysis of textile materials in a laboratory (simulating the captured outdoor atmospheric temperatures and those of the model's skin using the climatic chamber and the thermal manikin). The combination of different patternmaking, sewing and printing techniques in textile materials, along with the study of the camouflage environment and the human body's variation factors, as well as the introduction of biomimetic-inspired elements (chameleon's skin), enabled the creation of a clothing design process with innovative de-sign elements which allow us to thermally camouflage the human body and take clothing beyond the visible spectrum in a functional and artistic way.

Keywords: fashion design; textile material; textile pattern; thermal camouflage; biomimetic

1. Introduction

In the creation of clothing, several elements must be considered in the design process, in order to achieve a greater understanding of the concept that is intended to develop (e.g.,: the silhouette; the function; the proportion and line; the details; the printing and ornamentation; the color; the material; etc.) [1]. In this context of creation, it is possible to observe the purpose of unconventional materials [2], such as intelligent materials or materials that are changeable to stimuli (e.g. light, heat, etc.), adaptable to the environment in an interactive and intelligent way [3]. Collaborations between researchers in the areas of nanotechnology, biotechnology and digital technology allow the reaching of this textile innovation [4]. Clothing is also related to the communication component, the influence on individuals, without the use of words [5], allowing the transmission of innumerable meanings, as to suggest, to insinuate, to lie, among others [6].

The concealment and disclosure of the body can be associated to cultural and religious issues, such as the example of the use of the burqa [7]. However, in contemporary western society, aspects related to personal needs and desires reveal another reality, such as the problematic of technological surveillance of human activity, where visible and invisible worlds are observed, sensed and tested through technology. The exploration of a hypervisibility in clothing that allows to react against common sense visibility [8], it has been an emerging trend of "illusion wear" (optical illusions in clothing), and which allow not only the manipulation of the proportion/shape of the human body [9] such as changing one's identity. Some examples of projects related to optical camouflage, thermal camouflage and pattern manipulation in the textile material are observed in the development of clothing, namely: the "Invisibility cloak" by Susumu Tachi; the "Digital skins" by designer Nancy Tibury [10]; the conceptual project of clothing with different patterns visible in thermal image [11].

1.1. Biomimetic

Human beings are surrounded by geniuses on planet Earth [12], where to Nature is associated the role of final designer while coworker, providing diversified sources of inspiration, for the creation of new materials in order to obtain greater performance, to respond to the needs and capacity of modern industry [13]. Examples of commercial applications are the development of innovative nanomaterials and nanodevices based on bacteria, plants and animals [14]. Biomimetic inspiration is combined with the potential for innovation, the integration of technology with natural ecosystems, as well as the possibility of redesigning the way how humanity relates with and inhabits the planet [15]. Thus, the exploration of the biological structure of living beings allows new relationships between materiality, form and function, correlated with multidisciplinary approaches [16], being possible the exploration of the fusion of different concepts observable in Nature with the aim of creating multifunctional materials [17]. Therefore, the study of the functionality coming from the surface and structure of the animals is revealed as a field of observation and learning in great growth currently. Examples of surface studies are materials with optical properties, superhydrophobic, anti-wear, anti-fog and drag reduction, among others [18]. Consequently, the study of the structural component appears as a fundamental element in the addition of innovative properties to materials in a context where the fashion designer resembles an engineer, in an attempt to create portable environments in clothing [19]. Innovation with inspiration in Nature emerges as a support for the creation of new textile products adapted to various purposes, as illustrated by the example of Velcro®, inspired by structural principles from the burdock plant burrs [20]. In the animal world, we can observe the exploration of the adhesion mechanism in the foot of the gecko (in its ability to scale different surfaces vertically), in the development of the vast field of nanotechnology and modern engineering, being some examples the following scientific studies/articles [21–24]. Beyond reptile skin, the shells of molluscs are also examples of natural structures noticeable in Nature, associating animals such as octopus, squid, cuttlefish as examples of inspiration for functional products based on mimetic properties related to camouflage and color change [25]. Thus, in an aspect of adaptation to wearable interactive devices, the "electronic skin" (e-skin) project is observed, which allows interactive color changes, derived from the study of the capacity of alteration and color change visualized in the chameleon's skin [26]. Otherwise, the study of the structural colors present in some animals, derived from mechanisms with complex interactions between light and microstructures, have great potential for application in different industrial areas, such as textile [27], being an example the design of a transparent structural colored film, with inspiration in the optical transparency evident in the wings of the insect *Cephonodes hylas* that allows it to hide from enemies and camouflage in the environment [28], such as examples that link the nanotechnology and study/inspiration in structural color and optical properties present in the wings of the morpho butterfly [29–31] and the wings of the Rajah Brooke's birdwing butterfly [32]. In this situation, the structure is a relevant principle for application in engineering, which may enable the creation of new more benign ways of obtaining color [33]. In the modern textile industry and engineering, as explained previously, examples of projects and materials based on the structure are observed, namely: the clothing designed by Donna Sgro with the textile material "Morphotex" inspired by the morpho butterflies; the textile material "FastSkin" from the Speedo brand, inspired by shark's skin; the textile material "Geckskin", inspired by the gecko's feet; the "lotus effect" on textile surfaces, inspired by the lotus leaf. Yet, in the active camouflage aspect, the creation of dynamic soft textile surfaces is observed in a project led by Xuanhe Zhao, inspired by the behavior of the octopus's skin [16]. Otherwise, the exploration of biomimetics and its relationship with nature is also observed in contexts of fashion installation, sending the viewer to a new reality, such as the example of the "Biopiricy" (Autumn/Winter, 2014) and "Aeriform" (Autumn/Winter, 2017) collections, by designer Iris Van Herpen, also relating that the "negative space" of fashion, which surrounds clothing, also raises issues beyond the materiality of clothing. The film is considered as the future, in terms of a tool for fashion exhibition [34] and the fusion between technology and fashion will allow an improvement in terms of interactive and expressive possibilities while creating a poetic use/experience [35]. In addition, the effects of textile surface, of materials that change color or shape still have a character of exhibition that involves a theatrical and intimate dialogue, without the need for specific performance techniques [36].

1.2. Proposal

The main objective of this work refers to the development of a new clothing creation process with visible patterns in thermal images, in order to allow the thermal camouflage of the human body in an outdoor environment (diurnal and nocturnal), previously idealized. New design elements will be introduced in the clothing creation process, through the exploration of the behavior of colors in thermal image of the environment, human body and textile material, as well as the combination of manual and digital drawing, and the exploration of biomimetics, with the introduction of the study/inspiration of the chameleon's skin (animal associated with its facility of camouflage through the change of skin color, where the capacity appears initially in the literature through dermal chromatophores, and posteriorly by the adjustment active of a network of guanine nanocrystals that constitute the dermal iridophores [37,38]).

To sum up, it will be possible to observe that thermal vision and thermography can enable the development of several clothing design projects, such as the exploration of new paths in the creation process, in a functional aspect (e.g.,: military camouflage) or in an artistic and conceptual aspect (e.g.,: exhibition and performance contexts).

In this sense, the work is planned through the exploration of the thermal image, visible in different environments (outdoor and laboratory) through the use of a Testo 855 thermal imaging camera (Testo, Barcelona, Spain) in a spectrum range between 7.5–14 µm, with measurement option from −30 °C to 100 °C, emissivity value set to 0.95 (approximate emissivity value of the skin) and "Iron" color range selected.

For the design of the clothing in an outdoor environment, a rocky place by the sea was selected in order to reduce possible changes in the landscape over time (Carreço beach, Viana do Castelo, Portugal).

For the laboratory environment, a FITOCLIMA 24000 EDTU walk-in climatic chamber was selected. The value of 65% humidity was applied, and the atmospheric temperature values captured outdoors with a HI9564 thermal hygrometer (HANNA) were reproduced, in which for the nocturnal simulation 18 °C of atmospheric temperature and for the diurnal simulation 24 °C atmospheric temperature was applied. To replace the human figure in the outdoor environment, a Newton 34-zone sweating thermal mannekin) was used, which was programmed with a temperature of 34 °C, from the ISO 15831 standard (procedure used to measure the thermal insulation of clothing on a thermal mannekin).

Thus, with the comparison and study of the results obtained from various thermal images of the textile materials, the human body and the environment (indoor and outdoor), it was possible to develop the new process of creation of the thermal camouflage clothing, through the selection of specific materials (conventional knitted 100% polyester fabric and copper metallic pigment) and techniques (printing, patternmaking of small structures and sewing) that allow the development of prototypes of thermal camouflage and later of the model's thermal camouflage in the selected location, in a nocturnal environment and diurnal environment, in a moving and static way.

2. Materials and Methods

In a previous study carried out [39,40], thermal images were analyzed (in the "Iron" range of the thermal imaging camera), which allowed to elaborate different thermally visible color palettes were created, to assist the entire conceptual process, namely: the diurnal and nocturnal environment (with regard to the environment/body/clothing); and the laboratory environment (with regard to the environment/thermal manikin/textile material). The conceptual process was based on the perception of these resulting thermal colors in different selected environments (diurnal and nocturnal) in thermal image, in order to design clothing with finishing or structural surface treatments in the textile materials used according to these results to subsequently camouflage the best possible. Thus, the following materials with the following properties were selected for this study: (i) knitted fabric: jersey 100% polyester; white color; 153.39 g/m^2; 0.42 mm; 26.9 Tex; 15 wales/cm; 20 courses/cm; 0.42 mm; 0.76 ε (without printing); 0.50 ε (with printed copper pigment); (ii) printed copper pigment (formula: 80 g Hydra Clear for Metallic (clear base of the brand Virus®, Water Based Inks, Bergamo, Italy) with 20 g Copper Powder (metallic powder of the brand Virus®, Water Based Inks, Bergamo, Italy).

Therefore, considering that in the diurnal environment there is a greater predominance of yellow, orange and magenta colors, in contrast to the nocturnal environment where there is a greater predominance of orange, magenta and blue colors, the study of patterns continued with the study and analysis of the colors of different types of materials and different types of sewing techniques (which are present in the paper [40], Figure 11 and Table 3, namely: (i) 100% polyester knitted fabric to obtain the yellow color; (ii) 100% polyester printed knitted fabric with copper pigment to obtain the orange color; (iii) application of embossed structures with fully fusible interlining to obtain the magenta color; and (iv) application of embossed structures with partially fusible interlining to obtain effects in the magenta and orange colors). Following the study of the data observed in the laboratory (climatic chamber), as well as in the outdoor (in the place of future camouflage), it was possible to observe that the properties of the textile materials used/tested (e.g., textile structure, composition, yarn linear density, mass per square meter, thickness, emissivity, thermal resistance, thermal conductivity), by themselves, are not enough to present or to guarantee high performance of thermal camouflage functionality of clothing. However, it was observed the extreme relevance of manipulation, control and study of the AIR element present between the surface of the textile material and the skin surface of the model's body and the thermal mannequin, because this interferes with the interaction and stability of the color behavior of the image in thermal vision and the consequent performance of the textile material in thermal camouflage contexts.

Consequently, the functionality of thermal camouflage and the creation of illusory effects on clothing can only be acquired through the combination of the properties of the textile material and with an efficient and creative design process, which enhances the visual characteristics present in the textile material in thermal vision, in its interaction with the human body that uses it and the environment that surrounds it.

Thus, the study of design becomes extremely important for the textile material/clothing and textile pattern developed to perform its functionality in an efficient way, in the different perspectives that interact with the idealized thermal camouflage environment.

Therefore, the use of image overlay techniques (with the presence and absence of the human body), as well as the combination of a manual and digital drawing process, was essential for the understanding of the camouflage environment, the silhouette of the model (human body in a static and in a moving position) and the conceptualization of textile patterns which will possibly be applied to the materials used in the clothing to be created later.

The desired camouflage colors are observed in the thermal image with the presence of body temperature. However, it is only with the absence of the model's body in the thermal image that it becomes possible, by comparative method, to analyze all textures and patterns that can provide absence and presence of camouflage. The body presence in the thermal image allows, on the other hand, to project and design in the ideal proportion of the clothing, in order to obtain a better functionality.

2.1. Development of Patterns

In the course of the investigation, it was verified that for the creation of design process, the image and the materials' real color is not determinant in obtaining camouflage, being relevant another type of creation approach, more focused on the embossing of the material structure. In this sense, the study and inspiration of the chameleon's skin behavior was also interconnected in the combination of the colors, patterns and structures of the design process for the camouflage outfit. The inspiration/study of the chameleon, focused on the observation of the behavior of the visible color on the skin (colors and dimensions of patterns) as well as the small embossments and texture observed (irregular shapes and different dimensions). Some images that supported the basis of the conceptual process can be observed below in Figures 1–3.

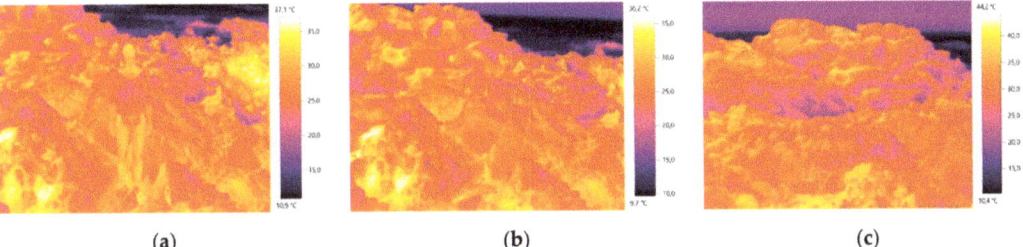

Figure 1. Thermal images of the base of the conceptual process of the diurnal prototype: (**a**) with the presence of the human body; (**b**) and (**c**) with the absence of the human body.

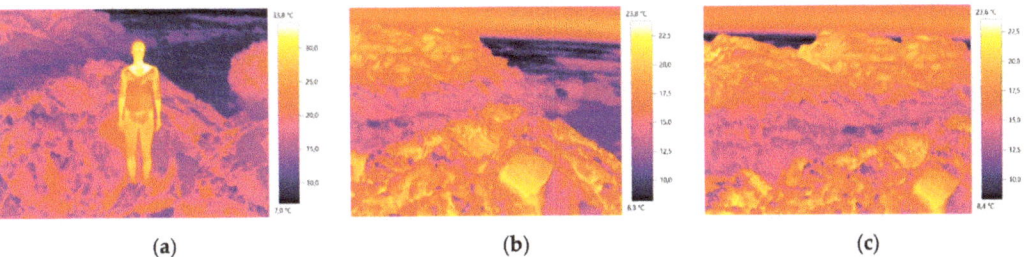

Figure 2. Thermal images of the base of the conceptual process of the nocturnal prototype: (**a**) with the presence of the human body; (**b**) and (**c**) with the absence of the human body.

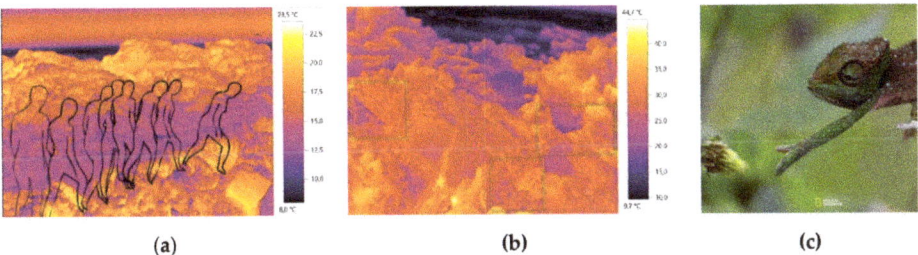

Figure 3. (**a**) Image of study of movement and silhouette; (**b**) Inspirational thermal image for the pattern modules M01 and M02; (**c**) Adapted chameleon-inspired photograph [41].

Among the various experiments carried out, different types of modules were performed, namely:

- M01 and M02, with measures of 28.50 cm × 30 cm, for the study of the manipulation of the air between the textile material and the human body using the printing and sewing of embossed structures techniques (1 cm height) with total and partially fusible interlining (to obtain the magenta color and effects in the magenta/orange color in thermal image). The technical drawings of the modules M01 and M02 can be observed in Figures 4 and 5. All small structures present an only vertical side seam, such as 0.5 cm of sewing value in all fittings. In addition, the fusible interlining was applied only on the side strip that raises and contours the shape (numbers: 2; 3; 8 and 9), as well as on the side strip and in the totality of the shape (numbers: 1; 4; 5; 6; 9; 10; 11; 12 and 13).

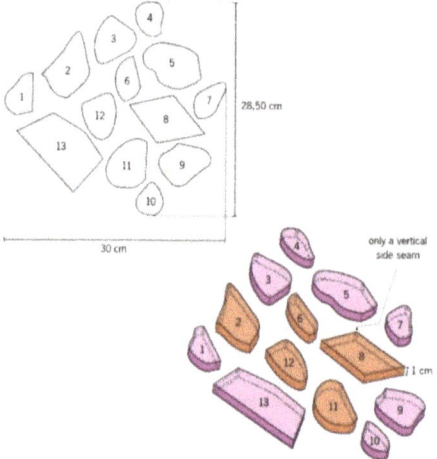

Figure 4. Technical drawing of the embossed structures of the pattern module M01.

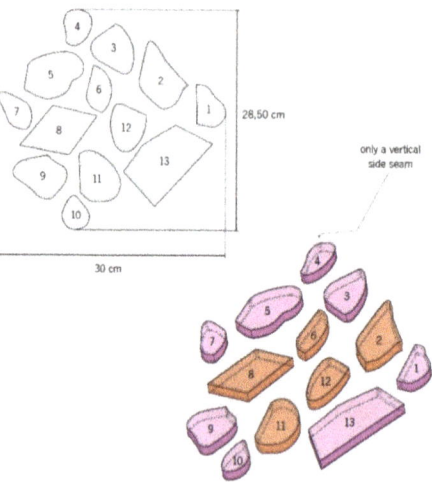

Figure 5. Technical drawing of the embossed structures of the pattern module M02.

Subsequently, the air behavior of the M01 module was observed in laboratory, similarly to previous experiments [39,40].

- M03, M04 and M05—based on M01 and M02 modules, were created in order to be placed tight to the body through the combination of the printed copper pigment technique and absent printed pigment areas (to obtain the yellow and orange colors in thermal image). Techniques of overlapping, rotating and moving of the shapes present in M01 module and M02 module, Figures 6 and 7, allowed to obtain a pattern with "pixelated" effect M03 (with measures of 50 cm × 50 cm), with abstract, asymmetric, irregular shapes and with different types of paint filling.

Figure 6. Assembly of different elements of the M03 module.

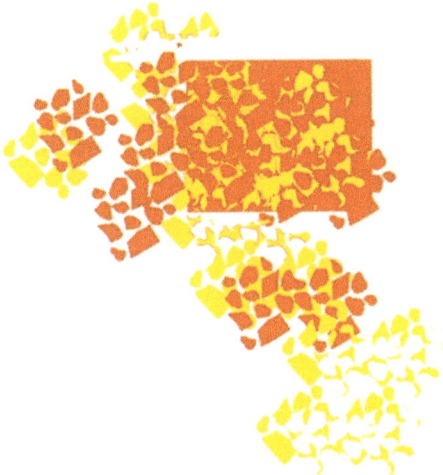

Figure 7. Final study of the M03 module.

Sequentially, the M03 module was reflected, giving rise to the M04 module (with measures of 50 cm × 50 cm). The M05 module (with measurements of 60 cm × 35 cm) was created with the intention of facilitating the subsequent process of sewing and patternmaking of the mask, keeping the original fitting of the M03 module on the front and creating another design for fitting with the M04 module, on the back.

- The M06 module was also created (with measurements of 50 cm × 50 cm), to be placed tight to the model's body using the printed copper pigment technique (to obtain the orange color in thermal image).

Once the modules were conceptualized, it was possible to develop two thermal camouflage prototypes (full-length overalls, in order to cover the body in its totality and be adaptable to the body movement): one for a diurnal environment and another for a nocturnal environment.

2.2. Development of Clothing Prototypes

In the diurnal prototype, the following pattern modules were applied: i) two pattern modules with small sewn embossed structures, with copper pigment, to provide orange and magenta colors in thermal image, namely: the "original" M01 placed on the front and the "reflected" M02 on the back of the prototype; (ii) three pattern modules with two colors visible in thermal image (yellow and orange), created through the absence and application of copper pigment on the knitted fabric, namely: the "original" M03 applied on the front of the prototype; the "reflected" M04 applied on the back; and the M05 of the mask that includes the front and the back; (iii) a pattern module with full color fill M06, with copper pigment to provide orange color in thermal image. Otherwise, in the nocturnal prototype, the following pattern modules were applied: (i) an "original" M01 pattern module with small sewn embossed structures, with copper pigment, to provide orange and magenta colors; (ii) and a pattern module with full color fill M06, with copper pigment to provide the orange color in thermal image.

Furthermore, in the conceptualization process, different drawing materials and software such as V 3.7 (Testo), Adobe Photoshop and Adobe Ilustrator were used. After the final conceptualization process, the printing, patternmaking and sewing processes of all prototypes were performed, with specific materials for both practices.

3. Results

For the conceptualization of clothing with thermal camouflage properties, the textile material and techniques chosen must adapt to the color present in the environment that will surround the model's body in the thermal image, as well as to the patterns and textures that surround it, and must present a similar color to the thermal result, if the aim is to camouflage. In contexts of thermal camouflage, the increase of the performance of the textile material and its thermal and emissive properties, is only acquired with the combination of the creative design process where the air component is extremely important in exploring the dynamics of the textile material with the body and the environment, since this element (AIR) allows the change and manipulation of colors on the body and silhouette of the model in thermal vision, emerging as a relevant and innovative element for design, in addition to temperature and thermal vision.

3.1. The Modules of the Conceptualized Patterns

3.1.1. Combination of the Printing Technique with Patternmaking of Embossed Structures

Figure 8 shows the final result of the illustration of the M01 (original) and M02 (reflected) module conceptualized, making it possible to observe the conjugation of the magenta color (derived from a fully fusible interlining structure) with effects of magenta and orange colors (derived from a partially fusible interlining structure).

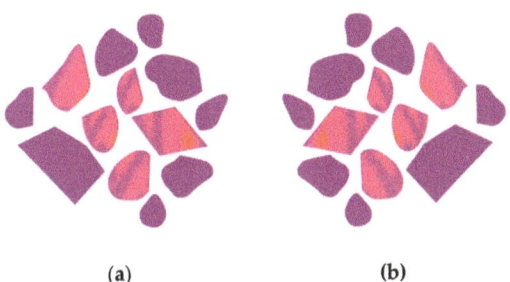

Figure 8. The pattern modules conceptualized with embossed structures and printed cooper pigment: (**a**) M01; (**b**) M02.

The color variations in the thermal image of the M01 module observed in the laboratory, are shown in Figure 9, as well as the module's sharpness and the ability to control and manipulate the air through small embossed structures. The desired colors for the Design conception of the camouflage prototype are observed in almost all structures, materials and techniques applied (100% polyester printed knitted fabric with copper pigment, with a combination of sewn embossed structures, fully and partially fusible interlining), in both diurnal and nocturnal simulations, previously tested.

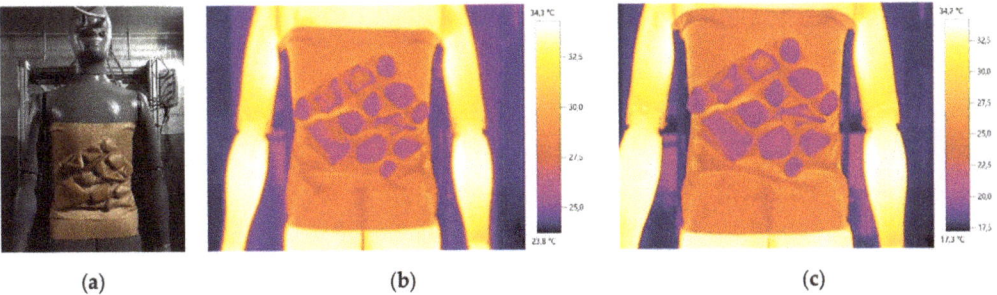

Figure 9. The M01 module tested in the laboratory: (**a**) Photography; (**b**) Thermal image of the diurnal simulation; (**c**) Thermal image of the nocturnal simulation.

3.1.2. Printing as Only Applied Technique

Figure 10 shows the final result of the conceptualized modules M03, M04, M05 with the combination of yellow and orange colors derived from the absence and presence of copper pigment, as well as the M06 module, only with a single color printing, in copper pigment.

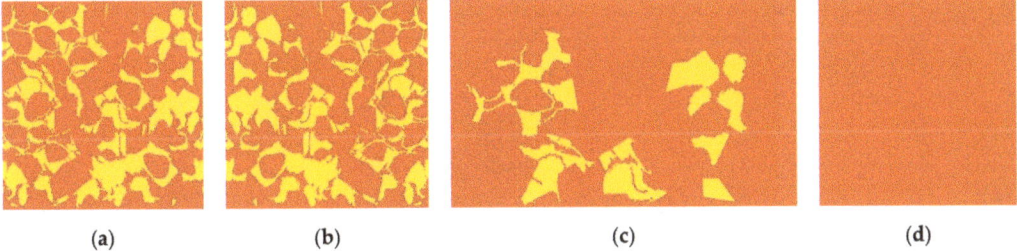

Figure 10. Pattern modules conceptualized with printed copper pigment: (**a**) M03; (**b**) M04; (**c**) M05; (**d**) M06.

3.1.3. Application of the Modules in the Final Thermal Camouflage Prototypes

Figures 11 and 12 show a thermal image of the model in the future camouflage environment and illustrations of the thermal camouflage prototypes for each environment: diurnal (with modules M01, M02, M03, M04, M05 and M06) and nocturnal (with modules M01 and M06).

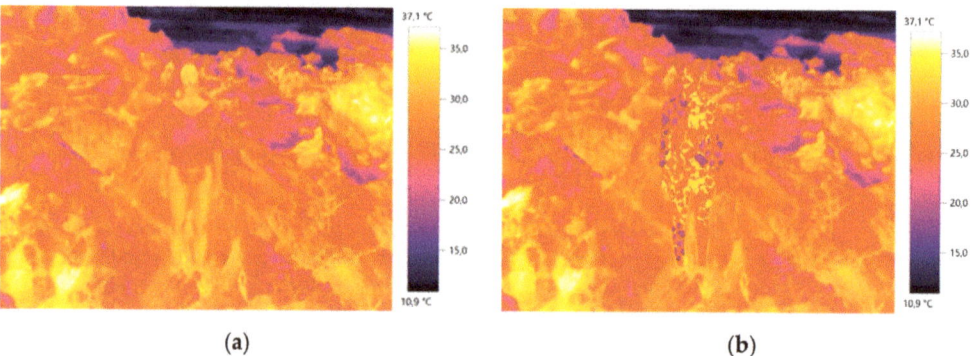

Figure 11. The diurnal thermal camouflage prototype conceptualized: (**a**) Base thermal image without the Design elaboration for the camouflage; (**b**) Base thermal image with the prototype illustration.

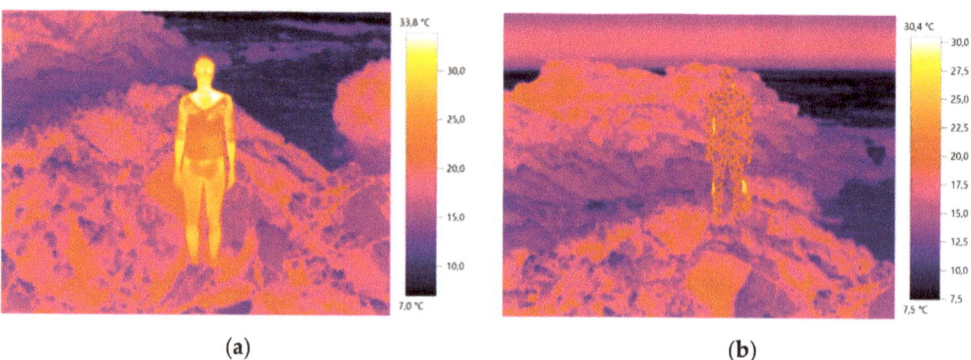

Figure 12. The nocturnal thermal camouflage prototype conceptualized: (**a**) Base thermal image without the Design elaboration for the camouflage; (**b**) Base thermal image with the prototype illustration.

Figure 13 shows the final illustrations of the front and back applied to the conceptualized thermal camouflage prototypes (diurnal and nocturnal) and respective pattern modules, based on the dimensions and silhouette of the model's body.

The technical drawings of the front and back of both conceptualized prototypes can be observed in Figure 14, relative to the diurnal thermal camouflage prototype and in Figure 15, relative to the nocturnal thermal camouflage prototype. Both prototypes feature a full-length overall tight to the body, with sleeves and long trousers, with a mask, gloves and socks incorporated. In the eyes, nose and mouth area, openings were applied. In the center of the back, a zipper was applied as well as an invisible zipper on both gloves and in the finishing of the sleeves. The cuts, printing and patternmaking techniques were combined in different way, in the prototypes to obtain diurnal and nocturnal thermal camouflage, as previously analyzed, in this article.

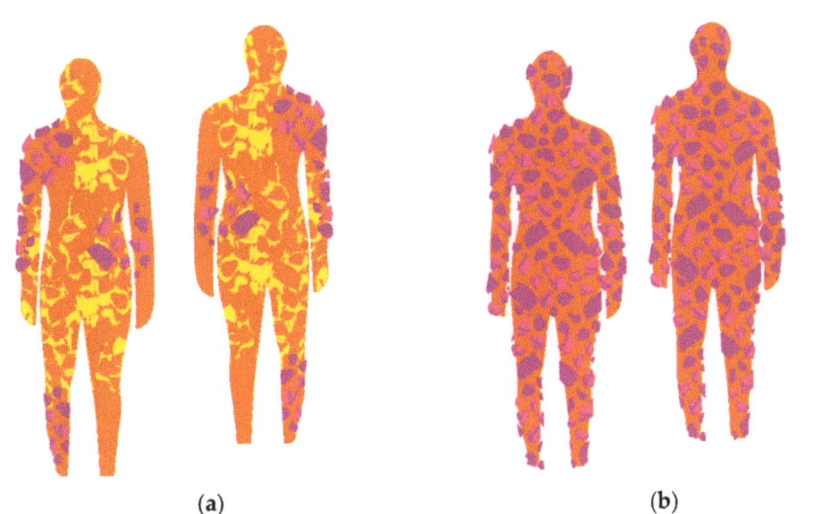

Figure 13. The conceptualized thermal camouflage prototypes: (**a**) Illustration of the front and back of the diurnal thermal camouflage prototype; (**b**) Illustration of the front and back of the nocturnal thermal camouflage prototype.

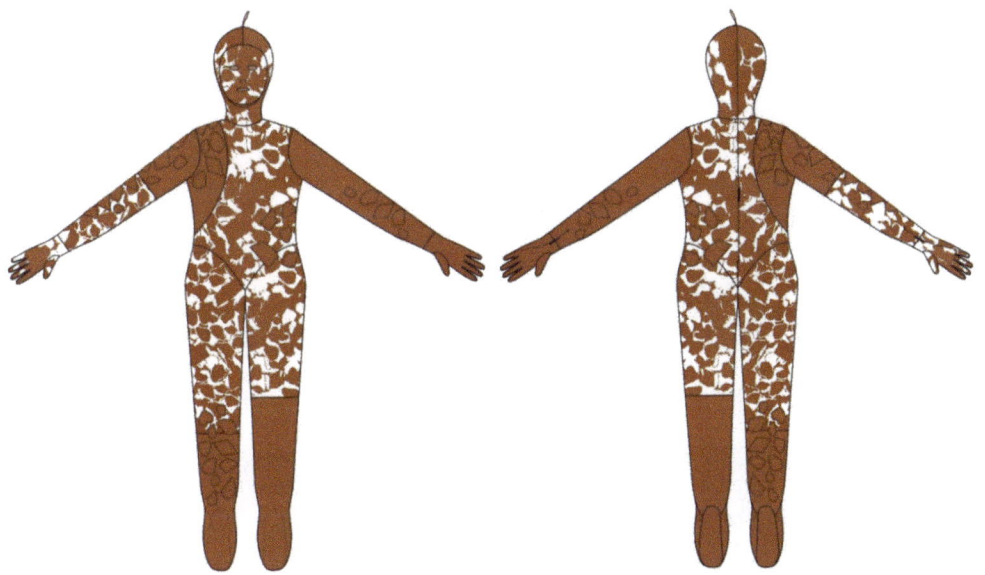

Figure 14. Technical drawing illustrated of the diurnal thermal camouflage prototype.

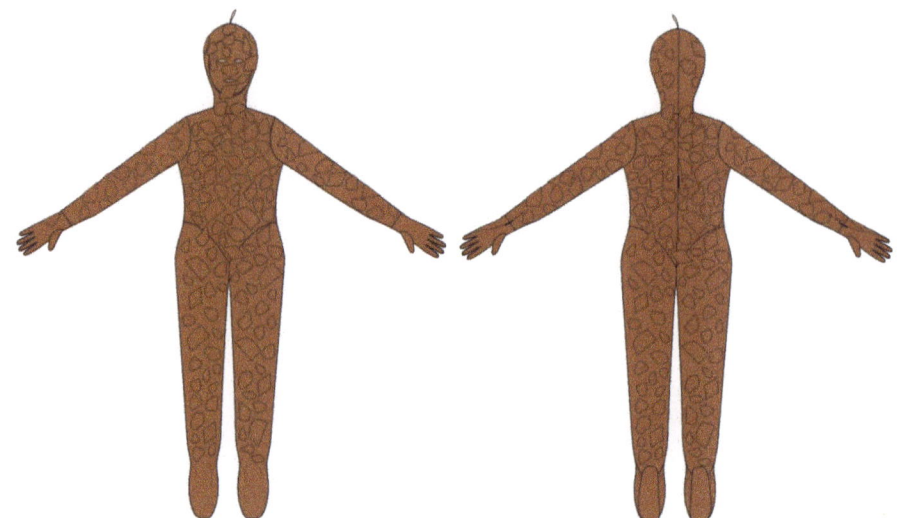

Figure 15. Technical drawing illustrated of the nocturnal thermal camouflage prototype.

To sum up, the results of the final thermal tests carried out in the idealized thermal camouflage environment, with the diurnal and nocturnal thermal camouflage prototypes can be observed bellow in Figures 16 and 17, taken from [42].

Figure 16. The conceptualized thermal camouflage prototype for the idealized diurnal environment [42].

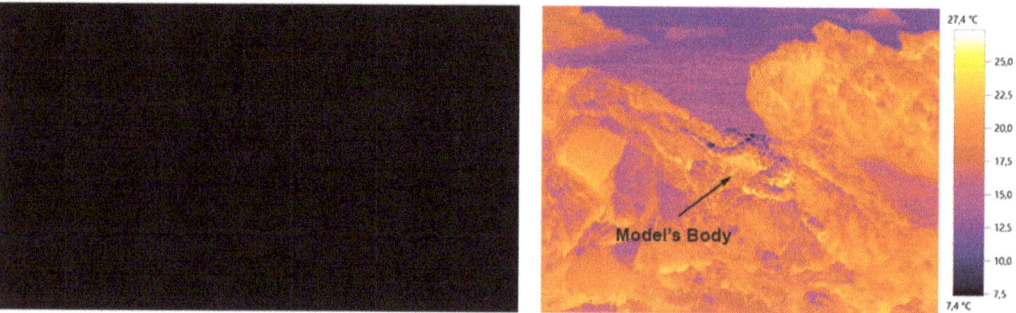

Figure 17. The conceptualized thermal camouflage prototype for the idealized nocturnal environment [42].

4. Discussion

The results obtained in the process of conceptualization of the prototypes and pattern modules allowed us to verify that it is possible to obtain and apply the thermal camouflage functionality in clothing, through the interconnection of the study of the different variants that involve the human body (female model), the environment (Carreço beach) and clothing (textile materials with different emissivity, combined with printing and patternmaking techniques). The increase in the performance and functionality of thermal camouflage of the textile material is obtained through the conjugation/application of the study and conceptual process of Design, exploring the clothing and the textile pattern in a global way, interconnected with the human body and the environment, where the air appears as a fundamental element in the control and stability of color in thermal image. Thus, the properties of textile materials (e.g., textile structure, composition, yarn linear density, mass per square meter, thickness, emissivity, thermal resistance, thermal conductivity), are not sufficient to enhance the functionality of thermal camouflage, since the material has to adapt to the surrounding visual environment, to its atmospheric conditions, as well as to the silhouette and temperature of the human body.

Therefore, in the conceptual design process for clothing with camouflage functionality, the following are relevant:

- The creation of a dynamic of colors in thermal image by the textile materials;
- The introduction of elements inspired by biomimetics in the design conception (e.g., chameleon skin);
- The study of the thermal image variation factors (atmosphere; human body; clothing/textile material; thermal imaging camera) and associated color palettes. In this case, the selection of another color palette (in addition to the "iron" range used in this work) will not alter the final result of thermal camouflage in clothing; however, it could provide less contrast and perceivability between the colors and temperatures captured by the thermal imaging camera in the selected color palette, which may make it difficult for the designer to work, when designing the clothing and, consequently, studying the color behavior visualized in the thermal image (either relative to the body, environment or textile material and techniques applied to it). Thus, in this parameter, it is up to the designer to reflect and choose the best method and color palette to apply.

In general terms, it was found that the thermal image with the presence of the model's body (with and without movement) should be used as a base model and as a start for the design conception, for presenting the colors and textures intended for the camouflage environment and compare the study of the thermal image with absence of the model's body, in order to verify all existing textures in the desired environment for the camouflage. In this way, it is essential to study the entire environment, all the representative details in the image (e.g., textures, materials, mechanisms ...), reflecting and envisioning the behavior of the human body at the time of camouflage, in a military or performance context. The combination of biomimetics, the study of the diversity of mechanisms present in nature, can reinforce the effectiveness of functionality, as was observed in this study with the interconnection of the chameleon skin structure.

Otherwise, the use of new design elements in the clothing creation process (e.g.,: thermal vision, color of the thermal image, temperature, air and thermal and emissive characteristics of the materials) such as the combination of conventional materials (polyester knitted fabric), metallic pigments (copper) and sewing and patternmaking techniques (embossed structures), solves the reach of a possible diurnal and nocturnal thermal camouflage, namely: for functional purposes (for example, for military users, with a view to the total camouflage of the human body) or for artistic purposes (in exhibition and performance contexts, with the exploration of new design elements freely, with a view to the creation of emotions and new perspectives in the visualization of clothing).

On the other hand, the impact of the varying textile patterns on the proposed clothing prototypes is very evident because the design and proportion of the pattern as well as the combination of different techniques (printing and sewn embossed structures) allow an interactive response between the temperature of the human body and the outdoor environment to which it is exposed, where the air and its absence, such as the combination of printing/pigment and its absence in the proposed pattern design enhances the response, the performance, the thermal camouflage functionality.

In this study, the 1 cm high embossed structures tested, in conventional knitted fabric printed with copper metallic pigment and with application of fusible interlining, potentiated another visible color in a thermal imaging camera (the magenta color), which without the introduction of this structure, the knitted fabric would present, adjusted to the human skin/body, the orange color (with printing) and the yellow color (without printing) in the thermal image. At this point, the structuring and manipulation of the air existing between the clothing and the human body/skin has a great impact on clothing, on the behavior of the textile material, on the conceptualized shapes and patterns and how these will later be seen in the thermal imaging camera. The introduction of different heights and dimensions in the textile structure, with the exploration of the elevation of the textile form, from the conceptual phase to the patternmaking and sewing phase, combining more or less irregular shapes, with greater or lesser dimensions, can and should be tested, in order to enable the study and control of more visible colors in thermal image, in search of creating new ways to achieve thermal camouflage. At this point, the introduction of the thermofixation technique, in a mechanical way, can help, facilitate and enable the development of a future structure.

Therefore, in addition to the application of surface design changes, the impact of varying textile patterns in the context of future work can be enhanced by using the introduction of new textile materials, which were not possible to be tested in this work, being an example the introduction and adaptation of smart materials and/or reactive materials, namely: thermochromic materials (which interact with temperature); photochromic materials (which interact with light intensity or brightness); hydrochromic materials (which interact with water); electrochromic materials (which interact with electrical current); materials with shape "memory" (which change their shape with different stimuli); between others. Thus, new visual effects can be obtained in the visible and invisible spectrum observed by the thermal imaging camera, enabling/allowing different interactions, illusions, movements, textile manipulations that can enhance the impact of the varying textile patterns in different contexts (artistic and military).

Otherwise, the conventional material (knitted fabric Jersey 100% Polyester; white color; 153.39 g/m^2; 0.42 mm; 26.9 Tex; 15 wales/cm; 20 courses/cm; 0.42 mm; 0.76 ε) and the metallic pigment tested (formula: 80 g Hydra Clear for Metallic (clear base of the brand Virus®, Water Based Inks, Bergamo, Italy) with 20 g Copper Powder (metallic powder of the brand Virus®, Water Based Inks, Bergamo, Italy), such as the printing and sewing technique applied in this work, proved to be a practical and economical solution to be reproduced in different textile products, however, it was also revealed as a starting point in the exploration of the textile surface with thermal camouflage functionality.

To sum up, there are no limits to the creation of new perspectives in clothing, combining the different visible and invisible spectra. The work and concept developed can be applied and studied in different aspects, such as in the exploration of different environments and camouflage backgrounds, in the exploration of different atmospheric conditions such as extreme adverse conditions, in the exploration of different types of bodies and silhouettes; in the exploration and creation of new prototypes and experimentation of new textile materials and textile manipulation techniques, in the exploration and interconnection with different arts and types of exhibition, in the exploration and interconnection with visual camouflage, envisaging a simultaneous thermal and visual camouflage functionality, with a view to developing the best possible camouflage.

Author Contributions: Conceptualization, C.P.; methodology, C.P.; investigation, C.P.; supervision, C.C.P. and R.F.; writing—original draft preparation, C.P., C.C.P. and R.F.; writing—review and editing, C.P., C.C.P. and R.F. All authors have read and agreed to the published version of the manuscript.

Funding: This research received no external funding.

Institutional Review Board Statement: Ethical review and approval were waived for this study, due to reason that it does not compromise the health of the people included.

Informed Consent Statement: Informed consent was obtained from all subjects involved in the study.

Data Availability Statement: No new data were created or analyzed in this study.

Conflicts of Interest: The authors declare no conflict of interest.

References

1. Seivewright, S. *Research and Design*; AVA Publishing SA: Lausanne, Switzerland, 2007.
2. Crane, D. Boundaries: Using cultural theory to unravel the complex relationship between fashion and art. In *Fashion and Art*; Geczy, A., Karaminas, V., Eds.; Bloomsbury Publishing: New York, NY, USA, 2012; pp. 99–110.
3. Lee, S. The shape-shifting skirt. In *Fashioning the Future: Tomorrow's Wardrobe*; Thames & Hudson: London, UK, 2005; pp. 109–126.
4. Seymour, S. Theoretical discourse. In *Fashionable Technology: The Intersection of Design, Fashion, Science and Technology*; Springer: Vienna, Austria, 2008; pp. 10–25.
5. Eco, U. O hábito fala pelo monge. In *Psicologia do Vestir*; Assírio e Alvim: Lisboa, Portugal, 1989; pp. 7–20.
6. Hollander, A. Dress. In *Seeing Through Clothes*; University of California Press: Berkeley, CA, USA, 1993; pp. 311–390.
7. Eicher, J. Body: The dressed body in fashion and art. In *Fashion and Art*; Geczy, A., Karaminas, V., Eds.; Bloomsbury Publishing: New York, NY, USA, 2012; pp. 77–86.
8. Quinn, B. Surveillance. In *Techno Fashion*; Berg: Oxford, UK, 2006; pp. 57–76.
9. Quinn, B. Future horizons. In *Fashion Futures*; Merrell: London, UK, 2012; pp. 198–229.
10. Braddock, S.; Harris, J. *Digital Visions for Fashion + Textiles*; Thames & Hudson: New York, NY, USA, 2012.
11. Pimenta, C. A Camuflagem Térmica no Design de Moda Conceptual. Master's Thesis, University of Beira Interior, Covilhã, Portugal, 2013.
12. Benyus, J. *Biomimicry: Innovation Inspired by Nature*; HarperCollins Publishers: New York, NY, USA, 1997.
13. Wang, Y.; Naleway, S.; Wang, B. Biological and bioinspired materials: Structure leading to functional and mechanical performance. *Bioact. Mater.* **2020**, *5*, 745–757. [CrossRef]
14. Bhushan, B. Biomimetics: Lessons from nature—An overview. *Philos. Trans. R. Soc. A Math. Phys. Eng. Sci.* **2009**, *367*, 1445–1486. [CrossRef]
15. Mackinnon, R.; Oomen, J.; Zari, M. Promises and presuppositions of biomimicry. *Biomimetics* **2020**, *5*, 33. [CrossRef]
16. Kapsali, V. *Biomimetics for Designers*; Thames & Hudson: London, UK, 2016.
17. Lui, K.; Jiang, L. Bio-inspired design of multiscale structures for function integration. *Nanotoday* **2011**, *6*, 155–175. [CrossRef]
18. Han, Z.; Mu, Z.; Yin, W.; Li, W.; Niu, S.; Zhang, J.; Ren, L. Biomimetic multifunctional surfaces inspired from animals. *Adv. Colloid Interface Sci.* **2016**, *234*, 27–50. [CrossRef] [PubMed]
19. Kapsali, V.; Dunamore, P. Biomimetic principles in clothing technology. In *Biological Inspired Textiles*; Abbott, A., Ellison, M., Eds.; Woodhead Publishing Limited: Cambridge, UK, 2008; pp. 117–135.
20. Wood, J. Bioinspiration in fashion—A review. *Biomimetics* **2019**, *4*, 16. [CrossRef]
21. Arzt, E.; Quan, H.; McMeeking, R.; Hensel, R. Functional surface microstructures inspired by nature—From adhesion and wetting principles to sustainable devices. *Prog. Mater. Sci.* **2021**, *119*, 100778. [CrossRef]
22. Li, X.; Li, Y.; Li, Y.; He, J. Gecko-like adhesion in the electrospinning process. *Results Phys.* **2020**, *16*, 102899. [CrossRef]
23. Gao, H.; Wang, X.; Yao, H.; Gorb, S.; Arzt, E. Mechanics of hierarchical adhesion structures of geckos. *Mech. Mater.* **2005**, *37*, 275–285. [CrossRef]
24. Wang, H.; Chen, S.; Liu, J. Adhesive evidence for gecko-inspired biomimetic fiber: Combination of experiments and modeling. In *Biomimetic Approaches in Engineering Practice*; Kolisnychenko, S., Ed.; Trans Tech Publications: Zurich, Switzerland, 2018; pp. 372–377.
25. Das, S.; Bhowmick, M.; Chattopadhyay, S.; Basak, S. Application of biomimicry in textiles. *Curr. Sci.* **2015**, *109*, 893–901. [CrossRef]
26. Chou, H.H.; Nguyen, A.; Chortos, A.; To, J.; Lu, C.; Mei, J.; Kurosawa, T.; Bae, W.G.; Tok, J.; Bao, Z. A chameleon-inspired stretchable electronic skin with interactive colour changing controlled by tactile sensing. *Nat. Commun.* **2015**, *6*, 8011. [CrossRef]
27. Kinoshita, S. *Structural Colors in the Realm of Nature*; World Scientific: Singapore, 2008.
28. Meng, Z.; Huang, B.; Wu, S.; Li, L.; Zhang, S. Bio-inspired transparent structural color film and its application in biomimetic camouflage. *Nanoscale* **2019**, *11*, 13377–13384. [CrossRef] [PubMed]
29. Butt, H.; Yetisen, A.; Mistry, D.; Khan, S.; Hassan, M.; Yun, S. Morpho butterfly-inspired nanostructures. *Adv. Opt. Mater.* **2016**, *4*, 497–504. [CrossRef]
30. Chen, Z.; Zhang, Z.; Wang, Y.; Xu, D.; Zhao, Y. Butterfly inspired functional materials. *Mater. Sci. Eng. R* **2021**, *144*, 100605. [CrossRef]
31. Liu, Y.; Huang, L.; Shi, W. Structural color bio-engineering by replicating Morpho Wings. In *Biomimetic Approaches in Engineering Practice*; Kolisnychenko, S., Ed.; Trans Tech Publications: Zurich, Switzerland, 2018; pp. 265–275.

32. Wilts, B.; Giraldo, M.; Stavenga, D. Unique wing scale photonics of male rajah brooke's birdwing butterflies. *Front. Zool.* **2016**, *13*, 36. [CrossRef]
33. Gebeshuber, I.; Macqueen, M. What is a physicist doing in the jungle? Biomimetics of the rainforest. In *Biomimetic Approaches in Engineering Practice*; Kolisnychenko, S., Ed.; Trans Tech Publications: Zurich, Switzerland, 2018; pp. 18–28.
34. Geczy, A.; Karaminas, V. *Fashion Installation: Body, Space and Performance*; Bloomsbury Visual Arts: London, UK, 2019.
35. Stead, L.; Goulev, P.; Evans, C.; Mamdani, E. The emotional wardrobe. *Pers. Ubiquitous Comput.* **2004**, *8*, 282–290. [CrossRef]
36. Birringer, J.; Danjoux, M. Wearable performance. *Digit. Creat.* **2009**, *20*, 95–113. [CrossRef]
37. Kreit, E.; Mäthger, L.; Hanlon, R.; Dennis, P.; Naik, R.; Forsythe, E.; Heikenfeld, J. Biological versus electronic adaptive coloration: How can one inform the other? *J. R. Soc. Interface* **2013**, *10*, 20120601. [CrossRef] [PubMed]
38. Teyssier, J.; Saenko, S.; Marel, D.; Milinkovitch, M. Photonic crystals cause active colour change in chameleons. *Nat. Commun.* **2015**, *6*, 6368. [CrossRef] [PubMed]
39. Pimenta, C.; Morais, C.; Fangueiro, R. The thermal colour and the emissivity of printed pigments on knitted fabrics for application in diurnal thermal camouflage garment. *Key Eng. Mater.* **2019**, *812*, 127–133. [CrossRef]
40. Pimenta, C.; Morais, C.; Fangueiro, R. The behavior of textile materials in thermal camouflage. In *Textiles Identity and Innovation: In Touch, Proceedings of the 2nd International Textile Design Conference (D_TEX 2019), Lisbon, Portugal, 19–21 June 2019*; CRC Press: London, UK, 2020; pp. 281–287. [CrossRef]
41. Goldberg, S. (Ed.) The race issue. *Off. J. Natl. Geogr. Soc.* **2018**, *233*, 1–156.
42. Pimenta, C.; Morais, C.; Fangueiro, R. Thermal camouflage clothing in diurnal and nocturnal environments. *Key Eng. Mater.* **2021**, *893*, 37–43. [CrossRef]

Article

Study on Effect of Leather Rigidity and Thickness on Drapability of Sheep Garment Leather

Hafeezullah Memon [1,*], Eldana Bizuneh Chaklie [2], Hanur Meku Yesuf [2,3] and Chengyan Zhu [1,*]

1. College of Textile Science and Engineering, International Institute of Silk, Zhejiang Sci-Tech University, Hangzhou 310018, China
2. Ethiopian Institute of Textile and Fashion Technology, Bair Dar University, Bahir Dar 6000, Ethiopia; elduyod@gmail.com (E.B.C.); 419005@mail.dhu.edu.cn (H.M.Y.)
3. College of Textiles, Donghua University, Shanghai 201620, China
* Correspondence: hm@zstu.edu.cn (H.M.); cyzhu@zstu.edu.cn (C.Z.)

Abstract: Understanding the performance and behavior of garment leathers provides valuable inputs for the design and production of leather garments. The drape is one of the important properties associated with garment fitness quality and appeal. This study aims to show how the independent variables flexural rigidity and thickness affect the dependent variable drapability. Nowadays, studies on the drape of garment leathers are scarce. In this work, the drape coefficient (DC) was measured for sheep garment leather, which influences the garment drapability, such as flexural rigidity in the range of 9.2 to 22 and thickness in the range of 0.64 to 0.96. The average DC was calculated in the range of 47.35 to 69.9% for the selected sheep leathers from four samples. The drapability of the garment leather was determined using the DC. Flexural rigidity and thickness have been shown to have a considerable influence on the DC, while they do bear a significant relationship to the DC. The results of this study can be used as an elementary tool for leather selection of appropriate materials for garments.

Keywords: leather; drape coefficient; leather rigidity; drapability; leather thickness

1. Introduction

Leathers for garments can be manufactured from almost all types of raw stock [1]. Nevertheless, the widely used raw material is sheepskins for their histological characteristics [2]. All the other skins such as goat skins, cow skins, and calf skins are also widely used, and the thickness range varies from 0.6 to 1.0 mm [3]. The qualities of the garment leathers are always required to be similar for their specific purpose and end use [4]. Some of the quality parameters and comfort parameters and the methodologies to be followed in leather processing are discussed below.

There are different properties that garment leather should have, such as lightweight, fastness (to rub, light, wash, etc.), durability, drapability, etc. [5]. On the other hand, the drape is the most significant garment feature, since it contributes to wear comfort [6]. This feature refers to the leather's capacity to fall in a similar manner to textile and conform to the body's shape when worn [7]. It is an inherent and essential quality for any leather to be called garment leather [8]. Whenever leather is having a cloth-like fall, it is said to have the quality of drape. The leathers require being as flexible as cloth with the utmost degree of softness without any firmness. Significantly less filling and more fat-liquoring will increase the drape quality of the leather [9]. Leather drapability is a morphological property that occurs when leather is hung down for gravity reasons. It is one of the essential indicators when measuring close-fitting clothing [6]. Therefore, good drapability is necessary for garment leathers [10]. The drapability of leather is closely related to the clothing classification [11]. There are two ways to examine leather drapes: using your senses or using testers (commonly used) [12]. A circular sample with a specific area is placed on a sample clamping plate, with their centers overlapping, causing the drape sample to droop along with the circular plate due to gravity [13].

According to the research done to date [6], the leather drape is determined mainly by the leather thickness, and rigidity/stiffness properties, with leather tensile and weight also of some importance. Therefore, to be able to determine the leather drape requires that the precise relationship between leather drape and leather rigidity /stiffness and thickness is known and that their effects on the leather physical property are also known [14]. In addition, leather garment drapes will be considered from the drape coefficient (DC) perspective only [15]. The bending length of the leather can determine the rigidity of the leather. For a better knowledge of drapability, it is essential to look into and grasp the fundamental bending behavior of leather [15]. The DC determines drapability; if the DC becomes between 30 and 80, the leather could be used for apparel [16]. It is well known that there are different hierarchical levels of leather within leather attach, and this is especially true for leather that has not been staked or fat-liquored [17]. The rigidity and stiffness of leather can be related to its bending length, as reported by [15]. The other main thing that must be noticed in the selection of leather is the effect of leather tanning material on the leather thickness and stiffness property [18]. Vegetable tanning material gives more thickness, fullness, roundness, etc. [19], whereas chrome gives empty, flexible, and soft leather [20].

This research aimed to show the actual effects of thickness and stiffness variability on leather drape quality difference by taking different leathers to measure all variables and analyze the result. Herein, leathers with vegetable and chrome tanned or re-tanned to reach each factor level in one sample leather have been studied. This helps set the optimum levels of independent variables thickness and flexural rigidity and optimize dependent variable drapability.

2. Experimental

2.1. Materials

Four types of sheep leather for the garment were reached from the EiTEX leather laboratory store. Among them, Leather 1 is chrome tanned leather; Leather 2 is vegetable tanned chrome re-tanned; Leather 3 has lower thickness higher rigidity; this leather is chrome tanned, vegetable re-tanned type of leather, and Leather 4 is vegetable-tanned.

Table 1 shows standard and measured values of independent variables (thickness and rigidity) that influence the dependent variable drape coefficient of the garment. If measured values from specimens in between their minimum and maximum limit levels of Indian standard IS 6490, the sheep leather samples can be used for apparel applications. Therefore, the specimens used for this research can be used for apparel applications.

Table 1. Variables and their levels.

Variables	Standard	Actual Measured Values
Thickness (mm)	0.6–1.0	0.64–0.96
Rigidity/Stiffness (mN/mm)	90–125	94–124
Drape Coefficient (%)	30–80	48–64

2.2. Method

Factorial designs are commonly employed in engineering experiments involving two factors where the combined effect of the factors on a response must be studied [21]. A factor's effect is defined as the change in reaction caused by a change in the factor's level [22]. The effects of leather thickness and rigidity on the separation force are next investigated using a factorial design [22]. A factorial design is commonly used to investigate the impacts of corresponding components, material qualities, and establish the best study circumstances, among other things. This includes ANOVA, regression analysis, model significance, model adequacy, and fit statistics using design expert software.

2.3. Research Design

This study follows an experimental research design (Full Factorial Design) because experimental research follows a scientific approach, where it includes a hypothesis, a variable that the researcher can manipulate, and the variable that can be evaluated, calculated, and compared. The study aims to distinguish the optimal effect of leather rigidity and thickness on garment drape property to select the appropriate leather property, which gives better results to recognizers obtained through experimental studies. These also required extensive experimentation, which involves examining the leather's physical properties through laboratory experimentation and giving better results, showing the effect of factors on leather drapes.

This testing works together with laboratory assistants. The conclusion of the cause-and-effect research reveals that two elements directly impact the leather's drapability. They are flexural rigidity and thickness of the material. Flexural rigidity is a measure of the stiffness of leather and is related to the bending length of the specimen due to gravitational force. Thickness is the distance through the leather, as distinct from width or height [23]. In a factorial experiment, these variables are assigned as independent variables.

Technical experience and understanding of operating and various experiments are required to set the factors in this experiment. It is known from their experiences that if all parameters are set too low, the drapability increases [24]. However, if these parameters are set too high, the drapability will be low because the DC will also be high. To avoid unfavorable outcomes in the experiment, proper samples of garment leather with the same thickness and rigidity are necessary. The expected ranges of factors are presented based on the experiment. Since all factors have two levels, a 22 factorial design is used. Each run is performed with three replicates at random, and the data are assessed at a significant level of 1 to achieve a very low type III error.

2.4. Thickness

The thickness of samples was measured with a digital thickness Gauge TF121C as leather has different thicknesses in different portions such as the shoulder (the thickest part), butt (the thicker area next to the shoulder area), belly (poor portion), and shank. Therefore, each leather sample was measured fifteen times to get a more accurate measurement, and finally, we used the average lower result of a thickness as a minimum and the average higher result as a maximum level.

Table 2 shows that the two samples (L1 and L3) have the same thickness and rigidity level, while the other two samples (L2 and L4) have the same thickness and rigidity levels. Due to these reasons, average measurements of two leather kinds, either L1 and L3 or L2 and L4, were used for minimum and maximum thickness levels.

Table 2. Measurements of leather thickness in mm.

Samples	1	2	3	4	5	6	7	8	9	10	11	12	13	14	15	Average
L1	0.75	0.62	0.58	0.61	0.89	0.63	0.54	0.55	0.56	0.57	0.67	0.58	0.65	0.69	0.65	0.635
L2	0.96	0.93	0.99	1.07	0.95	1.31	0.96	1.05	0.99	0.85	0.82	0.93	0.88	0.90	0.87	0.962
L3	0.71	0.68	0.59	0.60	0.80	0.65	0.55	0.54	0.61	0.58	0.65	0.58	0.65	0.59	0.77	0.6366
L4	0.92	0.97	0.99	1.0	0.93	1.09	0.95	1.15	0.95	0.92	0.93	0.88	0.94	0.91	0.851	0.9601

2.5. Bending Length

Bending length is the ability of the leather that can be bent somewhere [25]. The Shirley stiffness tester, which comprises a platform with smooth low friction and flat surfaces, was used to measure the bending length of samples. Specimens are made from sheep leather. Three specimens were cut from four kinds of leather (2.5 × 21.5 cm^2).

2.6. Flexural Rigidity

Flexural rigidity is a measure of the stiffness of leather and is related to the bending length of the specimen due to gravitational force [26]. It is usually attributed to the rigidity of collagen fibers in leather [27]. Flexural rigidity was determined according to the Indian standard IS 6490 test method [28]. The samples of dimensions 2.5 × 25 cm^2 were cut in parallel directions to the backbone of the leather, considering the size of the leather. The rectangular pieces were shorter in length than the specimen length specified in the standard, as it has been observed earlier that the deviation in the length of the samples up to 100 mm does not influence the flexural rigidity. The length of the slacker part of each sample (L) was measured with each side up, first at one end and then at the other, using the constant angle method. The mean value of L was measured for flexural rigidity (G) calculation [27], as is shown below in Equation (1).

$$G = W \times (L)^3 \tag{1}$$

where:
W = weight per unit area of leather mN/mm
Weight of the leather = mass (g)*gravity in m/s^2
L = Bending length in mm
G = Flexural rigidity in mN/mm

As shown in Table 3, L1 and L3, L2 and L4 have very similar thickness measurement results and took their result as an average minimum and maximum levels. As calculated, L1 and L2 have lower related rigidity, and L3 and L4 have related higher results, which means:

Table 3. Flexural rigidity of sample leathers.

Sample Leathers	Flexural Rigidity (G) in mN/mm
L1	93.96
L2	94.013
L3	124.419
L4	124.55

L1 has a lower thickness, lower rigidity, chrome-tanned.

L2 has a higher thickness, lower rigidity, vegetable tanned, and chrome re-tanned leather.

L3 has lower thickness higher rigidity; this leather is chrome tanned and vegetable re-tanned leather.

L4 has a higher thickness, higher rigidity, and is vegetable-tanned.

Vegetable tanning material gives more thickness, fullness, roundness, etc., whereas chrome gives empty and soft leather.

2.7. Drapability

Leather drapability, according to research, is a morphological property that occurs when leather is hung down for gravity [15]. It is one of the essential indicators when measuring close-fitting clothing [29]. The classification of leather as a garment, glove, upper, and so on has a strong influence on its drapability.

There are two ways to examine leather drapes: using your senses or using testers (commonly used). A circular sample with a specific area is placed on a sample clamping plate, with their centers overlapping, causing the drape sample to droop along with the circular plate due to gravity [13].

The drape sample is projected onto a white sheet, and light is used to create the draped figure of the sample, which may be obtained by a shaded area on the ring paper that represents leather drapability. The relevant indexes such as the DC can be obtained by weighing the ring paper and calculating the cut-shaded area. DC is commonly used to assess the drapability of leather and can be calculated in percentage as DC= [(W$_2$/W$_1$) * 100 percent], which is the projected area ratio to the original area; where W$_1$ = total weight of the ring paper (original area) and W$_2$ = weight of the shaded area of the ring paper (projected area).

The DC increases as the stiffness of the leather increases and vice versa [30]; thus, the drapability of leather by looking at the drape wave number and amplitude can be estimated. Three samples, each with a diameter of 30 cm and no crease on the surface, are prepared. Each sample should have two sides labeled "a" and "b", respectively. Three samples, each with a diameter of 30 cm and no crease on the surface, are prepared.

3. Results and Discussion

The DC and drapability of the material have an inverse relationship. As the DC becomes large, the drapability of the leather will decrease. As a standard, the garment leather drape should become between the accepted ranges 30–80, where DC below 30 is very limp leather and better to use for very small goods. The DC above 80 is very stiff leather, better to use for upper and safety gloves. The DC largely depends upon the flexural rigidity and thickness of the leather. The DC of sheep garment leather in different levels of the two factors is shown above. As the ANOVA indicates, all factors and their interactions are significant. The DC property between 47.75 and 69.9 largely depends on the thickness of the material. The thickness influences are significantly greater than the flexural rigidity value influences ranging from 94 to 124 sheep garment kinds of leather.

3.1. Drape Coefficient vs. Thickness

The thickness of an object is defined as the three descriptive measurements: height, width, and length. The thicknesses of the leathers were measured using a digital thickness tester. The design expert software analysis based on the given data shows thickness on leather drapability. Hence, as a thickness becomes higher, it leads to a higher DC. However, higher thickness in the garment, due to its resulting bulkiness, is often not preferred. The goal of this research was to minimize the thickness to optimize variables.

3.2. Drapability vs. Flexural Rigidity

Flexural rigidity is a measure of leather stiffness and is related to the bending length of the specimen. It is usually attributed to the rigidity of the collagen fibers in the leather. The flexural rigidity of garment sheep garment leathers was measured in parallel to the backbone direction. It is seen that there is a significant difference in the flexural rigidity values measured in the two levels ranging from 94 to 124 on DC results between 47 and 69. The DC and flexural rigidity have a positive or direct relationship. These values are significant. As the values of flexural rigidity become high, the DC will be higher, and the drapability will be lower. The plot of DC versus flexural rigidity (calculated from bending length values) reveals that there is a significant change.

3.3. Analysis of the Outcome

All key components and the interaction effect have a substantial impact on drapability, according to the analysis of variance (ANOVA), as discussed in Table 4. The 402.60 Model F-values indicate that the model is significant. An F-value of this magnitude has a 0.01% chance of occurring due to noise. While the F-values of rigidity, thickness, and interaction of both factors are significant, the F-values of rigidity, thickness, and interaction of both factors are not. Model terms with P-values less than 0.0500 are significant. Model reduction may improve the model if there are many nominal model terms (not including those required to support hierarchy).

The final model's projected R^2 of 0.9852 (98%) agrees reasonably well with the adjusted R^2 of 0.9910 (99%); that is, the difference is less than 0.2 (20%), implying that the final models can forecast with unpredicted changes of less than 2%, where enough precision measures the signal-to-noise ratio, as it is shown in Table 5.

Table 4. ANOVA for the factorial model of choice.

Source	Sum of Squares	Degree of Freedom	Square of Mean	F-Value	p-Value	Remarks
Model	770.14	3	256.71	402.60	<0.0001	Significant
A-Rigidity (mN/mm)	170.93	1	170.93	268.07	<0.0001	
B-Thickness (mm)	546.08	1	546.08	856.41	<0.0001	
AB	53.13	1	53.13	83.32	<0.0001	
Pure Error	5.10	8	0.6376			
Cor Total	775.24	11				

Table 5. Fit statistics for the model.

Std. Dev.	0.7985	R^2	0.9934
Mean	56.60	Adjusted R^2	0.9910
C.V. %	1.41	Predicted R^2	0.9852
		Adequate Precision	45.6374

3.4. Model Adequacy Checking

Before moving on to the next step, the residuals of the final models must be verified to see if they meet three criteria: first, they must be normally distributed; second, their variance must be constant; and third, they must be independent of the components. The normal probability plot of the residuals, the plot of residuals vs. fitted values, and the plot of residuals versus predicted values of the final models of the response variables, drapability, are shown in Figures 1 and 2. Table 6 shows all the analysis reports.

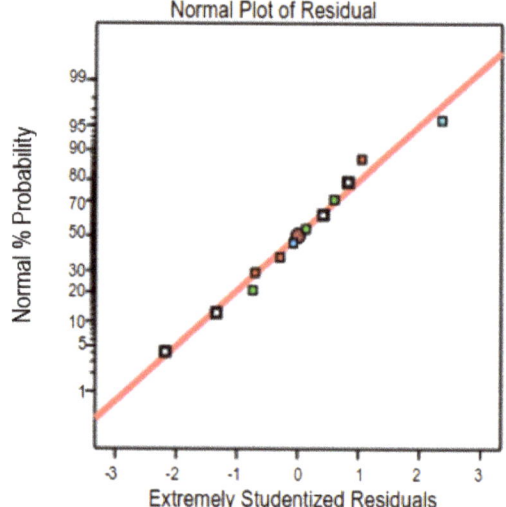

Figure 1. Data analysis for the normal plot of residuals.

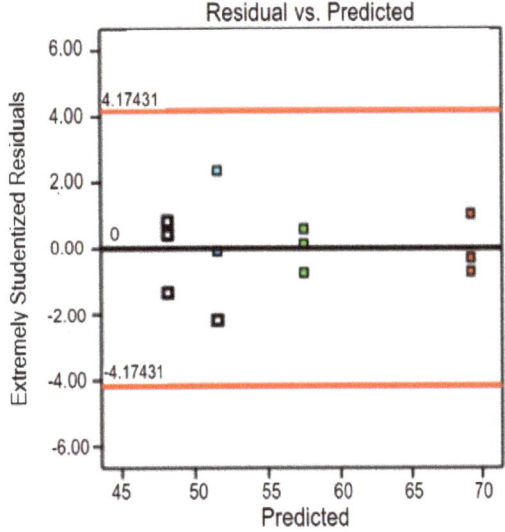

Figure 2. Residual vs. predicted analysis.

Table 6. Report of the design analysis.

Run Order	Thickness (mm)	Rigidity (mN/mm)	Actual Value of Drapability	Value Predicted of Drapability	Residual	Residuals That Have Been Externally Studentized	Standard Order
1	0.96	3.4	69.90	69.22	0.6767	1.044	12
2	0.96	1.3	56.97	57.47	−0.4967	−0.740	8
3	0.96	1.3	57.87	57.47	0.4033	0.593	7
4	0.96	3.4	69.02	69.22	−0.2033	−0.294	10
5	0.96	3.4	68.75	69.22	−0.4733	−0.703	11
6	0.64	1.3	48.73	48.18	0.5467	0.821	2
7	0.96	1.3	57.56	57.47	0.0933	0.134	9
8	0.64	1.3	47.35	48.18	−0.8333	−1.340	3
9	0.64	3.4	50.35	51.52	−1.17	−2.182	5
10	0.64	1.3	48.47	48.18	0.2867	0.416	1
11	0.64	3.4	52.75	51.52	1.23	2.357	6
12	0.64	3.4	51.47	51.52	−0.0533	−0.077	4

3.5. The Two Factors' Interaction

The interaction graph implies both factors have an interception at some point as the lines are not parallel, as Figure 3 shows. The graph shows the interaction; the more the thickness becomes, the more the DC. Therefore, the DC and thickness of the leather have a positive or direct relationship. The couture figure shown in Figure 4 represents whether individual factors or their interaction affect the response. Figure 4 represents how both factors' interaction influences the drapability. This can be identified by whether the couture is curved (has an effect) or a straight line (does not have an effect).

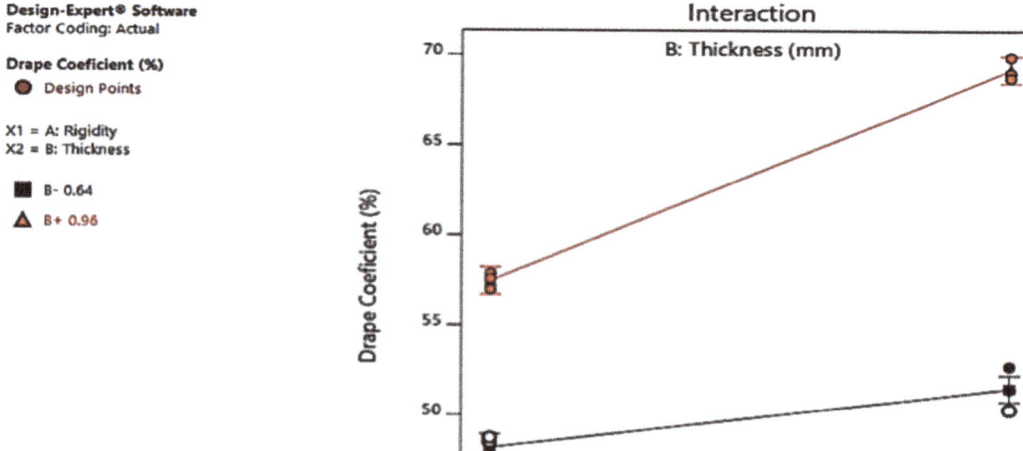

Figure 3. Interaction graph of the two independent variables.

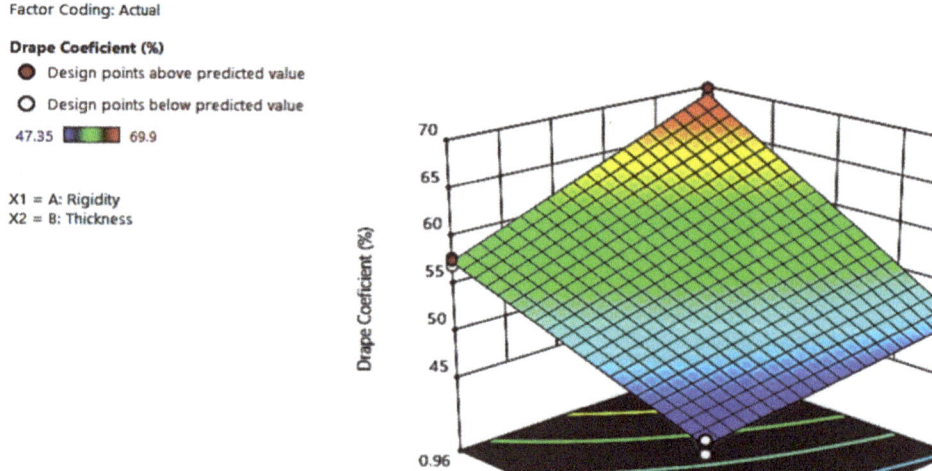

Figure 4. The three-dimensional interaction graph.

Furthermore, the interaction of both factors is important, as an ANOVA value indicates. The level of one factor affects the other factor. In addition, the two factors' interaction can be expressed by a three-dimensional graph, as Figure 4 shows.

3.6. Model of Regression

The link between the response and the factors can be described in Equation (2) using the final models (in terms of coded factors).

$$Y = B_0 + B_1X_1 + B_2X_2 + B_{12}X_1X_2 = 56.6 + 3.77X_1 + 6.75X_2 + 2.1X_1X_2 \quad (2)$$

Here, Y is the response (drapability), B_1 is factor A (thickness), B_2 is factor B (rigidity), B_1B_2 is the interaction of AB (both factors). Equation (2), in terms of coded factors, can be used to make predictions about the response for given levels of each factor. By default, the high levels of the factors are coded as +1, and the low levels are coded as −1. We obtained the regression model for drapability as A = +3.77, B = +6.75, AB =+2.10 with Intercept = +56.60. The coded equation helps identify the relative impact of the factors by comparing the factor coefficients. This equation can be useful to predict the drapeability level of sheep leather for garments with the minimum standard level of thickness and rigidity within the experimental range of thickness (0.64–0.96 mm) and rigidity (94–124 mN/mm).

3.7. Response Optimization

According to the response optimizer in Design-Expert software, the ideal factor setting is 0.64 mm stiffness and 94 mN/mm thickness. Therefore, the required drapability values in this study are set at 48.1833 DC, with lower and maximum bounds of 47.35 DC and 69.9 DC, respectively, as Figure 5 shows.

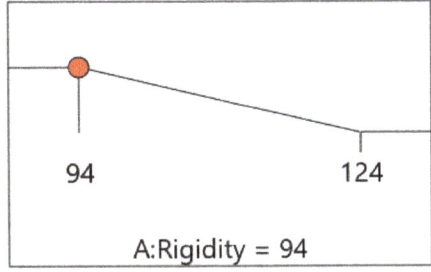
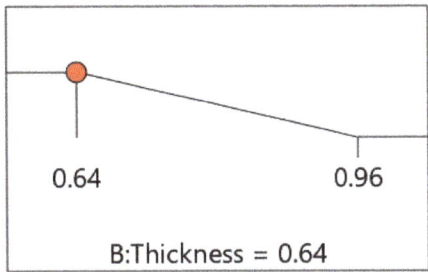
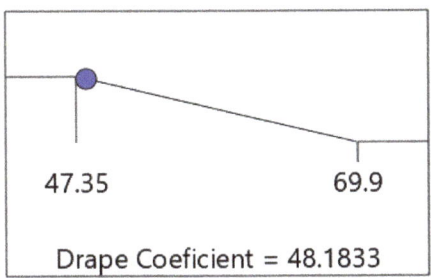

Figure 5. Variables optimization with lower and maximum bounds.

According to the findings of this study, the relationship between the parameters thickness and rigidity of a sample, and its resistance to drape deformation, is significantly intricate. The specimens sample leathers used in this research have a much smaller thickness and hardness than ideally upper leather (footwear leather) materials, resulting in a bigger drape when the sample is deformed in testing. As leather is not a homogeneous material, tested specimens show that the different portions of the samples have varying thicknesses and rigidities. This observation is highlighted in measuring different parts of the sample. Furthermore, when the four samples are examined at different levels, they reveal a substantial difference. Therefore, when the approximate pure drape model is applied to leather, significant deviations from the theoretical model are inevitable.

4. Conclusions

Previous studies studied thickness and rigidity, but they did not study its influence on the drape property of garment leather. This research explains the relationship between sample thickness, rigidity, and drape property and then extends the research by examining the influence of factors that could create a divergence from the ideal drape behavior. The 2^k full factorial design is used in this study to determine the appropriate thickness and hardness of garment leather. The results reveal that setting the thickness and rigidity to 0.64 mm and 94 mN/mm, respectively, is the best factor setting for the optimum level of DC of 48.1833%. The DC values can be regulated to be within the required range using this optimization. So, it is possible to conclude that leathers with 0.64 mm thickness and 94 mN/mm rigidity level show an excellent drape that hangs straight down in many little creases and folds. It clings to a bodily part or any other object when placed over it, revealing the form of whatever is beneath it. This study can be used as a benchmark for future studies and a standard reference for clothing makers. The publication has specified the reputability of assessed drapability to reproduce the experiment for other researchers. This study also has practical implications, such as assisting practitioners in comprehending numerous elements and selecting suitable material for garment firms and customers.

Author Contributions: Conceptualization, H.M., E.B.C. and H.M.Y.; methodology, H.M. and E.B.C.; software, H.M., E.B.C. and H.M.Y.; validation, H.M., E.B.C. and H.M.Y.; formal analysis, H.M.; investigation, H.M.; resources, H.M.; data curation, H.M., E.B.C. and H.M.Y.; writing—original draft preparation, H.M., E.B.C. and H.M.Y.; writing—review and editing, H.M., and C.Z.; visualization, E.B.C. and H.M.Y.; supervision, C.Z.; project administration, H.M.Y. and C.Z.; funding acquisition, C.Z. All authors have read and agreed to the published version of the manuscript.

Funding: This research received no external funding.

Informed Consent Statement: Not applicable.

Data Availability Statement: Data can be provided by corresponding author on request.

Acknowledgments: We would like to acknowledge EiTEX leather laboratory store.

Conflicts of Interest: The authors declare no conflict of interest.

References

1. Kumar, M.P.; Fathima, N.N.; Aravindhan, R.; Rao, J.R.; Nair, B.U. An Organic Approach for Wet White Garment Leathers. *J. Am. Leather Chem. Assoc.* **2009**, *104*, 113–119.
2. Mohamed, H.A.; Van Klink, E.G.; ElHassan, S.M. Damage caused by spoilage bacteria to the structure of cattle hides and sheep skins. *Int. J. Anim. Health Livest. Prod. Res.* **2016**, *2*, 39–56.
3. Kanagaraj, J.; Senthilvelan, T.; Panda, R.C.; Kavitha, S. Eco-friendly waste management strategies for greener environment towards sustainable development in leather industry: A comprehensive review. *J. Clean. Prod.* **2015**, *89*, 1–17. [CrossRef]
4. Krishnaraj, K.; Thanikaivelan, P.; Chandrasekaran, B. Mechanical properties of sheep nappa leather influencing drape. *J. Am. Leather Chem. Assoc.* **2008**, *103*, 215–221.
5. Ork, N.; Mutlu, M.M.; Yildiz, E.Z.; Pamuk, O. Sewability properties of garment leathers tanned with various tanning materials. *Ann. Univ. Oradea Fascicle Text. Leatherwork* **2016**, *XVII*, 197–202.
6. Choudhary, A.K.; Bansal, P. Drape measurement technique using manikins with the help of image analysis. In *Manikins for Textile Evaluation*; Woodhead Publishing Duxford, CAM: Oxford, UK, 2017; pp. 173–195.
7. Sasikala, L.; Ganesan, P.; Hariharan, S. Processing of Leather for Garments—An Overview. *Man-Made Text. India* **2007**, *50*, 356–360.
8. Eakanayake, E.; Jayamanne, S.; Wickramasinghe, W. Development of garment leather from Yellowfin tuna (Thunnusalbacares) skin. In Proceedings of the Research Symposium of UvaWellassa University, Technical Session-Sri Lanka: Aquatic Resources Technology, Badulla, Sri Lanka, 29–30 January 2015; pp. 33–35.
9. Işik, N.O.; Karavana, H.A. Determination of Some Physical Characteristics of Artificially Aged Chrome Tanned Garment Leather. *Text. Appar.* **2012**, *22*, 64–69.

10. Cui, Y.M.; Zhong, H.; Zhu, D.H. Classify Application and Development Trend of Children Clothing Material Modeling and Style. In *Proceedings of Advanced Materials Research*; Trans Tech Publications Ltd.: Bäch, Switzerland, 2011; pp. 534–538.
11. Sudha, T.B.; Thanikaivelan, P.; Aaron, K.P.; Krishnaraj, K.; Chandrasekaran, B. Comfort, Chemical, Mechanical, and Structural Properties of Natural and Synthetic Leathers Used for Apparel. *J. Appl. Polym. Sci.* **2009**, *114*, 1761–1767. [CrossRef]
12. Wang, X.; Liu, X.; Deakin, C.H. Physical and mechanical testing of textiles. In *Fabric Testing*; Woodhead Publishing Kidlington: Oxford, UK, 2008; pp. 90–124.
13. Hu, J.; Xin, B. 2-Structure and mechanics of woven fabrics. In *Structure and Mechanics of Textile Fibre Assemblies*, 2nd ed.; Schwartz, P., Ed.; Woodhead Publishing: Cambridge, MA, USA, 2008; pp. 27–60.
14. Ramasubramanian, M. Physical and mechanical properties of towel and tissue. In *Handbook of Physical Testing of Paper*; CRC Press: Boca Raton, FL, USA, 2001; pp. 683–896.
15. Cusick, G. 46—The dependence of fabric drape on bending and shear stiffness. *J. Text. Inst. Trans.* **1965**, *56*, T596–T606. [CrossRef]
16. Krishnaraj, K.; Thanikaivelan, P.; Phebeaardn, K.; Chandrasekaran, B. Effect of sewing on the drape of goat suede apparel leathers. *Int. J. Cloth. Sci. Technol.* **2010**, *22*, 358–373. [CrossRef]
17. He, X.; Wang, Y.N.; Zhou, J.F.; Wang, H.B.; Ding, W.; Shi, B. Suitability of Pore Measurement Methods for Characterizing the Hierarchical Pore Structure of Leather. *J. Am. Leather Chem. Assoc.* **2019**, *114*, 41–47.
18. Nasr, A.; Abdelsalam, M.; Azzam, A. Effect of tanning method and region on physical and chemical properties of barki sheep leather. *Egypt. J. Sheep Goat Sci.* **2013**, *8*, 123–130. [CrossRef]
19. Howes, F.N. Vegetable tanning materials. *Veg. Tann. Mater.* **1953**, *XI*, 325. [CrossRef]
20. Covington, A. The chemistry of tanning materials. In *Conservation of Leather and Related Materials*; Routledge: London, UK, 2005; pp. 44–57.
21. Jaynes, J.; Ding, X.; Xu, H.; Wong, W.K.; Ho, C.M. Application of fractional factorial designs to study drug combinations. *Stat. Med.* **2013**, *32*, 307–318. [CrossRef] [PubMed]
22. Box, G.; Bisgaard, S.; Fung, C. An explanation and critique of Taguchi's contributions to quality engineering. *Qual. Reliab. Eng. Int.* **1988**, *4*, 123–131. [CrossRef]
23. Haines, B.; Barlow, J. The anatomy of leather. *J. Mater. Sci.* **1975**, *10*, 525–538. [CrossRef]
24. Abdin, Y.; Taha, I.; El-Sabbagh, A.; Ebeid, S. Description of draping behaviour of woven fabrics over single curvatures by image processing and simulation techniques. *Compos. Part B Eng.* **2013**, *45*, 792–799. [CrossRef]
25. Zhou, N.; Ghosh, T.K. On-line measurement of fabric bending behavior: Part II: Effects of fabric nonlinear bending behavior. *Text. Res. J.* **1998**, *68*, 533–542. [CrossRef]
26. Xu, G.; Wilson, K.S.; Okamoto, R.J.; Shao, J.Y.; Dutcher, S.K.; Bayly, P.V. Flexural Rigidity and Shear Stiffness of Flagella Estimated from Induced Bends and Counterbends. *Biophys. J.* **2016**, *110*, 2759–2768. [CrossRef]
27. Phebe, K.; Thanikaivelan, P.; Krishnaraj, K.; Chandrasekaran, B. Factors Influencing the Seam Efficiency of Goat Nappa Leathers. *J. Am. Leather Chem. Assoc.* **2012**, *107*, 78–84.
28. Ukey, P.; Kadole, P. Characterizations of bottom wear fabrics for drapability. *Man-Made Text. India* **2020**, *48*, 15–17.
29. Maki, J.M.; Wetzel, J.E.; Wichiramala, W. Drapeability. *Discret. Comput. Geom.* **2005**, *34*, 637–657. [CrossRef]
30. Hunter, L.; Fan, J.; Chau, D. Fabric and garment drape. In *Engineering Apparel Fabrics and Garments*; CRC Press: Boca Raton, FL, USA, 2009; pp. 102–130.

Article

Analysis of Polygonal Computer Model Parameters and Influence on Fabric Drape Simulation

Slavenka Petrak [1,*], Maja Mahnić Naglić [1], Dubravko Rogale [1] and Jelka Geršak [2]

[1] Department of Clothing Technology, Faculty of Textile Technology, University of Zagreb, 10 000 Zagreb, Croatia; maja.mahnic@ttf.unizg.hr (M.M.N.); dubravko.rogale@ttf.unizg.hr (D.R.)
[2] Research and Innovation Centre for Design and Clothing Science, Faculty of Mechanical Engineering, University of Maribor, 2000 Maribor, Slovenia; jelka.gersak@uni-mb.si
* Correspondence: slavenka.petrak@ttf.unizg.hr

Abstract: Contemporary CAD systems enable 3D clothing simulation for the purpose of predicting the appearance and behavior of conventional and intelligent clothing in real conditions. The physical and mechanical properties of the fabric and the simulation parameters play an important role in this issue. The paper presents an analysis of the parameters of the polygonal computer model that affect fabric drape simulation. Experimental research on physical and mechanical properties were performed for nine fabrics. For this purpose, the values of the parameters for the tensile, bending, shear, and compression properties were determined at low loads, while the complex deformations were analyzed using Cusick drape meter devices. The fabric drape simulations were performed using the 2D/3D CAD system for a computer clothing design on a disk model, corresponding to real testing on the drape tester in order to allow a correlation analysis between the values of drape parameters of the simulated fabrics and the realistically measured values for each fabric. Each fabric was simulated as a polygonal model with a variable related to the side length of the polygon to analyze the influence of the polygon size, i.e., mesh density, on the model behavior in the simulation. Based on the simulated fabric drape shape, the values of the areas within the curves necessary to calculate the drape coefficients of the simulated fabrics were determined in the program for 3D modelling. The results were statistically processed and correlations between the values of the drape coefficients and the optimal parameters for simulating certain physical and mechanical properties of the fabric were determined. The results showed that the mesh density of the polygonal model is an important parameter for the simulation results.

Keywords: fabric; drape; physical and mechanical properties; polygonal model; 3D simulation

Citation: Petrak, S.; Mahnić Naglić, M.; Rogale, D.; Geršak, J. Analysis of Polygonal Computer Model Parameters and Influence on Fabric Drape Simulation. *Materials* 2021, *14*, 6259. https://doi.org/10.3390/ma14216259

Academic Editor: Philippe Boisse

Received: 31 August 2021
Accepted: 18 October 2021
Published: 21 October 2021

Publisher's Note: MDPI stays neutral with regard to jurisdictional claims in published maps and institutional affiliations.

Copyright: © 2021 by the authors. Licensee MDPI, Basel, Switzerland. This article is an open access article distributed under the terms and conditions of the Creative Commons Attribution (CC BY) license (https://creativecommons.org/licenses/by/4.0/).

1. Introduction

The behavior of textile materials from the aspect of mechanical properties, such as tensile, bending, shear, and compression properties, can be observed at two levels of load, i.e., at lower loads, to which textile materials are exposed during processing as well as in further use, and at higher loads. The mechanical properties of textile materials at lower loads can thus be used for process control and optimization, as well as for the development, construction and computer clothing design and the simulation of virtual conventional and intelligent clothing. Namely, the model takes into account the parameters of mechanical properties of textile materials as a complex textile structure, which allow the simulation of falling or drape [1]. Drape as a complex deformation of fabric can be generally defined from two viewpoints, as a two-dimensional or three-dimensional drape. The two-dimensional drape is associated with the deformation caused by gravity acting on a textile surface. Due to its own weight, the fabric bends in one plane, while the three-dimensional drape allows the fabric to be deformed into folds within more than one plane [2]. C.C. Chu defined 'Drape' and 'Drapeability' as terms for the property of textile materials which allows a fabric to orient itself into graceful folds or pleats when acted upon by the force gravity [3].

In practical terms, it means that the fabric has good drape qualities if the configuration is pleasing to the eye. From this point of view, the word drape is also a qualitative term [2]. Drape parameters can be used to predict the ability to shape and appearance of a garment on the human body and are usually tested using the Cusick drape meter, as a standard and well-known measurement equipment [1,4,5].

Given the increasing application of computer technologies in 2D/3D clothing design, more simple systems for determining the values of mechanical parameters of the fabric have recently been developed, whose parameters can be implemented in CAD systems for 3D prototype development and virtual testing of conventional and intelligent clothing [6–8]. However, the Kawabata Evaluation System (KES) is still the best known and most significant system for the objective evaluation of textiles. The main mechanical properties related to the fabric behavior during production and wearing are tensile, bending, shear, and compression properties [1]. Many authors have investigated the influence of the mechanical properties of fabrics on drape [9–14]. Geršak, who studied the influence of the elastic potential of fabrics which expresses fabric ability to recover, found that it directly influences the drapeability of the fabric. The results obtained from the relationship between the fabric elastic potential and its fit, i.e., drapeability, indicate that elastic potential influences the drape coefficient and the crease depth. Tensile elastic potential and bending elastic potential directly impact drape, while the influence of shear elastic potential is reflected indirectly through the value of shear hysteresis, known as $2HG5$ [15]. Jedda and associates analyzed the relationship between the drape coefficients and the mechanical properties determined by the FAST system and found that the bending and shear stiffness correlate with the drape coefficient better than the thickness [16]. Various prediction models have also been developed which, in addition to the drape coefficient, includes other parameters, such as the number, shape, and dimensions of the folds [13,17,18]. In his research, Jeong found a large variability in the number of folds in the same fabric, which he explains with the influence of the drape speed and the ratio of the sample size to the test disk [19,20]. The development of 3D clothing simulation methods began in the 1990s, with 2D cutting parts simulating their joining around a virtual body model, and surface deformability based on the values of the mechanical properties of the fabric being simulated [21]. The application of virtual 3D technology in the development of new clothing models contributes to greater precision and greatly reduces the cost of making trial models, given the virtual fit testing [22–24]. The precisely determined values of the parameters of textile material mechanical properties are a very important factor, i.e., the mathematical model on which the algorithm defining the behavior and deformability of the target material in the simulation is based [22,25], as well as the precise construction of 2D cutting parts and anthropometric characteristics of the virtual body model [24,26].

Clothing simulation models are mostly based on the particle mesh method and the finite element method [23,27]. Considering the complex anisotropic behavior of the textile material, the quadratic particles model and triangular springs mesh allow a better representation of the virtual fabric using objectively determined values of the mechanical properties of the target fabric. Most commercial clothing design and 3D simulation systems allow the entry of the data of mechanical properties determined by one of the two leading objective evaluation systems, KES and/or FAST [25,28–31]. The relevant parameters of physical and mechanical properties used for simulations are tensile elongations in warp and weft directions, bending rigidity in warp and weft direction, shear rigidity, thickness, and weight. If we compare the KES and FAST system, the KES system allows the determination of the partially and completely reversible deformations caused by small tensile, compression, shear, and bending stresses, while the FAST system measures similar low-stress fabric mechanical properties (compression, bending, extension, and shear), but does not measure recovery properties [1,32]. For example, shear properties measured using the KES system is based on the shear deformation of a specimen. As shown in Figure 1a, constant tension force F_{pt} is applied along the direction orthogonal to shear deformation and perpendicular to tensile force F, respectively. Shear rigidity is defined as the average slope of the curve F_G (γ), (Figure 1b) [1]:

$$G = tg\,\gamma = \frac{\Delta F_G}{\Delta \gamma} \qquad (1)$$

where shear force F_G is given by:

$$F_G = F - F_{pt} \cdot tg\,\gamma \qquad (2)$$

where F_G is the shear force per unit length, F_{pt} is the constant tensile force (9.807 cN per unit length), and γ is the shear angle ($\gamma = 8°$).

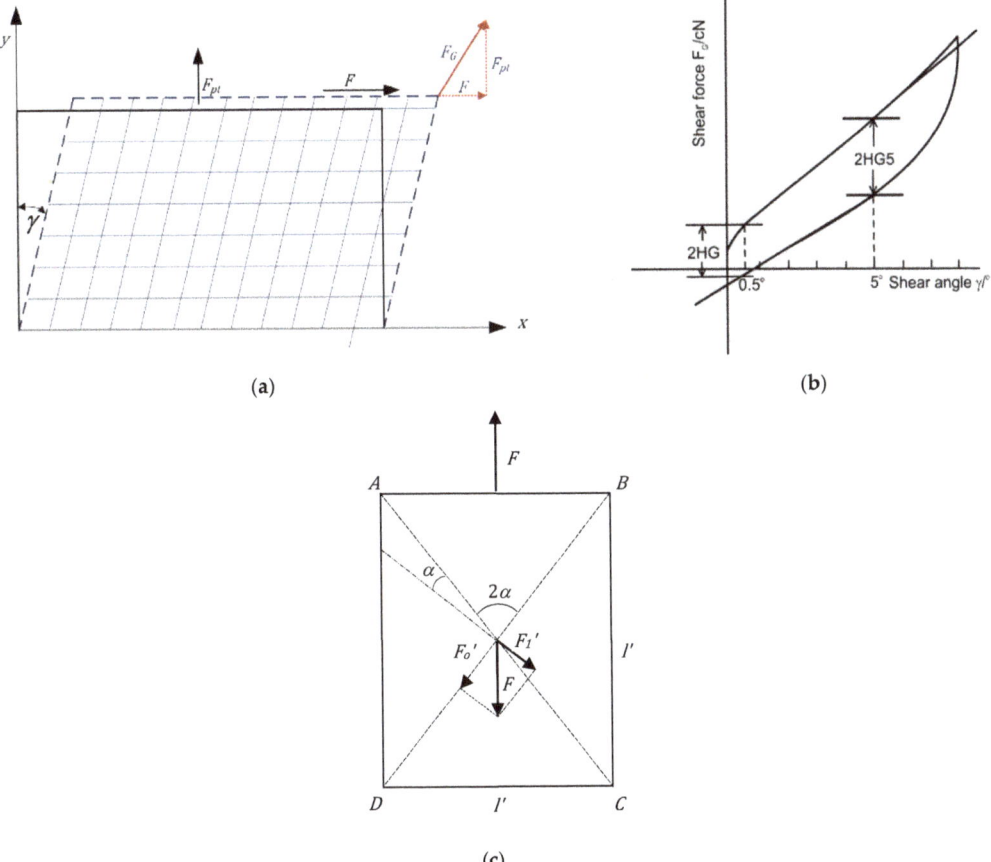

Figure 1. Principle of shear properties measurement: (**a**,**b**) Principles and diagram for KES and (**c**) FAST—Bias extension test.

In the case of the FAST system, it is a bias extension test, in which the warp and weft threads are parallel to the diagonals (Figure 1c) [1]. Applying a force F, e.g., in the vertical direction in the figure, the square is deformed into a rectangle, with sides l' and l'', and the threads incline at an angle $\pm\gamma$ compared with the vertical (Figure 1c). In the FAST measurement system, simplified formulae are used to calculate shear rigidity G, where the deformation is measured as the tangent of the shear angle and can be written as [1]:

$$G = \frac{F_G}{tg\,\gamma} = \frac{\frac{F}{2}}{2 \cdot e} = \frac{F}{4 \cdot e} = \frac{4.9 \cdot 100}{4 \cdot \varepsilon_{B5}} = \frac{123}{\varepsilon_{B5}} \qquad (3)$$

Thus, the application of data determined by the KES system for clothing simulation provides a better assessment of the appearance and fit of a virtual garment model [33]. The application of such a methodology makes it possible to assess the suitability of the target fabric for making a garment model with regard to the desired appearance and fit of the model on the virtual body [25,26].

The aim of the presented research is to determine the correlation between the density of the model polygonal mesh during the simulation and the deformation of the surface to which the mechanical and physical properties of the selected fabrics are assigned. The purpose of the research is to determine the optimal simulation parameters depending on the mechanical properties of the target fabric in order to gain insight into the algorithm that determines the mechanical behavior of the polygonal surface simulating the fabric in a CAD system.

2. Research Methodology

An objective evaluation of the parameters of physical and mechanical properties was performed on nine fabrics intended for the production of outerwear. The evaluation of the individual fabrics was carried out with two different systems, Fabric Assurance by Simple Testing (FAST) [34] and Kawabata Evaluation System (KES) [1,35], in order to examine the applicability of each system in the process of computer development and obtaining a realistic 3D model prototype. Complex deformations, i.e., drape of selected fabrics was examined using the Cusick drape meter device, and the values of the determined drape parameters of individual fabrics were used as reference values for evaluating the simulation results. In order to investigate the influence of the polygonal mesh density of digital cutting parts, for each fabric, four drape simulations were performed with a polygon side length variation of two, four, six, and eight millimeters. The obtained computerized drape models were used to analyze the shape and dimensions of the inside curve surface defined by the fabric drape. Model processing and surfaces calculations within the drape curves were performed using the CAD program for 3D modeling Rhinoceros [36]. Based on the determined areas, the drape coefficients were calculated for each simulated fabric in all four variations of the polygonal mesh density. Based on the determined values of the drape coefficients of the simulated fabrics, a correlation analysis was performed with the coefficients of real samples determined using the Cusick drape meter. In this way, the influence of mesh density on the result of 3D simulation was investigated, and the optimal polygon size was defined in relation to the physical and mechanical properties of the target fabric, which gives the results closest to the values of real samples.

2.1. Test Fabric Samples

Nine different fabrics intended for making conventional and intelligent outerwear were selected for the research. The selected fabrics differ according to the raw material composition, weave structure, thread density, and fabric weight, which is shown in Table 1. The thread density was determined on conditioned samples by counting the threads in the warp and weft direction using a magnifying glass according to ISO 7211/2-1984 [37]. The fabric weight was determined according to ISO 3801 [38], where five conditioned samples measuring 100 cm^2 were tested for each fabric, and the fabric mass is expressed as the mean value of individual measurements in g/m^2.

Table 1. Structure of selected fabrics according to raw material composition, wave, thread density, and fabric weight.

Fabric	Raw Material Composition	Wave	Density [Thread/cm]		Weight [g/m²]
			Warp	Weft	
F1	100% polyester fiber	Twill 3/1	112.2	41.2	135.2
F2	100% polyester fiber	Plane 1/1	72.8	36.3	83.7
F3	63% polyester fiber, 23% viscose fiber, 4% elastane fiber	Twill 2/1	35.7	28.2	286.1
F4	100% wool	Plane 1/1	38.9	35.9	161.8
F5	75% wool, 15% polyester fiber, 10% elastane fiber	Twill 2/1	23.7	24.1	178.7
F6	100% polyester fiber	Plane 1/1	19.6	18.0	159.8
F7	97% polyester fiber, 3% elastane fiber	Plane 1/1	23.2	20.8	178.1
F8	65% cotton, 31% polyester fiber, 4% elastane fiber	Twill 2/1	42.6	18.1	253.4
F9	65% polyester fiber, 35% viscose fiber	Plane 1/1	42.7	31.1	366.8

2.2. Determination of the Parameters of Mechanical Properties

Measuring devices of the FAST and KES systems were used to determine the values of the parameters the mechanical properties of the selected fabrics, as seen in Table 2.

Table 2. Values for the mechanical properties of fabrics determined by KES and FAST system (suffix 1 refers to warp direction and suffix 2 to weft direction).

Measurement system	Samples	EM-1 [%]	EM-2 [%]	B-1 [cNcm]	B-2 [cNcm]	G [cN/(cm°)$^{-1}$]	T0 [mm]
KES	F1	2.75	6.92	0.0195	0.0061	0.41	0.405
	F2	3.90	31.28	0.0085	0.0033	0.46	0.356
	F3	21.54	22.05	0.0297	0.0240	0.61	0.900
	F4	3.02	9.58	0.0324	0.0161	0.73	0.439
	F5	9.05	11.49	0.0214	0.0200	0.50	0.539
	F6	12.41	9.63	0.0362	0.3980	0.49	0.600
	F7	7.02	7.75	0.0532	0.4440	0.69	0.510
	F8	3.48	21.15	0.0743	0.0370	1.26	0.980
	F9	3.97	6.27	0.3354	0.0989	1.15	2.420

Measurement system	Samples	E100-1 [%]	E100-2 [%]	B-1 [µNm]	B-2 [µNm]	G [Nm^{-1}]	ST [mm]
FAST	F1	1.1	3.9	4.5	1.9	23	0.060
	F2	1.6	18.0	2.1	1.1	16	0.080
	F3	11.9	11.7	7.7	6.3	27	0.174
	F4	1.4	4.9	7.6	3.9	33	0.082
	F5	4.9	5.9	5.7	5.3	24	0.086
	F6	6.5	5.1	9.1	9.3	24	0.096
	F7	3.2	3.4	11.9	9.8	38	0.072
	F8	1.6	10.6	32.1	10.3	60	0.225
	F9	1.5	2.8	94.3	23.4	67	0.960

From the determined parameter sets of the systems, the following parameters were selected for 3D simulation: extension *EM* by maximum values of *F* (*Fm* = 490.35 cN per unit width) in the warp and weft direction (KES-FB system); extension *E100* by load of 98.07 cN per unit width in the warp and weft direction (FAST system); bending rigidity *B* in the warp and weft direction; shear rigidity *G*; thickness *T0* of fabric at maximum pressure 0.49 cN cm^{-2} (KES-FB system); and the surface layer thickness *ST* as the difference in thickness measured at the two loads (*ST* = *T2*–*T100*), where *T2* is thickness at 1.96 cN cm^{-2} and *T100* is thickness at 98.07 cNcm^{-2} as a compression property parameter (FAST system) [1,2], as seen in Table 2. The determined parameter values were translated into units suitable for input into a CAD clothing simulation system using the fabric converter [39].

2.3. Research on Fabric Drape

Fabric drape was investigated using a Cusick drape meter, (Figure 2), on conditioned 30-cm-diameter circular samples, where the measurement for each fabric was performed on five samples, and the results were expressed as the mean of the individual measurements for each fabric [1]. Table 3 presents the captured shaded surfaces of drape fabric samples obtained using Cusick drape meter, based on which the drape coefficients (K_d) were calculated for each fabric according to the Expression (4), where *A* [mm^2] is the projection area of the deformed fabric shape, R_1 [mm] is the radius of the horizontal disk (90 mm), and R_2 [mm] is the radius of the undeformed sample surface.

$$K_d = \frac{A - \pi R_1^2}{\pi (R_2^2 - R_1^2)} \quad (4)$$

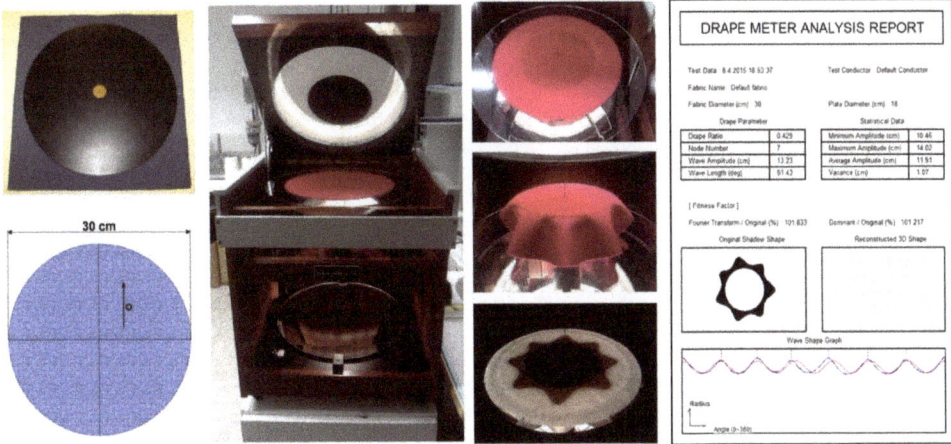

Figure 2. Fabric drape test on a Cusick drape meter.

Table 3. Measured values of drape coefficient for nine selected fabrics (F1–F9).

Samples	F1	F2	F3	F4	F5	F6	F7	F8	F9
K_d	0.222	0.227	0.285	0.319	0.348	0.403	0.442	0.548	0.647

2.4. Computer Simulation of Fabric Drape

An Optitex CAD system was used for computer simulations of fabric drape [40]. The method of performing simulations and determining the drape coefficients of the simulated samples was defined in accordance with the Cusick drape meter test method in real conditions. For this purpose, circular samples with a diameter of 30 cm were computer constructed (Figure 3a), and the simulations were performed on a solid disk model with a diameter of 18 cm, as found in the Cusick drape meter (Figure 3b). The parameter values of the physical and mechanical properties of the individual fabrics determined with the FAST and KES measuring systems were converted using the fabric editor converter and applied to circular samples for the simulation.

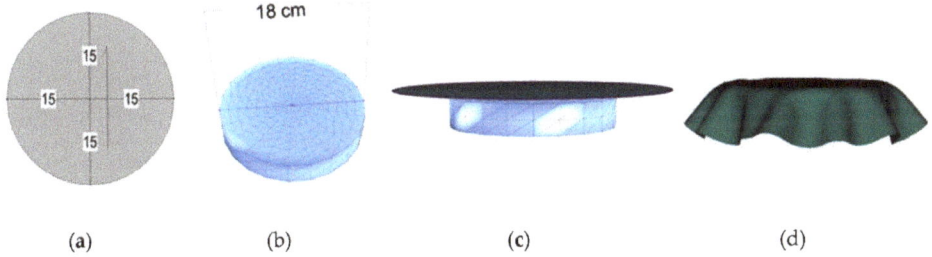

(a) (b) (c) (d)

Figure 3. The process of fabric drape computer simulation in the Optitex CAD system: (**a**) creation of a planar circular sample of given dimensions; (**b**) a solid disk model of specified dimensions; (**c**) positioning of the planar circular sample above the disk model; (**d**) computer simulation of fabric drape.

During the simulations, the circular samples are positioned exactly in the middle and just above the solid disk (Figure 3c), whereby a free fall of the sample onto the disk is performed and characteristic folds were formed as a shading curve of the fabric drape surface (Figure 3d). To determine the influence of polygonal mesh density on the simulation results, each fabric was simulated four times, with four different polygon dimensions, i.e., with a polygon side length of two, four, six, and eight millimeters, (Figure 4).

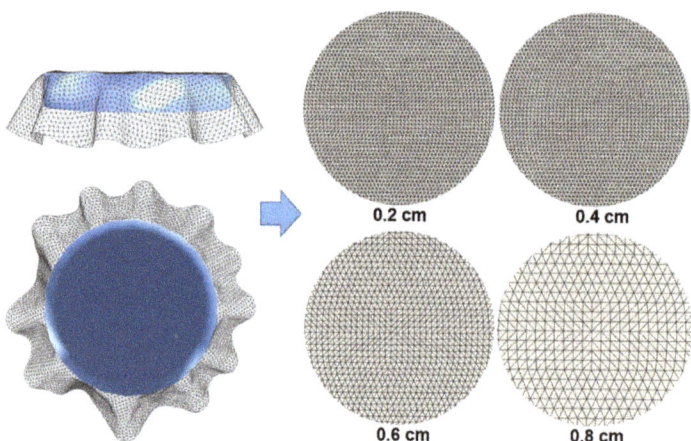

Figure 4. Computer circular fabric samples with different polygonal mesh densities, i.e., polygon side lengths.

Computer 3D models of drape samples were stored as surface models suitable for further analysis. The shape and surface analysis of the simulated samples was performed using the Rhinoceros 3D modeling program. In order to determine the drape coefficients on the simulated samples, it is primarily necessary to extract the characteristic drape curve that describes the shape and size of the projected area, and on the basis of which the coefficients are calculated. The edges of the simulated drape models were separated by computer processing, creating 3D curves which describe the edges of the simulated fabric models. The created 3D curves use a 3D flattening method mapped to the y plane and closed in the surface (Figure 5), which corresponds and is comparable to the method of testing and determining projected drape surfaces using the Cusick drape meter. Furthermore, computer measurements of surface areas defined by drape curves were performed, and based on the determined values, drape coefficients were calculated for each simulated sample according to the expression (4).

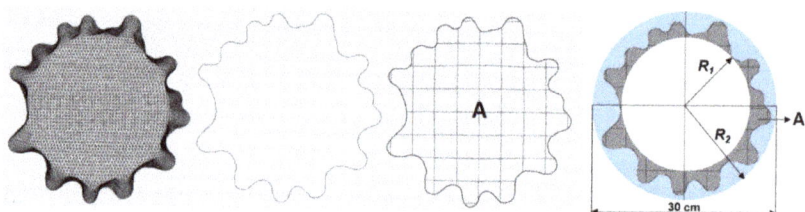

Figure 5. Computer processing of simulated drape samples in the Rhinoceros program.

3. Results and Discussion

From the results of the drape simulations performed, the influence of the model polygon mesh density on the properties of the simulated surface is visible. By increasing the dimensions of the polygon, i.e., decreasing the density of the polygonal surface, a change in the shape and size of the shaded drape surface can be seen from the presentation of the simulated samples shown in Table 4.

Table 4. Computer simulations of drape fabric samples with different polygonal mesh densities.

Samples	FAST				KES			
	0.2 mm	0.4 mm	0.6 mm	0.8 mm	0.2 mm	0.4 mm	0.6 mm	0.8 mm
F1								
F2								
F3								
F4								
F5								
F6								
F7								
F8								
F9								

The comparative analysis of the calculated drape coefficients of samples simulated using the values of the mechanical parameters determined by the FAST and KES system and the drape coefficient of the realistically measured samples showed a tendency towards proportionality between the drape parameters and the density parameters of the simulated polygonal surface in both cases. The samples simulated based on the mechanical parameters determined by the KES system showed a significantly higher sensitivity depending on the defined density of the model polygonal mesh compared to samples simulated with the parameters determined by the FAST system.

Figure 6 presents a scatter graph with 95 percent reliability and shows positive a linear correlation between the real (Kd) and simulated values of drape coefficients (Kd_0.2–Kd_0.8) with different side lengths. A high correlation was found between the values of the drape coefficient of the real and all simulated samples, with slight differences in values depending on the change in the parameters of the polygonal mesh. However, simulations performed using data determined by the KES system show higher values of the determination coefficient ($r^2 > 0.8$), which is expected, Table 5. Given the visible deviation of the individual samples on the graph (Figure 6), the relationships between the individual values of the drape coefficients and the mechanical properties of an individual fabric sample were further analyzed.

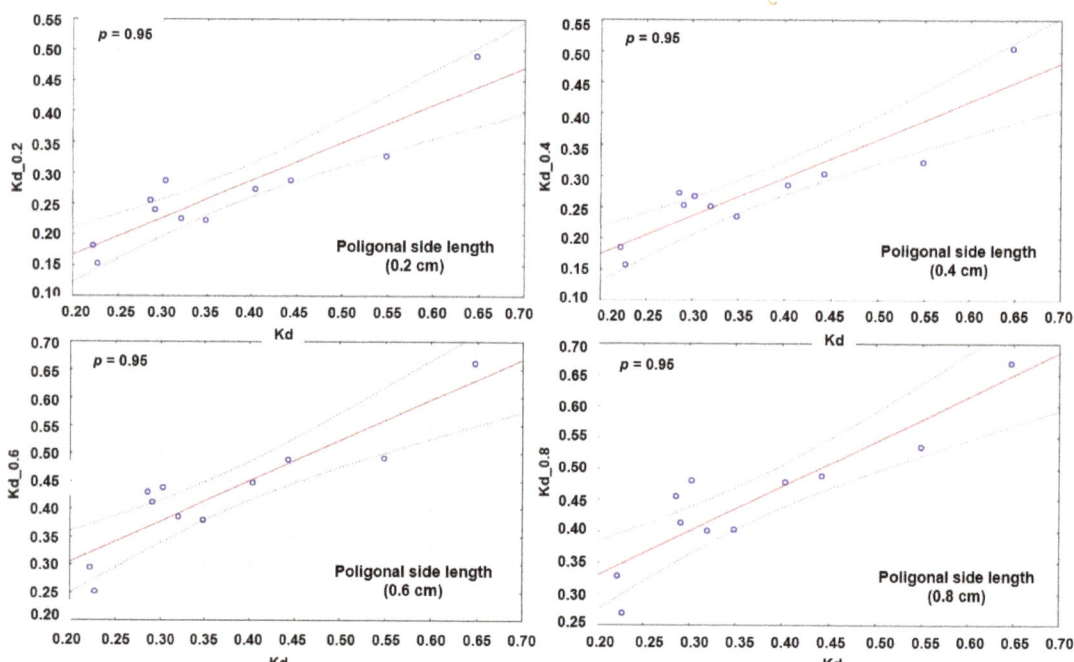

Figure 6. Linear correlation scatter graphs of real and simulated drape coefficients values with different polygonal mesh densities. p—reliability, Kd—drape coefficients of real measured fabrics, Kd_0.2—drape coefficients of fabrics simulated with polygonal side length of 2 mm, Kd_0.4—drape coefficients of fabrics simulated with polygonal side length of 4 mm, Kd_0.6—drape coefficients of fabrics simulated with polygonal side length of 6 mm, and Kd_0.8—drape coefficients of fabric simulated with polygonal side length of 8 mm.

Table 5. Linear correlation analysis between the values of the drape coefficients of real fabrics and fabrics simulated with different polygonal mesh densities.

Measurement System	FAST				KES			
Correlation parameters	K_d 0.2	K_d 0.4	K_d 0.6	K_d 0.8	K_d 0.2	K_d 0.4	K_d 0.6	K_d 0.8
r	0.868	0.861	0.897	0.915	0.910	0.912	0.900	0.899
r^2	0.754	0.742	0.805	0.839	0.828	0.831	0.810	0.809
t	5.250	5.090	6.093	6.841	6.591	6.649	6.197	6.186
p	0.001	0.001	0.000	0.000	0.000	0.000	0.000	0.000
a	0.037	0.026	0.272	0.305	0.046	0.052	0.159	0.187
b	0.234	0.946	0.750	0.747	0.607	0.612	0.726	0.712

r—Pearson correlation coefficient, r^2—determination coefficient, t—value to test the significance of the correlation coefficient, p—level of correlation significance, a—regression constant (value of the dependent variable when the value of the independent variable is equal to 0), and b—regression coefficient (average change in the dependent variable value for the unit change in the independent variable).

It is evident from the determined values of the drape coefficients of the samples simulated on the basis of mechanical parameters determined by the KES system with different polygonal mesh densities that the surface with higher density and thus higher mobility (polygon side size 0.4 mm) better corresponds to the samples with lower drape coefficient (Kd = 0.222–0.319), i.e., soft fabrics that have a nice drop. The lower density of the polygonal mesh of the simulated surface (polygon side size 0.6 mm) corresponds to samples with a higher drape coefficient (Kd = 0.348–0.442), i.e., stiffer fabrics that do not have such a nice drop, while for fabrics with a drape coefficient above 0.5 a polygon mesh with a polygon side size of 0.8 mm is most suitable, as seen in Table 6.

Table 6. Values of drape coefficients of simulated samples.

Samples	Drape K_d	Simulation FAST Data				Simulation KES Data			
		K_d 0.2	K_d 0.4	K_d 0.6	K_d 0.8	K_d 0.2	K_d 0.4	K_d 0.6	K_d 0.8
F1	0.222	0.113	0.319	0.492	0.530	0.182	0.186	0.295	0.329
F2	0.227	0.107	0.260	0.453	0.465	0.153	0.158	0.252	0.270
F3	0.285	0.080	0.258	0.434	0.466	0.256	0.273	0.431	0.457
F4	0.319	0.125	0.354	0.528	0.565	0.226	0.252	0.386	0.402
F5	0.348	0.108	0.315	0.507	0.556	0.224	0.235	0.380	0.403
F6	0.403	0.149	0.390	0.575	0.623	0.274	0.286	0.448	0.479
F7	0.442	0.141	0.411	0.611	0.642	0.289	0.304	0.489	0.490
F8	0.548	0.149	0.399	0.583	0.628	0.328	0.323	0.492	0.536
F9	0.647	0.203	0.785	0.840	0.853	0.491	0.505	0.662	0.669

Fabric samples F2 and F9 proved to be an exception to the observed behavior during the simulation. In the case of F2 fabric, the best simulation results were with a length of the polygon side of 0.6 cm instead of 0.4 cm, given the measured lower drape coefficient of the real fabric (Kd = 0.227). By looking at the values of the parameters of the mechanical properties of the fabric sample F2, one can notice a large value of fabric elongation at the maximum force in the weft direction ($EMT-2$ = 31.28%) and very small values of bending rigidity ($B-1$ = 0.0085 cNcm, $B-2$ = 0.0033 cNcm). The values obtained show that it is a very soft fabric with a large transverse stretch, and, from the aspect of performing numerical simulations, we can say that the fabric is extremely deformable. In this sense, the defined high density of the polygonal model mesh additionally contributes to the mobility of the surface and, as such, is not suitable for such samples because excessive deformations occur. The size of 0.6 cm was determined as the optimal value of the polygon size for the fabric sample F2, as presented in Table 6.

In addition, by considering the values of the parameters of the mechanical properties of fabric sample F9, a high value of fabric thickness ($T0$ = 2.42 mm) and higher values of bending rigidity ($B-1$ = 0.3354 cNcm, $B-2$ = 0.0989 cNcm) can be noticed. The obtained values show that it is a fabric with greater thickness and stiffness, and from the aspect of performing numerical simulations, we can say that the fabric is less deformable. In this sense, the optimal polygon size value for sample F9 was the polygon size of 0.6 cm, not 0.8, which was expected given the large value of the drape coefficient measured on a real fabric sample F9 (Kd = 0.647). The lower density of the polygonal surface of the model additionally contributes to the reduced deformability of the surface and, in the case of simulating fabrics with greater thickness and flexural stiffness, prevents real mobility. In this case, the defined smaller dimension of the polygon compensates for the missing surface mobility from the computer aspect and allows for more realistic simulation results (Table 6).

Since the drape is also influenced by other mechanical parameters in addition to the parameters of the bending properties, a correlation analysis was also carried out between the determined values of the mechanical parameters and the drape coefficients of the real and simulated fabric samples, as seen in Table 7.

Table 7. Linear correlation analysis of the parameters of physical and mechanical properties influences the value of the drape coefficient of real and simulated fabrics.

Drape Coefficient	Correlation Parameters	Mechanical Parameters from KES System						
		T0	B-1	B-2	EMT-1	EMT-2	G	W
Kd_{real}	r	0.740	0.849	0.916	−0.250	−0.411	0.884	0.642
	p	0.009	0.001	0.000	0.459	0.209	0.000	0.033
	a	0.292	0.292	0.212	0.406	0.458	0.117	0.147
	b	0.381	0.004	0.020	−0.010	−0.012	0.007	0.106
$Kd_{sim0.4}$	r	0.911	0.955	0.963	−0.352	−0.541	0.775	0.604
	p	0.000	0.000	0.000	0.288	0.086	0.005	0.049
	a	0.272	0.281	0.199	0.434	0.504	0.132	0.145
	b	0.514	0.005	0.023	−0.015	−0.017	0.007	0.110
$Kd_{sim0.6}$	r	0.882	0.932	0.928	−0.368	−0.557	0.756	0.552
	p	0.000	0.000	0.000	0.266	0.075	0.007	0.078
	a	0.399	0.406	0.333	0.541	0.604	0.278	0.302
	b	0.438	0.004	0.020	−0.014	−0.015	0.006	0.088

In this way, the simulated samples gained an insight into the algorithm that determines the mechanical behavior of the polygonal surface that simulates the fabric in the CAD system, which is protected by the manufacturer of the CAD system and is not available because it is a commercial program

For the values of the drape coefficients measured on real samples, a strong correlation was found with the parameters of bending rigidity in both directions and shear rigidity ($r > 0.8$), a medium strong correlation with the parameters of thickness and weight ($r > 0.6$), and a weak correlation with tensile elongation parameters in both directions. If we compare the established correlations of real samples with the correlations of drape coefficients and mechanical parameters of simulated samples, differences in the significance of individual parameters are visible, which can be related to the established differences in the drape visualization of real and simulated samples and determined values of the drape coefficients. The fabric thickness parameter ($T0$) shows a medium–strong correlation (r = 0.74) for the real samples and a strong correlation ($r = 0.91$) for the simulated samples with the drape coefficient, which leads to increased surface stiffness when simulating thicker fabrics, as in the case of the F9 fabric sample. Since the deformability of the surface in the simulation can be directly affected by changing the density parameter of the polygonal mesh, i.e., the length of the polygon side, a smaller optimal polygon size was determined for the F9 fabric sample in relation to the value proportional to the real drape coefficient. This reduced the surface stiffness caused by the excessive influence of the fabric thickness parameter.

In addition, although a weak correlation between drape and tensile parameters was found in both real and simulated samples, higher values of the correlation coefficient can be observed in the simulated samples. The increased influence of tensile parameters on the mechanical behavior of the polygonal surface in the simulation causes excessive mobility, and this may explain the formed shape differences between simulated fabric drape patterns and realistically determined values obtained using Cusick drape meter. An excessive influence of the parameters of tensile properties is especially visible in sample F2, which realistically has a significantly higher value of stretching in the weft direction than the other samples, and this higher value of stretching is further emphasized in relation to the model of mechanical behavior in the simulation, resulting in excessive deformability of the sample.

4. Conclusions

Simulations based on the physical and mechanical properties of fabrics allow prediction assessment of the appearance and behavior of products in the development process, thus saving time and costs required for the production of test samples, which is especially important in the development of new models of conventional and intelligent garments.

The mesh density of the polygonal model is a very important parameter to obtain the most relevant simulation results.

The investigation revealed that the optimal polygon size for fabrics with a drape coefficient of 0.20 to 0.35 is 0.4 mm, for fabrics with a drape coefficient of 0.35 to 0.50 is 0.6 mm, and for fabrics with a drape coefficient above 0.5 is 0.8 mm. However, it is also necessary to further analyze the individual parameters since significantly increased values of individual parameters can further affect the outcome of the simulation, as is the case with fabrics F2 and F9 from this study.

The research conducted using the CAD system Optitex determined the optimal parameters of simulations depending on the mechanical properties of the target fabric, which facilitates the process of computer prototype development and increases accuracy in analyzing and predicting the appearance and behavior of clothing before making a realistic test model. The use of 3D in the textile industry, especially in the fashion industry, is still too low due to a lack of understanding of the computer simulation evaluation process.

The simulated samples showed good topology results, meaning that the shaded areas are similar in size as the ones measured on real fabric samples, but there are visible shape deviations, where the number of folds differs from the number of folds determined by the Cusick drape meter. The correlation analysis of the projection area surface and the drape coefficient of real and simulated samples revealed a very strong linear correlation, which confirmed the influence of physical and mechanical properties on the fabric simulation, and the obtained results and differences can be used in further research to improve the system and simulation process for more realistic folding. In this sense, given the complexity of fabric drape, which is influenced by numerous parameters, future research will focus on the possibility of analyzing the effects of additional simulation parameters reflected in drape properties.

Author Contributions: Conceptualization, S.P. and M.M.N.; Data curation, M.M.N.; Formal analysis, S.P., M.M.N., D.R. and J.G.; Funding acquisition, D.R. and S.P.; Investigation, S.P. and M.M.N.; Methodology, S.P. and M.M.N.; Resources, S.P., D.R. and J.G.; Visualization, M.M.N.; Writing—original draft, S.P., M.M.N., D.R. and J.G.; Writing—review & editing, S.P., M.M.N., D.R. and J.G. All authors have read and agreed to the published version of the manuscript.

Funding: This work has been supported in part by the Croatian Science Foundation through the project IP-2018-01-6363 "Development and thermal properties of intelligent clothing (ThermIC)", as well as by the University of Zagreb trough research grant TP11/21 "Digital clothing for the creative industries".

Institutional Review Board Statement: Not applicable.

Informed Consent Statement: Not applicable.

Data Availability Statement: Data available in a publicly accessible repository.

Conflicts of Interest: The authors declare no conflict of interest.

References

1. Geršak, J. *Objektivno Vrednovanje Plošnih Tekstilija i Odjeće*; Tekstilno-tehnološki fakultet Sveučilišta u Zagrebu: Zagreb, Croatia, 2013.
2. Geršak, J. (Ed.) Study of the complex deformations of textile structure. In *Complex Fabric Deformations and Clothing Modelling in 3D*; (Complex Fabric Deformations & Modelling); LAP LAMBERT Academic Publishing: Saarbrücken, Germany, 2013; pp. 3–60.
3. Chu, C.C.; Cummings, C.L.; Teixeriara, N.A. Mechanics of Elastic Performance of Textile Material, Part V: A Study of the Factors Affecting the Drape of Fabric, Development of a Drape Meter. *Text. Res. J.* **1950**, *20*, 539–548. [CrossRef]
4. Cusick, G.E. The dependence of fabric drape on bending and shear stiffness. *J. Text. Inst.* **1965**, *56*, 596–606. [CrossRef]
5. Cusick, G.E. The measurement of fabric drape. *J. Text. Inst.* **1968**, *59*, 253–260. [CrossRef]
6. Kujipers, S.; Luible Bär, C.; Hugh Gong, R. The measurement of fabric properties for virtual simulation—A critical review. In *IEEE SA Industry Connections*; IEEE Standards Association (IEEE SA): New York, USA, 2020; pp. 8–26.
7. Strazdiene, E. Textiles Objective and Sensory Evaluation in Rapid Prototyping. *Mater. Sci.* **2011**, *17*, 407–412. [CrossRef]
8. Pavlinic, D.Z.; Gersak, J. Design of the System for Prediction of Fabric Behavior in Garment Manufacturing Processes. *Int. J. Cloth. Sci. Technol.* **2004**, *16*, 252–261. [CrossRef]

9. Peirce, F.T. The Handle of Cloth as a Measurable Quantity. *J. Text. Inst.* **1930**, *21*, 377–416. [CrossRef]
10. Abbott, N.J. The Measurement of Stiffness in Textile Fabrics. *Text. Res. J.* **1951**, *21*, 435–444. [CrossRef]
11. Niwa, M.; Seto, F. Relationship between Drapeability and Mechanical Properties of Fabrics. *J. Text. Mach. Soc. Jpn.* **1986**, *39*, 161–168. [CrossRef]
12. Jevšnik, S.; Geršak, J.; Gubenšek, I. The advance engineering methods to plan the behavior of fused panel. *Int. J. Cloth. Sci. Technol.* **2005**, *17*, 161–170. [CrossRef]
13. Robson, D.; Long, C.C. Drape Analysis using Imaging Techniques. *Cloth. Text. Res. J.* **2000**, *18*, 1–8. [CrossRef]
14. Kenkare, N.; Plumlee, T.M. Fabric drape measurement: A modified method using digital image processing. *J. Text. Appar. Technol. Manag.* **2005**, *4*, 1–8. Available online: https://textiles.ncsu.edu/tatm/wp-content/uploads/sites/4/2017/11/Plumlee_full_148_05.pdf (accessed on 10 July 2021).
15. Geršak, J. Study of relationship between fabric elastic potential and garment appearance quality. *Int. J. Cloth. Sci. Technol.* **2004**, *16*, 238–251. [CrossRef]
16. Jedda, H.; Ghith, A.; Sakli, F. Prediction of fabric drape using the FAST system. *J. Text. Inst.* **2007**, *93*, 219–225. [CrossRef]
17. Chu, C.C. *Determination of Factors Which Influence the Draping Properties of Cotton Fabrics*; Forgotten Books: London, UK, 2017; pp. 15–37.
18. Lo, W.M.; Hu, J.L.; Li, L.K. Modeling a Fabric Profile. *Text. Res. J.* **2002**, *72*, 454–463. [CrossRef]
19. Jeong, Y.J. A Study of Fabric-Drape Behavior with Image Analysis Part I: Measurement, Characterization, and Instability. *J. Text. Inst.* **1998**, *89*, 59–69. [CrossRef]
20. Jeong, Y.J.; Philips, D.G. A study of Fabric Drape Behavior with Image Analysis, Part II: The Effect of Fabric Structure and Mechanical Properties on Fabric Drape. *J. Text. Inst.* **1998**, *89*, 70–79. [CrossRef]
21. Sayem, A.S.M.; Kennon, R.; Clarke, N. 3D CAD systems for the clothing industry. *Int. J. Fash. Des. Technol. Educ.* **2010**, *3*, 45–53. [CrossRef]
22. Luible, C. Study of Mechanical Properties in the Simulation of 3D Garments. Ph.D. Thesis, Université de Genève, Genève, Switzerland, 2008.
23. Volino, P.; Magnenat-Thalmann, N. *Virtual Clothing: Theory and Practice*; Springer: Berlin/Heidelberg, Germany, 2000; pp. 35–60.
24. Kuijpers, A.A.M.; Gong, R.H. Virtual tailoring for enhancing product development and sales. In Proceedings of the 4th International Global Fashion Conference: Re-Thinking and Reworking Fashion, Ghent, Belgium, 20–21 November 2014; University College Ghent: Ghent, Belgium, 2014; pp. 1–27.
25. Ancutiene, K.; Strazdiene, E.; Lekeckas, K. Quality evaluation of the appearance of virtual close-fitting woven garments. *J. Text. Inst.* **2014**, *105*, 337–347. [CrossRef]
26. Petrak, S.; Mahnic, M.; Rogale, D. Impact of male body posture and shape on design and garment fit. *Fibres Text. East. Eur.* **2015**, *23*, 150–158. Available online: http://yadda.icm.edu.pl/yadda/element/bwmeta1.element.baztech-43775b65-d95a-4c8e-ad21-85f0a34a5ba0 (accessed on 7 July 2021). [CrossRef]
27. Rizzi, C.; Fontana, M.; Cugini, U. Towards virtual prototyping of complex-shaped multi-layered apparel. *Comput. Aided. Des. Appl.* **2004**, *1*, 207–216. [CrossRef]
28. Pandurangan, P.; Eischen, J.; Kenkare, N.; Lamar, T.A.M. Enhancing accuracy of drape simulation. Part II: Optimized drape simulation using industry-specific software. *J. Text. Inst.* **2008**, *99*, 219–226. [CrossRef]
29. Lim, H.S. Three Dimensional Virtual Try-on Technologies in the Achievement and Testing of Fit for Mass Customization. Ph.D. Thesis, North Carolina State University, Raleigh, North Carolina, 2009. Available online: https://repository.lib.ncsu.edu/handle/1840.16/3322 (accessed on 18 July 2021).
30. Wu, Y.Y.; Mok, P.Y.; Kwok, Y.l.; Fan, J.T.; Xin, J.H. An investigation on the validity of 3D clothing simulation for garment fit evaluation. In Proceedings of the IMProVe 2011 International Conference on Innovative Methods in Product Design, San Servolo, Venice, Italy, 15–17 June 2011; Libreria Internazionale Cortina Padova: Padova, Italy, 2011; pp. 463–468.
31. Power, J. Fabric objective measurements for commercial 3D virtual garment simulation. *Int. J. Cloth. Sci. Technol.* **2013**, *25*, 423–439. [CrossRef]
32. Luible, C.; Magnenat-Thalmann, N. Suitability of Standard Fabric Characterisation Experiments for the Use in Virtual Simulations. Available online: http://citeseerx.ist.psu.edu/viewdoc/download?doi=10.1.1.524.2550&rep=rep1&type=pdf (accessed on 21 July 2021).
33. Kenkare, N.; Lamar, T.A.M.; Pandurangan, P.; Eischen, J. Enhancing accuracy of drape simulation, Part 1: Investigation of drape variability via 3D scanning. *J. Text. Inst.* **2008**, *99*, 211–218. [CrossRef]
34. De Boos, A. Tester, D. *SiroFAST Fabric Assurance by Simple Testing. A System for Fabric Objective Measurement and Its Application in Fabric and Garment Manufacture*; Report No. WT97.02; SCIRO Division of Wool Technology: Geelong, Australia, 1994.
35. Kawabata, S. *The Standardisation and Analysis of Hand Evaluation*, 2nd ed.; The Hand Evaluating and Standardization Committee, The Textile Machinery Society of Japan: Osaka, Japan, 1980.
36. Becker, M.; Golay, P. *Rhino–Nurbs 3D Modeling*; New Riders: San Francisco, CA, USA, 1999.
37. ISO 7211/2-1984 Textiles–Woven Fabrics–Construction–Methods of Analysis, Part 2: Determination of Number of Threads per Unit Length. 1984. 2017. Available online: https://www.iso.org/standard/13842.html (accessed on 15 July 2021).
38. ISO 3801:1977 Textiles–Woven Fabrics–Determination of Mass per Unit Length and Mass per Unit Area. 1977. 2017. Available online: https://www.iso.org/standard/9335.html (accessed on 15 July 2021).

39. Fabric Converter. *Optitex Technical Documentation*; EFI Optitex: Rosh HaAyin, Izrael, 2012.
40. Optitex. Available online: https://optitex.com/ (accessed on 15 July 2021).

Article

Study Regarding the Kinematic 3D Human-Body Model Intended for Simulation of Personalized Clothes for a Sitting Posture

Andreja Rudolf [1,*], Zoran Stjepanovič [1] and Andrej Cupar [2,*]

1. Institute of Engineering Materials and Design, Faculty of Mechanical Engineering, University of Maribor, Smetanova 17, 2000 Maribor, Slovenia; zoran.stjepanovic@um.si
2. Mechanical Engineering Research Institute, Faculty of Mechanical Engineering, University of Maribor, Smetanova 17, 2000 Maribor, Slovenia
* Correspondence: andreja.rudolf@um.si (A.R.); andrej.cupar@um.si (A.C.)

Abstract: This study deals with the development of a kinematic 3D human-body model with an improved armature in the pelvic region, intended for a sitting posture (SIT), using Blender software. It is based on the scanned female body in a standing posture (STA) and SIT. Real and virtual measures of females' lower-body circumferences for both postures were examined. Virtual prototyping of trousers was performed to investigate their fit and comfort on the scanned and kinematic 3D body models and to make comparison with real trousers. With the switch from STA to SIT, real and virtual lower-body circumferences increase and are reflected in the fit and comfort of virtual and real trousers. In SIT, the increased circumferences are attributed to the redistribution of body muscles and adipose tissue around the joints, as well as changes in joints' shapes in body flexion regions, which are not uniformly represented on the kinematic sitting 3D body model, despite improved armature in the pelvic region. The study shows that average increases in waist, hip, thigh, and knee circumferences should be included in the process of basic clothing-pattern designs for SIT as minimal ease allowances, as should, in the future, armature designs that consider muscle and adipose tissues, to achieve realistic volumes for kinematic 3D body models in SIT.

Keywords: kinematic 3D human-body model; sitting posture; Blender software; virtual prototyping; personalized clothing

Citation: Rudolf, A.; Stjepanovič, Z.; Cupar, A. Study Regarding the Kinematic 3D Human-Body Model Intended for Simulation of Personalized Clothes for a Sitting Posture. *Materials* **2021**, *14*, 5124. https://doi.org/10.3390/ma14185124

Academic Editor: Dubravko Rogale

Received: 30 June 2021
Accepted: 2 September 2021
Published: 7 September 2021

Publisher's Note: MDPI stays neutral with regard to jurisdictional claims in published maps and institutional affiliations.

Copyright: © 2021 by the authors. Licensee MDPI, Basel, Switzerland. This article is an open access article distributed under the terms and conditions of the Creative Commons Attribution (CC BY) license (https://creativecommons.org/licenses/by/4.0/).

1. Introduction

In recent years, the use of computer-aided design (CAD) with various computational and analytical tools has played an important role in clothing design and customization. Three-dimensional design software has been mostly used to improve the garment design process through overcoming the limitations of ordinary 2D design methods by generating virtual 3D clothing prototypes. In addition, many researchers have attempted to develop 3D virtual garment prototypes using 3D body models and involving 3D human-body scanning in a static standing or dynamic posture, respectively, as the benefits of virtual prototyping have been recognized. All studies on scanning technologies for the standard scanning processes for effective data collection to study human body-size and -shape and defined protocols for automatic body measurements in static and dynamic anthropometry are focused on accurate 3D body models and the complexity of 3D human body-scan data modelling, with the purpose of measuring accurate body dimensions intended for sizing systems or the development of personalized clothes in a virtual environment [1–8]. The standard avatars in 3D CAD systems still have limitations, and cannot be exactly adapted to the many body figures, postures or body deformities required for personalized clothing prototyping, e.g., for the elderly, athletes, disabled people, pregnant women, workers, etc. Therefore, 3D scanning in the required body postures and generation of accurate

3D models of individuals is required [9–19]. Studies by Abtew M.A. et al. [20–23] focus on the development of comfortable and well-fitted bra patterns for customized female soft-body armour through the 3D-design process of adaptive bust on a virtual mannequin. These studies introduced a novel design technique for developing female adaptive bust volumes on a 3D female virtual mannequin. Their results showed that the proposed method reflected satisfactory fit and comfort, compared with 2D clothing-pattern designs. A co-design-based method for generating two-dimensional basic clothing-pattern designs, for physically disabled people with scoliosis and using three-dimensional virtual technology, was presented in work by Hong Y. et al. [24–26]. The design of functional garments for people with scoliosis or kyphosis, using computer simulation techniques, kinematic 3D body models and the CASP methodology, has been investigated in other studies [27,28]. CASP means curvature, acceleration, symmetry and proportionality. This methodology can be applied for the analysis of geometrical surfaces and their evaluation. An early stage of this analysis methodology was used in a study by Cupar et al. [29]. A recent study, regarding the 3D digital adaptive thorax modelling of people with spinal disabilities, intended for performance-clothing design applications, shows that the developed adaptive thorax model helps to design a basic bodice-pattern design, adapted to the patient's evolving morphology, by recognizing the anthropometric points from certain parts of the skeleton [30,31].

All the approaches described, of which there are many more, are based on the recognition that static 3D models of the human body are not suitable for the construction of sports, medical and protective clothing. Indeed, clothing that conforms to a static shape can be very uncomfortable when performing daily tasks, such as walking, sitting, or reaching. In studies on virtual prototyping of clothing for people with limited body abilities, design considerations made by using a scanned 3D body model in a sitting posture ensure the ergonomic comfort of the garments when worn in a sitting posture, and take into account the functional requirements imposed by strength and movement limitations, such that they do not lead to additional health problems for paraplegics [32,33].

With the development of computer graphics in the field of human body kinematics and the animation of virtual garments, rigged human-body models also began to emerge, from the very beginning [34,35], to improve deformable virtual humans [36]. For animation purposes, an automatic adaptation of existing general models to scanned 3D body-model mesh was developed to be used for kinematic body models [37]. That there is a growing interest in designing personalized garments in sitting or other postures by using a kinematic 3D body model is reflected in the research [38–41]. Kozar et al. [38] designed an adaptive 3D body model intended for the development of clothing for people with limited body capabilities using Blender software. The studies by Zhang D. and Krzywinski S. [40,41] highlight that there is currently no kinematic 3D human-body model for representing realistic body deformations during movement, especially in the elbow, knee, and hip joints. Therefore, they investigated four methods, linear-blend skinning in the simulation software Clo3d, auto-rigging of 3D scans on the online service Maximo and a skinned multi-person linear human model and anatomical simulation using the plugin Ziva Dynamics, which also indicates improper deformation of the body mesh during bending of the elbow, knee, and hip joint regions. In recent research by Klepser A. and Pirch K. [9], a new process for generating 3D body models using Blender 3D software outside of 3D garment simulation software was developed, resulting in a parametric and rigged 3D body model that can be used across platforms. From this research, it is evident that, despite carefully performed scanning and mesh processing, difficulties remained due to incorrect mesh deformations during posture adjustments.

Without a realistic 3D human body shape and natural human postures, it is difficult to create properly fitting clothing. In order to achieve a comfortable wear comfort and reduce development time, clothing should be designed based on specific body postures. In a study by Gill and Parker [7], they researched standing scanning postures (scan posture with feet 40 cm apart and a natural, relaxed standing posture with legs closer together) were found to have a significant effect on hip girth and could cause an average change of 2 cm, which can consequently impact garment fit. While moving the body and performing different body postures, body dimensions may change; especially, an increase of lower body girths has been observed in a sitting posture [5,13,19]. Therefore, these changes in body dimensions must also be considered when developing special clothing pattern designs intended for performing different movements or to be worn in specific body postures. In the work of Delph S. L. et al. [42], it is pointed out that the dynamic simulation of movements enables the software that integrates models describing the anatomy and physiology of the elements of the neuromusculoskeletal system and the mechanics of multi-joint movement. Recent trends in 4D-scanning technology predict a major breakthrough in building a kinematic 3D body model that will provide accurate 3D body-mesh deformation and meet the human need for soft-tissue deformations that mirror real humans [43,44]. In the research of Pons-Moll G. et al. [43] two technologies to capture and process the data of full bodies in motion, 4D scanning and 4D-mesh alignment, were used to analyse over 40,000 scans of ten subjects. Analysis of how soft-tissue motion causes mesh triangles to deform relative to a base 3D body model was performed and a Dyna model was created that approximates soft-tissue deformation and relates the subspace coefficients to the changing pose of the body.

The present study focuses on the development of a kinematic 3D human-body model intended for the development of personalized clothes for a sitting posture, using virtual prototyping. A prerequisite for 3D human-body motion is the kinematic structure of a 3D body model, adapted to the anatomical and anthropometric characteristics of the subjects. Therefore, an improved armature in the pelvic region was modelled, using Blender software, and based on a scanned female body in a standing and a sitting posture. Real and virtual measures of female lower-body circumferences for both postures and their changes when moving between postures were investigated to compare changes of the kinematic 3D body model's mesh during sitting posture adaption. Virtual simulations of personalized- and real-trouser prototypes were investigated to evaluate their fit on standing, sitting and kinematic-sitting 3D body models.

2. Methods

2.1. Measuring Lower-Body Dimensions in the Standing and Sitting Postures of Real Persons

Measurements of the lower-body dimensions in a standing and a sitting posture were carried out to evaluate changes in lower-body dimensions between the observed postures. A case group of twenty-two female subjects voluntarily participated in the study, ranging in age from 20 to 24 years, with different body heights (BH), body weights (BW) and body mass indexes (BMI). Informed consent was obtained from all subjects involved in the study for the use of body measurements for research. During measurement, participants wore leggings, and measurements were rounded up to the nearest half centimetre. The basic average data of the measured participants are presented in Table 1.

Table 1. Basic data of the measured participants.

Age and Body Attributes	Symbol	$\bar{x}$ (x_{min}; x_{max})	SD (cm)	CV (%)
Age (years)	A	21.82 (20; 24)	1.26	5.77
Body height (cm)	BH	168.00 (158.00; 175.00)	0.06	3.36
Body weight (kg)	BW	67.73 (49.00; 95.00)	13.57	20.03
Body mass index	BMI	24.07 (16.37; 32.87)	4.69	19.50

Examined were those body circumferences that impact the development of clothes for a sitting posture and the lower part of the body (trousers). The four body dimensions, measured according to the standard ISO 8559-1 [45], were: waist circumference, hip circumference, thigh circumference and knee circumference in a standing and sitting posture, as shown in Figure 1. During manual measurements, using a measuring tape, locations of the body dimensions, anthropometric landmarks and the standard procedure for measuring the human body, were considered according to the standard ISO 8559-1 [45]. All lower-limb measurements were taken from the right limbs. In sitting posture, hip circumference was measured transversely as the maximum circumference around the buttocks, while knee circumference was measured at a 90-degree angle with knees bent. Manual measurements were rounded up to the nearest half centimetre.

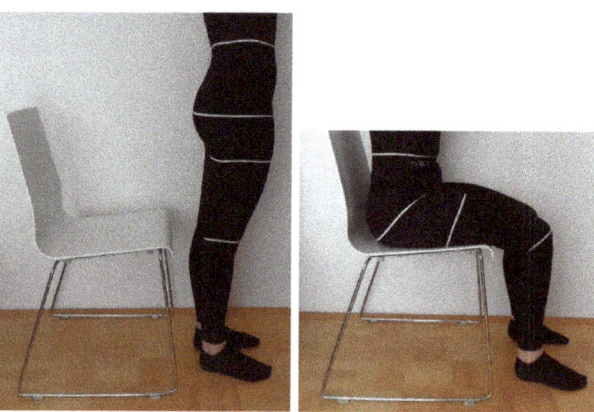

Figure 1. Measurement of real body dimensions.

Data Analysis

The mean values ($\bar{x}$), standard deviations (SD) and coefficients of variation (CV) were calculated for the measurements of body dimensions in a standing (MSTA) and measurements in a sitting posture (MSIT). Differences (D) were calculated between the mean values of MSTA and MSIT. A positive value (+) means that the MSTA is greater than the MSIT and a negative value (−) vice versa.

The Pearson correlation coefficient was used to analyse the relationship between basic data of the participants and their body measurements, as well as the relationship between the basic data of the participants and differences in body measurements between the standing and sitting postures. The aim was to evaluate if there were any influences of the basic data of the participants in this case study on twenty-two female subjects' body height (BH), body weight (BW) and body mass index (BMI) on body measurements for a standing posture and on differences in body measurements between standing and sitting postures ($D_{MSTA-MSIT}$).

2.2. Construction of the Kinematic 3D Body Model

The main idea of the kinematic 3D body model is to take a 3D scan, in a standing posture, and to then pose it to a required posture. Since the human body is a complex bundle of bones, muscles, other soft tissues and skin, such a kinematic model must be complex enough to simulate these real structures. Some authors have created virtual models with most of these structures [39]. In this study, scans of the real female body were used, and they were attached to an armature. When building the armature, focus was mainly on the lower parts of the body (the pelvis region, hip joints and knee joints) due to our goal of exploring the sitting posture. To build a correct kinematic 3D body model, two 3D scans of the same person were used, one in a standing and one in a sitting posture. The standing scan was transformed into a kinematic 3D body model by adding an armature that allowed placing the model in a sitting posture, using the sitting scan as a reference.

2.2.1. Three-Dimensional Scanning

To construct the virtual skeleton for the kinematic 3D body model, scanning of female participants was performed in a standard standing posture and in a sitting posture. Informed consent was obtained from the woman who participated in the study.

An ATOS II 400 3D optical scanning system (Gom GmbH, Braunschweig, Germany), with a measurement volume of 1200 mm × 1200 mm, was used for scanning. During scanning, the woman stood and sat with knees bent at 90 degrees and hands slightly raised, breathing normally, and wearing a thin bodysuit. Scanning, with arms raised and with space between the thighs, was carried out for the purpose of the virtual prototyping of clothing. Optical scanning was performed from different angles and heights to digitize a complete person. A real-time surface mesh of the individual scan was acquired using Atos V6.0.2-6 software (Gom GmbH, Braunschweig, Germany). In addition, 3D human-body-mesh modelling techniques and surface-reconstruction techniques, such as manual mesh-holes filling, mesh cleaning and an automatic algorithm called Poisson reconstruction of mesh, which creates a new watertight mesh over the old one, were used to obtain 3D body models in standing and sitting postures. Finally, some manual mesh sculpting and adjustments were performed in critical areas such as the crotch, where the scanner was unable to capture the surface, to fulfil and reliably present the 3D scans, which method is presented in detail in [38].

2.2.2. Armature as a Virtual Skeleton

In this study the open-source creation suite Blender 3D [46–48] was used to construct the kinematic 3D human-body model, based on the scanned female body in standing and sitting postures. The whole process was interactive, with much trial and error.

In this study, our focus was on the pelvic region of the torso, as this is a critical region for mesh deformation during sitting-posture adjustment. In the first part of the study, the armature was performed with a simple model consisting of only two bones, one on each side, to connect the spine with the thigh bone. This solution was insufficient because the 3D model deformed inappropriately when changing posture from standing to sitting. Placing the bones of the standing 3D body model in position to achieve sitting position, adequate to scanned body models, lead to missing mesh volume in the buttocks area; moving bones to the back side of the model, to achieve a more fulfilled shape of the buttock, resulted to an overall wrong kinematic model, where too-short thighs appear in the sitting posture. These findings lead to the implementation of two pairs of additional bones, so-called helper bones, in the pelvic region, as shown in Figures 2 and 3.

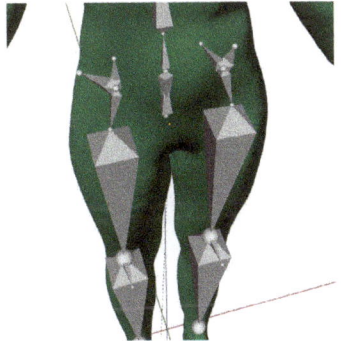

Figure 2. Blender 3D—improved armature of the lower part of the kinematic 3D body model.

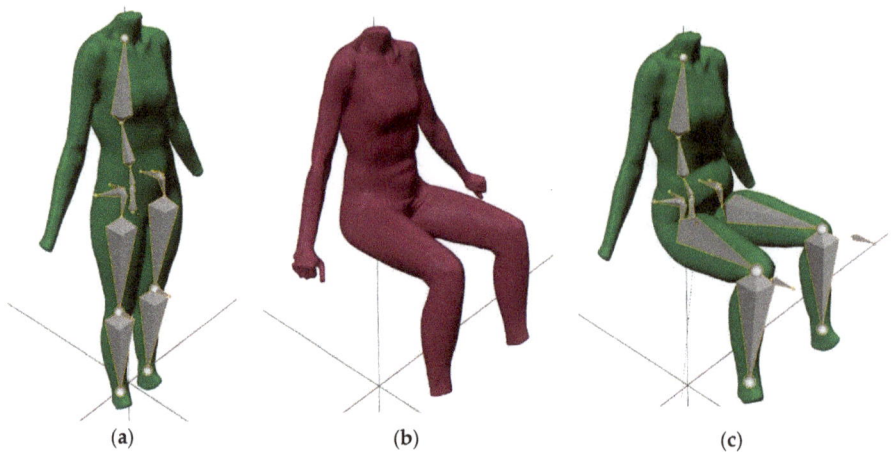

Figure 3. Virtual body models used in this study: (**a**) standing scanned 3D body model with an improved armature, (**b**) sitting scanned 3D body model and (**c**) standing scanned 3D body model, posed with an improved armature, herein called the kinematic sitting 3D body model.

The built armature's absolute x, y and z coordinates for each bone's head and tail in Blender 3D, for a standing posture of a scanned person, are summarized in Table 2, and for a sitting posture in Table 3. Our armature for a sitting posture is based on the reference scan and scanned 3D body model of the sitting posture, which allowed us to accurately compare and evaluate the observed circumferences of the lower-body parts (Figure 1). Blender 3D does not support automatic skeleton adjustment to the 3D body model, therefore manual adjustment was required. The study by [43] made progress in this direction using a 4D scanning technique, in which the additional dimension is time. This means that the change of posture can be accurately examined at each intermediate stage, and that body shape is correctly captured and presented throughout the entire movement sequence.

Table 2. Armature's absolute x, y and z coordinates for every bone's head and tail in Blender 3D, for the standing posture of a scanned person.

Bone Part Standing Bone Name	Bone Head Location			Bone Tail Location			Parent Bone
	x	y	z	x	y	z	
Bone.002	−6.25	61.55	1107.00	−4.64	26.30	1333.00	Bone.003
Bone.003	−7.33	27.08	978.80	−6.25	61.55	1107.00	Bone.001
Bone.001	−6.81	48.95	817.00	−7.33	27.08	978.80	Pelvis Master
Pelvis Master	−6.81	48.97	722.40	−6.81	48.95	817.00	-
Pelvis Son	−6.81	48.95	817.00	−6.81	48.97	722.40	Pelvis Master
Leg L Base	79.78	−87.49	872.89	80.15	−86.11	791.70	Pelvis Son
Leg L	80.15	−86.11	791.70	86.63	−59.00	420.60	Leg L Base
Leg L.001	86.63	−59.00	420.60	77.81	−33.04	30.17	Leg L
Buttocks L	79.77	−87.49	872.89	89.68	72.03	720.10	Leg L Base
Buttocks L.001	79.77	−87.49	872.89	137.90	1.46	835.90	Buttocks L
Leg R Base	−99.35	−87.49	871.29	−114.93	−87.49	792.91	Pelvis Son
Leg R	−114.93	−87.49	792.91	−106.22	−59.00	419.18	Leg R Base
Leg R.001	−106.22	−59.00	419.18	−82.70	−33.04	31.69	Buttocks R
Buttocks R	−99.35	−87.49	871.29	−108.00	71.90	719.70	Leg R Base
Buttocks R.001	−99.35	−87.49	871.29	−145.33	1.80	854.00	Buttocks R

Table 3. Armature's absolute x, y and z coordinates for every bone's head and tail in Blender 3D, for the sitting posture of a sitting posture.

Bone Part Sitting Bone Name	Bone Head Location			Bone Tail Location			Parent Bone
	x	y	z	x	y	z	
Bone.002	−6.87	67.53	1112.00	−3.75	22.40	1336.00	Bone.003
Bone.003	−7.42	51.64	980.30	−6.87	67.53	1112.00	Bone.001
Bone.001	−6.81	48.97	817.00	−7.42	51.64	980.30	Pelvis Master
Pelvis Master	−6.81	48.95	722.40	−6.81	48.97	817.00	-
Pelvis Son	−6.81	48.97	817.00	−6.81	48.95	722.40	Pelvis Master
Leg L Base	79.78	−87.50	872.90	80.15	−86.11	791.70	Pelvis Son
Leg L	80.15	−86.11	791.70	179.80	−446.10	783.40	Leg L Base
Leg L.001	179.80	−446.10	783.40	206.70	−482.80	396.10	Leg L
Buttocks L	79.78	−87.50	872.90	117.50	81.44	735.40	Leg L Base
Buttocks L.001	79.78	−87.50	872.90	151.00	−7.02	862.90	Buttocks L
Leg R Base	−99.34	−87.50	871.30	−94.96	−85.95	791.60	Pelvis Son
Leg R	−94.96	−85.95	791.60	−187.30	−447.90	783.30	Leg R Base
Leg R.001	−187.30	−447.90	783.30	−202.10	−485.20	394.70	Buttocks R
Buttocks R	−99.34	−87.50	871.30	−158.80	80.16	741.80	Leg R Base
Buttocks R.001	−99.34	−87.50	871.30	−158.80	−4.57	872.80	Buttocks R

2.2.3. Rigging

Rigging is a procedure whereby a scanned mesh and armature are bundled into a kinematic 3D body model. The purpose of building a kinematic 3D body model is to adapt the body to the different postures needed for developing personalized garments, and it is performed in a virtual environment, using CAD 3D software (version).

To investigate the kinematic sitting posture, standing and sitting 3D body models of a given participant were used as reference and for comparison (Figure 3a,b). Virtual models were positioned in the space overlapping the upper part of the body. With the kinematic 3D body model (Figure 3a,c; green), we tried to achieve the posture of a scanned sitting 3D body model (Figure 3b; purple).

When building our kinematic 3D body models, we focused on the lower part of the body, containing the hip joints in the pelvic region and the knee joints. For the rigging of the 3D body models' mesh and armature in Blender 3D, the Parent with Automatic Weights tool was used. The armature enabled us to pose any 3D body-model mesh by selecting the armature and manually moving/rotating the appropriate bones into a desired pose that best fit the sitting 3D body model.

2.2.4. Measuring Scanned 3D Body Models

Measurements of the scanned virtual 3D body models, in standing and sitting postures and in the kinematic sitting 3D body model, were carried out with the aim of exploring whether any changes in the observed circumferences (WC, HC, TC, KC) occurred with the change of the body posture from standing to sitting and between the scanned 3D body model and the kinematic 3D body model in a sitting posture.

All measurements can be referred directly to real body measurements, therefore, the measurement positions are the same as shown in Figure 1. Measurements were performed in Blender 3D using the section plane with the Modifier Boolean Difference tool. Newly created hole circumferences in section planes were selected and evaluated using the Add-in Measure-It tool. All measurements of circumferences were repeated five times. The mean values ($\bar{x}$), standard deviations (SD) and coefficients of variation (CV) were calculated for the measurements of 3D bodies' dimensions in a scanned standing posture (3DMSTA), scanned sitting posture (3DMSIT) and kinematic sitting posture (K3DMSIT). Differences (D) were calculated between the mean values of 3DMSTA and 3DMSIT, 3DMSTA and K3DMSIT, 3DMSIT and K3DMSIT. A positive value (+) means that the 3DMSTA is greater than the 3DMSIT or K3DMSIT and a negative value (−) vice versa.

2.3. Virtual Simulation of Personalized-Trouser and Real-Trouser Prototypes

Construction of the basic trouser-pattern design was performed for the scanned female, whose body height was 165.0 cm and whose virtually measured waist, hip, thigh and knee circumferences from the scanned 3D body model, in standing posture, were recorded (WC = 69.0 cm, HC = 97.5 cm, TC = 54.5 cm, KC = 35.5 cm), using the rules and equations of the construction system M. Müller & Sohn [49] for the calculation of proportional measures (trousers outside length, front and back width, sitting depth, knee height), where the width at the knee line and bottom edge was 20.0 cm. A regular trouser style, made of woven fabric and which fit the body without pressure, was used for this research. Continuous sitting, which is unavoidable for people with immobile lower extremities, the elderly and, often, in many professions, such as drivers, secretaries, machine operators etc., or in continuous work behind a computer for several hours (which we have witnessed rise in the last two years, due to COVID-19), may affect vascular endothelial function in the lower extremities. Indeed, it has been shown that prolonged flexion of the hip and knee joints, as occurs during sitting, and the associated low blood flow caused by arterial bending, is disadvantageous for leg vascular health [50]. In addition, analysis of the effect of different sitting postures on skin temperature of the lower extremity has shown that sitting postures cause a decrease in blood flow volume to the lower extremities, resulting in a decrease in lower-extremity temperature. Especially, sitting with the legs crossed affects the circulation of blood-flow volume much more than normal sitting with uncrossed legs, on a chair [51].

In order to investigate the influence of the body-posture change from standing to sitting on changes in the lower-body dimensions, and thus on the fit of trousers to the body in different postures, the first basic trouser-pattern design was constructed for the standing posture, without ease allowances in waist and hip circumferences, and the second basic trouser-pattern design was constructed with minimal ease allowances, based on investigated differences of circumferences between the mean values of the MSTA and MSIT. Since the dimensions of the thigh circumference and knee circumference are smaller than the dimensions in the regular basic trouser-pattern design, we did not add any ease allowance in these construction lines.

The construction of the personalized trouser-pattern designs was performed using the OptiTex PDS system. Three-dimensional simulations of the trouser prototypes and analysis of their fit on the standing, sitting and kinematic sitting 3D body models were performed using the 3D and Tension modules of the OptiTex PDS V11 system. A 3D Tension Map and the Tension XY tool were used to measure the highest tension of the fabric of the trousers in both the warp (X) and weft (Y) direction, and together (XY), which are also presented individually for the back part. Fit evaluation, based on the visual assessment of colour maps, can be subjective, because of small, nuanced differences in colour, and because colour maps represent the maximum value in red and the minimum value in blue for each garment, regardless of absolute values [52–54]. Some researchers [54–57] have included the numerical values provided by colour maps in their analyses, since the tension maps of two different fabrics may look similar while their maximum tension values may differ. The amount of virtual tension influencing the fabric is given by the OptiTex PDS V11 system in units of gf/cm (i.e., forces acting in a unit of length), which has here been converted into SI units N/cm. The colour band on the tension scale ranges from red to yellow, through green, to blue, where red indicates the maximum values and blue the minimum values of virtual tension. By default, the Tension Map module displays the highest value of tension found in the garment.

A 100% cotton plain-weave fabric with a surface mass of 154.94 g m^{-2} was used for the trouser simulations and the sewing of real-trouser prototypes for a comparison of their fitting. The linear density for the warp was 32 tex and, for the weft, was 38 tex. The fabric density in the warp direction was 25 threads/cm and, in the weft direction, was 22 threads/cm.

The low-stress mechanical parameters of the used fabric (extensibility, bending rigidity, shear rigidity, surface thickness) were measured using the FAST measuring system, Table 4, and considered for trouser virtual simulations by using Fabric Converter of OptiTex software, which converts extensibility (%) in resistance to stretch (gf/cm), bending rigidity (μNm) in resistance to bend (dyn·cm), shear rigidity (N/m) in resistance to shear (dyn·cm), thickness (cm) and surface mass (gm^{-2}).

Table 4. Mechanical parameters of the used fabric measured by FAST measuring system.

Mechanical Parameters	Direction	Unit	Measured Value
Extensibility (E100)	warp	%	0.93
	weft	%	5.60
Bending rigidity (B)	warp	μNm	19.08
	weft	μNm	7.46
Shear rigidity (G)	-	Nm^{-1}	189.23
Surface thickness (ST)	-	mm	0.151
Surface mass (W)	-	gm^{-2}	154.94

3. Results with Discussion

3.1. Comparison of Real Human Body Circumferences between Standing and Sitting Postures

The mean values ($\bar{x}$), Standard deviations (SD) and Coefficients of variation (CV) for both the measurements in a standing posture (MSTA) and measurements in a sitting posture (MSIT) of the real female bodies are presented in Table 5. In addition, the minimum and maximum values of each body dimension, the calculated differences (D) between the mean values of the MSTA and MSIT (D$_{MSTA-MSIT}$) and THE percentages of these differences are given in Table 5.

Table 5. Body measurements in standing and sitting postures and the differences between the mean values of the MSTA and MSIT.

Body Dimensions	Symbol	MSTA			MSIT			D$_{MSTA-MSIT}$	
		$\bar{x}$ (cm) x_{min} (cm) x_{max} (cm)	SD (cm)	CV (%)	$\bar{x}$ (cm) x_{min} (cm) x_{max} (cm)	SD (cm)	CV (%)	$\overline{Dx}$ (cm)	$\overline{Dx}$ (%)
Waist circumference	WC	77.25 61.00 102.50	10.15	13.14	77.98 62.50 104.00	10.64	13.64	−0.73	−0.94 [1]
Hip circumference	HC	106.61 87.00 130.00	11.17	10.47	109.84 90.00 135.00	11.58	10.54	−3.23	−3.03
Thigh circumference	TC	61.95 50.00 74.00	7.30	11.78	62.77 50.00 75.00	7.53	11.99	−0.82	−1.32
Knee circumference	KC	39.84 33.00 47.00	4.00	10.04	41.32 36.00 48.50	4.13	10.00	−1.48	−3.71

[1] Note: the minus sign means that the body dimension increased when the body changed its posture from standing to sitting.

Based on the measured body dimensions, it was found that changes occur in all measured circumferences when subjects change their posture from standing to sitting. The average waist circumference (WC), which was 77.25 cm in the standing position, increased to 77.98 cm or an average of 0.73 cm (0.94%). In the sitting posture, the hip circumference (HC) also increased by an average of 3.23 cm (3.03%), thigh circumference (TC) by an average of 0.82 cm (1.32%) and knee circumference (KC) by an average of 1.48 cm (3.71%), Table 5.

The analyses of measured circumferences show that, with increasing body weight (BW) and body mass index (BMI), WC, HC, TC and KC increase, while no effect of body height (BH) on the measured circumferences was been observed for a standing (Figure 4) or sitting (Figure 5) posture.

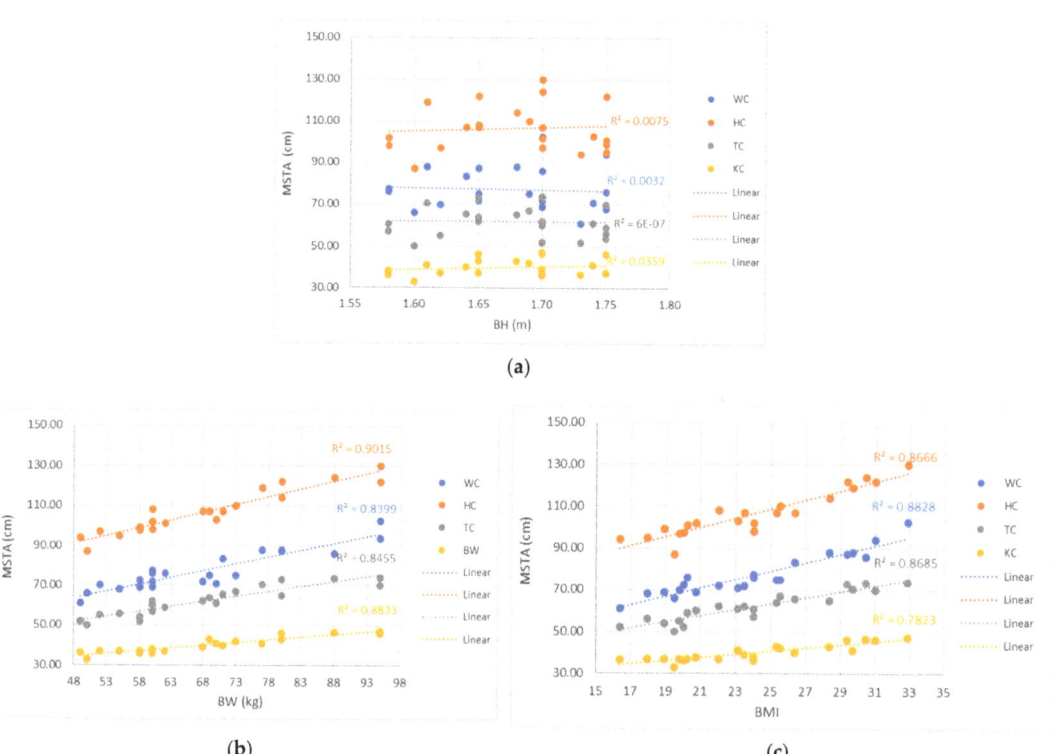

Figure 4. Influence of participants' basic data on observed measurements of body circumferences in the standing posture (MSTA): (**a**) body height, (**b**) body weight, (**c**) body mass index.

High linear trends of the increase in the measured circumferences of the tested participants with the increase in their BW (R^2 between 0.8399 to 0.9015 for MSTA and between 0.8260 to 0.8934 for MIST) and BMI (R^2 between 0.7823 to 0.8828 for MSTA and between 0.6719 to 0.8863 for MIST), were observed for both the seated and standing postures (Figures 4 and 5). The greatest correlation was found between the increase in hip circumference with increasing body weight and BMI. Measuring the hip circumference in the construction of trouser-pattern design is extremely important for worn comfort, especially when sitting.

In Figures 4 and 5, more than half of the study participants had a normal BMI, i.e., between 18.5 and 24.9 (50.00%), BMI below 18.5 or underweight had only 9.09%, pre-obesity 27.27% and obesity class I 13.64%, the categories of which are addressed in the source [58] for adults over 20 years.

In this case study of twenty-two female subjects aged between 20 and 24 years, a wide range of tested participants was measured, with respect to their physical characteristics, as shown in Table 1 and Figures 4 and 5. Therefore, the relationships between the basic data of participants (BH, BW, BMI) and body measurements (WC, HC, TC, KC) were examined using Pearson correlation coefficients (Table 6).

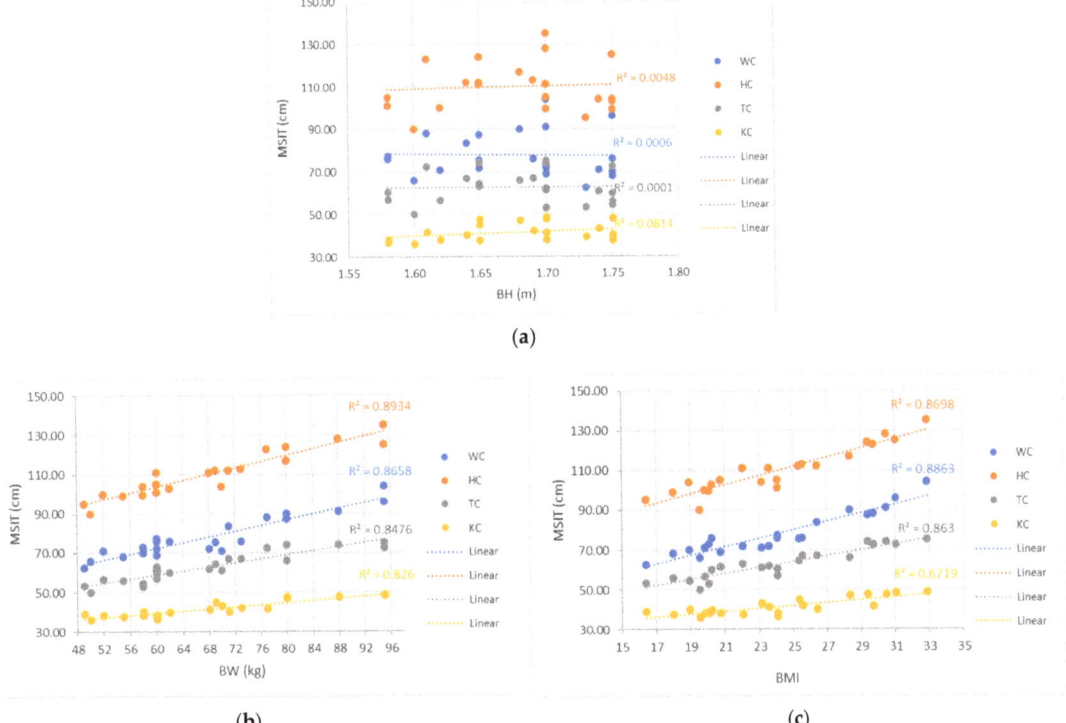

Figure 5. Influence of participants' basic data on observed measurements of body circumferences in a sitting posture (MSIT): (**a**) body height, (**b**) body weight, (**c**) body mass index.

Table 6. Pearson correlation coefficients between participants' basic data and their body measurements.

Basic Data/Body Measurements	WC	HC	TC	KC
BH	−0.057	0.086	−0.001	0.189
BW	0.916	0.949	0.920	0.940
BMI	0.940	0.931	0.932	0.884

The Pearson correlation coefficients clearly show and support the above findings that all measured body circumferences are influenced by BMI, which is influenced to a greater extent by body weight compared with body height. The Pearson correlation coefficients indicated a very high and strong association of BW and BMI with all measured body circumferences, Table 6.

No obvious trends were found between the basic data of the participants studied and the calculated differences in the measured circumferences. The Pearson correlation coefficients also show a slight-to-weak association of the studied data and relationships between the basic data of the participants and the differences (D) of the body measurements between the standing and sitting postures ($DWC_{MSTA-MSIT}$, $DHC_{MSTA-MSIT}$, $DTC_{MSTA-MSIT}$, $DKC_{MSTA-MSIT}$), Table 7.

Table 7. Pearson correlation coefficients between the basic data of the participants and differences of body measurements between the standing and sitting posture.

Basic Data/Body Measurements	$DWC_{STA-SIT}$	$DHC_{STA-SIT}$	$DTC_{STA-SIT}$	$DKC_{STA-SIT}$
BH	−0.287	0.135	0.118	−0.377
BW	−0.501	−0.286	−0.301	0.004
BMI	−0.401	−0.340	−0.264	0.135

It was found that the average increases in measured circumferences in the sitting posture (WC, HC, TC and KC) did not depend on BH, BW or BMI. Therefore, it can be concluded that there is a difference in the measured circumferences when the posture is changed from standing to sitting, due to the change in the shape of the joints and the redistribution of the muscles and adipose tissue surrounding the joint, resulting in an increase of the observed circumferences.

Based on the comparison of real female-body circumferences between standing and sitting postures, it can be concluded that the measured body circumferences are mainly influenced by body weight, regardless of body height. Moreover, when changing postures from standing to sitting, the measured body circumferences increased on average regardless of body height: WC = 0.73 cm, HC = 3.23 cm, TC = 0.82 cm, and KC = 1.48 cm. Thus, it can be pointed out that, in the process of construction of basic clothing-pattern designs, intended for garments of a woven fabric, ease allowances must be added to the body dimensions to ensure the unhindered movement of the person in the garment. When determining clothing dimensions, body dimensions are constant and ease allowances variable, which is discussed in detail in the paper by [59,60]. The correct amount of ease allowance provides a well-fitting and functional garment. Generally, ease allowance is divided into two types [61]. First, the wearing ease allowance, which refers to the amount of added fabric allowed over and above the body dimensions to ensure the comfort, mobility and drape of clothing, and secondly, design ease allowance, which refers to the amount of added fabric to achieve a particular design and shape. Ng R. et al. [62] classified ease allowances according to three different functions: (a) basic movement, breathing and sitting, which require static ease allowances and also known as standard ease; (b) nonstandard body shapes (fat, thin, big hip, strong leg, ...) and their movements (walking, jumping, running, etc.), which require dynamic ease allowances; and (c) styling ease, which is the extra spacing required to conform the desired shape. Chen Y. et al. [63] also point out importance of the fabric ease allowance, which considers the influence of the mechanical properties of clothing fabric on the positive or negative ease allowance. The properties of stretchy knitted fabrics require special garment-pattern design and negative ease allowance.

When constructing basic pattern designs for regular trousers intended for woven fabric, standard ease allowances are usually added to the waist and hip circumference, while the circumference of the trouser leg is adjusted at the thighs and knees, according to the clothing's model. In the study by McKinney et al. [64], the body-to-pattern measurement and shape relationships in trouser patterns, drafted by two methods, were explored. It was found that their body-to-pattern measurement and shape relationships were inconsistent between and within the methods, making them unsuitable for use in computer-aided custom patternmaking. A comparative study of four trouser-pattern-making methods (Aldrich, Armstrong, Bunka, and ESMOD) was carried out on a standard standing and five dynamic postures of a female subject, aged 29, and with a BH of 167.7 cm, BW of 68 kg and BMI of 24.1 [65]. All pattern-making methods provided different amounts of ease allowances for the waist and hip girths, where the Armstrong method added the greatest amount (8.88 cm in the waist and 6.34 cm in the hips), while other methods suggested at least 3.0 cm in the waist and 4.0 cm at the hips. All pattern-making methods provided construction of trousers for the woven fabric, where slight differences between construction methods and the suggested amounts of ease allowances influenced the clothing's fit and overall shape. Specifically, dissatisfaction increased slightly when the subject changed poses, and their discomfort increased when changing postures (stepping at walking pace, sitting 90°, stooping 90°, climbing the stairs, squatting on hams). Studies on the fit of women's trousers by measuring the differences between the trousers and body scans (ease) for twenty-four subjects, aged 35 to 55 years, representing three body-shape groups, showed no clear dependence on body shape [66]. Research on trouser fit was conducted for women aged 55 years and older and tested 176 participants in five states, the results of which highlighted that variations in body size, shape, proportion, and posture create difficulties for effective ready-to-wear sizing systems with a practical number of sizes [67]. Therefore, for a ready-to-wear sizing system, a comprehensive measurement of the population has been carried out in last decades, using 3D human-body scanning [1,6]. In the research by Petrak S. and Mahnić Naglić M. [5], two dynamic body postures (standing, sitting) were investigated in 80 female subjects, aged 20 to 30 years. It was found that with a change of the body posture from standing to sitting, hip width increased in the range of 0.7 to 6.0 cm, and it can be assumed that hip girths also increased. Similarly, it was also found in case studies on the development of personalized clothing in smaller number of female and male test subjects that, with a change of the body posture from standing to sitting, hip girth increased [13,32].

Based on the results of the case studies of women described in this section, it can be assumed that when constructing basic pattern-designs for regular female trousers intended for woven fabric, it is necessary to add a minimal ease allowances on measures of the waist circumference and hip circumference of WC = 0.73 cm, HC = 3.23 cm, and for trousers with narrower trouser legs, ease allowances on the thigh and knee circumference of minima for TC = 0.82 cm and KC = 1.48 cm.

3.2. Virtual 3D Body Models in Sitting Posture

Creating a kinematic 3D body model requires much skill and knowledge in and of 3D modelling software. Most methods are manual, and there is usually a lot of trial and error.

A simple armature in the pelvic region [38–41,68,69] was used in the preliminary study, which produced too-short legs and thus incorrect knee position. If the femur bone was moved, the buttocks were unnaturally curved, with a lack of volume, as shown in Figure 6. Subsequently, we introduced and tested additional bones in the pelvic area. With the implementation of so-called helper bones, the kinematic 3D body model retained the appropriate kinematics of the tibia, femur, and retained the shape of the buttocks and thighs in the folded area of a sitting posture. In Figure 6b, an improved armature structure with helper bones in the pelvic region is shown, which enabled more suitable kinematics for moving between the standing and sitting postures of the kinematic 3D body model, which featured corrected positions for the knees and buttocks, as well as buttock shape.

When analysing 3D body models in sitting posture, there were some shape differences between the scanned sitting 3D body mesh (purple) and kinematic sitting 3D body mesh (green) in Figure 7. Due to transparent representation of the scanned sitting 3D body mesh and kinematic sitting 3D body mesh in Figure 7a,b, green areas at the crotch and, on the other side, purple areas at hips, are noticeable.

The smaller crotch in the kinematic sitting 3D body mesh can be explained by mesh filling and repairing in the scan-preparation phase. When 3D scanning the human body, there are some places that are difficult to capture in a standing posture, such as the crotch or the armpits, where the scanner cannot capture the surface. Holes and gaps appear in the 3D scan, and they must be filled for a correct virtual garment pattern design. This reconstruction can also produce an inaccurate mesh that is visible in a sitting posture. Wider hips on the scanned 3D sitting mesh can be interpreted as the fact that in a real human body in sitting posture, the soft tissues in the sitting area are pressed and stretched around the hips as shown in Figure 7c.

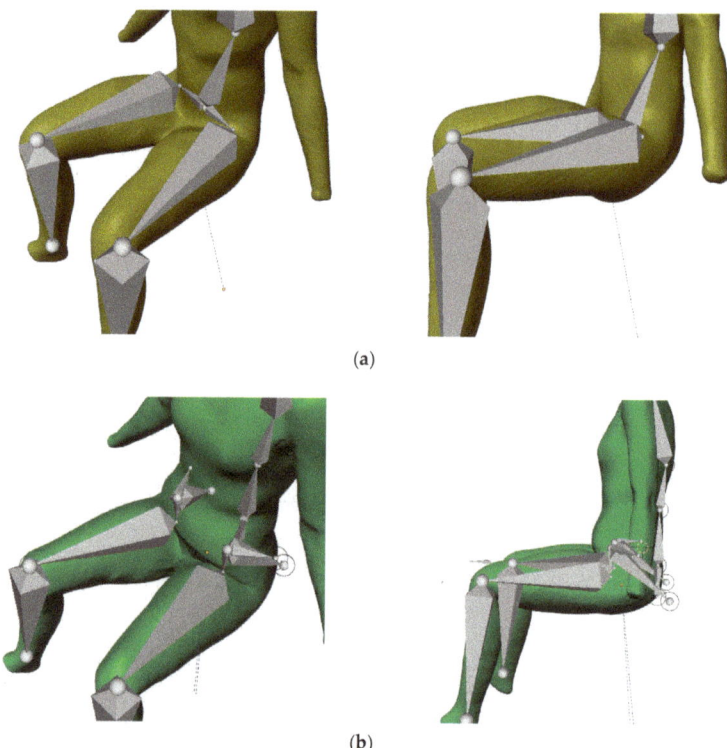

Figure 6. Armature structure of kinematic 3D body model: (**a**) simple armature in pelvic region, (**b**) improved armature with helper bones in pelvic region.

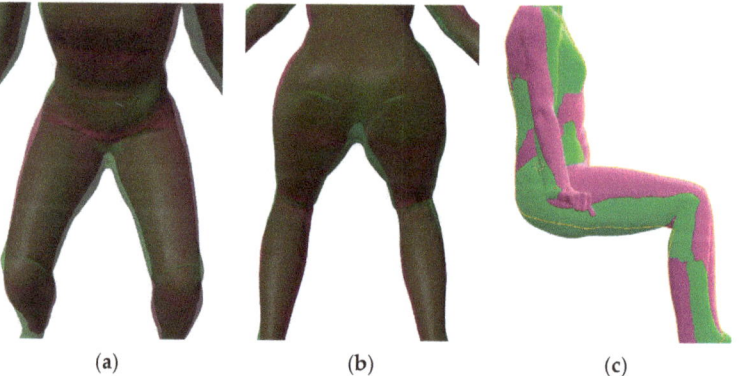

Figure 7. Overlapping meshes of a sitting 3D body model (3DBMSIT) (purple) and kinematic sitting 3D body model (K3DBMSIT) (green): (**a**) the upper front view, (**b**) the lower bottom view, (**c**) side view.

The standing posture requires muscular activity and tension, therefore the geometry of the body is different in the standing and sitting postures. These differences can be seen very well in Figure 8 and will be evaluated in Section 3.2.1 with measured circumferences of the lower body parts. Even with the improved armature in the pelvic region, differences between the scanned and kinematic sitting 3D body models in the flexion of the hip joints in the pelvis, which unnaturally moves the soft tissues towards the abdomen, can be clearly seen in Figure 8. A pronounced unnatural fold under the bended knees can be also seen. In general, the real human body is not a uniform solid model, but is built of different tissues and structures. For this reason, building a kinematic 3D body model that will simulate real-body behaviour becomes more challenging. Researchers [9,39–41] have had the same issues of inappropriate 3D body mesh bending in regions, where the bending angle is large.

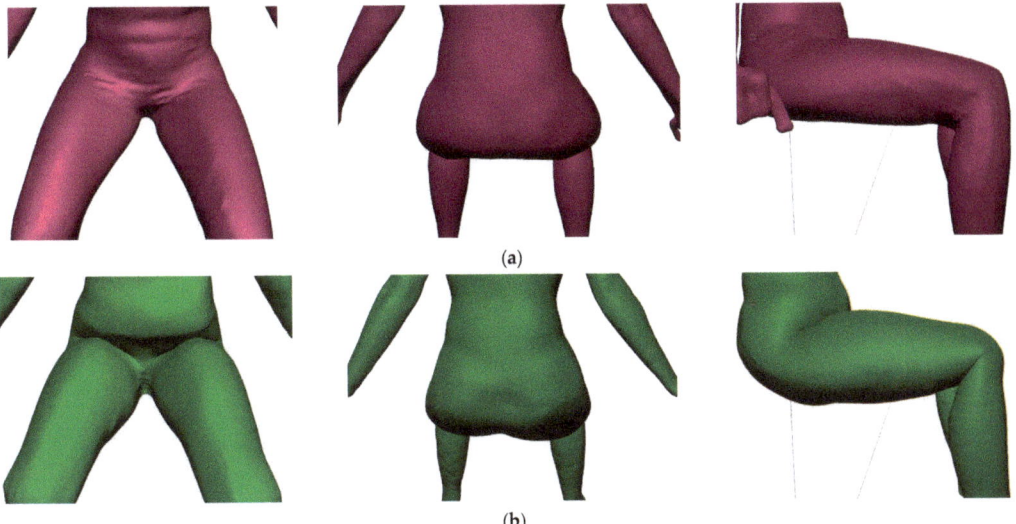

Figure 8. Comparison of joints regions: (**a**) scanned sitting 3D body model (purple), (**b**) kinematic sitting 3D body model (green).

There are some services already available, wherein virtual characters can be downloaded, but they are intended for animation [40]. These characters consist of a 3D body mesh and a simple inner armature, suitable for animation. Additionally, for 3D animation the inappropriate 3D body-mesh bending can be hidden with some additional geometry or post-processing effects, but for clothing-pattern design, a clean and correct geometry is required. For this purpose, an improved armature in pelvic region and additional helper bones for proper flexion of hips joints in pelvis, respectively, were added to the armature of the standing kinematic 3D body model in this study, to simulate the posture, size, and shape of kinematic model as closely as possible to the scanned sitting 3D body model.

The current solution for 3D body models intended to simulate clothing is to manually adjust and reconstruct the mesh. However, these adjustments are not possible without a 3D scan of the intended wearer in a sitting posture, so some predictions and guidelines are suitable at this point. In future research, incorporating as many simulated tissues as possible is likely to be the preferred method for creating kinematic 3D body models for virtual clothing-pattern design. Some available solutions are intended for modelling, simulation, and analysis of the neuromusculoskeletal system of the human body as used in [42,43].

3.2.1. Comparison of Scanned 3D Body-Models' Measurements between the Standing (3DMSTA) and Sitting (3DMSIT) Postures and the Kinematic 3D Body Model in Sitting Posture (K3DMSIT)

The mean values ($\bar{x}$), standard deviations (SD) and coefficients of variation (CV) were calculated for measurements of 3D body dimensions in scanned standing posture (3DMSTA), scanned sitting posture (3DMSIT) and kinematic sitting posture (K3DMSIT), Table 8. The mean values of circumferences are rounded up to the nearest half centimetre. Differences (D) were calculated between the mean values of 3DMSTA and 3DMSIT, 3DMSTA and

K3DMSIT, 3DMSIT and K3DMSIT, Table 9. A positive value (+) means that the 3DMSTA is greater than the 3DMSIT or K3DMSIT and a negative value (−) vice versa.

Table 8. Three-dimensional body measurements of standing (3DMSTA), sitting (3DMSIT) and kinematic sitting (K3DMSIT) postures.

Measurements of 3D Body Models	Symbol	3DMSTA			3DMSIT			K3DMSIT		
		$\bar{x}$ (cm)	SD (cm)	CV (%)	$\bar{x}$ (cm)	SD (cm)	CV (%)	$\bar{x}$ (cm)	SD (cm)	CV (%)
Waist circumference	WC	69.00	0.27	0.40	71.00	0.0	0.00	68.50	0.55	0.80
Hip circumference	HC	97.50	0.65	0.67	103.00	1.44	1.40	96.00	0.71	0.74
Thigh circumference	TC	54.50	0.35	0.65	57.50	0.57	1.00	57.50	0.89	1.57
Knee circumference	KC	35.50	0.42	1.18	39.00	0.35	0.91	39.50	0.89	2.29

Table 9. Differences between mean values of the 3DMSTA and 3DMSIT, 3DMSTA and K3DMSIT and between 3DMSIT and K3DMSIT.

Measurements of 3D Body Models	Symbol	$D_{3DMSTA-3DMSIT}$		$D_{3DMSTA-K3DMSIT}$		$D_{3DMSIT-K3DMSIT}$	
		$\overline{Dx}$ (cm)	$\overline{Dx}$ (%)	$\overline{Dx}$ (cm)	$\overline{Dx}$ (%)	$\overline{Dx}$ (cm)	$\overline{Dx}$ (%)
Waist circumference	WC	−2.20 [1]	−3.20	0.40	0.58	2.60	3.66
Hip circumference	HC	−5.40	−5.54	1.40	1.46	6.80	6.61
Thigh circumference	TC	−2.80	−5.14	−2.60	−4.55	0.20	0.35
Knee circumference	KC	−3.60	−10.17	−3.70	−9.46	0.10	−0.26

[1] Note: The minus sign means that the body dimension increased when the body changed posture from a standing to a sitting.

Standing postures require muscle activity and tension, therefore the observed circumferences differ between scanned 3D standing body model, scanned 3D sitting body model and kinematic 3D sitting model. The analysis of the measurement results of the 3D body models shows circumferential differences when changing posture from standing to sitting, with respect to the real person's measurements. In sitting postures, all measured circumferences show an increased value. The highest increase was seen in the waist and knee circumferences, which is consistent with the results of the real measurements in Tables 5 and 9. The waist and hip circumferences of the kinematic sitting 3D body model decreased, as compared with the standing and sitting 3D body models, while the circumferences of the thigh and knee.

The differences between the mean values of the standing and sitting 3D body models ($D_{3DMSTA-3DMSIT}$) showed an increase in circumferences, which were for WC of 2.20 cm (3.20%), HC of 5.40 cm (5.54%) TC of 2.80 cm (5.14%) and KC of 3.60 cm (10.17%) (Table 9). The differences between the mean values of the standing 3D body model and kinematic sitting 3D model ($D_{3DMSTA-K3DMSIT}$) indicated a decrease in the waist and hips circumferences of the kinematic sitting 3D body model, while its thigh and knee circumferences increased. Significant differences between mean values are observed between the two sitting 3D body models ($D_{3DMSIT-K3DMSIT}$) (Table 9). Especially the hip circumference of the kinematic sitting 3D body model is 6.8 cm smaller than that of the sitting 3D body model, which can also be observed in Figure 8. The large HC difference can be explained by the lack of 3D program tissue simulation ability, while in Blender 3D, automatic weights were used for the armature—mesh interference. With some mesh weights, better adaptation and redistribution results can be expected. A large, triangular 3D scan was used for the kinematic 3D body model in this case, moreover, researchers in [40] used the quadratic simplified 3D mesh model, Skinned Multi-Person Linear (SMPL), which promises easier handling, but it still has a lack of proper bending at joints with large angular rotations.

3.3. Virtual Simulations of Personalized Trousers on Standing, Sitting and Kinematic Sitting 3D Body Models and Real-Trouser Prototypes

Virtual simulations of the basic trouser-pattern design were carried out for a female participant with a body height of 165.0 cm, and her virtually measured waist, hip, thigh, and knee circumferences on the scanned 3D body model in a standing posture were: WC = 69.0 cm, HC = 97.5 cm, TC = 54.5 cm and KC = 35.5 cm. The real-trouser prototypes were sewn for the same person. The first pair of trousers were constructed without ease allowances (Figure 9) and the second pair of trousers with minimal ease allowances (Figure 10). The researched average differences in circumferences between standing and sitting postures (Table 5) were rounded up to the nearest half centimetre and used as standard ease allowances. Therefore, the ease allowances were 1.0 cm for the waist circumference and 3.5 cm for the hip circumference. The thigh and knee circumferences were not considered, because regular-fit trousers have a larger circumference of the trouser leg than the measured thigh and knee circumferences, even if we add ease allowances for the TC of 1.0 cm and for the knee KC of 1.5 cm. For the virtual simulation used 100% cotton woven fabric, had low extensibility in the warp direction (0.93%) and high shear rigidity (189.23 Nm^{-1}), while its extensibility in the weft direction was 5.6%, and bending rigidity of 19.08 μNm in the warp direction and 7.46 μNm in the weft direction.

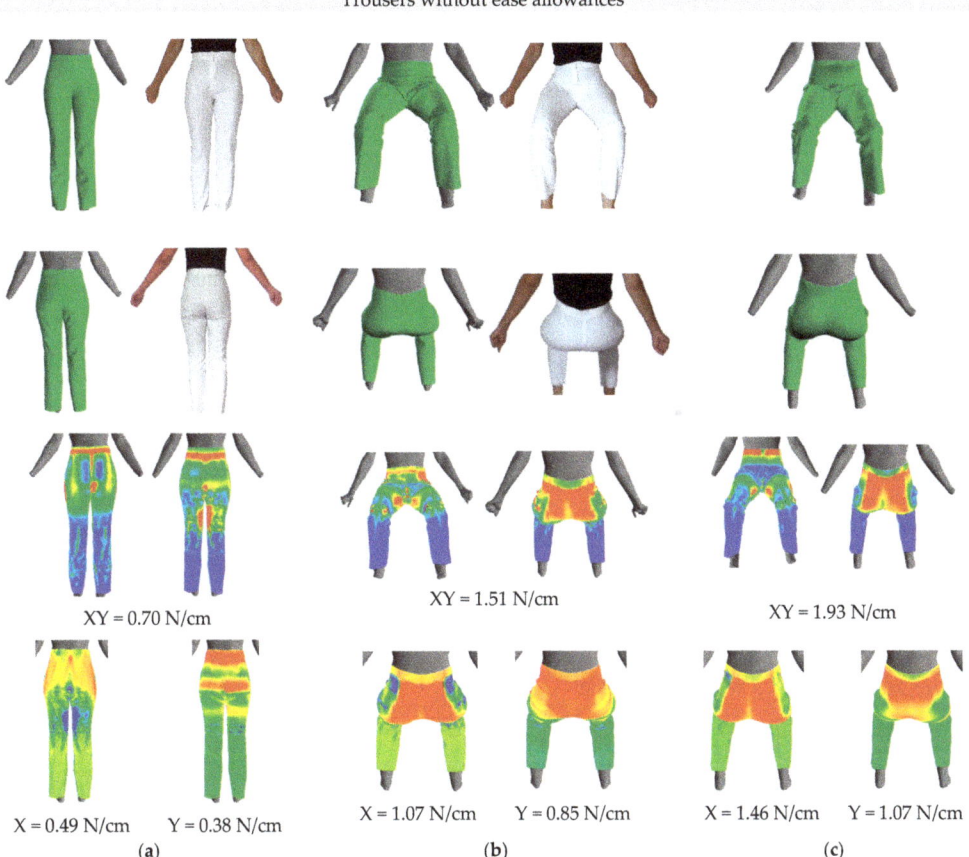

Figure 9. Virtual and real-trouser prototypes without ease allowances and tension in trouser fabric: (**a**) standing 3D body model and real person, (**b**) sitting 3D body model and real person, (**c**) kinematic sitting 3D body model.

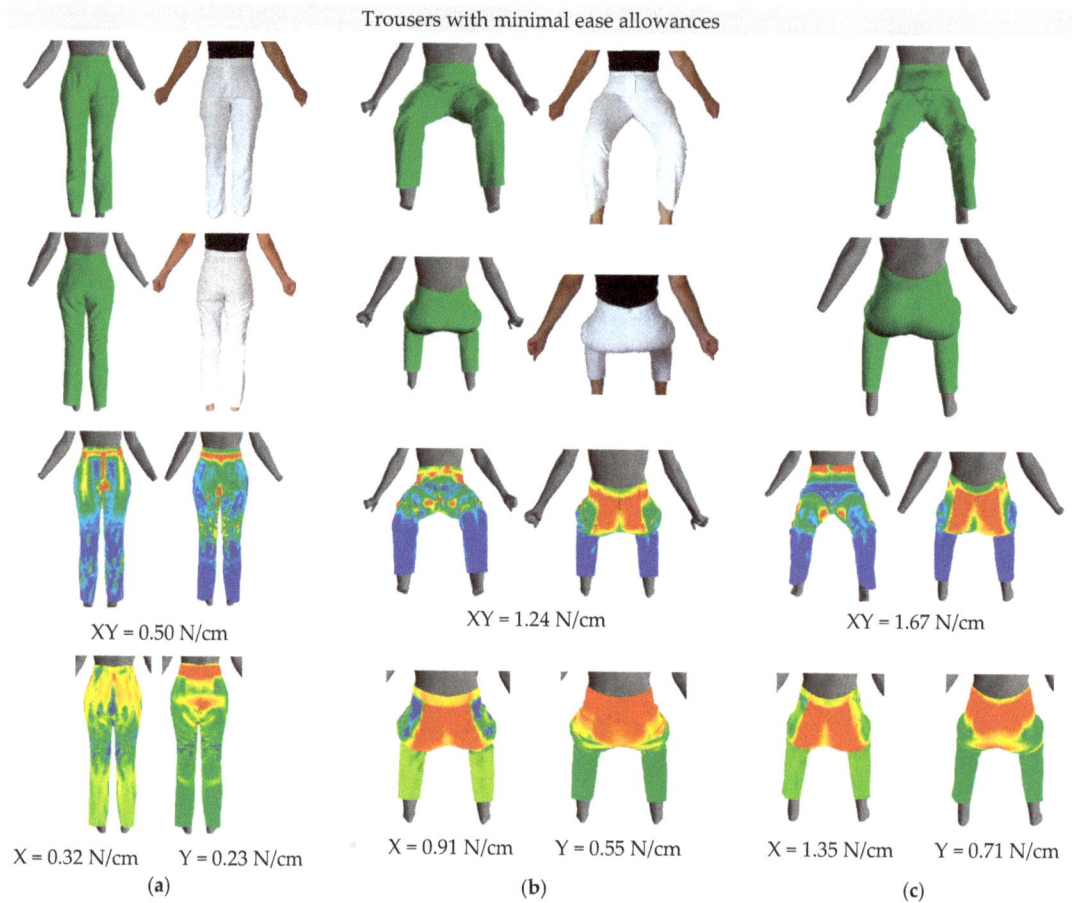

Figure 10. Virtual and real-trouser prototypes with minimal ease allowances and tension in trouser fabric: (**a**) standing 3D body model and real person, (**b**) sitting 3D body model and real person, (**c**) kinematic sitting 3D body model.

Different fits of the virtual trouser prototypes, with and without minimal ease allowances, were between different 3D body models. A perfect fit of the trousers without ease allowances is observed on the standing 3D body model (Figure 9a), while, considering minimal ease allowances, a slightly wider virtual prototype of the trousers can be seen (Figure 10a). The latter is also reflected in the tension of the fabric used for the trousers in this study. In a standing posture, the maximum tension XY in trousers without ease allowances was 0.70 N/cm (X = 0.49 N/cm, Y = 0.38 N/cm) and for trousers with minimal ease allowances it decreased to 0.50 N/cm (X = 0.32 N/cm, Y = 0.23 N/cm). Tension in the fabric was observed mainly in the waist region and at the hips and knees (red), which decreased when simulating trousers with minimal ease allowances around the waist and at the hips and knees. The latter can also be seen in the fit of the real pants to the participant. When observing the maximum tension in the warp and weft directions, a decrease in maximum tensions were found for trousers with minimal ease allowance, and the colour maps thereof showed a decrease in tension, especially of the hip-girth region in the weft direction.

On the sitting 3D body model, the trousers with and without minimal ease allowances in the buttocks region were quite tight (Figures 9b and 10b), which is consistent with an increase of the measured real and virtual circumferences in the sitting posture (Tables 5 and 8) and reflects an increase of the tension in virtual and real trousers. The maximum tension, XY, in the trouser fabric was higher in a sitting posture, compared with a standing posture, and applies for trousers without ease allowances 1.51 N/cm (X = 1.07 N/cm, Y = 0.85 N/cm), while for trousers with minimal ease allowances, it decreased to 1.24 N/cm (X = 0.91 N/cm, Y = 0.55 N/cm) for a sitting posture. The same can be seen for the real trousers, where a small fold of fabric appears in the region between the waist and hips, especially on the right side. The highest tension was observed around the buttocks (red), which, on the one hand, is attributable to the increase in circumference around the waist and hips, and, on the other hand, to the regular trouser-pattern design, which is intended for a standing posture. The decrease in virtual tension of the fabric on the colour map is more difficult to detect. The increase in virtual fabric tension can also be attributed to low fabric extensibility in the warp direction and high shear rigidity (Table 4), which is exacerbated by a sitting posture with restraint of the trousers at the waist and buttocks. In the sitting posture, the trousers' height at the back area is too low, therefore, back trouser-pattern pieces should be raised to the waistline to reduce the trousers' tension in the back area and to increase comfort when sitting in them. It should be noted that it is extremely difficult to photograph virtual and real-trouser prototypes from the same angle. Therefore, the body postures and details of fitting the trousers to the test subject and her 3D body model may differ slightly.

Sitting 3D body models differ in measured circumferences and in the shape of the buttocks and abdomen. Although the circumferences around the waist and hips of the kinematic sitting 3D body model are smaller, Figures 9c and 10c compared to these circumferences of the sitting 3D body model, Figures 9b and 10b, the tension in these trousers increased. This is attributed to the bending of the kinematic 3D body model from a standing to a sitting posture in the pelvic region and unrealistic tissue redistributions around the waist and hips. For a kinematic sitting 3D body model, the maximum tension (XY) in trousers without ease allowances was 1.93 N/cm (X = 1.46 N/cm, Y = 1.07 N/cm) and, for trousers with minimal ease allowances, decreased to 1.67 N/cm (X = 1.36 N/cm, Y = 0.71 N/cm), which is consistent with a decrease in virtual tension of the trousers with minimal ease allowance on a sitting 3D body model. This decrease in the virtual tension of the fabric on the colour map is more difficult to detect.

The findings of this study agree with those of a study wherein the influence of different ease allowances on the virtual fit of basic bodice-pattern designs was investigated [57]. It was found that, as the ease allowance increases (0/3/6/9/12 cm), the maximum virtual tension in the woven fabric decreases, from a value of 0.61 N/cm for a zero ease allowance to 0.15 N/cm for an ease allowance of 12 cm.

Based on the results in this part of our research, we conclude that, with the virtual simulation of trousers and real-trouser prototypes designed according to the dimensions of a standing person, it is necessary to consider at least minimal ease allowances on trouser-construction lines for comfort while sitting. On the other hand, we have found that our kinematic sitting 3D body model is not yet reliable enough to develop personalized clothing pattern designs, since, so far, the 3D body mesh does not follow real changes in circumferences and shapes in a sitting posture, despite a carefully planned anatomical skeleton.

4. Conclusions

Virtual prototyping of personalized clothing has been a current and reliable tool for some time. Typically, this involves the use of scanned 3D body models, in both standard and specific postures, that are important for the development of personalized clothing. By building a reliable kinematic 3D body model, we could shorten, for the needs of personalized clothing development, the usual process of scanning the body in different postures as a process of fitting and prototyping of real personalized clothing.

This study focused on the development of a kinematic 3D human-body model intended for a sitting posture by using Blender software. It was based on a scanned female body in standing and sitting postures. In the construction of the armature, the focus was mainly on the lower body parts of the pelvic region, hip, and knee joints, due to our exploration of the sitting posture. The study's participants were female, real and virtual measures of lower-body circumferences in both postures. Virtual prototyping of trousers and sewing of real-trouser prototypes were performed to investigate their fit and comfort on scanned and kinematic 3D body models, as well as on a real body. Namely, to examine prolonged flexion of the hip and knee joints that occurs during sitting and the associated low blood flow caused by arterial flexion, as such are unfavourable for leg vascular health.

It was found that by changing the body posture from standing to sitting, the real and virtual body circumferences changed, which affected the fit and comfort of the virtual and real trousers. Our analysis showed that the measured circumference of the lower body parts increases with body weight and BMI, while an effect on circumferential difference when changing posture from a standing to a sitting was not observed. Therefore, it could be supposed that in a sitting posture, the increase in circumferences of the lower body parts reflects, on average, all tested persons regardless of their basic physical characteristics. The findings on the case group in this research highlight that research is needed for the production of ready-to-wear clothing, as are anthropometric surveys to determine the ease allowances for various construction systems, body postures and textile materials. An increase in measured circumferences was found also for the sitting 3D body model and is attributed to the redistribution of body muscles and adipose tissue surrounding the joints, as well as changes in joint shapes in the regions of body flexion, which are not represented equally on the kinematic sitting 3D body model, regardless of our improved armature in the pelvic region. This part of the research also provides guidelines for software developers and for upgrading kinematic 3D-body-model-mesh deformation when changing body postures.

Based on this study's measured results, it can be concluded that average increases in waist, hip, thigh, and knee circumferences should be included in the process of designing basic female clothing patterns for a sitting posture, as minimal ease allowances when using a woven fabric, and, in the future, armatures should be constructed with muscles and adipose tissues to achieve realistic circumferences and shapes at the observed regions of a kinematic 3D body model in a sitting posture.

Author Contributions: Conceptualization, A.R., A.C. and Z.S.; methodology, A.R., A.C. and Z.S.; software, A.R., A.C. and Z.S.; formal analysis, A.R., A.C. and Z.S.; investigation, A.R., A.C. and Z.S.; writing—original draft preparation, A.R., A.C. and Z.S.; writing—review and editing, A.R., A.C. and Z.S. All authors have read and agreed to the published version of the manuscript.

Funding: The research was founded by Slovenian Research Agency (Research Programme P2-0123: Clothing Science, Comfort and Textile Materials), as well as the project of EC Programme Erasmus+, project OptimTex.

Institutional Review Board Statement: Not applicable.

Informed Consent Statement: Informed consent was obtained from all subjects involved in the study regarding the use of body measurements for the research.

Data Availability Statement: Not applicable.

Conflicts of Interest: The authors declare no conflict of interest. The funders had no role in the design of the study; in the collection, analyses, or interpretation of data; in the writing of the manuscript, or in the decision to publish the results.

References

1. Bragança, S.; Arezes, P.; Carvalho, M.; Ashdown, S. Current state of the art and enduring issues in anthropometric data collection. *DYNA* **2016**, *83*, 22–30. [CrossRef]
2. Parker, C.J.; Gill, S.; Harwood, A.; Hayes, S.G.; Ahmed, M. A method for increasing 3D body scanning's precision: Gryphon and consecutive scanning. *Ergonomics* **2021**, 1–21, ahead of print. [CrossRef]
3. Lapkovska, E.; Dāboliņa, I. Sizing for a Special Group of People: Best Practice of Human Body Scanning. In *Environment. Technology, Resources, Proceedings of the 12th International Scientific and Practical Conference, Rezekne, Latvia, 20–22 June 2019*; Rezekne Academy of Technologies: Rezekne, Latvia, 2019; Volume 1, pp. 136–141.
4. Špelić, I.; Petrak, S. Complexity of 3D Human Body Scan Data Modelling. *Tekstilec* **2018**, *61*, 235–244. [CrossRef]

5. Petrak, S.; Mahnić, M. Dynamic anthropometry—Defining protocols for automatic body measurement. *Tekstilec* **2017**, *60*, 254–262. [CrossRef]
6. Gill, S. A review of research and innovation in garment sizing, prototyping and fitting. *Textile Prog.* **2015**, *47*, 1–85. [CrossRef]
7. Gill, S.; Parker, C.J. Scan posture definition and hip girth measurement: The impact on clothing design and body scanning. *Ergonomics* **2017**, *60*, 1123–1136. [CrossRef]
8. Parker, C.J.; Gill, S.; Hayes, S. 3D Body Scanning has Suitable Reliability: An Anthropometric Investigation for Garment Construction. In Proceedings of the 8th International Conference and Exhibition on 3D Body Scanning and Processing Technologies, Montreal, QC, Canada, 11–12 October 2017.
9. Klepser, A.; Pirch, C. Is this real? Avatar Generation for 3D Garment Simulation. *J. Text. Appar. Technol. Manag. (JTATM)* **2021**, *12*, 1–11.
10. Spahiu, T.; Shehi, E.; Piperi, E. Personalized avatars for virtual garment design and simulation. *UNIVERSI—Int. J. Educ. Sci. Technol. Innov. Health Environ.* **2015**, *1*, 56–63.
11. Jevšnik, S.; Pilar, T.; Stjepanović, Z.; Rudolf, A. Virtual prototyping of garments and their fit to the body. In *DAAAM International Scientific Book 2012*; Katalinić, B., Ed.; DAAAM International Publishing: Vienna, Austria, 2012; pp. 601–618.
12. Stjepanović, Z.; Rudolf, A.; Jevšnik, S.; Cupar, A.; Pogačar, V.; Geršak, J. 3D virtual prototyping of a ski jumpsuit based on a reconstructed body scan model. *Bul. Inst. Politeh. Din Iaşi. Secţia Text. Pielărie* **2011**, *57*, 17–30.
13. Bogović, S.; Stjepanović, Z.; Cupar, A.; Jevšnik, S.; Rogina-Car, B.; Rudolf, A. The Use of New Technologies for the Development of Protective Clothing: Comparative Analysis of Body Dimensions of Static and Dynamic Postures and its Application. *AUTEX Res. J.* **2019**, *19*, 301–311. [CrossRef]
14. Nakić, M.; Bogović, S. Computational design of functional clothing for disabled people. *Tekstilec* **2019**, *62*, 23–33. [CrossRef]
15. Cupar, A.; Stjepanović, Z.; Olaru, S.; Popescu, G.; Salistean, A.; Rudolf, A. CASP methodology applied in adapted garments for adults and teenagers with spine deformity. *Ind. Text.* **2019**, *70*, 435–446. [CrossRef]
16. Bruniaux, P.; Cichocka, A.; Frydrych, I. 3D Digital Methods of Clothing Creation for Disabled People. *Fibres Text. East. Eur.* **2016**, *24*, 125–131. [CrossRef]
17. Cieśla, K.; Frydrych, I.; Krzywinski, S.; Kyosev, Y. Design workflow for virtual design of clothing for pregnant women. *Commun. Dev. Assem. Text. Prod.* **2020**, *1*, 148–159.
18. Olaru, S.; Popescu, G.; Anastasiu, A.; Mihăilă, G.; Săliştean, A. Innovative concept for personalized pattern design of safety equipment. *Ind. Text.* **2020**, *71*, 50–54. [CrossRef]
19. Naglić, M.M.; Petrak, S.; Stjepanović, Z. Analysis of 3D construction of tight fit clothing based on parametric and scanned body models. In Proceedings of the 7th International Conference on 3D Body Scanning Technologies, Lugano, Switzerland, 30 November–1 December 2016; pp. 302–313.
20. Abtew, M.A.; Bruniaux, P.; Boussu, F. Development of adaptive bust for female soft body armour using three dimensional (3D) warp interlock fabrics: Three dimensional (3D) design process. *IOP Conf. Ser. Mater. Sci. Eng.* **2017**, *254*, 052001. [CrossRef]
21. Abtew, M.A.; Bruniaux, P.; Boussu, F.; Loghin, C.; Cristian, I.; Chen, Y. Development of comfortable and well-fitted bra pattern for customized female soft body armor through 3D design process of adaptive bust on virtual mannequin. *Comput. Ind.* **2018**, *100*, 7–20. [CrossRef]
22. Abtew, M.A.; Bruniaux, P.; Boussu, F.; Loghin, C.; Cristian, I.; Chen, Y.; Wang, L. A systematic pattern generation system for manufacturing customized seamless multi-layer female soft body armour through dome-formation (moulding) techniques using 3D warp interlock fabrics. *J. Manuf. Syst.* **2018**, *49*, 61–74. [CrossRef]
23. Abtew, M.A.; Bruniaux, P.; Boussu, F.; Loghin, C.; Cristian, I.; Chen, Y.; Wang, L. Female seamless soft body armor pattern design system with innovative reverse engineering approaches. *Int. J. Adv. Manuf. Technol.* **2018**, *98*, 2271–2285. [CrossRef]
24. Hong, Y.; Zeng, X.; Bruniaux, P.; Liu, K. Interactive virtual try-on based three-dimensional garment block design for disabled people of scoliosis type. *Text. Res. J.* **2017**, *87*, 1261–1274. [CrossRef]
25. Hong, Y.; Zeng, X.; Bruniaux, P.; Liu, K.; Chen, Y.; Zhang, X. Collaborative 3D-To-2D Tight-Fitting Garment Pattern Design Process for Scoliotic People. *Fibres Text. East. Eur.* **2017**, *5*, 113–117. [CrossRef]
26. Hong, Y.; Bruniaux, P.; Zeng, X.; Liu, K.; Curteza, A.; Chen, Y.; Cedex, R. Visual-simulation-based personalized garment block design method for physically disabled people with scoliosis (PDPS). *AUTEX Res. J.* **2018**, *18*, 35–45. [CrossRef]
27. Stjepanović, Z.; Cupar, A.; Jevšnik, S.; Stjepanović, T.K.; Rudolf, A. Construction of adapted garments for people with scoliosis using virtual prototyping and CASP method. *Ind. Text.* **2016**, *67*, 141–148.
28. Rudolf, A.; Stjepanović, Z.; Cupar, A. Designing the functional garments for people with physical disabilities or kyphosis by using computer simulation techniques. *Ind. Text.* **2019**, *70*, 182–191.
29. Cupar, A.; Kaljun, J.; Pogačar, V.; Stjepanović, Z. Methodology framework for surface shape evaluation. In Proceedings of the International Conference on Mechanical Engineering (ME 2015), Viena, Austria, 15–17 March 2015; pp. 58–65.
30. Mosleh, S.; Abtew, M.A.; Bruniaux, P.; Tartare, G.; Xu, Y.; Chen, Y. 3D Digital Adaptive Thorax Modelling of Peoples with Spinal Disabilities: Applications for Performance Clothing Design. *Appl. Sci.* **2021**, *10*, 4545. [CrossRef]
31. Mosleh, S.; Abtew, M.A.; Bruniaux, P.; Tartare, G.; Chen, Y. Developing an Adaptive 3D Vertebrae Model of Scoliosis Patients for Customize Garment Design. *Appl. Sci.* **2021**, *11*, 3171. [CrossRef]
32. Rudolf, A.; Cupar, A.; Kozar, T.; Stjepanović, Z. Study regarding the virtual prototyping of garments for paraplegics. *Fibers Polym.* **2015**, *16*, 1177–1192. [CrossRef]

33. Rudolf, A.; Görlichová, L.; Kirbiš, J.; Repnik, J.; Salobir, A.; Selimović, I.; Drstvenšek, I. New technologies in the development of ergonomic garments for wheelchair users in a virtual environment. *Ind. Text.* **2017**, *68*, 83–94.
34. Boulic, R.; Magnenat-Thalmann, N.; Thalmann, D. A global human walking model with real-time kinematic personification. The Visual Computer. *Int. J. Comput. Graph.* **1990**, *6*, 344–358.
35. Jung, M.; Badler, N.I.; Noma, T. Animated Human Agents with Motion Planning Capability for 3D-Space Postural Goals. *J. Vis. Comput. Animat.* **1994**, *5*, 225–246. [CrossRef]
36. Kalra, P.; Magnenat-Thalmann, N.; Moccozet, L.; Sannier, G.; Aubel, A.; Thalmann, D. Real-time animation of realistic virtual humans. *IEEE Comput. Graph. Appl.* **1998**, *18*, 42–56. [CrossRef]
37. Seo, H.; Magnenat-Thalmann, N. An automatic modelling of human bodies from sizing parameters. In Proceedings of the 2003 Symposium on Interactive 3D Graphics, SI3D 2003, Monterey, CA, USA, 28–30 April 2003; pp. 19–26.
38. Kozar, T.; Rudolf, A.; Cupar, A.; Jevšnik, S.; Stjepanovič, Z. Designing an adaptive 3D body model suitable for people with limited body abilities. *J. Text. Sci. Eng.* **2014**, *4*, 1–13.
39. Leipner, A.; Krzywinski, S. 3D Product Development Based on Kinematic Human Models. In Proceedings of the 4th International Conference on 3D Body Scanning Technologies, Long Beach, CA, USA, 19–20 November 2013.
40. Zhang, D.; Krzywinski, S. Development of a Kinematic Human Model for Clothing and High Performance Garments. In Proceedings of the 3DBODY.TECH 2019—10th International Conference and Exhibition on 3D Body Scanning and Processing Technologies, Lugano, Switzerland, 22–23 October 2019; pp. 68–73.
41. Zhang, D.; Krzywinski, S. Development of a Kinematic Human Model for Clothing Design/simulation. In Proceedings of the AUTEX2019—19th World Textile Conference on Textiles at the Crossroads, Ghent, Belgium, 11–15 June 2019.
42. Delp, S.L.; Anderson, F.C.; Arnold, A.S.; Loan, P.; Habib, A.; John, C.T. OpenSim: Open-Source Software to Create and Analyze Dynamic Simulations of Movement. *IEEE Trans. Biomed. Eng.* **2007**, *54*, 1940–1950. [CrossRef] [PubMed]
43. Pons-Moll, G.; Romero, J.; Mahmood, N.; Black, M.J. Dyna: A model of dynamic human shape in motion. *ACM Trans. Graph.* **2015**, *34*, 120. [CrossRef]
44. Klepser, A.; Morlock, S.; Loercher, C.; Schenk, A. Functional measurements and mobility restriction (from 3D to 4D scanning). In *Anthropometry, Apparel Sizing and Design*, 2nd ed.; The Textile Institute Book Series; Woodhead Publishing: Sawston, UK, 2020; pp. 169–199.
45. ISO 8559-1:2017. *Size Designation of Clothes—Part 1: Anthropometric Definitions for Body Measurement*; ISO: Geneva, Switzerland, 2017.
46. Blender 3D. Available online: https://www.blender.org/download/ (accessed on 25 May 2021).
47. Blender Help. Available online: https://docs.blender.org/manual/en/latest/animation/armatures/introduction.html (accessed on 25 May 2021).
48. Blender Help. Available online: https://docs.blender.org/manual/en/latest/animation/armatures/bones/structure.html (accessed on 25 May 2021).
49. System, M.; Müller, S. Damenhose im Jeans-Stil. In *Konstruktionen für Röcke und Hosen*; Deutsche Bekleidungs-Akademie München, Rundschau Verlag: München, Germany, 1996; pp. 82–84.
50. Walsh, L.K.; Restaino, R.; Martinez-Lemus, L.; Padilla, J. Prolonged leg bending impairs endothelial function in the popliteal artery. *Physiol. Rep.* **2017**, *5*, e13478. [CrossRef]
51. Namkoong, S.; Shim, J.; Kim, S.; Shim, J. Effects of different sitting positions on skin temperature of the lower extremity. *J. Phys. Ther. Sci.* **2015**, *27*, 2637–2640. [CrossRef]
52. Power, J. Fabric objective measurements for commercial 3D virtual garment simulation. *Int. J. Cloth. Sci. Technol.* **2013**, *25*, 423–439. [CrossRef]
53. Sayem, A.S.M. Objective analysis of the drape behaviour of virtual shirt, part 1: Avatar morphing and virtual stitching. *Int. J. Fash. Des. Technol. Educ.* **2017**, *10*, 158–169. [CrossRef]
54. Brubacher, K.; Tyler, D.; Apeagyei, P.; Venkatraman, P.; Brownridge, A.M. Evaluation of the Accuracy and Practicability of Predicting Compression Garment Pressure Using Virtual Fit Technology. *Cloth. Text. Res. J.* **2021**. First Published. [CrossRef]
55. Allsop, C.A. An Evaluation of Base Layer Compression Garments for Sportswear. Master's Thesis, Manchester Metropolitan University, Manchester, UK, 2012.
56. Sayem, A.S.M. Objective analysis of the drape behaviour of virtual shirt, part 2: Technical parameters and findings. *Int. J. Fash. Des. Technol. Educ.* **2017**, *10*, 180–189. [CrossRef]
57. Sayem, A.S.M.; Bednall, A. A novel approach to fit analysis of virtual fashion clothing. In Proceedings of the 19th Edition of the International Foundation of Fashion Technology Institutes Conference (IFFTI 2017—Breaking the Fashion Rules), Amsterdam, The Netherlands, 28–30 March 2017; The Amsterdam Fashion Institute (AMFI): Amsterdam, The Netherlands, 2017.
58. World Health Organization. Body Mass Index—BMI. Available online: https://www.euro.who.int/en/healthtopics/disease-prevention/nutrition/a-healthy-lifestyle/body-mass-index-bmi (accessed on 13 May 2021).
59. Gill, S. Improving garment fit and function through ease quantification. *J. Fash. Mark. Manag.* **2011**, *15*, 228–241. [CrossRef]
60. Gill, S.; Chadwick, N. Determination of ease allowances included in pattern construction methods. *Int. J. Fash. Des. Technol. Educ.* **2009**, *2*, 23–31. [CrossRef]
61. Shan, Y.; Huang, G.; Qian, X. Research Overview on Apparel Fit. In *Soft Computing in Information Communication Technology*; Luo, J., Ed.; Advances in Intelligent and Soft Computing, AISC 161; Springer: Berlin/Heidelberg, Germany, 2012; pp. 39–44.

62. Ng, R.; Cheung, L.; Yu, W. Dynamic ease allowance in arm Raising of Functional Garments. *Sen'I Gakkaishi* **2008**, *64*, 52–58. [CrossRef]
63. Chen, Y.; Zeng, X.; Happiette, M.; Bruniaux, P.; Ng, R.; Yu, W. A new method of ease allowance generation for personalization of garment design. *Int. J. Cloth. Sci. Technol.* **2008**, *20*, 161–173. [CrossRef]
64. McKinney, E.; Gill, S.; Dorie, A.; Roth, S. Body-to-Pattern Relationships in Women's Trouser Drafting Methods: Implications for Apparel Mass Customization. *Cloth. Text. Res. J.* **2017**, *35*, 16–32. [CrossRef]
65. Lim, H.-W.; Cassidy, T. A comparative study of trouser pattern making methods. *J. Text. Eng. Fash. Technol.* **2017**, *1*, 189–196. [CrossRef]
66. Petrova, A.; Ashdown, S.P. Three-Dimensional Body Scan Data Analysis: Body Size and Shape Dependence of Ease Values for Pants' Fit. *Cloth. Text. Res. J.* **2008**, *26*, 227–252. [CrossRef]
67. Schofield, N.A.; Ashdown, S.P.; Hethorn, J.; LaBat, K.; Salusso, C.J. Improving Pant Fit for Women 55 and Older through an Exploration of Two Pant Shapes. *Cloth. Text. Res. J.* **2006**, *24*, 147–160. [CrossRef]
68. Pirch, C.; Klepser, A.; Morlock, S. Using 3D Scanning to Create 4D Motion Data for Clothing Simulation. In Proceedings of the 3DBODY.TECH 2020—11th International Conference and Exhibition on 3D Body Scanning and Processing Technologies, Online Conference, 17–18 November 2020; Available online: https://www.3dbody.tech/cap/papers2020.html (accessed on 13 May 2021).
69. Zhang, D.; Wang, J.; Yang, Y. Design 3D garments for scanned human. *J. Mech. Sci. Technol.* **2014**, *28*, 2479–2487. [CrossRef]

Article

A New Method for Testing the Breaking Force of a Polylactic Acid-Fabric Joint for the Purpose of Making a Protective Garment

Slavica Bogović [1,*] and Ana Čorak [2]

1 Department of Clothing Technology, Faculty of Textile Technology, University of Zagreb, 10000 Zagreb, Croatia
2 Primary School Dr. Franjo Tudjman, 53520 Korenica, Croatia; ana.corak2@skole.hr
* Correspondence: slavica.bogovic@ttf.unizg.hr

Abstract: 3D printing is a technology that is increasingly used in the individualization of clothing, especially in the construction of garments for people with disabilities. The paper presents a study on the use of 3D printed knee protectors intended for wheelchair users. Due to the specific purpose of this 3D printed object, the breaking force of the polylactic acid (PLA) combined with 100% cotton and 100% polyester fabric was investigated. This paper will also describe a new method for testing the breaking force of a 3D printed polymer (PLA) combined with an incorporated fabric. Test samples were made, and the input parameters used in 3D printing were defined for testing purposes. A 3D knee protector for wheelchair users was developed based on a digitized model of the human body. The durability of the shape of the 3D printed shield was also tested after washing at temperatures of 40 °C, 50 °C and 60 °C. A clothing model that provides adequate user protection was proposed based on the conducted research. A construction solution has been proposed that enables the application of a 3D printed individualized garment element.

Keywords: 3D printing; textile applications; polylactic acid (PLA); breaking force; knee protector

Citation: Bogović, S.; Čorak, A. A New Method for Testing the Breaking Force of a Polylactic Acid-Fabric Joint for the Purpose of Making a Protective Garment. *Materials* **2022**, *15*, 3549. https://doi.org/10.3390/ma15103549

Academic Editor: Dubravko Rogale

Received: 1 April 2022
Accepted: 11 May 2022
Published: 16 May 2022

Publisher's Note: MDPI stays neutral with regard to jurisdictional claims in published maps and institutional affiliations.

Copyright: © 2022 by the authors. Licensee MDPI, Basel, Switzerland. This article is an open access article distributed under the terms and conditions of the Creative Commons Attribution (CC BY) license (https://creativecommons.org/licenses/by/4.0/).

1. Introduction

The application and development of individualized products based on 3D printing have been on the rise in the recent years. In the production of clothing, 3D printing is used to make entire garments, in which cases its segments are not integrated in the garment, or for the printing of patterns or incorporated elements on textile materials [1–4]. The additional function of the garment can be achieved in terms of design, the functionality of the garment (e.g., protection) or for the purpose of incorporating electronic components into the garment [5,6]. Different 3D printing technologies are used for 3D printing: fused deposition modeling (FDM) printers, stereolithography (SLA) printers, selective laser sintering (SLS) or polyjet modeling (PJM). For all of the above mentioned 3D printing technologies, it is necessary to create a digital 3D model in stereolithography file (STL) format that is converted into G-code, containing all the data needed for 3D printing [7].

Acrylonitrile butadiene styrene (ABS), polyethylene terephthalate (PET), polylactic acid (PLA), nylon and thermoplastic polyurethane (TPU) are polymers most commonly used for the 3D printing of garments, clothing segments and patterns or incorporated elements on textile materials. The choice of polymer depends on the final application of the garment, as well as the properties of the polymer and textile materials. The adhesion of different polymers to textile materials has been investigated in previous studies. The influence of textile surface properties on the adhesion strength of plastic flexible polymers is also investigated, taking into account the mechanical/physical and chemical mechanisms of adhesion using FDM technology [2,7–9].

To achieve higher adhesion strength, tests were conducted on the influence of mechanical and thermal parameters of 3D printing, such as the distance of layers from the material along the z-axis [10–12], chemical factors [6] and the treatment of textile material with plasma [9,13]. The treatment of textile surfaces with polymeric materials showed an increase in adhesive strength [8,14]. The bond strength of textile material and polymers is examined by varying a large number of 3D printing parameters, taking into account the raw material composition of textile material, the material density and the type of textile fabric [2,15].

Maintenance plays a major role in the functionality of the garment. Therefore, adhesion studies of 3D printed polymers after washing are performed regarding the number of repetitions of the wash cycle. This enables the collection of relevant data that can be used for the correct choice of polymer and textile material regarding its density and structure and their combinations. Based on these studies, it can be determined which combinations of polymers and materials are suitable for long-term or short-term use, as well as the combinations suitable only for prototyping [6,16].

PLA is a polymer that is increasingly used in FDM-based 3D printing due to its properties. It has sufficient mechanical strength, is biodegradable and can be adapted to different purposes by varying the parameters of the 3D print and the 3D modeling process. In its purest form, it has properties that are not suitable for the long-term use of 3D printed objects. PLA polymer composites are used for 3D printing. Such polymer composites directly affect the solubility and mechanical properties of the polymer, as well as the properties of the 3D printed object [17,18].

Clothes for wheelchair users [19] should be made according to individual measures and the possibilities of using these garments to ensure adequate protection. The use of 3D technology can therefore be a key element in ensuring adequate protection. A 3D body scanner is used to digitize the human body, take measurements and define the shape of the human body with all its specifics in a short time. By using the digitized model of the human body, it is possible to make a customized garment and define measures and forms of protective elements for each person specifically [19,20]. Therefore, the possibilities of applying 3D technology can be further explored for the development of individualized clothing for wheelchair users that differs from standard clothing. The shape and properties of the 3D printed protective elements of a garment are also investigated. In order to ensure the functionality of such garments and to facilitate their handling and use, it is necessary to examine the possibility of incorporating textile material into a polymer that is shaped according to the part of the body for which protection is provided [20]. It is necessary to test the strength of the bond of the polymer and textile material in order to fix the 3D printed protection element on the garment, which is shaped and positioned exactly according to the shape of the user's body.

When it comes to clothes for wheelchair users, it is necessary to develop functional clothing in a studious way in order to enable easier dressing in addition to protection, given the limited movements of the users [20].

A method of testing the breaking force of the bond between the polymer and textile material was developed with that purpose. Past research was based on testing the strength of the bond where the polymer was applied to the textile material. The samples were defined according to the standards that were the foundation for the mentioned research [2,8,9,11,13,15,18,21].

2. Materials and Methods

This research on the application of 3D printed polymer PLA for the purpose of making protective elements that are an integral part of clothing was conducted in three steps to find the best 3D print parameters that affect the quality and usability properties of 3D printed individualized knee protectors for wheelchair users.

The breaking force of the bond between the polymer and textile material on the samples was tested on samples where the textile material is located between the layers of 3D printed polymers of different fillings. Knee protectors were made based on a 3D scan of the human body. The stability of the shape of the 3D printed knee protectors of different wall thicknesses [20] was tested at different washing temperatures.

2.1. Materials

Biodegradable polymer PLA filament (produced by Devil Design Ryszka Mateja Company, Poland) was used for the experiment. The filament has a diameter of 1.75 mm, a density of 1.24 g/cm^3, an extrusion temperature of 200–250 °C and a heated bed of 50–60 °C. The basic parameters of 3D printing were defined, as shown in Table 1.

Table 1. Basic parameters of the 3D print polylactic acid (PLA) used to test the breaking strength of the bond with 100% cotton and 100% polyester fabric.

	Filament Data			3D Print Settings				
	Extrusion Temp. [°C]	Diameter [mm]	Speed [mm/s]	Density [g/cm^3]	Layer Height [mm]	Bed Temp. [°C]	Nozzle Size [mm]	Fill Angle
PLA	210	1.75	30	1.24	0.3	60	0.4	45°

The textile materials for which the tensile strength tests were performed are 100% cotton and 100% polyester. The characteristics are shown in Table 2.

Table 2. Characteristics of textile materials bonded with polylactic acid (PLA) for determining the breaking force.

Fabric	Raw Material Composition	Density [Thread/cm]		Weave	
		Warp	Weft		
1	100% cotton	25	24		
2	100% polyester	18	22		

2.2. Methods

3D printing is a method of creating objects based on a digital 3D model. The method of 3D printing using FDM technology is based on layering a polymer which is heated and melted at certain temperatures. The melted polymer passes through the nozzle above the bed of the 3D printer. The nozzle can be of different diameters depending on the object and polymer used for 3D printing. It moves along the x and y axes, leaving behind a thin strand of polymer. The polymer is transported toward the nozzle using rollers, which prevents the clogging of the nozzle. After one pass, the mechanism holding the nozzle rises along the z axis. In this way, the model is created layer by layer, attaching itself to the preheated bed in the first pass and to itself in all of the subsequent passes.

The process takes place according to the G-Code generated from a stereolithography file of a 3D object and the printing parameters. The parameters are defined according to the final use of the 3D object and the polymer in use. Depending on the properties of the polymer, it is necessary to define the temperature of the nozzle and bed, as well as the manner in which the polymer is to be fed to the nozzle. Parameters of 3D printing such as wall thickness (number of layers), layer height, object infill (density and shape), support, the speed of printing and others are determined depending on the use of the 3D object. The printing support must be defined for parts of the object that do not have a previous layer on which they could be printed. It is required to construct the 3D digital object shape in a CAD program, taking into account all of the features and limitations of the 3D printing method. 3D construction and G-Code preparations can reduce the need for support, which reduces the time of printing and polymer use.

Based on the described method of 3D printing, samples were made to test the breaking force of the PLA/textile/PLA bonds of samples that were supposed to be incorporated into the garment.

2.3. Preparation of the 3D Sample Model

Investigation of the influence of 3D printing parameters with FDM technology on the breaking force of a sample that has an integrated fabric between the polymer layers was performed by adapting the sample to the standards of HRN EN ISO 13934-1: 2013 (Textiles—Tensile properties of fabrics—Part 1: Determination of maximum force and elongation at maximum force using the strip method (ISO 13934-1: 2013; EN ISO 13934-1: 2013)) [22].

For the investigation of the breaking force, test samples were made in accordance with the method of testing prescribed by the stated norm. To test the breaking strength of the textile material between the layers of the 3D printed polymer, a 50 × 200 mm sample model was constructed, as shown in Figure 1. The dimensions of the central part of the sample into which the fabric is integrated were 50 × 100 mm. The end sides of the test sample were used to secure the test sample to the dynamometer clamps.

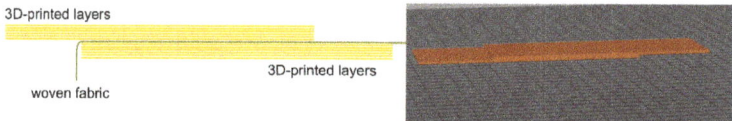

Figure 1. 3D sample model for breaking force test.

Test samples for measuring the breaking force were prepared using a Creality CR-10 Max 3D printer. The bed of the 3D printer is 450 mm × 450 mm in depth and width and 470 mm in height. Six types of samples were prepared, varying the fill density, as shown in Table 3. The 3D printing of the samples was performed with filling densities of 20%, 60% and 100% for 100% cotton and 100% polyester fabric.

Table 3. Defining 3D printed pattern fillings.

Infill Density	Preparation for 3D Printing	3D Printing of Model Infill
20%		
60%		
100%		

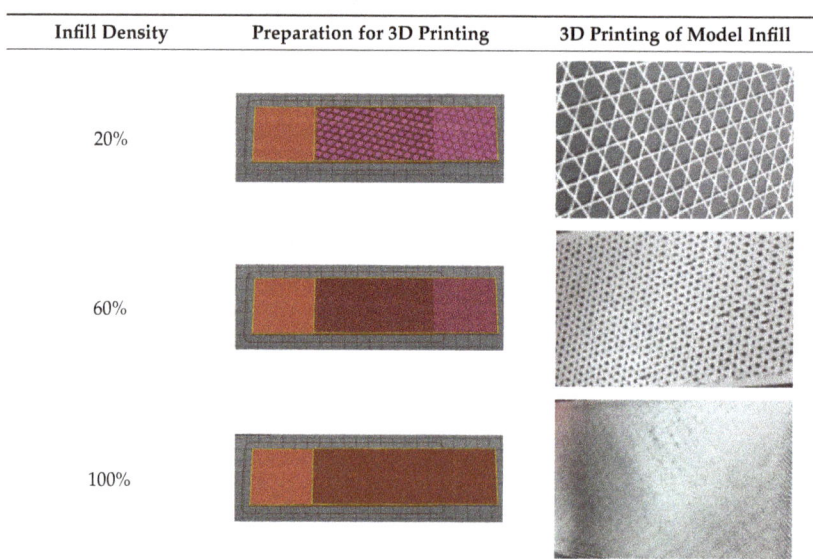

3. Results and Discussion
3.1. Breaking Force

The breaking force of the PLA/textile/PLA bond was tested and measured on the prepared samples on a MesdanLab Strength Tester dynamometer. Figure 2 shows a protector sample clamped in the dynamometer. The sample was pre-tensioned with 0.5 N.

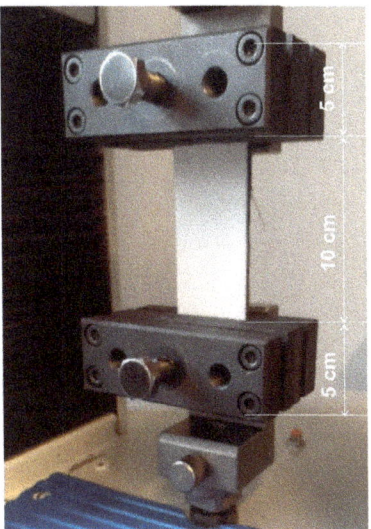

Figure 2. Sample testing on a MesdanLab Strength Tester dynamometer.

Based on the prepared samples for testing the breaking strength of the PLA/textile/PLA bond, the results were obtained for fabrics of raw material compositions of 100% cotton and 100% polyester (PES). Different 3D printed samples of 20, 60 and 100% density were used, and the results are shown in Figure 3. Figure 4 shows a graph of the elongation of the samples.

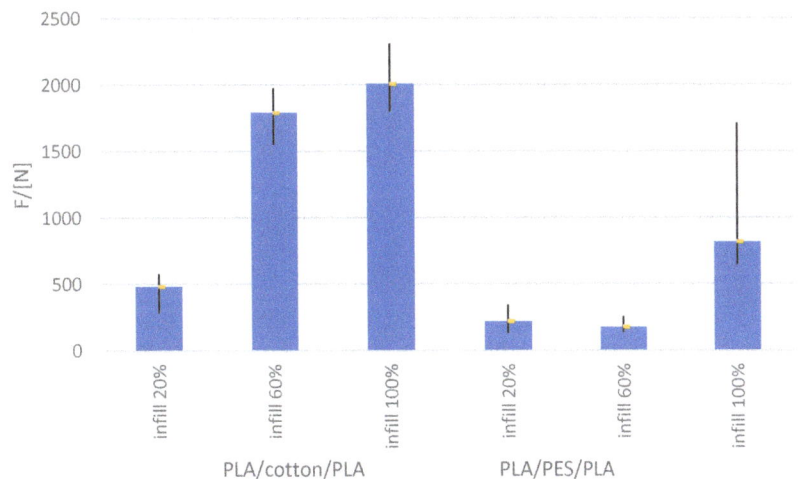

Figure 3. Test results of the breaking strength of PLA/cotton/PLA and PLA/polyester/PLA with fillings of 20, 60 and 100%.

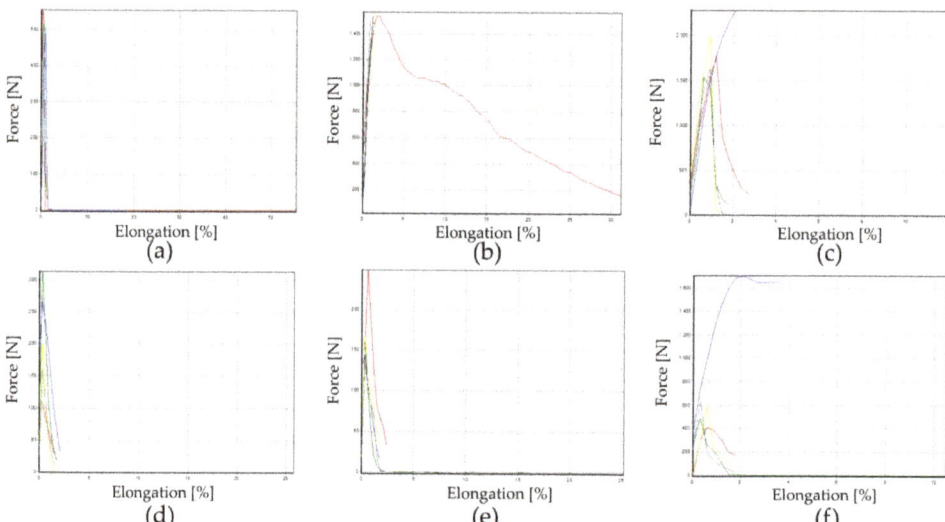

Figure 4. Test results of the elongation of PLA/cotton/PLA with fillings 20% (**a**), 60% (**b**) and 100% (**c**); and PLA/polyester/PLA with fillings 20% (**d**), 60% (**e**) and 100% (**f**).

Based on the presented results of measurements performed under the same conditions, it can be observed that there is a deviation in the amount of breaking force depending on the filling density of the 3D printed samples and the type of fabric used. The fabrics used in the test are of approximate density. The deviation in the breaking force values for PLA/cotton/PLA and PLA/polyester/PLA for samples of the same filling density is significant.

According to the obtained results, it can be concluded that the raw material composition of textile material has a significant influence. At the same percentage of filling (60%), the breaking force of the 3D printed element with incorporated cotton material is many times higher than when incorporating polyester fabric. In the case of cotton fabric with 100% filling, the breaking force is twice as high as in the case of the polyester sample at the same filling value. Previous studies on the adhesion of polymers to fabrics have also shown the better adhesion of polymers to cotton materials than polyester, although the polymer was applied only on the upper side of the fabric [21]. The difference between the breaking force in the test of cotton material and fillings of 60 and 100% is not negligible, and the presented results show that the breaking force does not increase linearly. Besides the 3D printing parameters, the sequencing of the layers also has an effect on the breaking force. The sequence is defined by the G-Code and is automatically generated after setting the 3D printing parameters. The 3D printing of higher density objects takes longer, which allows more time for the previous layer to cool down and harden. If textile is placed between two such layers, inconsistency in surface filling occurs.

The results show that there is a discrepancy between the values of the elongation and the breaking force of the same group of samples. The dispersion of data might be caused by the unevenness of the textile material or the unevenness of the 3D printing sample.

During the test of the breaking force, it was noticed that the separation of the test sample occurred between the textile material and the lower layer of the polymer, while the fabric remained adhered to the upper layer of the polymer. Since these are low density fabrics, the polymer has passed through the fabric. The aforementioned facts indicate that there is a need for new discoveries in the application of 3D printed objects onto textile materials. The strength of the bond can be further increased by reducing the fabric density or by using construction methods by defining openings at the PLA/fabric/PLA bonds.

3.2. Application onto the Garment

The stability of the 3D model of the protector was investigated after testing the breaking force of the polymer/fabric/polymer bonds. The 3D model of the protector is based on the 3D model of the human body. The positions and shapes of the protective elements of the garment were determined by analyzing the 3D human model. Point cloud segments of the human body serve as a foundation for 3D shield modeling, while body measurements serve to develop and construct the garment into which the shields are incorporated (Figure 5). The construction of clothing and protectors is carried out in two different ways. The trousers are constructed two-dimensionally, while the protectors are constructed using a 3D modeling software package.

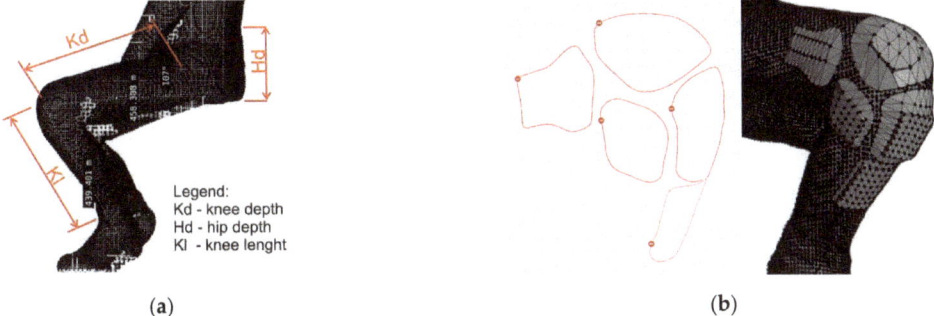

Figure 5. Defining body measures (**a**) and protector shape (**b**) according to the 3D human model.

To create a 3D element that is integrated into a garment, it is necessary to take into account a number of parameters that affect the final shape and function of the garment. The comfort and adherence of the 3D printed element are crucial, because solid protective elements cannot be subsequently adapted to the body shape (Figure 6). Knowledge of 3D printing technology is also very important, because it significantly affects the 3D design of objects. Thus, with FDM 3D print technology, it is important to take into account the polymer layering and the angle of inclination of the walls, wall thickness, etc. Since this is a human body with all its specifics, it is necessary to shape the protective elements according to the body. Parallel to the design of the 3D printed element, it is necessary to carry out the construction of the garment, which is additionally shaped according to the 3D printed element in order to finally connect two different elements and ensure the function for which it is intended (Figure 7). To bond two different materials of polymer and textile material, a regular plate is placed as a base into which the textile material is integrated [20].

Figure 6. Defining the intersection of the protector and knee (**a**); Defining the comfort and adherence of the 3D printed element and knee (**b**).

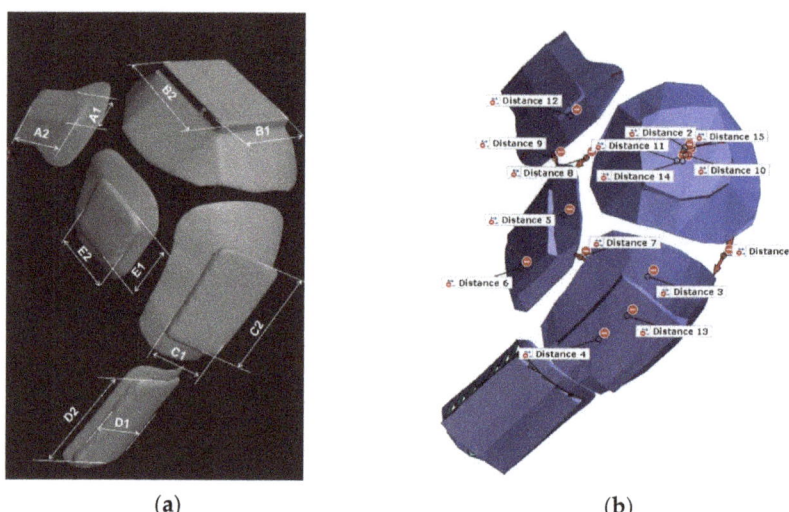

Figure 7. Determining the dimensions of the protector applied to the material (**a**); and the distance between protector elements (**b**).

Since textile materials of higher strength are commonly used in the production of protective garments resulting from the greater thickness and density of the textile material, it is necessary to consider the possibility of applying design solutions that allow for the better incorporation of textile material into 3D printed individualized protective elements. The application of 3D printed shapes in garments can thus be increased. The construction solution shown in Figure 8 is proposed for this purpose. The construction of the garment is adapted to the body in the sitting position. The entire process of garment construction is also adapted to the technical performance of 3D printing.

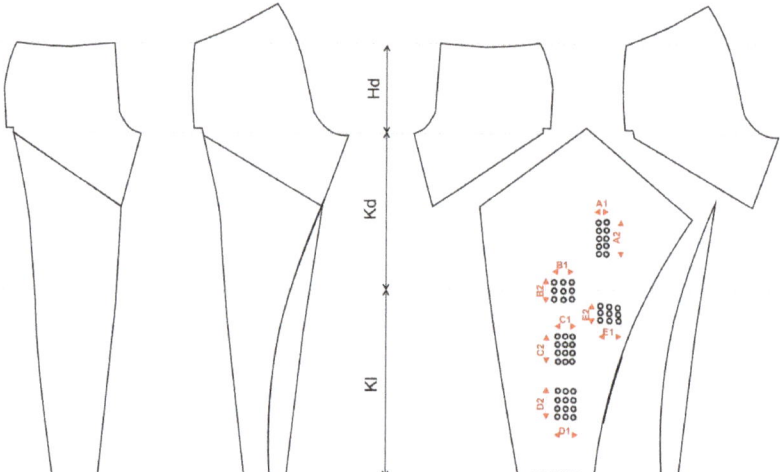

Figure 8. Cutting parts for 3D printing of protectors by incorporating textile material into the polymer.

The body size shown in Figure 4a was used to adapt the cut of the trousers for a person in a sitting position. Measures for knee depth (Kd), hip depth (Hd) and knee length (Kl) were defined, as they have an effect on the fitting of the garment. The measurement also defines the exact positions of the protector incorporated into the garment. The 3D constructed protectors have a rectangular base that allows for easy integration of the fabric. The dimensions of the base, as well as all of the parts of the protector and the distance between the protectors, are shown in Figure 6. The positions of the protectors were exactly calculated based on the above mentioned measurements (Figure 8). This modeled individualized cutting part of the trousers provides the possibility of integrating a flat cutting part into 3D models of the shield.

Two protectors of different wall thicknesses (7 and 12 3D printed layers) were made to investigate the stability of the shape of the protector in the washing process at different temperatures (Figure 9).

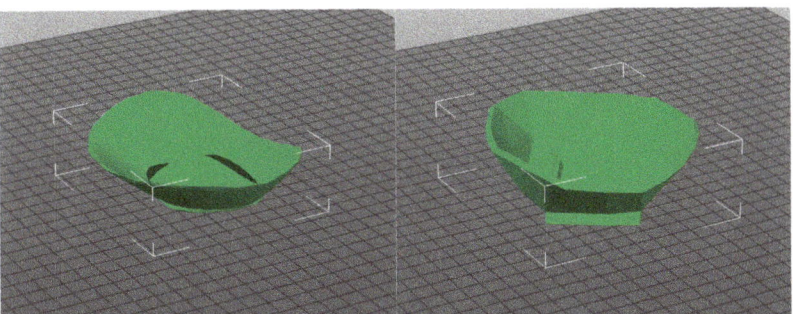

Figure 9. Knee protector made according to a 3D scan of the human body with a wall thickness of 7 and 12 3D printed layers [20].

Since the combination of the polymer and textile itself is not the only factor that indicates the functionality and usable properties of the finished garment, it is necessary to conduct tests of shape stability. To test the stability of the shape, the protectors were subjected to a washing process at temperatures of 40, 50 and 60 °C. The washing procedure was repeated five times for those elements that did not change shape in the previous wash cycle. During the washing process, the samples were left for 45 min in the aforementioned temperatures.

The samples of a 3D printed element with a wall thickness of seven layers of 3D printed PLA were washed at different washing temperatures (40, 50 and 60 °C), where deformations of the shape are clearly visible. Table 4 shows the results obtained after testing the stability of the shape of a 3D printed object with different wall thicknesses at different washing temperatures.

The protective element with a smaller wall thickness (seven layers) was deformed at washing temperatures of 50 and 60 °C after the first wash cycle, while its shape remained unchanged at a temperature of 40 °C. Therefore, the undeformed protector was subjected to further washing cycles at a temperature of 40 °C.

The thick-walled knee protector retained its shape even after five wash cycles at all of the tested wash temperatures.

After five washing cycles, it was noticed that the surface of the 3D fastened elements became rougher, while the edge became uneven with the shield of smaller wall thickness. The reason for this is the dissolution of the polymer in water, which indicates the reduced durability of the built-in 3D printed elements in the garment. Based on the presented results, it can be concluded that the wall thickness of the 3D printed element affects the stability of the shape in the washing process. This research is necessary because it indicates the applicability of 3D printed elements incorporated into a garment. The optimal wall thickness, in addition to ensuring the stability of the shape in the washing process, will ensure adequate protection of the part of the body for which the protective element is intended.

Table 4. Results of shape stability testing at different washing temperatures.

Washing Temperature	After a Single Wash	After 5 Washes
40 °C		
50 °C		
60 °C		

4. Conclusions

Based on the presented results, it can be concluded that there is a need for further research related to the application of 3D printed elements that are incorporated into garments in order to achieve additional functionality that is primarily concerned with protection. Previous research on the adhesion of polymers to textile materials conducted by the method of 3D printing on the material, along with the presented research, indicates the possibility of incorporating textile material into 3D printed polymers by developing a new method of testing the breaking force of the polymer/textile/polymer. By varying the parameters of 3D printing, it is possible to determine the optimal parameters of 3D printing and the type of textile product that will ensure the adequate application of the garment. It is also evident from the above that the construction of clothing and the 3D modeling of the elements integrated into the garment are interdependent and that design solutions can be found to provide adequate and reusable garments, especially for sensitive target groups such as people with disabilities.

Author Contributions: Conceptualization, S.B.; Data curation, S.B. and A.Č.; Formal analysis, S.B. and A.Č.; Funding acquisition, S.B.; Investigation, S.B. and A.Č.; Methodology, S.B. and A.Č.; Resources, S.B. and A.Č.; Visualization, S.B. Writing—original draft, S.B.; Writing—review & editing, S.B. and A.Č. All authors have read and agreed to the published version of the manuscript.

Funding: The financial support was provided within the framework of the Support for Scientific Research 2021 "Microbial barriers of textiles as a basis for functional design of specific purpose clothing", financed by the University of Zagreb.

Institutional Review Board Statement: This research does not require ethical approval.

Informed Consent Statement: Written informed consent has been obtained from the person to publish this paper.

Acknowledgments: The authors express their gratitude to Maja Somogyi Škoc for her help. The authors also thank Dario Bogović for the translation of the paper and Antonia Treselj for proofreading the paper.

Conflicts of Interest: The authors declare no conflict of interest.

References

1. Kozior, T.; Blachowicz, T.; Ehrmann, A. Adhesion of three-dimensional printing on textile fabrics: Inspiration from and for other research areas. *J. Eng. Fibers Fabr.* **2020**, *15*, 1558925020910875. [CrossRef]
2. Korger, M.; Glogowsky, A.; Sanduloff, S.; Steinem, C.; Huysman, S.; Horn, B.; Ernst, M.; Rabe, M. Testing thermoplastic elastomers selected as flexible three-dimensional printing materials for functional garment and technical textile applications. *J. Eng. Fibers Fabr.* **2020**, *15*, 1558925020924599. [CrossRef]
3. Milošević, P.; Bogović, S. 3D Technologies in Individualized Chest Protector Modelling. *Text. Leather Rev.* **2018**, *2*, 46–55. [CrossRef]
4. Sitotaw, D.B.; Ahrendt, D.; Kyosev, Y.; Kabish, A.K. Additive Manufacturing and Textiles-State of the Art. *Appl. Sci.* **2020**, *10*, 5033. [CrossRef]
5. Spahiu, T.; Piperi, E.; Grimmelsmann, N.; Ehrmann, A.; Shehi, E. 3D Printing as a New Technology for Apparel Designing and Manufacturing. In Proceedings of the Interantional Textile Conference, Dresden, Germany, 24–25 November 2016.
6. Pei, E.; Shen, J.; Watling, J. Direct 3D Printing of Polymers onto Textiles: Experimental, Studies and Applications. *Rapid Prototyp. J.* **2015**, *21*, 556–571. [CrossRef]
7. Grothe, T.; Brockhagen, B.; Storck, J.L. Three-dimensional printing resin on different textile substrates using stereolithography: A proof of concept. *J. Eng. Fibers Fabr.* **2020**, *15*, 1558925020933440. [CrossRef]
8. Meyer, P.; Döpke, C.; Ehrmann, A. Improving adhesion of three-dimensional printed objects on textile fabrics by polymer coating. *J. Eng. Fibers Fabr.* **2019**, *14*, 1558925019895257. [CrossRef]
9. Korger, M.; Bergschneider, J.; Lutz, M.; Mahltig, B.; Finsterbusch, K.; Rabe, M. Possible Applications of 3D Printing Technology on Textile Substrates. In Proceedings of the 48th Conference of the International Federation of Knitting Technologists (IFKT), Materials Science and Engineering, Moenchengladbach, Germany, 8–11 June 2016; Volume 141. [CrossRef]
10. Mpofu, N.S.; Mwasiagi, J.I.; Nkiwane, L.C.; Githinji, D.N. The use of statistical techniques to study the machine parameters affecting the properties of 3D printed cotton/polylactic acid fabrics. *J. Eng. Fibers Fabr.* **2020**, *15*, 1558925020928531. [CrossRef]
11. Spahiu, T.; Al-Arabiyat, M.; Martens, Y.; Ehrmann, A.; Piperi, E.; Shehi, E. Adhesion of 3D printing polymers on textile fabrics for garment production. In Proceedings of the Aegean International Textile and Advanced Engineering Conference (AITAE 2018), 2018 IOP Conference Series: Materials Science and Engineering, Lesvos, Greece, 5–7 September 2018; Volume 459. [CrossRef]
12. Bochnia, J.; Blasiak, M.; Kozior, T. A Comparative Study of the Mechanical Properties of FDM 3D Prints Made of PLA and Carbon Fiber-Reinforced PLA for Thin-Walled Applications. *Materials* **2021**, *14*, 7062. [CrossRef] [PubMed]
13. Kozior, T.; Döpke, C.; Grimmelsmann, N.; Junger, I.J.; Ehrmann, A. Influence of fabric pretreatment on adhesion of three-dimensional printed material on textile substrates. *Adv. Mech. Eng.* **2018**, *10*, 1687814018792316. [CrossRef]
14. Unger, L.; Scheideler, M.; Meyer, P.; Harland, J.; Görzen, A.; Wortmann, M.; Dreyer, A.; Ehrmann, A. Increasing Adhesion of 3D Printing on Textile Fabrics by Polymer Coating. *Tekstilec* **2018**, *61*, 265–271. [CrossRef]
15. Narula, A.; Pastore, C.M.; Schmelzeisen, D.; El Basri, S.; Schenk, J.; Shajoo, S. Effect of knit and print parameters on peel strength of hybrid 3-D printed textiles. *J. Text. Fibrous Mater.* **2018**, *1*, 2515221117749251. [CrossRef]
16. Martensa, Y.; Ehrmanna, A. Composites of 3D-Printed Polymers and Textile Fabrics. In Proceedings of the ICMAEM-2017, IOP Conference Series: Materials Science and Engineering, Secunderabad, India, 3–4 July 2017; Volume 225, p. 012292. [CrossRef]
17. Tümer, E.H.; Erbil, H.Y. Extrusion-Based 3D Printing Applications of PLA Composites: A Review. *Coatings* **2021**, *11*, 390. [CrossRef]
18. Chalgham, A.; Ehrmann, A.; Wickenkamp, I. Mechanical Properties of FDM Printed PLA Parts before and after Thermal Treatment. *Polymers* **2021**, *13*, 1239. [CrossRef] [PubMed]
19. Nakić, M.; Bogović, S. Computational Design of Functional Clothing for Disabled People. *Tekstilec* **2019**, *62*, 23–33. [CrossRef]
20. Čorak, A. Application of 3D Technologies in the Manufacture of Men's Trousers Intended for Rugby Players in Wheelchairs. Graduate Thesis, University of Zagreb Faculty of Textile Technology, Zagreb, Croatia, 20 October 2020.
21. Mpofu, N.S.; Mwasiagi, J.I.; Nkiwane, L.C.; Njuguna, D. Use of regression to study the effect of fabric parameters on the adhesion of 3D printed PLA polymer onto woven fabrics. *Fash Text.* **2019**, *6*, 24. [CrossRef]
22. HRN EN ISO 13934-1:2013; Textiles—Tensile Properties of Fabrics—Part 1: Determination of Maximum Force and Elongation at Maximum Force Using the Strip Method (ISO 13934-1:2013; EN ISO 13934-1:2013). Croatian Standards Institute: Zagreb, Croatia, 2013.

Article

Reactive Printing and Wash Fastness of Inherent Flame Retardant Fabrics for Dual Use

Martinia Glogar *, Tanja Pušić, Veronika Lovreškov and Tea Kaurin

Department of Textile Chemistry and Ecology, Faculty of Textile Technology, University of Zagreb, 10000 Zagreb, Croatia; tanja.pusic@ttf.unizg.hr (T.P.); veronika.lovreskov@ttf.unizg.hr (V.L.); tea.kaurin@ttf.unizg.hr (T.K.)
* Correspondence: martinia.glogar@ttf.unizg.hr

Abstract: The possibility of reactive printability on protective flame—resistant fabrics, varied in composition of weft threads and weave was investigated. In addition, the wash fastness of printed samples was analyzed. The functional properties of fabrics were assessed by measuring of the Limiting Oxygen Index (LOI). Printing was performed with two printing pastes varied in thickeners and two dyestuff concentrations. The samples were analyzed by microscopic imaging using digital microscope and spectrophotometric measurement before and after the five washing cycles. The results confirmed the printability of FR inherent fabrics specified through fine colored effects and optimal wash fastness.

Keywords: FR fabric; printing; wash fastness; spectrophotometry

Citation: Glogar, M.; Pušić, T.; Lovreškov, V.; Kaurin, T. Reactive Printing and Wash Fastness of Inherent Flame Retardant Fabrics for Dual Use. *Materials* **2022**, *15*, 4791. https://doi.org/10.3390/ma15144791

Academic Editor: Ricardo J. C. Carbas

Received: 10 May 2022
Accepted: 30 June 2022
Published: 8 July 2022

Publisher's Note: MDPI stays neutral with regard to jurisdictional claims in published maps and institutional affiliations.

Copyright: © 2022 by the authors. Licensee MDPI, Basel, Switzerland. This article is an open access article distributed under the terms and conditions of the Creative Commons Attribution (CC BY) license (https://creativecommons.org/licenses/by/4.0/).

1. Introduction

Non-flammability of textiles has been the subject of intensive research since the 1940s, although there are also studies published as early as 1735 and 1821 on finishing of cellulosic textiles using alum, ferrous sulphate and borax, which laid the foundation for a modern approach to the research and development of non-flammable textiles [1–3]. There are two main directions in current research in the field of non-flammable textiles—research of flame-retardants of innovative structure and pre-treatment methodologies, and research into inherently non-flammable fibers, yarns and textiles. The most common flame retardants formulation developed in the second part of the 20th century was based on phosphorous, nitrogen or halogen derivate. For cotton, formulations based on organophosphorus compounds, such as tetrakis (hydroxymethyl) phosphonium chloride (THPC), hydroxyl functional organophosphorus oligomer (HFPO), dimethyloldihydroxyethyleneurea (DMD-HEU) as well as some formulations based on monoguanidine dihydrogen phosphate (MGHP) and 3-aminopropylthoxysilane (APS), in some combinations with phosphoric acid were used [4–13]. The characteristics of such finishes applied on cellulose-based textiles can be changes in the comfort and physical-mechanical properties, as well as their negative environmental profile (e.g., release of free formaldehyde) [14–16]. The dynamics of changes in these properties depends on the characteristic of textile materials, process conditions, the composition of the bath, the method of application of functional agents (e.g., plasma, impregnation, LbL-layer by layer etc.) drying temperature and condensation [14]. In order to overcome such shortcomings, the direction of research is moving towards environmentally more acceptable and economically justified formulations with reduced amount of formaldehyde being released from the treated fabric, using butane tetra carboxylic acid (BTCA) as the binding agent [4]. In addition, some halogen-free phosphorus–nitrogen-based flame retardants were developed to promote more char formation during burning of the cellulosic substrate [17–19].

In 2011, Horrocks published a comprehensive study of innovative FR functionalization based on the application of nano-ZnO or TiO_2, various clay compounds, and polycarboxylic acids for application to cotton [20,21]. In studies of low-pressure plasma pre-treatment in developing nano-coatings, Tsafack et al. reported an FR efficiency in grafting of phosphorus-containing acrylate monomers to polyacrylonitrile fabrics [22,23]. Jama et al. studied the possibility of using low pressure plasma in deposition of silicon-based nano-layer for improving the flame retardancy of polyamide 6 [24,25].

In the scope of inherent FR yarns and fabrics, intended for protective clothing, mixtures of fibers (synthetic and natural origin) are primarily applied, in order to achieve optimal non-combustibility, but also meet certain comfort standards. Sonee et al. tested the burning behavior of FR viscose and meta-aramid blended in three different ratios, as well as FR viscose with polyamide 6.6 and meta-aramid blended in two different ratios, confirming the best behavior of viscose/meta-aramid blends in ratio 30:70 [26,27]. One of the problems with blends with the purpose of achieving inherently non-combustible fabric is the nonlinearity of the final properties of the burning behavior to the structure of the blend [28]. Due to the high hydrophobicity and crystallinity of individual components of such blends, dyeing and printing is complex and requires a systematic approach to the selection of dyestuff, but also optimizing a key process parameter in dyeing or printing. Manyukov et al. have published an interesting study about dyeing of thermostable para/meta aramid by pre-treatment in a solvent-water-swelling system, and subsequent dyeing by the depletion process with a mixture of disperse and cationic dye [29].

Regarding printing on inherently flame-retardant fabrics, a review of the literature revealed a research gap and a relatively small number of publications. This was exactly the challenge and incentive in the design of the experiment, to examine, among other things, the characteristics of coloration that can be achieved given the limited content of components that can be dyed. The research is part of the project of the development of a functional fabric that has the properties of inherent flame retardancy and high comfort, whose visual and functional properties can be improved through printing.

2. Materials and Methods

Research was performed on five non-commercial fabrics whose composition and constructional properties have been designed and developed in cooperation of Croatian textile factory Cateks and University of Zagreb faculty of Textile Technology. Two fabrics were in twill 2/2, two in twill 3/1 and one in ripstop. The warp thread of these fabrics are the same, blend of meta-aramid, m-AR (95%) and para-aramid, p-AR (5%), while the weft threads are different in composition, polyamide (PA), FR viscose (CV FR) am meta-aramid (m-AR), Table 1.

Table 1. Composition of fabrics.

	Fabric 1	**Fabric 2**	**Fabric 3**	**Fabric 4**	**Fabric 5**
Weave	Ripstop	Twill 2/2	Twill 2/2	Twill 3/1	Twill 3/1
Warp	95% m-AR 5% p-AR	95% m-AR 5% p-AR	95% m-AR 5% p-AR	95% m-AR 5% p-AR	95% m-AR 5% p-AR
Weft	2% PA 20% PA 6.6 38% CV FR 40% m-AR	2% PA 20% PA 6.6 38% CV FR 40% m-AR	20% PA6.6 40% CV FR 40% m-AR	2% PA 20% PA 6.6 38% CV FR 40% m-AR	20% PA6.6 40% CV FR 40% m-AR

Detailed characterization of structural and mechanical properties of fabrics is shown in Table 2.

Samples were woven on sample weaving machine Fanyuan Instrument DW598 (Hefei Fanyuan Instrument Co. Ltd., Hefei, Anhui, China), fully automated loom for weaving patterns with a woven rod. The loom is characterized with the maximum width of the base of 50 cm; number of wefts per minute of 30–60 and with maximum number of sheets of 20.

Table 2. Constructional and mechanical characteristics of fabrics.

	Mass [g/m^2]	Thickness 0.5 kPa [mm]	Thickness 1 kPa [mm]	Warp Density [Threads/cm]	Weft Density [Threads/cm]
Fabric 1	203	0.66	0.63	38	20
Fabric 2	197	0.78	0.75	37	20
Fabric 3	223	0.80	0.78	37	20
Fabric 4	195	0.84	0.82	36	20
Fabric 5	215	0.91	0.88	37	20

2.1. LOI (Limiting Oxygen Index)

The flame resistance of the new fabrics was tested by LOI—Limiting Oxygen Index. measured using Limiting Oxygen Index Apparatus (Concept Equipment Ltd., Arundel, UK) according to EN ISO 4589-2: 2017 Plastics—Determination of burning behavior by oxygen index—Part 2: Ambient—temperature test. Samples in size 140 mm × 53 mm were conditioned (RH 65%, 24 h) according to standard ISO 139:2008/A1 Textiles—Standard Atmospheres for Conditioning and Testing.

2.2. Micro Cone Calorimeter (MCC)

Microscale combustion calorimeter (MCC) from Govmark, Farmingdale, NY, USA was applied for the thermal characterization of five FR fabrics according to ASTM D7309-2007. The sample in mass of few milligrams was heated to a specified temperature using a linear heating rate of 1 °C/s in a nitrogen stream, flow rate of 80 cm^3/min. The thermal degradation products were mixed with a 20 cm^3/min oxygen stream.

2.3. Fabric Surface pH Measuring

Contact pH measurement of functional fabrics was tested using a contact electrode InLab Surface Pro-ISM® device SevenCompact ™ Duo S213, Mettler Toledo, Greifensee, Switzerland.

2.4. Screen Printing

Printing with reactive dyes was performed by hand screen procedure. After drying and fixing of the prints, wash fastness was tested. Reactive dye and two thickeners were used for the preparation of printing pastes, each in two different viscosities defined by the ratio of water and dry matter. The first thickener was CHT-Alginat MV (CHT Group and Co) and the second Alkagum NS (Diamalt AG München 13, München, Germany). Both were prepared in 4% and 9% of a thickening agent (dry matter), so different viscosities were tested. The reactive dyestuff anthraquinone structure used was Brilliantblau V-R spez, Bezema (C.I. Reactive Blue 19, C.I. 61,200), in two concentrations. Pastes with a lower concentration of dye are marked with "a", and pastes with a higher concentration of dye with "b". The composition of the printing pastes is shown in Table 3. Quantities of components in printing pastes are shown per 100 g of printing paste. The samples were printed by hand screen procedure, dried and fixed with steam for 10 min.

2.5. Wash Fastness Testing and Spectrophotometric Measurement

Printed samples were tested in a laboratory apparatus Polycolor, Mathis, Oberhasli, Swiss. The test was performed according to standard ISO 105-C06:2010 (A2S) Textiles—Tests for color fastness—Part C06: Colour fastness to domestic and commercial laundering, using 5 g/L standard detergent (ECE Non phosphate detergent without optical brightener agent), with bath ratio 1:8, temperature 40 ± 2 °C, time 30 min, through 5 cycles. The samples were air dried between each cycle.

Table 3. Composition of printing pastes.

Paste	Thickener		Dyestuff	Urea	Na$_2$CO$_3$
1a	CHT-Alginat MV (4%)	50 g	1.26 g	20 g	4 g
2a	CHT-Alginat MV (9%)	50 g	1.26 g	20 g	4 g
3a	Alkagum NS (4%)	50 g	1.26 g	20 g	4 g
4a	Alkagum NS (9%)	50 g	1.26 g	20 g	4 g
1b	CHT-Alginat MV (4%)	50 g	7.5 g	20 g	4 g
2b	CHT-Alginat MV (9%)	50 g	7.5 g	20 g	4 g
3b	Alkagum NS (4%)	50 g	7.5 g	20 g	4 g
4b	Alkagum NS (9%)	50 g	7.5 g	20 g	4 g

The wash fastness of prints was assessed by spectral evaluation of samples before and after washing, using remission spectrophotometer DataColor 850 (Datacolor AG, Lucerne, Switzerland), with constant instrument aperture, standard light D65 and d/8°geometry. The results are shown in terms of color difference values calculated according to CIE76 formula.

2.6. Microscopic Imaging

The microscopic examination of a printed samples was performed using DinoLite AM7013 under magnification of 50×. The imaging was performed before and after the 5th cycle of washing and drying. The microscopic imaging has been performed with the following parameters: magnification: 50×/1.3 MP; unit: mm; horizontal FOV [accuracy]: 9.564 mm [+/−0.192 mm]; one pixel increment (one keyboard arrow press): ~7.4 µm).

The scanning of the fabrics was performed with resolution of 1200 dpi.

3. Results

FR properties of fabrics before printing in warp and weft direction by LOI-Limiting Oxygen Index were examined according to a standard protocol are presented in Table 4.

Table 4. LOI of fabrics.

Fabric Sample	LOI [%]			
	Warp	σ [%]	Weft	(σ) [%]
Fabric 1	33.6	0.266	33.55	0.266
Fabric 2	34.2	0.191	33.96	0.266
Fabric 3	34.4	0.191	34.37	0.151
Fabric 4	34.1	0.110	34.10	0.261
Fabric 5	34.68	0.151	34.68	0.110

The results of LOI in Table 4 indicate on high resistance to burning, so all samples meet the criteria specified for the inherently flame-resistant fabrics with LOI > 26%. Comparison of LOI values did not show significant differences within variations in weft and weave.

MCC as a method for evaluation of the fire retardancy of materials using only a few milligrams of sample showed small differences between samples, Table 5.

Tested fabrics can be grouped according to values of MCC parameters as follows. The heat release capacity (ηc) of fabrics ranges from 39.00 $[J(g\cdot K)^{-1}]$ for fabric 2 to 51.00 $[J(g\cdot K)^{-1}]$ for fabric 5. Other fabrics are specified between these values. Fabric 1, Fabric 2 and Fabric 4 possessed more optimal values than Fabric 3 and 5. Peak heat release rates (Qmax) for selected fabrics 1, 2 and 4 fully follow the values of the heat release capacity. Specific heat release of samples (hc) fabric 1 and fabric 3 is less than 7.00 $[kJ\cdot g^{-1}]$, while to fabric 2 and fabric 5 possessed the same value, 7.20 $[kJ\cdot g^{-1}]$, and finally the fabric 5 to which belongs the highest value of 8.00 [kJ·g^{-1}]. Pyrolysis residues of all tested fabrics are higher than 40%. Finally, according to MCC criteria small differences were found between tested fabrics, which can be attributed to structural features, primarily in the fabric embroidery. Nevertheless, all fabrics can be specified as fire retardant.

Table 5. MCC parameters.

Parameters	Fabric 1	Fabric 2	Fabric 3	Fabric 4	Fabric 5
ηc (J[g·K]$^{-1}$)	43.00	39.00	50.00	44.0	51.00
Qmax [W·g^{-1}]	42.68	38.71	50.73	41.56	51.72
hc [kJ·g^{-1}]	6.60	7.20	6.40	8.00	7.20
hc, gas [kJ·g^{-1}]	12.30	12.90	11.46	13.74	12.19
Tmax [°C]	281.9	276.0	281.1	279.0	279.3
Residue [%]	46.33	44.17	44.16	41.79	40.93

The surface of fabrics was characterized by pH value, which may indicate on character of residual substances, Table 6.

Table 6. pH evaluated by contact electrode.

Sample	pH	(σ)	T [°C]
Fabric 1	4.57	0.246	25.5
Fabric 2	4.73	0.380	25.7
Fabric 3	4.59	0.249	25.7
Fabric 4	4.51	0.264	24.9
Fabric 5	4.40	0.452	24.5

The values obtained indicate acidity of the surface which may be caused by yarn preparations. The acidic surrounding can affect the reduced bonding of the reactive dye with the fiber, since it requires alkaline conditions to induce the reaction of the dye with the fiber. So samples were before printing washed in a mild detergent composed from anionic and non-ionic surfactants. After the analyses of FR properties of the fabrics along with the characterization of the surface pH, the process of printing with reactive dye was approached.

Figures 1–5 shows microscopic images of printed fabrics. It was observed that the binding of dyes in the process of printing and fixing occurs exclusively on the viscose components of the yarn. The difference in the depth of blue coloration, that is visually noticeable in samples Fabric 1, Fabric 2 and Fabric 4, compared to samples Fabric 3 and Fabric 5, stems from the difference in the basic color of fabrics, which has a grey component in weft yarn. The images of the samples serve the visual orientation of the appearance of the achieved coloration. In addition, the quality of the print is, in general, assessed by the sharpness of the print and the achieved color characteristics.

For Fabric 1, for samples printed with Paste 2a, 4a and 4b, satisfactory print sharpness was obtained (Figure 1).

The result indicates that a thickener with a higher dry matter content is suitable for this type of fabric with a hydrophobic fiber content (aramids and PA), as in the case of the indicated pastes, regardless of the type of thickener. The microscopic images clearly show the component of aramid yarn that remained uncolored, and the proportion of the same is visible due to the characteristic of the weaving. Since the appearance of the color depends on the interaction of simultaneous remission from the dyed and undyed components of the fabric, an influence of the uncolored component on obtained color appearance can be expected.

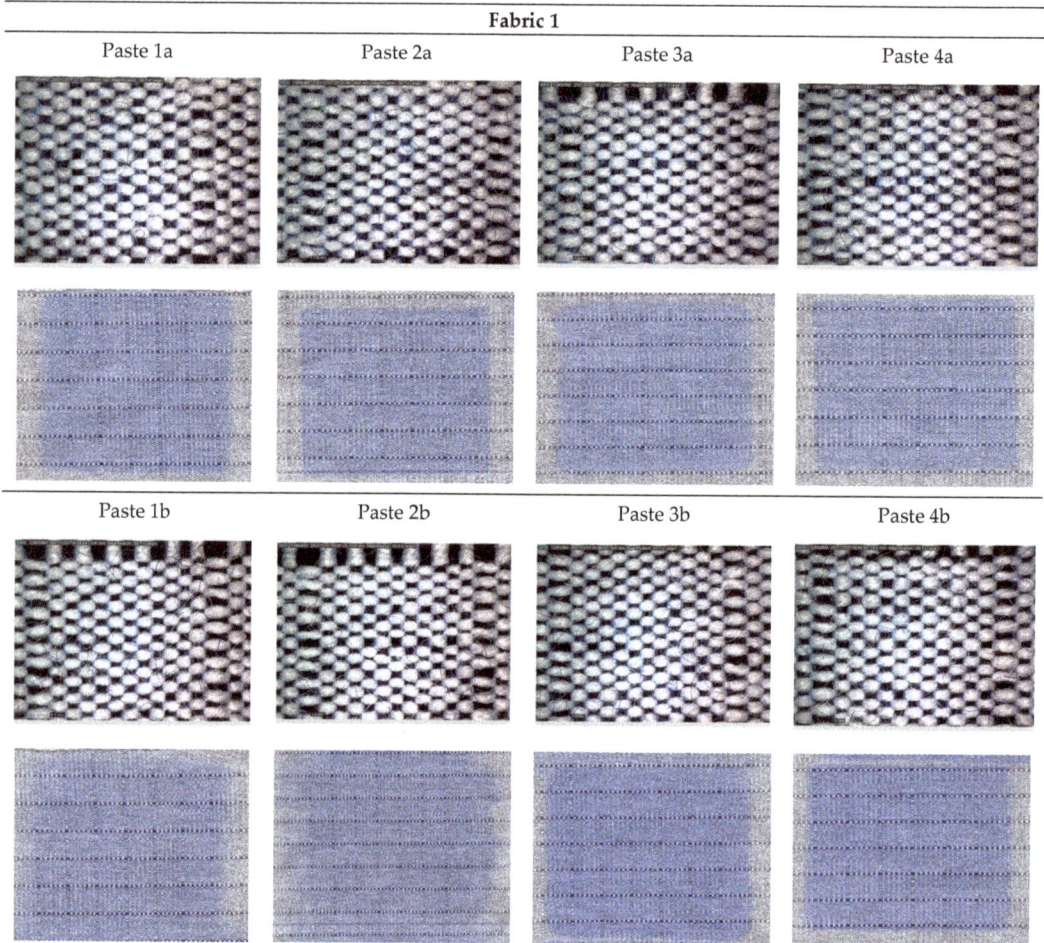

Figure 1. DinoLite microscopic and scanned images of printed Fabric 1.

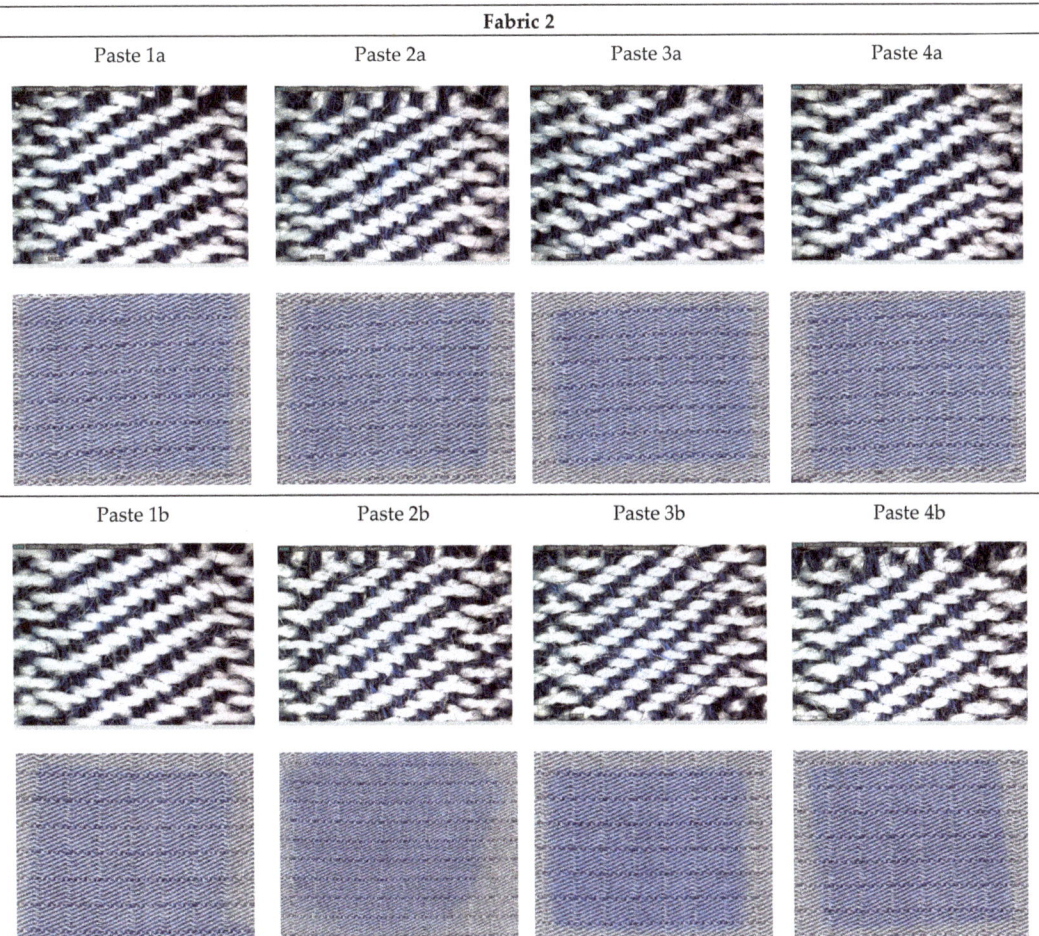

Figure 2. DinoLite microscopic and scanned images of printed Fabric 2.

Fabric 2 (Figure 2) differs from Fabric 1 in the type of weaving, which also affects the behavior of the printing paste.

It can be noticed that for Paste *a*, regardless of the type of thickener and the amount of dry matter, satisfactory print sharpness was obtained, but for Paste *b* where a higher amount of dye is present (recipes in Table 3), some printing paste spreading outside the pattern cotour occurred for thickener CHT-Alginate MV. A starch-based thickener (Alkagum NS) proved to be more suitable for printing Fabric 2 with higher dye concentration.

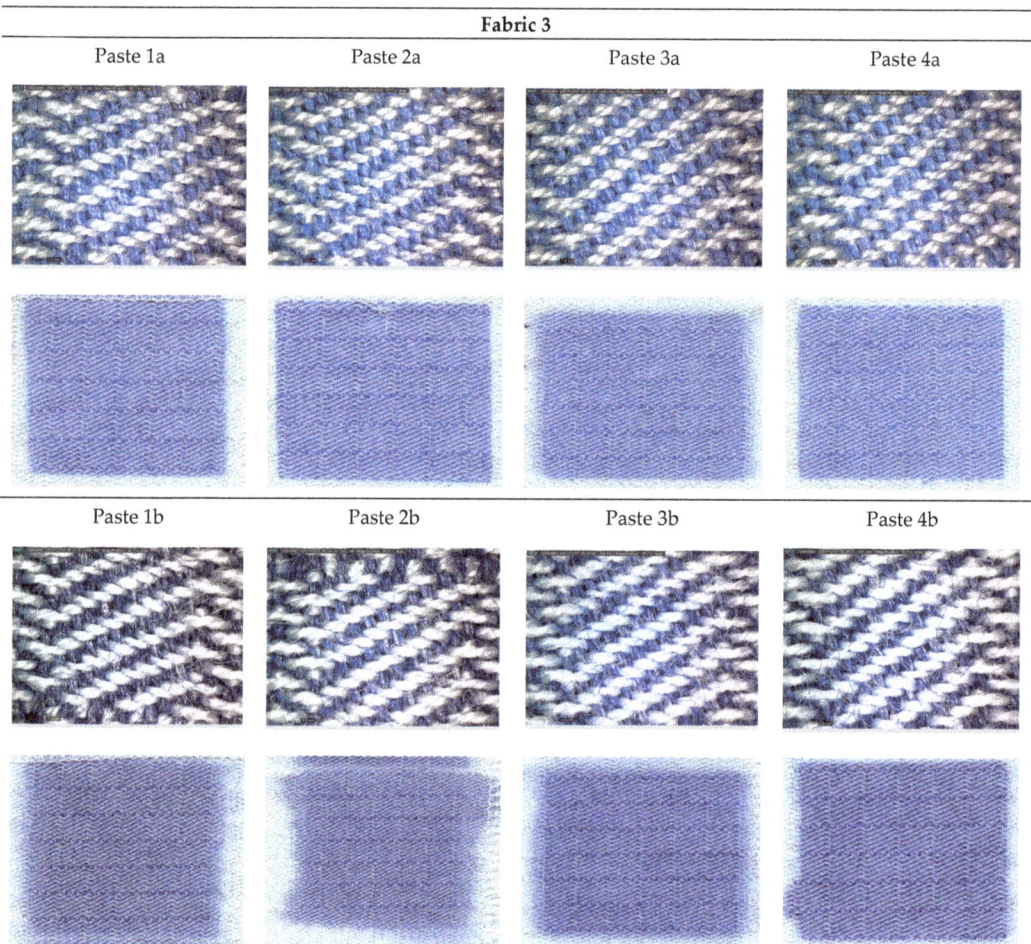

Figure 3. DinoLite microscopic and scanned images of printed Fabric 3.

Fabric 3 (Figure 3) is of the same weave as Fabric 2, differing in the proportion of PA fibers (Table 1). Furthermore, in contrast with Fabrics 1 and 2, Fabric 3 has completely uncolored components in its composition and does not already contain the basic gray shade. That is why a clearer blue color was obtained.

A similar trend of differences is observed as for Fabric 2. For Paste b, which is characterized by a higher proportion of dyestuff, spreading of dye outside the print contour is observed. On microscopic images it is observed for samples printed with the same Paste b, a darker color of the dye-binding component, which is associated with the occurrence of dye capillary spreading. For the component that has an affinity for the reactive dye, which is viscose, there was saturation, i.e., maximum dye binding, and the excess dye caused capillary spillage.

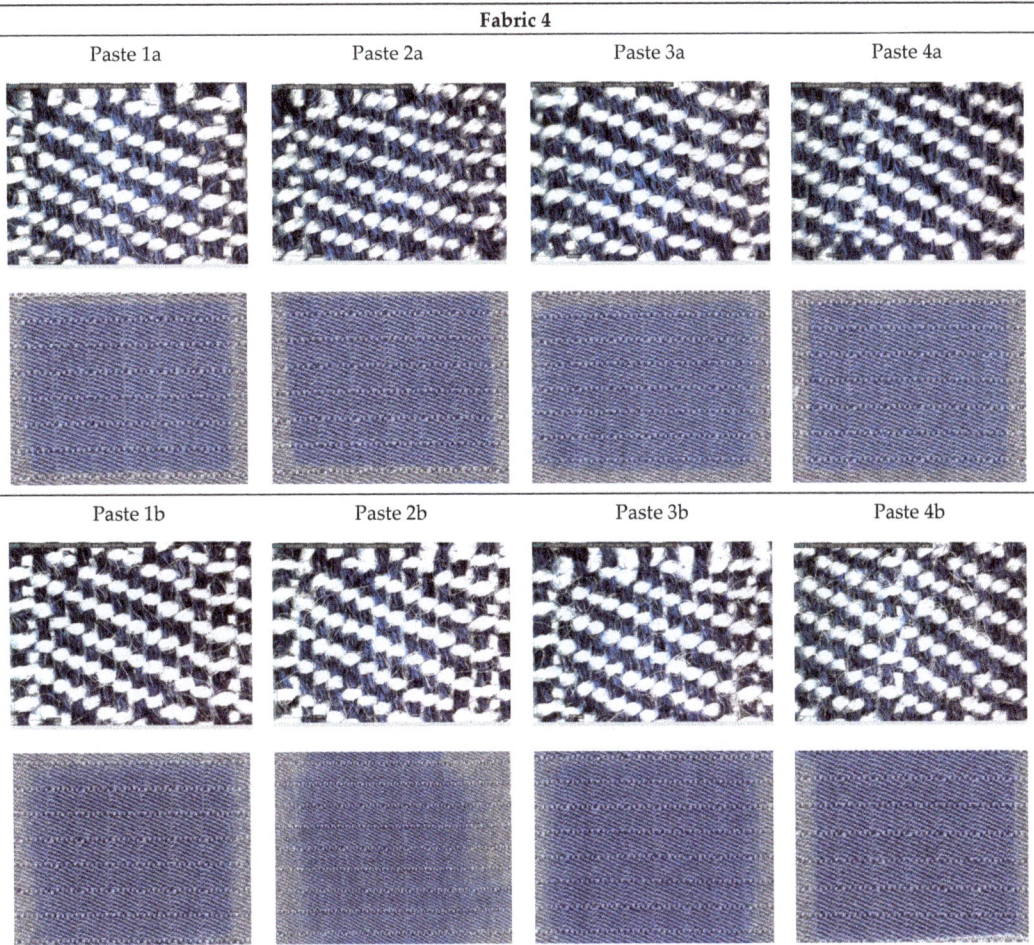

Figure 4. DinoLite microscopic and scanned images of printed Fabric 4.

For fabric 4 (Figure 4), printed with Paste *a*, with a lower dye content, satisfactory print sharpness was obtained for recipes with a higher viscosity thickener meaning a higher dry matter content (Paste 2a and 4a), regardless of the type of thickener.

As for Pastes *b* with a higher dye content, again a satisfactory print sharpness is obtained for printing pastes 3 and 4 with a starch-based thickener (alkagum NS). Microscopic images clearly show the difference in weaving structure, which results in a difference in the ratio of dyed and undyed components in the fabric content, and the impact of such yarn ratio on the subjective color appearance as well as on the objective measurements that follow, is expected.

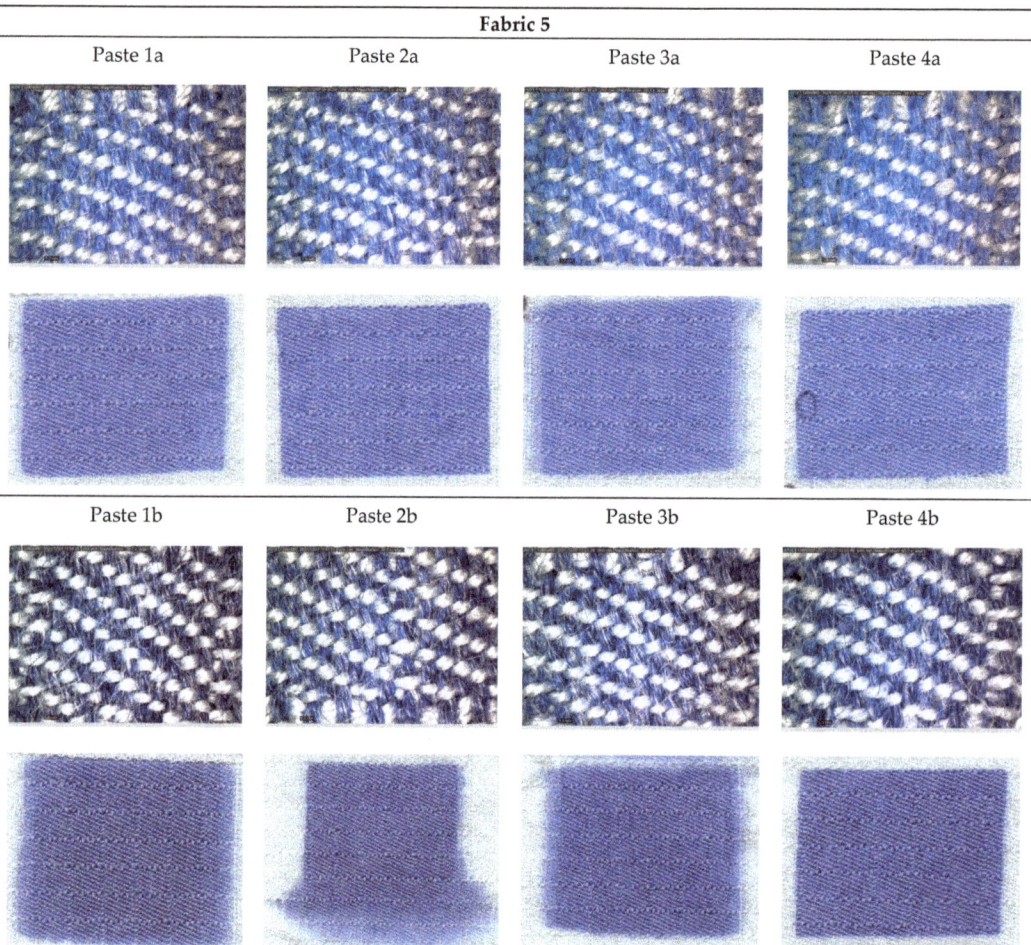

Figure 5. DinoLite microscopic and scanned images of printed Fabric 5.

For Fabric 5 (Figure 5), for Pastes *a*, with a lower content of dye, satisfactory, given the sharpness of the print contour, proved to be pastes with a thickener of higher viscosity, ie a higher proportion of dry matter (2a, 4a), regardless of the type of thickener. When printing with pastes with a higher proportion of dye, capillary spillage occurs, with the exception of the sample printed with Paste 4b (paste with a starch-based thickener, higher viscosity).

Paste 1a showed the stability in viscosity and homogeneity of the structure and sharp, equal prints were achieved without spillage and capillary migration. The same effects were achieved for pastes 2a, 3b and 4b. Pastes 1b, 3a and 4a did not gave satisfactory print sharpness, although the coverage of surface with printing paste is of optimal uniformity. In addition, during printing and fixing, capillary spreading of the paste occurred, although there were no changes in the viscosity and homogeneity of the paste during the process. Paste 2b also did not achieve satisfactory print sharpness, although the coverage of the surface with printing paste was optimal here as well.

Comparative analysis is given of the objective values of color strength (K/S) and the ratio of lightness (L*) and chroma (C*) as a definition of color intensity. The results of K/S are shown graphically in Figure 6a–e, and the L*/C* values are shown in Table 7.

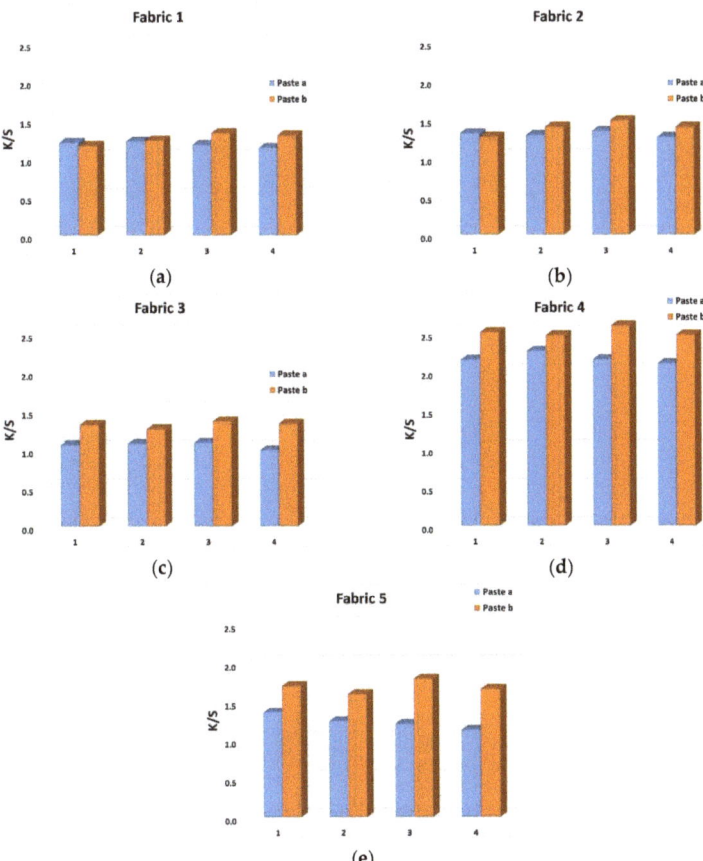

Figure 6. K/S objective values of color of printed surfaces: Fabric 1 (**a**), Fabric 2 (**b**), Fabric 3 (**c**), Fabric 4 (**d**) and Fabric 5 (**e**).

Objective K/S values show a relatively low achieved color strength. As expected, slightly higher values were obtained for printing pastes "b" with a higher concentration of dyestuff. Such lower K/S values were to be expected, given that only a partial ratio of components contained in the yarn composition is capable of bonding with dyestuff (the cellulosic part contained in the viscose yarn component).

The Kubelka-Munk coefficient, which objectively evaluates the color strength (K/S), is defined by the amount of dye that binds to the fiber in the printing process. The higher the color strength coefficient K/S, the greater the color depth itself, which estimates the coverage of the surface with color in the printing process. The printing process in which dyes are applied is actually a dyeing process from small, concentrated baths that takes place in the fixing phase. In this process, a certain amount of dye binds to the fiber, which depends on the conditions created by the thickener and other components in the printing paste during the fixing phase.

However, the K/S color intensity data alone does not provide information on the appearance of the achieved color. The K/S value needs to be analyzed in the context of the ratio of lightness (L^*) and chroma (C^*) for a given tone ($h°$). Namely, the same values of color strength will not have the same interpretation for different color hues and will depend on the understanding of the nature of color. According to its nature, blue belongs to naturally darker colors, which means that maximum chroma is achieved at lower lightness, expressed in objective CIE values, at lightness levels below 50. Therefore, the results indicated a lower color intensity, given the obtained lightness values above 50 (except for fabric 4) and relatively low saturation values. Such ratio of lightness (L*) and chroma (C*) is in accordance with the K/S values. The specific ratio of lightness and chroma, in general, is the definition of color intensity and depends, in addition to the parameters of the dyestuff and the textile material, also on the color itself.

Table 7. Lightness (L*) and chroma (C*) relationship for printed samples. as definition of color intensity.

	L*	C*		L*	C*
Fabric 1 Paste 1a	59.83	10.11	Fabric 1 Paste 1b	59.79	9.83
Fabric 1 Paste 2a	59.73	9.87	Fabric 1 Paste 2b	59.29	9.18
Fabric 1 Paste 3a	59.94	10.36	Fabric 1 Paste 3b	56.80	12.48
Fabric 1 Paste 4a	60.71	9.92	Fabric 1 Paste 4b	57.86	11.56
	L*	C*		L*	C*
Fabric 2 Paste 1a	58.61	8.47	Fabric 2 Paste 1b	59.68	7.34
Fabric 2 Paste 2a	58.57	8.68	Fabric 2 Paste 2b	57.75	7.86
Fabric 2 Paste 3a	57.82	9.48	Fabric 2 Paste 3b	55.65	10.94
Fabric 2 Paste 4a	58.94	8.53	Fabric 2 Paste 4b	57.32	9.09
	L*	C*		L*	C*
Fabric 3 Paste 1a	60.22	17.76	Fabric 3 Paste 1b	56.84	13.28
Fabric 3 Paste 2a	60.56	18.34	Fabric 3 Paste 2b	57.97	13.86
Fabric 3 Paste 3a	59.90	18.94	Fabric 3 Paste 3b	55.48	17.70
Fabric 3 Paste 4a	61.45	18.95	Fabric 3 Paste 4b	56.42	16.53
	L*	C*		L*	C*
Fabric 4 Paste 1a	48.58	11.85	Fabric 4 Paste 1b	46.28	10.41
Fabric 4 Paste 2a	47.90	11.68	Fabric 4 Paste 2b	47.05	10.07
Fabric 4 Paste 3a	48.40	12.06	Fabric 4 Paste 3b	44.93	13.03
Fabric 4 Paste 4a	48.66	12.30	Fabric 4 Paste 4b	46.20	11.85
	L*	C*		L*	C*
Fabric 5 Paste 1a	55.27	20.02	Fabric 5 Paste 1b	51.51	14.24
Fabric 5 Paste 2a	56.20	21.19	Fabric 5 Paste 2b	53.02	15.13
Fabric 5 Paste 3a	56.60	21.10	Fabric 5 Paste 3b	50.41	17.96
Fabric 5 Paste 4a	57.48	22.08	Fabric 5 Paste 4b	51.53	18.26

In print characterization, color reproduction quality control is performed by evaluating the spectrophotometric parameters. In dye printing, achieving a satisfactory intensity and brilliance of coloration of printed parts of the fabric is very complex. Precisely due to the chemical bonding of the dye with the fiber, the obtained coloration becomes an integral part of the fabric structure, which provides more satisfactory properties of fastness, but lower brilliance of the dye. It is the color intensity as well as the brilliance of the print coloration that can be defined by the specific ratio of lightness (L*) and chroma (C*). In the samples tested in this paper, the lower intensity of the obtained coloration is contributed by the specific composition of the fabric in which more than half of the content are fibers that cannot be dyed by the process of dye-fiber chemical bonding.

The spectrophotometrically measured values of samples after the 5th wash cycle were compared with the values of unwashed samples and the analysis of the change in the value of K/S (color strength) and the specific relationship of lightness (L*) and chroma (C*) was performed. The results for K/S are shown graphically on Figure 7 and the results for lightness (L*) and chroma (C*) are given in Table 8.

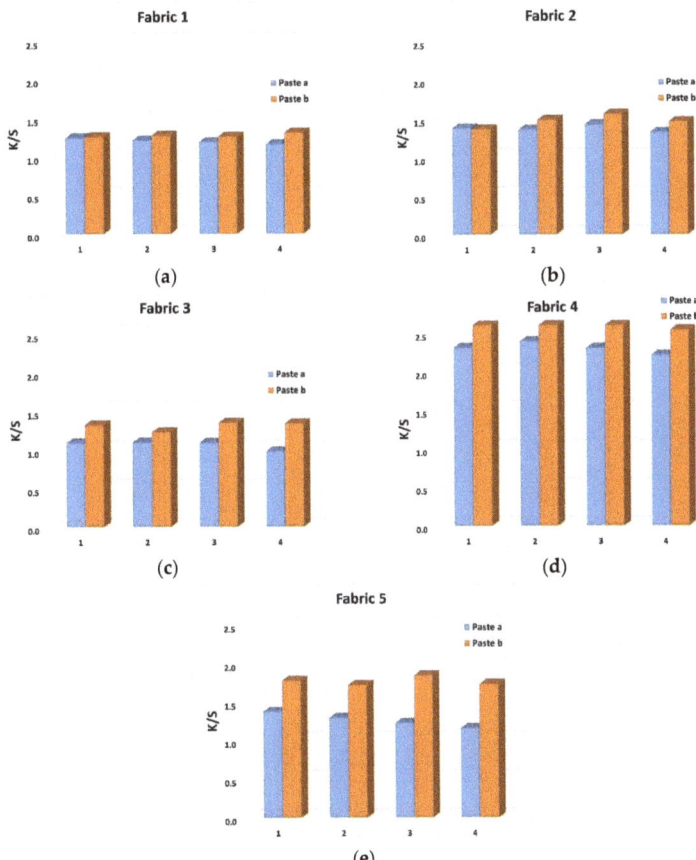

Figure 7. K/S objective values of color of printed surfaces after 5 washing cycles: Fabric 1 (**a**), Fabric 2 (**b**), Fabric 3 (**c**), Fabric 4 (**d**) and Fabric 5 (**e**).

For K/S values, a minimal change was obtained, after the washing cycles, which indicate that there were no evident release of dye. Slightly higher differences in K/S values after washing were obtained for Fabric 4 printed with Paste b, regardless of the type and viscosity of the thickener. It can be said that, in this sample, the amount of bound dye was lower and there was a higher release of dye from during washing.

For the ratio of lightness (L*) and chroma (C*), in general for all samples (Fabrics 1 to 5), it was observed that there was an increase in lightness (L*) and a decrease in chroma (C*). This result of the lightness-chroma (L*-C*) ratio is in line with the K/S values and the indication of a slight dye release. An increase in lightness and a decrease in chroma indicates a shift in coloration towards a lighter, more pastel color shade, indicating albeit slight but still fading coloration in the wash.

In further analysis, the color differences are calculated based on objective spectrophotometric measurement of samples before and after the 5 washing cycles, taking the values of printed unwashed samples as the standard (reference) values. The results are shown in terms of color parameters differences (lightness difference dL*, chroma difference dC* and hue difference dh) as well as in terms of total color difference value (dE), presented graphically in Figure 8.

Table 8. Lightness (L*) and chroma (C*) relationship for printed samples after 5 washing cycles.

	L*	C*		L*	C*
Fabric 1 Paste 1a	60.66	8.77	Fabric 1 Paste 1b	59.48	8.36
Fabric 1 Paste 2a	60.61	8.58	Fabric 1 Paste 2b	59.72	7.74
Fabric 1 Paste 3a	60.67	9.21	Fabric 1 Paste 3b	58.68	10.71
Fabric 1 Paste 4a	61.28	8.59	Fabric 1 Paste 4b	58.66	9.96
	L*	**C***		**L***	**C***
Fabric 2 Paste 1a	58.69	7.66	Fabric 2 Paste 1b	59.34	6.64
Fabric 2 Paste 2a	58.75	7.69	Fabric 2 Paste 2b	57.71	7.01
Fabric 2 Paste 3a	57.70	8.63	Fabric 2 Paste 3b	55.56	9.91
Fabric 2 Paste 4a	59.10	7.65	Fabric 2 Paste 4b	57.50	8.08
	L*	**C***		**L***	**C***
Fabric 3 Paste 1a	60.40	16.76	Fabric 3 Paste 1b	57.62	11.73
Fabric 3 Paste 2a	60.84	17.40	Fabric 3 Paste 2b	59.08	12.45
Fabric 3 Paste 3a	60.59	17.50	Fabric 3 Paste 3b	56.22	16.44
Fabric 3 Paste 4a	62.31	17.41	Fabric 3 Paste 4b	57.03	15.14
	L*	**C***		**L***	**C***
Fabric 4 Paste 1a	48.03	11.01	Fabric 4 Paste 1b	46.28	9.65
Fabric 4 Paste 2a	47.53	10.85	Fabric 4 Paste 2b	46.64	9.39
Fabric 4 Paste 3a	47.90	11.18	Fabric 4 Paste 3b	45.20	12.13
Fabric 4 Paste 4a	48.46	11.33	Fabric 4 Paste 4b	46.37	10.94
	L*	**C***		**L***	**C***
Fabric 5 Paste 1a	55.44	19.61	Fabric 5 Paste 1b	51.26	13.67
Fabric 5 Paste 2a	55.92	21.25	Fabric 5 Paste 2b	52.15	15.17
Fabric 5 Paste 3a	56.72	20.82	Fabric 5 Paste 3b	50.37	17.51
Fabric 5 Paste 4a	57.52	21.78	Fabric 5 Paste 4b	51.36	17.80

For the parameters of lightness (L*) and hue (h°), the differences (dL* and dh) were for all samples within the ranges of tolerances (tolerances for dC* = 0.8–1.5; dL* = 1.2–2; dh = 0.5–0.8). The differences obtained for chroma parameter (dC*) were higher and outside the tolerances. However, although the individual differences in chroma (dC*) were out of the tolerances, the values of the total color difference (dE), taking into account the obtained minimum differences in lightness and hue, were also mostly within the tolerance limits. More pronounced values of the total color difference (dE), outside the range of tolerance (dE < 1.5), were obtained only for Fabric 1, Paste 3b. The results of the differences are acceptable and indicate a satisfactory color wash fastness.

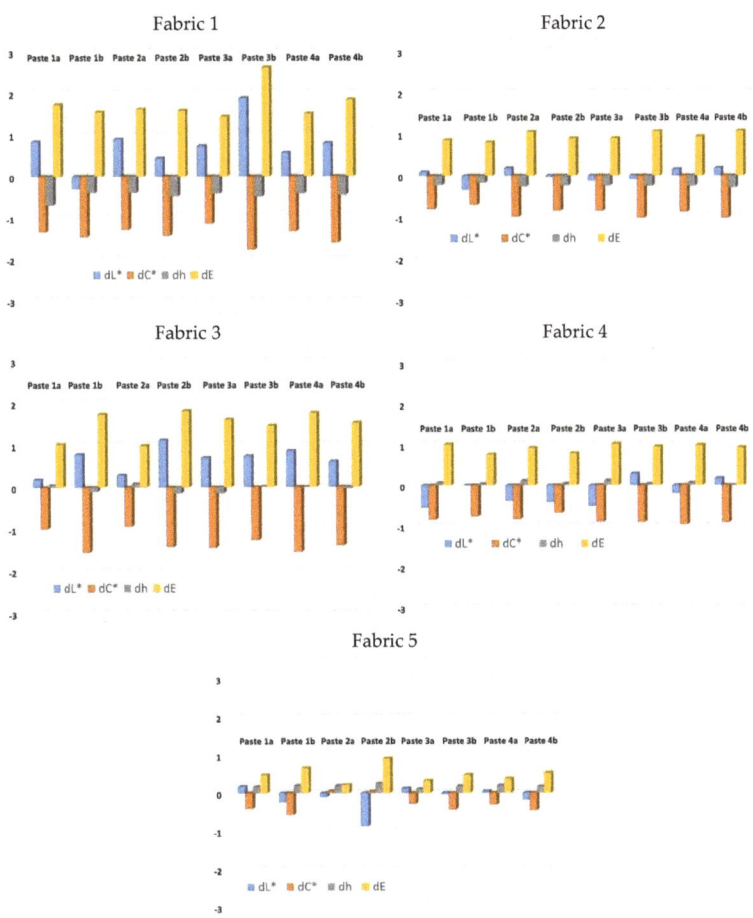

Figure 8. Color differences values calculated according to CIE76 formula. comparing values of unwashed samples with values obtained after 5th wash cycle.

4. Conclusions

Based on the analysis, it can be confirmed that it is possible to achieve a certain level of coloration in fabrics that contain a high proportion of aramid fibers (95% in warp and 40% in weft yarn composition), if a certain ratio of cellulose component is contained (30–40% in weft yarn). The weft yarn contained a component of viscose (38–40%) that has the ability to bond reactive dye, and coloration was achieved even in samples that had a certain basic color i.e., were not completely uncolored (Fabrics 1, 2 and 4). Although, by objective evaluation of the color strength, lower K/S values were obtained (1.1 to 2.6), by analyzing the relationship between the color lightness parameters (L*) and chroma (C*), a satisfactory color intensity was achieved (maximum values L* = 62.31 and C* = 21.78).

The results confirmed the optimal wash fastness in the process of five washing cycles, more emphasized for Fabric 5, where for all printing conditions (sort and viscosity of the thickener as well as dyestuff content) the lowest differences values have been obtained (dE_{CIE} = 0.32–0.91).

This research is part of the comprehensive research of the characteristics of FR fabrics and the possibility of their finishing in the processes of dyeing and printing.

Author Contributions: Conceptualization was performed by T.P. and M.G.; T.P. and T.K. performed LOI measurement and analysis; printing was performed by V.L. and T.K. performed washing; V.L. performed spectrophotometric measurement and microscopic analyses. M.G. and V.L. performed colorimetric analysis; All authors participated in writing-original draft preparation. All authors have read and agreed to the published version of the manuscript.

Funding: This work has been supported by the European Union from the European Regional Development Fund under the project KK.01.2.1.02.0064 Development of multifunctional non-flammable fabric for dual use.

Institutional Review Board Statement: Not applicable.

Informed Consent Statement: Not applicable.

Data Availability Statement: Data available in a publicly accessible repository.

Acknowledgments: The research was performed on equipment purchased by K.K.01.1.1.02.0024 project "Modernization of Textile Science Research Centre Infrastructure" (MI-TSRC). "A paper recommended by the 14th Scientific and Professional Symposium Textile Science and Economy, The University of Zagreb Faculty of Textile Technology".

Conflicts of Interest: The authors declare no conflict of interest.

References

1. Alongi, J.; Carosio, F.; Horrocks Richards, A.; Malucelli, G. *Update on Flame Retardant Textiles: State of the Art Environmental Issues and Innovative Solutions*; A Smithers Group Company: Shawbury, UK, 2013; pp. 1–14.
2. Horrocks, A.R. flame retardant finishes and finishing. In *Textile Finishing*; Heywood, D., Ed.; Society of Dyers and Colourists: Bradford, UK, 2003.
3. Bourbigot, S. Flame retardancy of textiles: New approaches. In *Advances in Fire Retardant Materials*; Horrocks, A.R., Price, D., Eds.; Woodhead Publishing: Cambridge, UK, 2008; pp. 9–40.
4. Uddin, F. Recent flame retardant consumption: Textiles. *Int. J. Sci. Eng.* **2019**, *10*, 805–819.
5. Uddin, F. Concerns of brominated flame retardant. *Ind. Fabr. Bull.* **2003**, *3*, 55–56.
6. Gaan, S.; Salimova, V.; Rupper, P.; Ritter, A.; Schmid, H. flame retardant functional textiles. In *Functional Textiles for Improved Performance. Protection and Health*; Pan, N., Sun, G., Eds.; Woodhead Publishing: Cambridge, UK, 2011; pp. 98–130.
7. Schartel, B.; Kebelmann, K. Fire testing for the development of flame retardant polymeric materials. In *Flame Retardant Polymeric Materials*; Hu, Y., Wang, X., Eds.; Routledge Handbooks Online: Boca Raton, USA, 2019.
8. Wakelyn, P.J.; Rearick, W.; Turner, J. Cotton and flammability-overview of new developments. *Am. Dyest. Rep.* **1998**, *87*, 13–21.
9. Wu, W.; Yang, C.Q. Comparison of DMDHEU and Melamine-Formaldehyde as the binding agents for a hydroxy-functional organophosphorus flame retarding agent on cotton. *J. Fire Sci.* **2004**, *22*, 125–142. [CrossRef]
10. Yang, C.Q.; Qiu, X. Flame-retardant finishing of cotton fleece fabric: Part I. The use of a hydroxy-functional organophosphorus oligomer and dimethyloldihydroxylethyleneurea. *Fire Mater.* **2007**, *31*, 67–81. [CrossRef]
11. Wu, W.; Yang, C.Q. Comparison of different reactive organophosphorus flame-retardant agents for cotton. Part II. Fabric flame resistant performance and physical properties. *Polym. Deg. Stab.* **2007**, *92*, 363–369. [CrossRef]
12. Lecoeur, E.; Vroman, I.; Bourbigot, S.; Lam, T.M.; Delobel, R. Flame retardant formulations for cotton. *Polym. Deg. Stab.* **2001**, *74*, 487–492. [CrossRef]
13. Lecoeur, E.; Vroman, I.; Bourbigot, S.; Delobel, R. Optimization of monoguanidine dyhidrogen phosphate and aminopropylethoxysilane based flame retardant formulations for cotton. *Polym. Deg. Stab.* **2006**, *91*, 1909–1914. [CrossRef]
14. Alongi, J.; Horrocks, R.A.; Carosi, F. *Update on Flame Retardant Textiles: State of the Art, Environmental Issues and Innovative Solutions*; Smithers Rapra Technology: Shawbury, UK, 2013; pp. 148–152.
15. Gupta, D. Softening treatments for technical textiles. In *Advances in the Dyeing and Finishing of Technical Textiles*; Gulrajani, M.L., Ed.; Woodhead: Oxford, UK, 2013; pp. 154–163.
16. Sun, L.; Wang, H.; Li, W.; Zhang, J.; Zhang, Z.; Lu, Z.; Zhu, P.; Dong, C. Preparation, characterization and testing of flame retardant cotton cellulose material: Flame retardancy, thermal stability and flame-retardant mechanism. *Cellulose* **2021**, *28*, 3789–3805. [CrossRef]
17. Schindler, W.D.; Hauser, P.J. (Eds.) Flame retardant finishes. In *Chemical Finishing of Textiles*; Woodhead Publishing Limited: Boca Raton, FL, USA, 2004; pp. 98–116.
18. Banerjee, S.K.; Day, A.; Ray, P.K. Fire proofing jute. *Text. Res. J.* **1985**, *56*, 338–343. [CrossRef]
19. Kandola, B.K.; Horrocks, A.R.; Price, D.; Coleman, G.V. Flame retardant treatments of cellulose and their influence on the mechanism of cellulose pyrolysis. *J. Macromol. Sci.* **1996**, *36*, 794–796. [CrossRef]
20. Horrocks, A.R. Flame Retardant Challenges for Textiles and Fibres: New Chemistry Versus Innovatory Solutions. *Polym. Degrad. Stab.* **2011**, *96*, 377–392. [CrossRef]

21. Bourbigot, S.; Flambard, X. Heat Resistance and Flammability of high Performance Fibre: A Review. *Fire Mater.* **2002**, *26*, 155–168. [CrossRef]
22. Tsafack, M.J.; Levalois-Grützmacher, J. Plasma-induced Graft-polymerization of flame retardat monomers onto PAN fabrics. *Surf. Coat. Technol.* **2006**, *200*, 3503–3510. [CrossRef]
23. Tsafack, M.J.; Levalois-Grützmacher, J. Towards multifunctional surface using the plasma-induced graft-polymerization (pigp) process: Flame and waterproof cotton textiles. *Surf. Coat. Technol.* **2007**, *201*, 5789–5795. [CrossRef]
24. Errifai, I.; Jama, C.; Le Bras, M.; Delobel, R. Fire retardant coating using cold plasma polymerization of a fluorinated acrylate. *Surf. Coat. Technol.* **2004**, *180*, 297–301. [CrossRef]
25. Quede, A.; Jama, C.; Supiot, P.; Le Bras, M.; Delobel, R.; Dessaux, O.; Goudmand, P. Elaboration of fire retardant coatings on pplyamide-6 using a cold plasma polymerization process. *Surf. Coat. Technol.* **2002**, *67*, 424–428. [CrossRef]
26. Sonee, N.; Arora, C.; Parmar, M. Burning behaviour of aramid and FR viscose blended fabrics. *Indian J. Fibre Text. Res.* **2019**, *44*, 238–243.
27. Sonee, N.; Arora, C.; Parmar, M.S. The flame-retardant performances of blending fabrics of flame-retardant viscose and nylon 6. 6 fiber with different blending ratio. *Int. J. Eng. Res. Appl.* **2017**, *7*, 87–91. [CrossRef]
28. Wolter, N.; Beber, V.C.; Haubold, T.; Sandinge, A.; Blomqvist, P.; Goethals, F.; van Hove, M.; Jubete, E.; Mayer, B.; Koschek, K. Effects of flame-retardant additives on the manufacturing. mechanical and fire properties of basalt fiber-reinforced polybenzoaxine. *Polym. Eng. Sci.* **2020**, *1*, 551–561.
29. Manyukov, E.A.; Sadova, S.F.; Baeva, N.N.; Platonov, V.A. Study of dyeing of thermostable para/meta-aramid fibre. *Fibre Chem.* **2005**, *37*, 55–58. [CrossRef]

Article

Method of Predicting the Crimp of Jacquard-Woven Fabrics

Eglė Kumpikaitė [1,*], Eglė Lapelytė [1] and Stasė Petraitienė [2]

[1] Department of Production Engineering, Faculty of Mechanical Engineering and Design, Kaunas University of Technology, Studentų Str. 56, LT-51424 Kaunas, Lithuania; eglelap@gmail.com
[2] Department of Applied Mathematics, Faculty of Mathematics and Natural Sciences, Kaunas University of Technology, Studentų Str. 50, LT-51424 Kaunas, Lithuania; stase.petraitiene@ktu.edu
* Correspondence: egle.kumpikaite@ktu.lt

Abstract: The aim of this study was to investigate the distribution of crimp in new jacquard fabric structures (in which one-layer and two-layer weaves are combined) in the fabric width and to create a method of crimp prediction. It was established that crimp was around 18.80% and changed within the limits of errors, i.e., a range of only ~4%, in the fabric width. It can therefore be said that the warp crimp was constant in the fabric width. Because the warp crimp of jacquard fabric changed insignificantly (within the limits of errors), it can be stated that the fabric-setting parameters and structural solutions were chosen and matched correctly, and such fabric can be woven on any jacquard weaving loom.

Keywords: linen fabric; warp crimp; calculated crimp; experimental crimp

1. Introduction

There are several definitions of crimp. Percentage crimp is defined as the mean difference between the straightened thread length and the distance between the ends of the thread while in the cloth, expressed as a percentage [1]. The shortening of yarn length in a fabric is known as crimp [2]. When the warp and weft yarn interlace in a fabric, they follow a wavy path. This waviness of yarn is called crimp [3]. According to the definition of crimp, two values must be known: the length of cloth from which the yarns are removed and the straightened length of the thread. In order to straighten the thread, sufficient tension must be applied to remove all the kinks without stretching the yarn. In practice, it is seldom possible to remove all the crimp before the yarn itself begins to stretch. From these two values, the crimp percentage can be calculated with Equation (1):

$$\text{Yarn Crimp} = 100 \cdot \frac{\text{Straightened Yarn length} - \text{Yarn length in fabric}}{\text{Fabric Length}} \quad (1)$$

Warp and weft crimp percentage are two factors that have an influence on a fabric's abrasion resistance, shrinkage, and fabric behaviour during strength testing [3].

Factors that affect crimp include the following:
1. Physical properties such as elasticity, rigidity, bending behaviour, etc. of fibres and yarns;
2. Count of warp and weft threads;
3. Setting of the threads;
4. Tension on the threads during weaving;
5. Yarn and fabric structure;
6. Physical and chemical treatment of the fabric after weaving [3].

Crimp frequency, amplitude, crimp stability, crimp elongation, and crimping point are some of the important properties that determine crimp. A fibre's crimp characteristics have a strong influence on the processing performance of the fibres [3].

Citation: Kumpikaitė, E.; Lapelytė, E.; Petraitienė, S. Method of Predicting the Crimp of Jacquard-Woven Fabrics. *Materials* **2021**, *14*, 5157. https://doi.org/10.3390/ma14185157

Academic Editors: Dubravko Rogale and Philippe Boisse

Received: 29 July 2021
Accepted: 6 September 2021
Published: 8 September 2021

Publisher's Note: MDPI stays neutral with regard to jurisdictional claims in published maps and institutional affiliations.

Copyright: © 2021 by the authors. Licensee MDPI, Basel, Switzerland. This article is an open access article distributed under the terms and conditions of the Creative Commons Attribution (CC BY) license (https://creativecommons.org/licenses/by/4.0/).

The construction of the warp and weft yarns are determined by various factors, such as the type and setting of weaving loom, the weaving conditions (moisture amount, temperature, yarn tension), the fabric weave, linear densities, the raw material of the threads, the flexibility of the warp and weft, and the pressure strength of the warp and weft threads in the positions of the floats. Thread crimp changes during weaving, fabric relaxation, and finishing [4]. Recently, it has been noticed that the thread diameter in the float positions and beat-up force can be attributed to these parameters [5].

Warp and weft crimp is important because it has an influence on weaving technology, warp and weft expenditure, fabric shrinkage during weaving and fabric properties (warp and weft breaking force, elongation at break, shape stability) [6].

During warping, it is important that all warp threads warped on the warp beam are equally taut. This is to ensure that the fabric's properties during weaving do not change and remain the same throughout the entire fabric length. Without ensuring the same warp system tension across the whole width of the fabric, unequal fabric properties will be obtained. This not only affects the weaving process but can also negatively influence the weaving loom. Uneven tension in the thread system unequally affects their crimp and its distribution in the fabric [7].

Because warp and weft threads are oriented in two directions in the fabric, stretching fabric in the directions of warp and weft lengthens or widens the fabric in the direction of tension and shrinks it in the opposite direction. Elongation of the fabric is directly related to the decrease in crimp shrinkage in the same direction. Threads lengthen in the direction of load after removal of crimp, and the opposite thread system gains higher crimp. This will increase until one thread system does not interfere with the other [8].

Topalbekiroglu and Kaynak performed research [9] in which they investigated how fabric weaves influence a fabric's stability. At first, they established that fabrics that have short floats in their weaves are more stable. If the weaves have longer floats, applying higher crimp can avoid instability in the fabric shape. This means that to have stable fabric, correct choice of certain weaves is needed, as well as ensuring optimal crimp [9].

Crimp also influences the fabric's strength. Any modification of the fabric structure, which spreads the friction over a larger area improves abrasion resistance. Abrasion resistance increases when the friction load affects a larger number of threads and fibre removal or transfer from the fabric during friction decreases. Of course, floating threads, which are higher in comparison with the fabric surface, will suffer the most from abrasion load. In order to save the fabric's tensile properties, fabric abrasion resistance should be increased; fabric appearance and/or mass loss are secondary factors. By increasing fabric crimp, abrasion resistance decreases. It is thought that this occurs because the crimp is lower when length of floats is higher. Fabric weaves and high crimp values, which would make some floats higher than others, thus concentrating abrasion loads, are not wanted. The locations of these floats would become the main weak links in the whole fabric [6].

The aim of Stig's and Hallström's research [10] was to find out how fabric weave and crimp influence rigidity and strength in 3D textile structures. It was established that rigidity and strength decrease non-linearly when crimp increases. The final strength of specimens with straight threads and of specimens with waved threads differed and noticeable stresses were significantly higher in threads that are waved by crimp. Specimens which contained crimped threads are weaker than those which are not crimped. Inserting perfectly straight threads into 3D fabric increases composite stiffness and strength, but some of the fibre volume also increases. However, the inserted threads are straight and stiffer than the warp threads, which are not waved by crimp. Because of this, they reach maximal stresses under low loads [10]. This phenomenon influences the decrease in 3D fabric strength in the longitudinal direction and the increase in stiffness in the longitudinal direction.

The aim of this paper was to investigate the distribution of crimp in jacquard fabrics across the fabric width and to create a method of predicting crimp in jacquard fabrics.

2. Experimental Procedure

New jacquard fabric structures, in which one-layer and two-layer weaves are combined, were woven by the joint-stock company Lincasa on a P1 rapier weaving loom (Lindauer Dornier GmbH, Lindau, Germany). The warp and weft yarns were linen; their linear density was 50 tex. They were single-ply yarns with 523 m^{-1} twist in the Z direction, the breaking force was 5.286 N and the elongation at break was 1.7%. The warp setting was 21 ends/cm and the weft setting was 16 picks/cm. In the weft direction, every second thread was bleached and every other second thread was of a natural colour. The warp was bleached. The number of jacquard machine hooks was 3360. The fabric width in the reed was 160 cm. The initial warp tension was 7.43 mN/tex, the healds' cross-advance was at 10 degrees of the main loom shaft. Fabrics were woven in seven different designs (Figure 1).

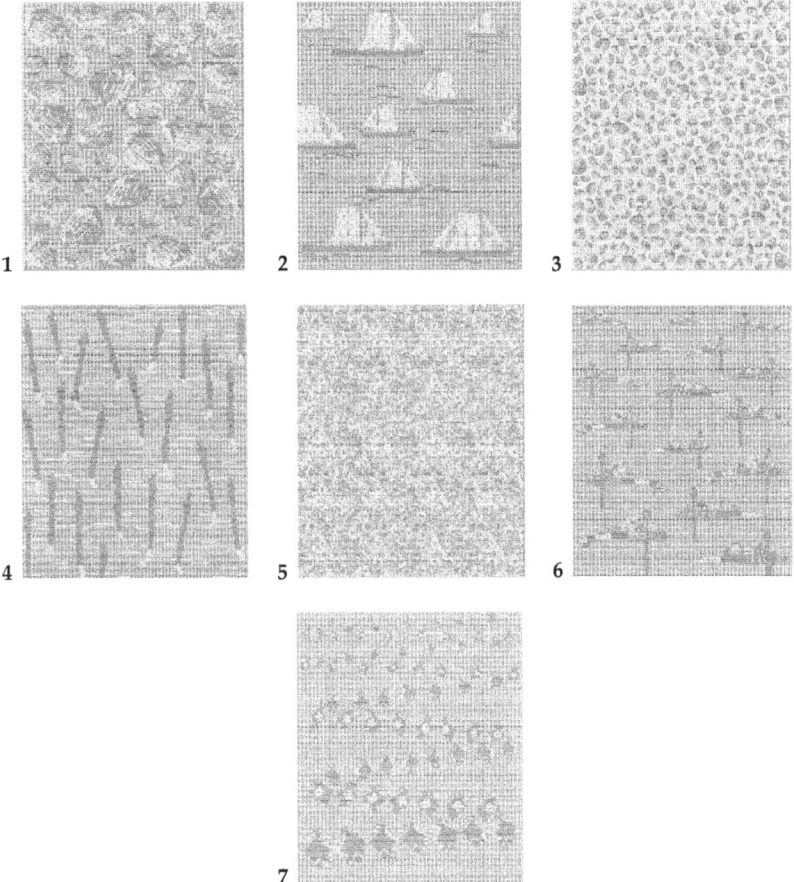

Figure 1. Designs of the jacquard fabrics used for the experiments: **1**—big shells; **2**—ships; **3**—small shells; **4**—birds; **5**—crystals of sand; **6**—weathervanes; **7**—dried fish.

The dimensions of all designs were 1.5 m × 2.0 m. All the designs were woven using five different weaves, which are presented in Figure 2. Weaves 1 and 2 used dark grey and greyish in all designs. These weaves were similar but they began from different threads. For this reason, the bleached weft dominated in Weave 1 and the natural-colour weft dominated in Weave 2. The background of all designs was woven in Weave 4. Weaves 3 and 5 were used for pattern highlighting and correspond to two other shades of grey in the designs.

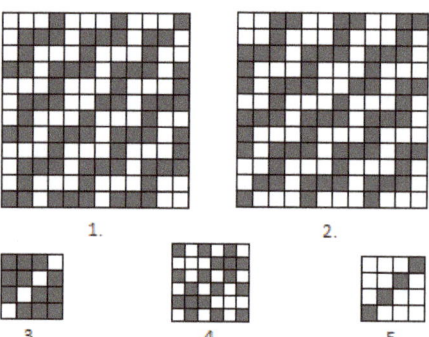

Figure 2. Weaves of used Jacquard fabrics: **1**—weave of dark grey colour in fabric design; **2**—weave of greyish colour in fabric design; **3**—weave of light grey colour in fabric design; **4**—weave of background in fabric design; **5**—weave of grey colour in fabric design.

Examples of fragments of jacquard fabric Design 1 from the right (Figure 3a) and wrong (Figure 3b) sides are presented in Figure 3. To show a general view of the design would make no sense, because the dimensions of the design are very big and the differences between different weaves and colours do not emerge in pictures of the entire design. The same situation occurs with the other six designs.

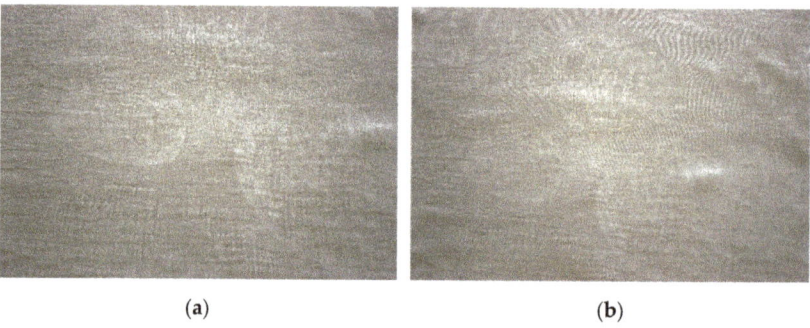

Figure 3. Fragment of jacquard fabric Design 1: (**a**) from the right side; (**b**) from the wrong side.

A known problem of all jacquard fabrics' weavability is that the weaving design has to be such that uniform warp yarn crimp over the entire width of the fabric can be achieved. This problem exists in every jacquard fabric but this question is particularly relevant when combining one-layer and two-layer weaves in one fabric. In order to distribute one-layer and two-layer weaves evenly in the fabric's width, warp crimp was established in different places of the fabric width. A random part of the fabric length was chosen and a segment of a certain length was measured. We tried to ensure that this place was not the beginning of the repeat and that the fabric repeat did not finish in the whole segment. The fabric was divided into six approximately equal parts, distributed equally across the fabric width. When the fabric had been divided into separate specimens, 10 warp threads were unpicked in the right side of the specimen. Their length was measured. Equation (2) was used for calculating the crimp:

$$a = \frac{l_s - l_{aud}}{l_s} \cdot 100\% \tag{2}$$

where: l_s is the length of the warp thread and l_{aud} is the length of fabric specimen.

3. Results and Discussion

When weaving jacquard fabrics, it is very important that the warp crimp in different places in the fabric's width should be the same or differ insignificantly. In such a case, the jacquard fabric's weavability will be good. In other cases, some warp yarns or their groups will be stretched and other ones will be loose. This fact is especially important in fabrics in which one-layer and two-layer weaves are combined. However, the investigation of fabric crimp across the width is very relevant to all jacquard fabrics.

In order to investigate crimp distribution in the fabric's width, warp crimp in various parts of the fabric was established. A random location in the fabric length was chosen and a segment 40 cm in length was measured. The fabric was divided into six parts, which were distributed across the fabric width as follows: 25 cm, 20 cm, 20 cm, 20 cm, 20 cm, 20 cm, and 25 cm. A schematic view of specimen partitioning is shown in Figure 4.

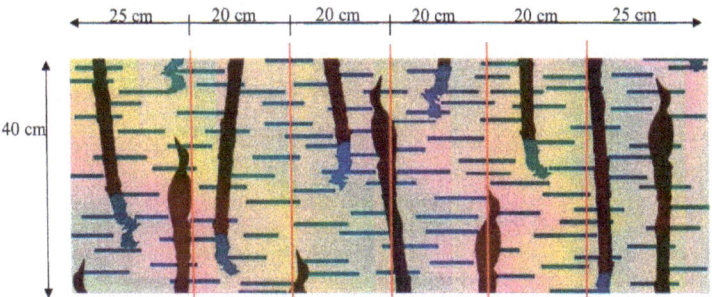

Figure 4. Division of the fabric section.

After dividing the fabric into separate specimens, 10 warp threads were unpicked from the right side of each specimen. The length of these 10 threads was measured. The crimp of the warp threads was calculated according to Equation (2).

The distribution of crimp across the fabric width is demonstrated in Figure 5. It can be seen that the differences in crimp in different places across the fabric width are insignificant: they differ by up to 4% (18.40–19.20%), i.e., they vary within the error margin, because the variation coefficient is up to 5.4%. Thus, it can be stated that the warp crimp is almost equal in different places of the fabric. The conclusion can be drawn that the fabric design has been made correctly, as the locations of different weaves were distributed evenly throughout the entire fabric repeat.

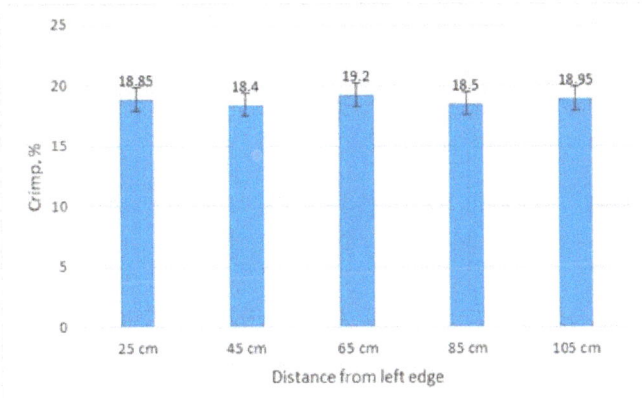

Figure 5. Results of fabric crimp calculation across the fabric width.

In order to explain how crimp changes in parts of fabric woven in different weaves, an original method is suggested. By applying this method, the crimp was calculated in certain segments. Specimens 10 cm in length were manufactured in which sections woven with different weaves were found. In the case of new Jacquard fabric structures, the parts woven in one-layer and two-layer weaves were analysed separately, because the float distribution inside these groups was similar. Ten different warp threads were pulled out from four different specimens and the crimp was calculated according to Equation (2) separately for one-layer and two-layer fabric sections.

The crimp of one-layer weaves should be the same because the floats' lengths and their distribution are the same in one-layer weaves; only the character of the floats differs. The same situation also occurs with two-layer weaves because the floats are only shifted through a single thread and the floats' distribution in the weave are the same. The results of calculating the crimp for one-layer and two-layer weaves are presented in Table 1. In the case of another jacquard fabric structure, the number of columns in the table will correspond to the number of different weaves in the fabric. The dispersion of the results is very low, i.e., the coefficient of variation for one-layer weaves is just 0.17% and for two-layer weaves, it is 0.19%. Thus, the accuracy of crimp calculation is very high.

Table 1. Results of calculating the crimp for one-layer and two-layer weaves.

Number	Crimp of One-Layer Weaves%	Crimp of Two-Layer Weaves%
1	15.25	21.88
2	23.08	18.70
3	23.08	28.57
4	16.67	18.70
5	20.00	25.92
6	23.08	26.47
7	16.67	20.63
8	24.24	21.88
9	19.35	20.00
10	18.03	20.00
11	19.35	19.35
12	25.37	25.37
13	19.35	19.35
14	18.03	18.03
15	16.67	16.67
16	18.70	18.70
17	20.63	20.63
18	18.03	18.03
19	18.70	18.70
20	11.50	11.50
Average	19.29	20.45
Standard deviation	3.32	3.85
Coefficient of variation	0.17	0.19
Histogram	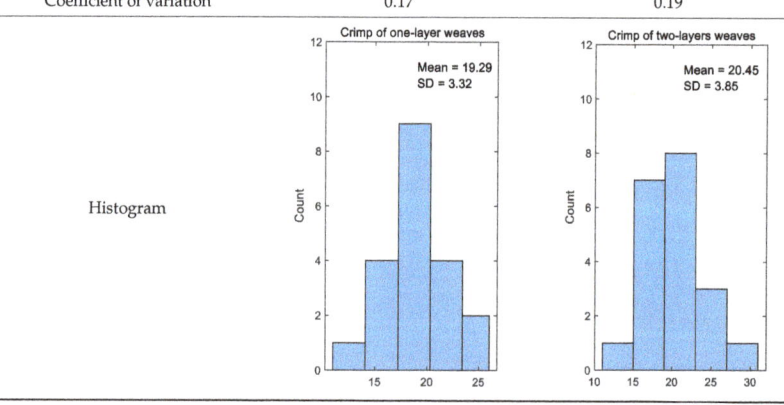	

In order to calculate the crimp, we determined the percentage of warp threads interwoven in one-layer and two-layer weaves along their length in the specimen. In the case of another jacquard fabric structure, the percentage of warp threads interwoven in each different weave will be calculated. Adobe Photoshop CS6 software (version 13.1x, Adobe Systems Incorporated, San Chose, CA, USA), which allowed us to visualise the created fabric at the pixel level, was used for calculating this parameter. It was established during this investigation that one pixel equals one warp thread. Next, a newly created jacquard fabric design was divided across the entire fabric width into the same six parts as the real fabric (25 cm, 20 cm, 20 cm, 20 cm, 20 cm, and 25 cm), the length of which is 40 cm. Ten pixel columns from the specimen's right side were calculated. The proportion of one-layer and two-layer weaves in separate warp threads is presented in Table 2.

Table 2. Proportion of one-layer and two-layer weaves.

Number	25 cm from Edge		45 cm from Edge		65 cm from Edge		85 cm from Edge		105 cm from Edge	
	1-l. We. Part%	2-l. We. Part%	1-l. We. Part%	2-l. We. Part%	1-l. We. Part%	2-l. We. Part%	1-l. We. Part%	2-l. We. Part%	1-l. We. Part%	2-l. We. Part%
1	66.2	33.8	97.6	2.4	45.2	54.8	62.2	37.8	100	0
2	73.8	26.2	97.6	2.4	44.2	55.8	63.1	36.9	100	0
3	79.0	21.0	97.9	2.1	47.2	52.8	63.8	36.2	100	0
4	82.1	17.8	97.9	2.1	46.8	53.2	64.0	36.0	100	0
5	88.1	11.9	98.1	1.9	45.0	55.0	65.0	35.0	100	0
6	92.1	7.8	98.1	1.9	47.4	52.6	65.4	34.6	100	0
7	93.4	6.6	98.2	1.8	48.6	51.4	65.8	34.2	100	0
8	94.9	5.1	98.3	1.7	49.4	50.6	66.3	33.7	100	0
9	95.2	4.8	98.5	1.5	46.5	53.5	67.4	32.6	100	0
10	98.2	1.4	98.7	1.3	46.0	54.0	67.4	32.6	100	0

When proportion of one-layer and two-layer weaves in the chosen warp threads is known, the crimp of chosen warp threads can be calculated according to Equation (3):

$$a_{sk} = \frac{d_{1-sl}a_{1-sl} + d_{2-sl}a_{2-sl}}{100} \quad (3)$$

where d_{1-sl} is the proportion of one-layer weaves in a certain warp thread (%), a_{1-sl} is the crimp of one-layer weaves (from Table 2), d_{2-sl} is the proportion of two-layer weaves in a certain warp thread (%) and a_{2-sl} is the crimp of two-layer weaves (from Table 2).

Moreover, d_{1-sl} and d_{2-sl} were calculated according to Equations (4) and (5):

$$d_{1-sl} = \frac{n_{1-sl}}{n_{sum}} \cdot 100\% \quad (4)$$

$$d_{2-sl} = \frac{n_{2-sl}}{n_{sum}} \cdot 100\% \quad (5)$$

where n_{1-sl} is the number of pixels in the column corresponding to number of threads woven in one-layer weaves along the warp thread in the jacquard pattern draft, n_{2-sl} is the number of pixels in the column corresponding to number of threads woven in two-layer weaves along the warp thread in the draft jacquard pattern, and n_{sum} is the number of pixels in one column of the analysed part of the draft jacquard pattern.

General warp crimp consists of a few one-layer and a few two-layer weaves, and their proportions in a jacquard fabric are evaluated in Equation (3), when crimps of each separate weave are known. Thus, the warp crimp of each separate warp thread in jacquard fabric can be predicted according to Equation (3) just by looking at the draft fabric pattern.

If the jacquard fabric is woven using another weave and structure, Equations (3)–(5) can be rewritten as:

$$a_{skb} = \frac{d_1a_1 + d_2a_2 + \cdots + d_na_n}{100} \quad (6)$$

$$d_1 = \frac{n_1}{n_{sum}} \cdot 100\% \tag{7}$$

$$d_2 = \frac{n_2}{n_{sum}} \cdot 100\% \tag{8}$$

$$d_n = \frac{n_n}{n_{sum}} \cdot 100\% \tag{9}$$

where d_1 is the proportion of the first weave in a certain warp thread (%), a_1 is the crimp of the first weave, d_2 is the proportion of the second weave in a certain warp thread (%), a_2 is the crimp of the second weave, d_n is the proportion of the nth weave in a certain thread (%), a_n is the crimp of the nth weave, n_1 is the number of pixels in the column corresponding to number of threads woven in the first weave along the warp thread in the draft jacquard pattern, n_2 is the number of pixels in the column corresponding to the number of threads woven in the second weave along the warp thread in the draft jacquard pattern, n_n is the number of pixels in the column corresponding to number of threads woven in the nth weave along the warp thread in the draft jacquard pattern, and n_{sum} is the number of pixels in one column of the analysed part of the draft Jacquard pattern.

The results of calculating the crimp are demonstrated in Table 3. The coefficients of variation of the experiments, which vary from 0.14% to 0.16%, show the high accuracy and low dispersion of the calculated crimp.

Table 3. Results of calculating the crimp.

Number	25 cm from Edge%	45 cm from Edge%	65 cm from Edge%	85 cm from Edge%	105 cm from Edge%
1	17.49	15.41	18.89	17.75	15.25
2	21.93	22.98	20.64	21.46	23.08
3	24.23	23.20	25.98	25.07	23.08
4	17.03	16.71	17.75	17.40	16.67
5	20.70	20.11	23.26	22.07	20.00
6	23.35	23.15	24.86	24.25	23.08
7	16.93	16.74	18.71	18.02	16.67
8	24.12	24.20	23.05	23.44	24.24
9	19.38	19.36	19.70	19.56	19.35
10	18.07	18.06	19.09	18.67	18.03
Average	20.32	19.99	21.19	20.77	19.95
Standard deviation	2.94	3.22	2.87	2.86	3.26
Coefficient of variation	0.14	0.16	0.14	0.14	0.16
Histogram					

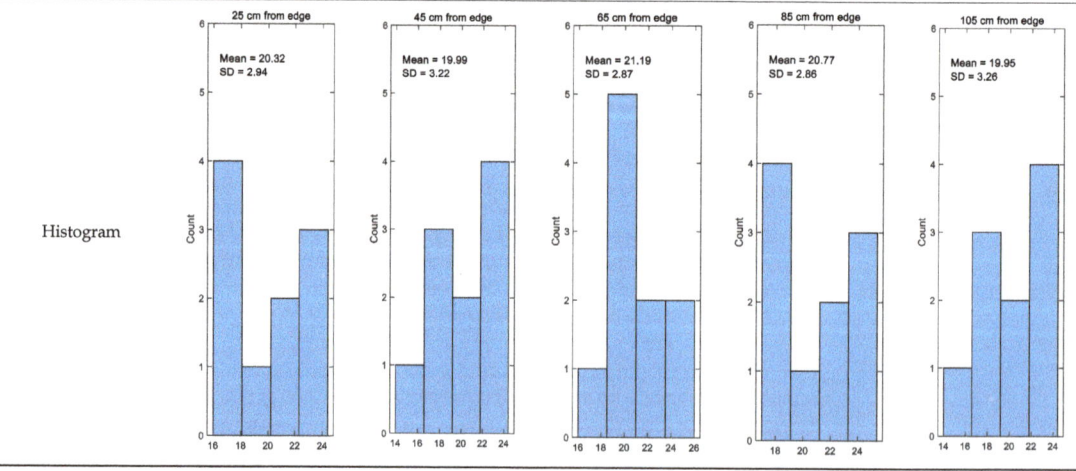

In a comparison between the experimental and calculated crimp results, comparable crimp results are shown in Figure 6. As can be seen from the diagram, the calculated crimp is higher than the experimental crimp in all cases by 5% for 105 cm distance from the fabric edge to 10% for 65 cm distance from the fabric edge. Such results were obtained because the crimp of separate weaves was established from shorter segments of the warp threads, which were woven in such weaves. Because of that, the accuracy of crimp could have been established differently.

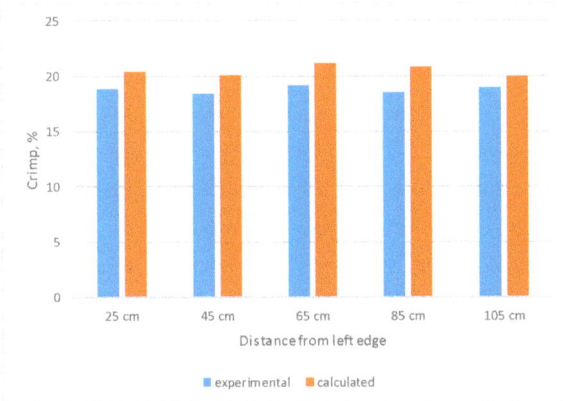

Figure 6. Comparison of experimental and calculated crimp.

The results showed that warp crimp changed within the limits of errors in the jacquard fabric's width. Thus, it can be stated that the fabric's structural parameters and structural solutions had been chosen correctly.

The same investigation of crimp was performed with seven different jacquard fabrics patterns woven with the same weaves but with different distributions in the fabric. The results are presented in Table 4. The relative error of the experimental crimp varied from 7.86% to 12.35% and the relative error of the calculated crimp varied from 6.06% to 12.41%. Thus, it can be stated that the accuracy of the results is sufficient.

Table 4. Results of the experimental and calculated crimp of different fabrics.

Design Number	Experimental Crimp%	Calculated Crimp%
1	19.04 ± 2.23	20.05 ± 1.32
2	18.58 ± 2.04	19.96 ± 1.21
3	19.12 ± 1.40	20.14 ± 2.50
4	19.08 ± 2.35	21.10 ± 1.87
5	18.44 ± 1.45	20.57 ± 1.44
6	18.63 ± 1.89	20.92 ± 2.34
7	18.78 ± 2.32	20.24 ± 2.41

As can be seen from Table 4, the distributions of warp crimp in fabrics with different distributions of weaves are similar. Thus, it can be stated that the suggested method of predicting warp crimp in the weave structure described is correct. It can be used for jacquard fabrics of any structure.

4. Conclusions

After evaluation of jacquard fabrics' crimp in the warp direction, it was established that the crimp was 18.80% and changed within the limits of errors, i.e., a range of only 4% percent, in the fabric width. Thus, it can be said that the warp crimp was constant across the fabric width.

Because the warp crimp of Jacquard fabric changed insignificantly (within the limits of errors), it can be stated that the fabric's setting parameters and structural solutions were chosen and matched correctly and such a fabric could be woven on any Jacquard weaving loom.

After a comparison of the experimental and calculated warp crimps, it was established that calculated crimp in all cases was higher than the experimental crimpy by 5–10% across the fabric width. The reason for this may be the lower accuracy of establishing the calculated crimp, because the crimp of different weaves was established from shorter segments of the warp threads.

Author Contributions: Conceptualisation; E.K.; methodology; E.K. and E.L.; formal analysis; S.P. and E.L.; investigation; E.L. and E.K.; writing—original draft preparation; S.P. and E.K. All authors have read and agreed to the published version of the manuscript.

Funding: This research received no external funding.

Institutional Review Board Statement: Not applicable.

Informed Consent Statement: Not applicable.

Data Availability Statement: Data are contained within this article.

Conflicts of Interest: The authors declare no conflict of interest.

References

1. Sarkar, P. What Is Crimp% in Fabric and How to Measure Warp and Weft Crimp%. Available online: https://www.onlineclothingstudy.com/2014/06/what-is-crimp-in-fabric-and-how-to.html (accessed on 16 July 2020).
2. Kiron, M.J. Crimp Percentage/Determination of Crimp Percentage Textile Learner. Available online: https://textilelearner.blogspot.com/2012/01/crimp-percentage-determination-of-crimp.html (accessed on 16 July 2020).
3. Al Mahfuj, A. Crimp & Crimp Interchange. Available online: https://www.slideshare.net/mahfujsms/crimp-crimp-interchange-71266447 (accessed on 16 July 2020).
4. Mertova, I.; Neckar, B.; Ishtiaque, S.M. New Method to Measure Yarn Crimp in Woven Fabric. *Text. Res. J.* **2015**, *86*, 1–13. [CrossRef]
5. Afroz, F.; Siddika, D.A. Effect of Warp Yarn Tension on Crimp in Woven Fabric. *Eur. Sci. J.* **2014**, *10*, 202–207.
6. Kovar, R. Length of the yarn in plain-weave crimp wave. *J. Text. Inst.* **2011**, *102*, 582–597. [CrossRef]
7. Ozkan, G.; Eren, R. Warp Tension Distribution over the Warp Width and Its Effect on Crimp Distribution in Woven Fabrics. *Int. J. Cloth. Sci. Technol.* **2010**, *22*, 272–284. [CrossRef]
8. Shahabi, N.E.; Mousazadegan, F.; Hosseini Varkiyani, S.M.; Saharkhiz, S. Crimp Analysis of Worsted Fabrics in the Terms of Fabric Extension Behaviour. *Fibers Polym.* **2014**, *15*, 1211–1220. [CrossRef]
9. Topalbekiroğlu, M.; Kaynak, H.K. The Effect of Weave Type on Dimensional Stability of Woven Fabrics. *Int. J. Cloth. Sci. Technol.* **2008**, *20*, 281–288. [CrossRef]
10. Stig, F.; Halstrom, S. Effects of Crimp and Textile Architecture on the Stiffness and Strength of Composites with 3D Reinforcement. *Adv. Mater. Sci. Eng.* **2019**, *2019*, 8439530. [CrossRef]

Article

The Influence of Finishing on the Pilling Resistance of Linen/Silk Woven Fabrics

Eglė Kumpikaitė [1,*], Indrė Tautkutė-Stankuvienė [1], Lukas Simanavičius [1] and Stasė Petraitienė [2]

[1] Department of Production Engineering, Faculty of Mechanical Engineering and Design, Kaunas University of Technology, Studentų str. 56, LT-51424 Kaunas, Lithuania; indre.stankuviene@gmail.com (I.T.-S.); lukas@klt.lt (L.S.)

[2] Department of Applied Mathematics, Faculty of Mathematics and Natural Sciences, Kaunas University of Technology, Studentų str. 50, LT-51424 Kaunas, Lithuania; stase.petraitiene@ktu.lt

* Correspondence: egle.kumpikaite@ktu.lt

Abstract: The pilling resistance of fashion fabrics is a fundamentally important and frequently occurring problem during cloth wearing. The aim of this investigation was to evaluate the pilling performance of linen/silk woven fabrics with different mechanical and chemical finishing, establishing the influence of the raw material and the peculiarities of dyeing and digital printing with different dyestuff. The pilling results of the dyed fabrics were better than those of the grey fabrics and even a small amount of synthetic fiber worsened the pilling performance of the fabric. Singeing influenced the change in the pilling resistance of the linen/silk fabrics without changing the final pilling resistance result. Singeing had a stronger influence on the fabrics with a small amount of synthetic fibers. The pilling resistance of printed fabrics was better than that of grey and dyed fabrics without and with singeing. The pilling resistance of pigment-printed fabrics was better than that of the reactive-printed fabrics.

Keywords: linen/silk fabric; pilling resistance; singeing; pigment and reactive printing

1. Introduction

It is known within the practice of textile manufacturing that the finishing of woven fabrics, such as abrasion and pilling resistance, has a particularly strong influence on their end-use properties. Thus, it is highly relevant to solve the issue of decreasing the pilling resistance of woven fabrics before their use in everyday wearing.

Fabric pilling is considered to be a performance and aesthetic property of a woven product that determines its quality [1].

Pilling is a fabric surface defect that develops due to fiber movement or the slippage of yarns caused by abrasion and wear. Pilling occurs in four steps: fuzz formation, entanglement, growth, and wear-off. The formation of fuzz and pills influences the aesthetics and durability of the fabric, as well as the demand of consumers [2]. Abrasion and pilling resistance can be established using the Martindale abrasion tester, which generates a movement according to the Lissajous curve and can test several samples simultaneously [3].

The abrasion and pilling resistance of both woven and knitted fabrics have been investigated by a number of scientists. The properties of these fabrics are influenced by such factors as the raw material (the composition of the fibers, the amount of synthetic fiber in the yarn, etc.), the structure (woven and knitted fabrics, different weaves, yarn structure parameters, etc.), and the finishing (mechanical and chemical finishing, different coatings, etc.).

Different scientists have analyzed the influence of the raw material on abrasion and pilling resistance. It was established that a larger amount of polyester fiber in a cotton/PES blend reduces the pilling resistance of knitted fabrics [4]. The addition of PES, PA fibers, or elastane filaments to the structure of socks enhanced the abrasion resistance of the garment [5,6]. The percentage of the mass loss was higher for knitted samples from wool than that for cotton samples [7]. The pilling resistance of polyester/wool woven fabrics has also been investigated. The number of polyester fibers migrated on the surface of a yarn increased with an increase in the polyester content of the blend and, hence, the pilling increased. The converse was true for the wool fibers used in the blend [8,9]. Fabrics comprising 100% cotton had better pilling performance than blended cotton/PES fabrics but their abrasion resistance was the lowest [10]. In summary, it can be stated that the content of synthetic fibers has a significant influence on the abrasion resistance and pilling performance of fabrics. Blends of natural and synthetic fibers have been investigated widely, whereas blends of two natural fibers of a different nature—linen (cellulose fiber) and natural silk (protein fiber), for example—have not been studied. This is a new blend of two natural fibers in one yarn and it has not been studied in earlier research.

The pilling and abrasion resistance also depend on the yarn structure. The length and fineness of the component fibers, as well as the yarn twist, density, and type of the weave of the fabric, influence pilling resistance [11,12]. The use of yarns with a higher linear density increased the abrasion resistance of knitted socks [5], polyester/wool [8], and polyester/cotton [13] woven fabrics. When the twist of the fabrics is higher, the relative slippage between the fibers decreases and the hairiness floating on the surface of the fabric is less. Therefore, the tendency to pill gradually decreases [12]. The abrasion and pilling resistance of fabrics produced from compact yarns were higher than those of fabrics produced from ring yarns [14], and they were lower in fabrics from carded yarn than in fabrics from combed yarn [15]. Pilling grades, from low to high, are ring-spun fabric, siro-spun fabric, compact-spun fabric, and siro–compact-spun fabric, respectively [12]. Yarn structure is also an important parameter when pilling performance is studied. In this research, the fiber composition and structure of the fabric of the yarn were constant.

The structure of the fabric is another important factor involved in pilling and abrasion resistance. Knitted fabrics of a different structure have been frequently analyzed by scientists. The abrasion resistance and pilling performance of interlock fabrics were higher than those of jersey fabrics [14]. Single jersey knitted fabrics showed improved resistance compared to rib and moss stitch structures [7]. Abrasion resistance depends on the structure of knitted fabric parameters, such as the stitch length: the abrasion resistance decreases when the stitch length increases [15,16]. Woven fabrics exhibited better pilling performance than knitted ones [7]. The pilling performance of a woven fabric depends on the float length, i.e., weaves with shorter floats have better pilling resistance than those with long floats [8,13,17]. Fabric woven at a proper loom setting or warp yarn tension showed higher strength, less pilling, and a superior abrasion tendency when compared with fabrics woven at various other levels of warp yarn tension [18]. The structure of the woven fabrics used was always the same—the raw material, structure, and linear density of the warp and weft, warp and weft settings, and weave—in all investigated fabrics.

The greatest amount of research has analyzed the influence of finishing on abrasion and pilling resistance. Processes such as singeing, cropping, and heat setting significantly reduce the tendency to pill [8,11]. Mechanically singed samples exhibited a better pilling grade than samples without singeing [4,8,9,12,19]. The abrasion resistance and pilling performance of dyed fabrics were higher than those of loom state fabrics [6,14,19]. Pilling resistance significantly increased with the wool content in polyester/wool fabrics and the heat setting temperature [20]. The tendency to pill of a CVC knitted fabric could be reduced by singeing and the heat setting [21]. Different coatings also affect the improvement of abrasion resistance and pilling performance [22–24]. A crease-resistant finishing applied to apparel fabric improved the usage characteristics of clothes and enhanced the pilling performance of woven fabrics [25]. Different mechanical and chemical means were used for the woven fabrics investigated in the article. One of these finishing methods was new and another had already been used by other researchers but, on the whole, the focus was on investigating the pilling performance of fabrics. Loom state, dyed, singed, and printed woven fabrics are investigated in this paper. The pilling performance of printed fabrics has not been investigated in any earlier scientific work; thus, the behavior of the fabric during the pilling test is important, especially the comparison of active and pigment printing. These two methods differ by their nature and technology, so it is relevant to establish and compare their pilling performance.

One more novelty of this investigation is the use of a mathematical analysis to establish the start of the pilling process, a new approach in textile research.

To deal with the information gap, the aim of the present article is to investigate the pilling performance of linen/silk woven fabrics with different mechanical and chemical finishing, establishing the influence of the raw material and the peculiarities of dyeing and digital printing with different dyestuff.

2. Experiment

2.1. Object of the Investigation

The object of the investigation was a woven fabric from a single blended yarn, 26 tex, 70% linen/30% silk in the warp and in the weft. The warp setting was 220 dm^{-1} and the weft setting was 223 dm^{-1}. Another fabric used for comparison was woven from linen 28 tex yarn in warp and cotton/PES twisted 28 tex threads in weft. The fibrous composition of this fabric was 86% linen/12% cotton/2% PES. Both fabrics were woven using a double-layer weave with a one-layer weave. The warp and weft repeats of the weave were large and a general plan of the weave could cover a large area in the article. For this reason, the schemes of drawing-in and the cards of the fabric are shown separately. The weave was chosen because of the existence of one-layer (dense) and two-layer (rare) places in one fabric. Such a fabric structure allowed for the evaluation of the pilling performance in rarer and denser parts of the fabric. The fabric drawing-in scheme is shown in Figure 1 and the cards of the woven fabrics are presented in Figure 2. The fabrics were woven by the textile company Klasikinė tekstilė (Lithuania).

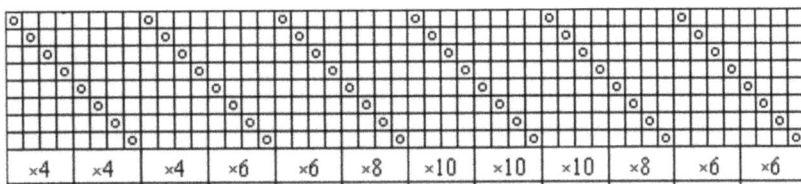

Figure 1. Drawing-in scheme of the woven fabric.

The fabrics were treated with different finishings. At first, grey fabrics and dyed fabrics without any additional finishing operations were investigated. After this, grey and dyed fabrics after singeing were analyzed. Two types of printed fabric (pigment and reactive printing) were also researched.

2.2. Finishing Materials and Technologies

All the finishing procedures (washing, dyeing, rinsing, softening, and drying) were performed in a BRONGO 100 (Brongo srl, Florence, Italy) machine. The fabrics were washed for 10–15 min at a temperature of 65 °C and dyed for 75–120 min at a temperature of 60 °C. Active dyestuff *Everzol* (Everlight Chemical, Taipei, Taiwan) was used. The fabrics were rinsed in cold water twice and in hot water twice after dyeing. One session of rinsing took 5 min. The softening was performed in an acid environment using a *Perustol CCF* (Rudolph Group, Gerestried, Germany) softener.

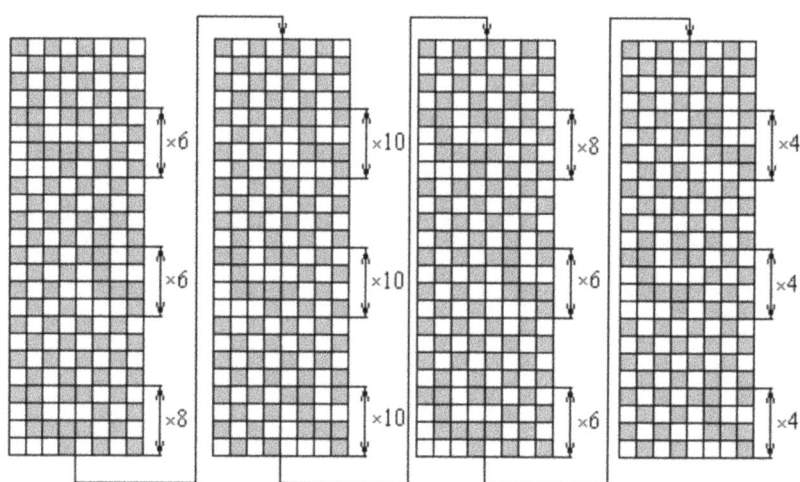

Figure 2. Cards of the woven fabric.

The singeing of the loom state fabrics was performed in a *Vollenweider* gas singeing machine (Xetma Vollenveider, Waedenswil, Switzerland) with an open flame. Two different sets of burners were used for singeing both sides of the fabric by threading the fabric suitably. In the gas singeing machine, the fabric in an open width passed at a speed of 55 m/min.

Both pigment and reactive digital printing were achieved with a piezoelectric DOD ink head. The piezoelectric material was placed in an ink-filled chamber behind each nozzle. When a demand (impulse) is applied, this special material changes shape, which generates a pressure pulse. This effect allows the ink fluid to exit the nozzle. This means that any type of ink can be used for printing with the same print head. Pigment printing was performed in a *Mimaki* printing machine (Mimaki Engineering Co., Ltd., Tomi, Japan) and reactive printing was conducted in an *Mtex500* printing machine (Techno Fashion World, Milano, Italy).

For the experiment, we used both pigment dyestuff and reactive dyestuff. The pigment and reactive dyestuffs were manufactured by *Mimaki* (Mimaki Engineering Co., Ltd., Tomi, Japan). The difference between the two inks is fundamental. Pigment dyestuff contains a binder in the dyestuff and thus the fabric does not need any special preparation before printing commences. Reactive dyestuff inks do not possess this feature. Fabrics need to be specially prepared for the print; they are soaked in a combination of chemicals that controls their color intensity, background color, and line sharpness. All the ingredients are fundamentally important but urea plays a key role in the soaking process. Urea is responsible for the color intensity and it must fulfill a high set of requirements for the drying process. The other major difference is the fixation process. Printed fabrics with pigments need a completely dry fixation and do not require washing afterwards; when using reactive dyestuff, a steaming process is required and washing afterwards is a necessary step.

2.3. Weather Conditions of the Experiment

The samples were laid on a plain horizontal surface. Air could pass through the fabric. The samples were conditioned for at least 24 h in standard weather conditions (Standard LST EN ISO 139: 2005/A1: 2011) before testing, i.e., the temperature was 20 ± 2 °C and the relative humidity was set at 65 ± 4%.

2.4. Methods of Establishing Pilling Resistance

The pilling resistance tests were performed using a MESDAN-LAB, Code 2561E (SDL Atlas, Rock Hill, UK) Martindale abrasion and pilling tester according to Standard ISO 12945-2:2000 "Determination of fabric propensity to surface fuzzing and to pilling—Part 2: Modified Martindale method". A picture of the tester is shown in Figure 3.

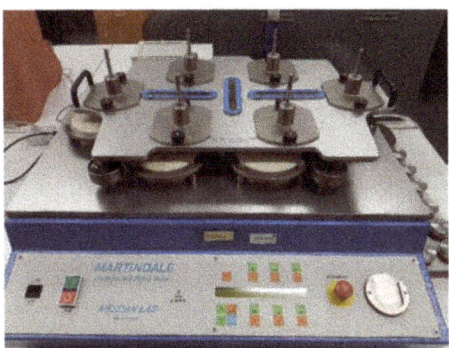

Figure 3. MESDAN-LAB, Code 2561E Martindale abrasion and pilling tester.

Six circular samples, from which three were placed on holders and another three were placed on the pilling table, were cut from the investigated fabrics. Each sample was evaluated by three experts according to an organoleptic evaluation after a certain number of cycles as specified in the standard (125, 500, 1000, and 2000 cycles). The marks of pilling of each sample were recorded and the average result of all the evaluations was established after the evaluation of the sample appearance. The evaluation of the pilling marks is described in Table 1.

Table 1. Evaluation of the pilling marks.

Mark	Description
5	Appearance does not change.
4	Slight fuzzing on the surface and (or) partially formed pills.
3	Medium fuzzing on the surface and (or) medium pilling. Pills of different magnitude and density partially cover the fabric surface.
2	Significant fuzzing and (or) significant pilling. Pills of different magnitude and density cover a large part of the fabric surface.
1	Extremely significant fuzzing on the surface and (or) extremely significant pilling. Pills of different magnitude and density cover the whole fabric surface.

2.5. Mathematical Analysis of the Results

A mathematical analysis was performed using MATLAB software.

3. Results and Discussion

At first, grey and finished woven fabrics without any additional finishing were tested whilst seeking to establish the influence of the finishing on the fabric pilling resistance. The results are shown in Figure 4. The number of cycles was set during the experiment and the pilling and abrasion tester stopped after a set number of cycles. For this reason, a statistical analysis of the number of cycles cannot be provided. As can be seen, the results of the grey and dyed linen/silk fabrics were similar. Only at the beginning of the pilling test (125 cycles) was the mark of the dyed linen/silk fabric slightly (by 0.5) higher than that of the grey fabric. In the middle of the pilling test (after 500 and 1000 cycles), the marks were essentially the same, i.e., the score was 3.5 for linen/silk fabric. At the end of the test, the pilling mark was the same; it was equal to 2.5 for both the grey and dyed linen/silk fabrics. It can be seen from the diagram that the mark of the grey fabric changed gradually after each abrasion period. The character of the dyed fabric in the diagram was slightly different, i.e., after 125 abrasion cycles, the changes in the fabric surface were not significant. After 500 cycles, the mark decreased significantly (by 1 point) and during the next abrasion period (1000 cycles), it did not change again. At the end of the test (after 2000 cycles), a significant change (by 1 point) could be seen again. The pilling marks of the dyed linen/silk fabrics were higher than those of the grey fabrics. The results of the linen/cotton/PES fabric differed from the results of the linen/silk fabrics, i.e., the pilling performance of grey fabric was better than that of linen/silk fabric. However, the pilling marks of dyed linen/cotton/PES fabric were significantly lower than those of linen/silk fabric. Thus, it can be stated that even a small amount of synthetic fiber worsens the pilling performance of the fabric. The result of investigation [9] also showed the same tendency, i.e., the pilling performance of the dyed fabrics was higher than that of the loom state fabrics. Only investigations of cotton (cellulose fiber) [4,5,7,21,25] and wool (protein fiber) [6,23,24] and their blends with synthetics were found in the scientific literature. The tendencies of the results were similar because the raw material of the analyzed fabrics was of a similar nature (linen, cellulose fiber; natural silk, protein fiber). The reason for these results could be that the dyestuff seemed to adhere the formed fuzzes and pills to the surface of the fabric; the pilling resistance of the dyed fabric then improved. The used dyestuff may have had an influence on the pilling performance of the analyzed fabrics because reactive dyes form covalent bonds with the fabric. This situation was not analyzed in this article because it is not the object of textile engineering. A statistical analysis of the results cannot be provided because the result of the evaluation was marked; it could only be 1, 1.5, 2, 2.5, 3, 3.5, 4, 4.5, or 5, and no errors or other statistical parameters could be calculated for the pilling marks.

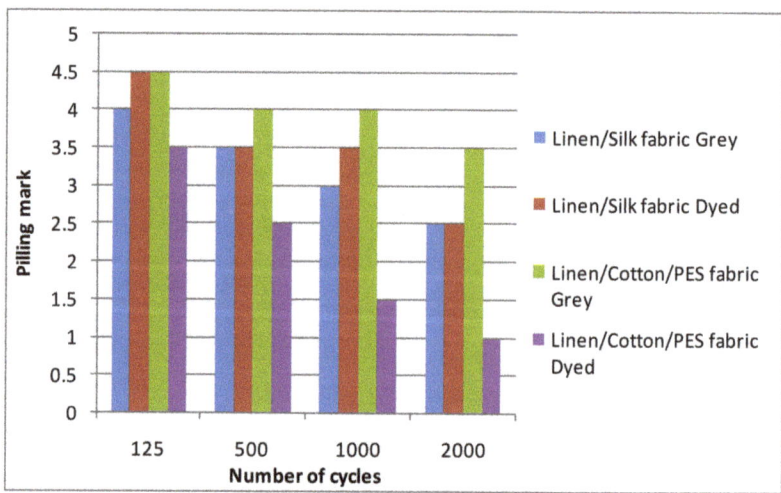

Figure 4. Pilling marks of grey and dyed linen/silk fabrics without additional mechanical finishing.

When seeking to find a method for better pilling resistance, a type of additional mechanical finishing—singeing—was performed on the grey fabric. The singeing process improves the surface of fabric and its pilling resistance by removing protruding fibers from it [4,8,14]. The fabric was also dyed after singeing. The diagrams of the pilling resistance of the grey and finished linen/silk fabrics after singeing are shown in Figure 5. According to reference [9], the pilling performance of dyed linen/silk fabrics was higher than that of grey fabrics. Comparing the given results with the results of linen/cotton/PES fabrics, it can be seen that results of grey fabrics were better than those of linen/silk fabrics, but they remained the same as the results without singeing. The pilling performance was almost the same as the results of linen/silk fabrics, but they were much better than the results before singeing. Thus, it can be stated that singeing had a greater influence on the fabric with a small amount of synthetic fibers. The pilling resistance of different raw material fabrics after singeing differed. Most investigations are related to woven and knitted fabrics from cotton (cellulose fiber) [4,5,7,21,25] and wool (protein fiber) [6,23,24] and their blends with PES, PA, and elastane. It was established that the pilling performance of 100% natural fiber fabrics was higher than that of blends with synthetic fibers [4–7,21,23–25]. A blend of two natural fibers (linen and natural silk; cellulose and protein fibers) was analyzed in this article. The pilling performance influenced by the singeing process differed from the results of the blends with synthetic fibers. In summary, the results showed that singeing did not have a significant influence on the pilling resistance of blends of two natural fibers. This was also confirmed by the ANOVA statistical method using MATLAB software (Figure 5). If 0.8164 > 0.06, the hypothesis about the equality of averages was accepted and averages of groups did not differ statistically.

ANOVA Table

Source	SS	df	MS	F	Prob>F
Columns	0.03125	1	0.03125	0.06	0.8164
Error	3.1875	6	0.53125		
Total	3.21875	7			

Figure 5. ANOVA results of singed fabrics.

As can be seen from the diagram in Figure 6, the nature of the pilling diagrams for both the grey and finished fabrics was almost the same, i.e., almost all the marks were the same after a certain number of abrasion cycle periods. In comparison, the diagrams depicting the experiments without and with rinsing highlighted that the nature of the diagram was different. At first, the mark reached the level of mark 4 and the level did not change after 125 and 500 abrasion cycles. In the next two abrasion periods, the appearance of both the grey and dyed fabrics changed significantly. The final result of the pilling test of the fabrics after the singeing treatment was the same as that of the fabrics without singeing. It can be stated that the results of pilling resistance improved because the nature of the diagrams was better for the fabrics after singeing, i.e., those fabrics preserved a higher mark of pilling resistance for longer during the abrasion cycles. In addition, it could be seen that better results of pilling resistance were obtained for the grey fabric after singeing than for the dyed fabric. According to reference [8], these results are because the dyed fabric suffered a greater number of mechanical effects during the dyeing process, which worsened the general pilling resistance of the dyed fabric. The fabric weave can also have an influence on the pilling performance of the fabric. The two-layer (rare) parts form pills quicker than one-layer parts because the length of the floats is longer in two-layer parts. These results correspond with earlier investigations [9,23–25], which state that shorter thread floats result in the better pilling performance of the fabric. The fabric was woven from rotor-spun yarn and the length of the fiber was short. Due to this, shorter fibers fuzzed on the surface of the dyed fabric after the dyeing process and this influenced the pilling performance of the dyed fabric. Thus, singeing exerted an influence on the pilling resistance of the fabrics but it was not significant. In a comparison of the results of the fabrics without and with singeing, it could be seen that singeing improved the pilling performance of both the grey and dyed fabrics but the influence of the dying process was more significant because the results showed that the grey and dyed fabrics changed in nature. In conclusion, this type of mechanical finishing needs to be used as an additional finishing for better pilling resistance. Using the ANOVA analysis for the data from Figure 6, it was established that 0.03 > 8, and it could be concluded that the hypothesis about the equality of averages was canceled. This means that the results differed (Figure 7).

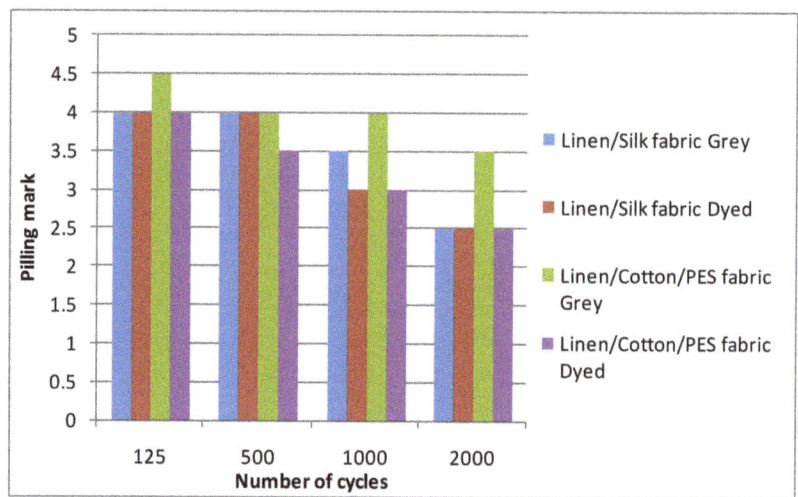

Figure 6. Pilling marks of grey and dyed linen/silk fabrics with additional mechanical finishing.

```
                          ANOVA Table
Source      SS      df      MS      F     Prob>F
-------------------------------------------------
Columns    0.5       1      0.5     8      0.03
Error      0.375     6      0.0625
Total      0.875     7
```

Figure 7. ANOVA analysis results for printed fabrics.

Finishing such as digital printing has become popular in textile finishing. Thus, the pilling resistance of two types of digital printing—pigment printing and reactive printing—was analyzed in this study. The diagrams of the pilling resistance of both types of printing are presented in Figure 8.

As can be seen in Figure 6, the mark of the pilling resistance changed significantly (by one mark for the pigment-printed fabric and by 1.5 marks for the reactive-printed fabric). After this, the mark remained constant until the end of the pilling test. This may have been caused by the different method of printing and the use of a different dyestuff. In pigment printing, the dyestuff distributes only on the surface of the fabric and it does not soak into the fabric inside, i.e., the fabric pattern can be seen only on the right side of the fabric. With reactive printing, the fabric is soaked and the active dyestuff is absorbed into the fabric. The printed pattern slightly saturates into the wrong side of the fabric. The result of the pigment-printed fabric was better because the areas where the pigment dyestuff was on the surface of the fabric were more resistant than those where there was a lack of the pigment dyestuff. The final pilling marks for the fabrics printed using both methods were significantly better than those of the dyed and singed fabrics. No references to the pilling performance of printed fabrics have been found in the existing scientific literature; thus, it can be stated that this research is new and important. The dyestuff did not absorb into the wrong side of the fabric during digital printing; the dyestuff absorption was more superficial, in contrast to the dyed fabrics. Thus, the dyestuff formed a cover on the surface of the fabric and this improved the pilling performance of the fabrics. The pigment-dyed fabrics showed particularly improved pilling resistance because the fabric absorbed the pigment dyestuff less than when the active method was used.

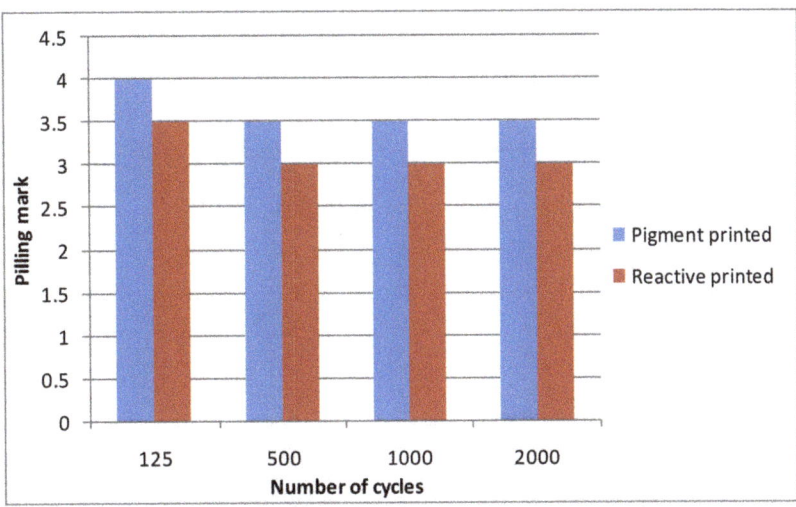

Figure 8. Pilling resistance of linen/silk fabrics after pigment and reactive printing.

The dynamics of the changes in appearance of all the fabrics with different finishing are presented in Table 2. It can be seen from the pictures in the table that the appearance of all the fabrics already began to change after 125 abrasion cycles. The surfaces of the fabrics fuzzed and pills partially started to form. The appearance of the fabrics worsened gradually after each number of cycles; fuzzing increased on the surface of the fabrics and the number of the pills rose, whereas the fabrics without and with singeing achieved a pilling mark of 2.5 after 2000 abrasion cycles. The appearance of the fabrics printed using both methods changed less. These fabrics showed only moderate pilling of their surfaces (marks 3 or 3.5).

Table 2. Appearance of linen/silk fabrics during the pilling test.

Appearance before Test	Appearance after 500 Cycles	Appearance after 2000 Cycles
	Grey fabric without singeing	
	Finished fabric without singeing	
	Grey fabric with singeing	

Table 2. *Cont.*

Appearance before Test	Appearance after 500 Cycles	Appearance after 2000 Cycles
	Finished fabric with singeing	
	Pigment-printed fabric	
	Reactive-printed fabric	

Thus, in summary, it can be stated that the pilling resistance of fabrics and the nature of their changing depends on the finishing used for them.

4. Conclusions

Generally, the pilling results of the dyed fabrics were better than those of the grey fabrics. The reason for this may have been that the dyestuff adhered the fuzzes and pills to the surface of the fabric and this led to the better pilling resistance of the dyed fabric. The investigation's results also showed that even a small amount of synthetic fibers worsens the pilling performance of the fabric.

Singeing influenced the nature of the change in the pilling resistance of the linen/silk fabrics without changing the final pilling resistance mark. These results could have been influenced by the raw material of the fabric, in which two natural fibers (linen and natural silk) of a different nature were blended. Singeing had a greater influence on the fabric with a small amount of synthetic fibers.

The pilling resistance of the printed fabrics was better than that of the grey and dyed fabrics without and with singeing. The reason was that, during dyeing, the entire fabric was immersed in the dye solution and the dyestuff was absorbed into the fabric. During printing, the dye was applied only to the surface of the fabric, improving the pilling resistance of the fabric.

The pilling resistance of pigment-printed fabrics was better than that of the reactive-printed fabrics because of the peculiarities of the type of dyestuff and its penetration into the fabric, i.e., the fabric absorbed the pigment dyestuff less than the reactive one. The reactive dye formed covalent bonds with the fabric so, during dyeing with the pigment dyestuff, it did not create any chemical reactions with the fabric.

The general recommendations will be to use mechanical finishing combined with digital printing for blends of two natural fibers and to use singeing and dyeing for blends of natural and synthetic fibers. These combinations of finishing will be optimal for the given raw materials.

Future studies may include different, new, and unexpected blends of textile fibers, such as linen/wool, linen/alpaca, their blends with other natural and synthetic fibers, etc. They can include such fabrics' end-use properties, such as pilling and abrasion resistance, and search for means of improving of these properties.

Author Contributions: Conceptualization: E.K.; methodology: L.S. and I.T.-S.; formal analysis: S.P. and I.T.-S.; investigation: E.K. and L.S.; writing—original draft preparation: E.K. and S.P. All authors have read and agreed to the published version of the manuscript.

Funding: This research received no external funding.

Institutional Review Board Statement: Not applicable.

Informed Consent Statement: Not applicable.

Data Availability Statement: The data presented in this study are available on request from the corresponding author.

Conflicts of Interest: The authors declare no conflict of interest.

References

1. Eldessouki, M.; Hassan, M. Adaptive neuro-fuzzy system for quantitative evaluation of woven fabrics pilling resistance. *Expert Syst. Appl.* **2015**, *42*, 2098–2113. [CrossRef]
2. Basit, A.; Latif, W.; Ashraf, M.; Rehman, A.; Iqbal, K.; Maqsood, H.S.; Jabbar, A.; Baig, S.A. Comparison of Mechanical and thermal Comfort Properties of Tencel Blended with Regenerated Fibers and Cotton Woven Fabrics. *AUTEX Res. J.* **2019**, *19*, 80–85. [CrossRef]
3. Jerkovic, I.; Pallares, J.M.; Capdevila, X. Study of the Abrasion Resistance in the Upholstery of Automobile Seats. *AUTEX Res. J.* **2010**, *10*, 14–20.
4. Smiriti, S.A.; Islam, M.A. An Exploration on Pilling Attitudes of Cotton Polyester Blended Single Jersey Knit Fabric After Mechanical Singeing. *Sci. Innov.* **2015**, *3*, 18–21. [CrossRef]
5. El-Dessouki, H.A. A Study on Abrasion Characteristics and Pilling Performance of Socks. *Int. Des. J.* **2010**, *2*, 229–234.
6. Busiliene, G.; Lekeckas, K.; Urbelis, V. Pilling Resistance of Knitted Fabrics. *Mater. Sci.* **2011**, *17*, 297–301. [CrossRef]
7. Coldea, A.M.; Vlad, D. Study Regarding the Physical-Mechanical Properties of Knits for Garments—Pilling Performance. In *MATEC Web of Conferences, Proceedings of the 8th International Conference on Manufacturing Science and Education (MSE 2017): Trends in New Industrial Revolution, Sibiu, Romania, 7–9 June 2017*; EDP Sciences: Les Ulis, France, 2017; Volume 121.
8. Mominul Alam, S.M.D.; Katun Sela, S.; Nayab-Ul-Hossain, A.K.M. Mechanical Attribution in Improving Pilling Properties. Available online: https://www.textiletoday.com.bd/mechanical-attribution-improving-pilling-properties/ (accessed on 17 September 2021).
9. Hamouda Elshakankery, M. Pilling Resistance of Blended Polyester/Wool Fabrics. Available online: https://www.fibre2fashion.com/industry-article/3259/pilling-resistance-of-blended-polyester-wool-fabrics (accessed on 17 September 2021).
10. Rahman, M. Influence of Stitch Length and Structure on Selected Mechanical Properties of Single Jersey Knitted Fabrics with Varying Cotton Percentage in the Yarn. *J. Text. Eng. Fash. Technol.* **2018**, *4*, 189–196. [CrossRef]
11. Sivakumar, V.R.; Pillay, K.P.R. Study of Pilling in Polyester/Cotton Blended Fabrics. *Indian J. Text. Res.* **1981**, *6*, 22–27.
12. Wang, R.; Xiao, Q. Study on Pilling Performance of Polyester/Cotton Blended Woven Fabrics. *J. Eng. Fibers Fabr.* **2020**, *15*, 1558925020966665. [CrossRef]
13. Amin, R.M.; Rana, R.I.M. Analysis of Pilling Performance of Different Fabric Structures with Respect to Yarn Count and Pick Density. *Ann. Univ. Oradea Fascicle Text. Leatherwork* **2021**, *72*, 9–14.
14. Akaydin, M.; Can, Y. Pilling Performance and Abrasion Characteristics of Selected Basic Weft Knitted Fabrics. *Fibres Text. East. Eur.* **2010**, *2*, 51–54.
15. Ayesha, S.; Nasir, U.M.; Mohammad, A.J.; Nur, N.A.; Kowshik, S. Effects of Carded and Combed Yarn on Pilling and Abrasion Resistance of Single Jersey Knit Fabric. *IOSR J. Polym. Text. Eng.* **2017**, *2*, 39–43.
16. Shiddique, N.A.; Repon, R.; Al Mamun, R.; Paul, D.; Akter, N.; Shahria, S.; Islam, S. Evaluation of Impact of Yarn Count and Stitch Length on Pilling, Abrasion, Shrinkage and Tightness Factor of 1 × 1 Rib Cotton Knitted Fabrics. *J. Text. Sci. Eng.* **2018**, *8*. [CrossRef]
17. Jahan, I. Effect of Fabric Structure on the Mechanical Properties of Woven Fabrics. *Adv. Res. Text. Eng.* **2017**, *2*, 10–18. [CrossRef]
18. Sayed, U.; Jhatial, R.A. Influence of Warp Yarn Tension on Cotton Greige and Dyed Woven Fabric Properties. *Mehran Univ. Res. J. Eng. Technol.* **2013**, *1*, 125–132.
19. Shakhawat Hossain, M.D.; Dey Naimul Hasan, M.I.S.C. An Approach to Improve the Pilling Resistance Properties of Three Thread Polyester/Cotton Blended Fleece Fabric. *Heliyon* **2021**, *7*, e06921. [CrossRef]
20. Abdel-Fattah, S.H.; El-Katib, E.M. Improvement of Pilling Properties of Polyester/wool Blended Fabrics. *J. Appl. Sci. Res.* **2007**, *3*, 1206–1209.

21. Shakhawat, H. Effect of Singeing and Heat Setting on Pilling Properties of CVC Single Jersey Knit Fabric. *Int. J. Curr. Eng. Technol.* **2017**, *7*, 266–271.
22. Zhichao, H.; Wenxing, C. Preparation and Characterization of Hot Melt Copolyester (PBTI) Ultrafine Particles and Their Effect on the Anti-Pilling Performance of Polyester/Cotton Fabrics. *Polymers* **2018**, *10*, 1163. [CrossRef]
23. Rombaldoni, F.; Mossotti, R.; Montarsolo, A.; Songia, M.B.; Innocenti, R.; Mazzuchetti, G. Thin Film Deposition by PECVD Using HMDSO-O_2-Ar Gas Mixture on Knitted Wool Fabrics in Order to Improve Pilling Resistance. *Fibers Polym.* **2008**, *5*, 566–573. [CrossRef]
24. El-Sayed, H.; El-Khatib, E. Modification of Wool Fabric Using Ecologically Acceptable UV-assisted Treatments. *J. Chem. Technol. Biotechnol.* **2005**, *80*, 1111–1117. [CrossRef]
25. Can, Y.; Akaydin, M.; Turhan, Y.; Ay, E. Effect of Wrinkle Resistance Finish on Cotton Fabric Properties. *Indian J. Fibre Text. Res.* **2009**, *34*, 183–186.

Article

Bibliometric Analysis of Artificial Intelligence in Textiles

Habiba Halepoto [1], Tao Gong [1,2,*], Saleha Noor [3] and Hafeezullah Memon [4]

1. Engineering Research Center of Digitized Textile and Fashion Technology, Donghua University, Shanghai 201620, China; 317111@mail.dhu.edu.cn
2. College of Information Science and Technology, Donghua University, Shanghai 201620, China
3. School of Information Science and Engineering, East China Science and Technology University, Shanghai 200237, China; saleha.noor@yahoo.com
4. College of Textile Science and Engineering, Zhejiang Sci-Tech University, Hangzhou 310018, China; hm@zstu.edu.cn
* Correspondence: taogong@dhu.edu.cn

Abstract: Generally, comprehensive documents are needed to provide the research community with relevant details of any research direction. This study conducted the first descriptive bibliometric analysis to examine the most influential journals, institutions, and countries in the field of artificial intelligence in textiles. Furthermore, bibliometric mapping analysis was also used to examine diverse research topics of artificial intelligence in textiles. VOSviewer was used to process 996 articles retrieved from Web of Science—Core Collection from 2007 to 2020. The results show that China and the United States have the largest number of publications, while Donghua University and Jiangnan University have the highest output. These three themes have also appeared in textile artificial intelligence publications and played a significant role in the textile structure, textile inspection, and textile clothing production. The authors believe that this research will unfold new research domains for researchers in computer science, electronics, material science, imaging science, and optics and will benefit academic and industrial circles.

Keywords: bibliometric analysis; textiles; research trend; artificial intelligence; Web of Science

Citation: Halepoto, H.; Gong, T.; Noor, S.; Memon, H. Bibliometric Analysis of Artificial Intelligence in Textiles. *Materials* **2022**, *15*, 2910. https://doi.org/10.3390/ma15082910

Academic Editor: Dubravko Rogale

Received: 20 March 2022
Accepted: 13 April 2022
Published: 15 April 2022

Publisher's Note: MDPI stays neutral with regard to jurisdictional claims in published maps and institutional affiliations.

Copyright: © 2022 by the authors. Licensee MDPI, Basel, Switzerland. This article is an open access article distributed under the terms and conditions of the Creative Commons Attribution (CC BY) license (https://creativecommons.org/licenses/by/4.0/).

1. Introduction

Artificial intelligence has changed people's lives, facilitating the performance of repeated tasks with maximized accuracy. Like in everyday life, artificial intelligence also finds its application in the field of textiles. Textile materials are characterized by flexibility, fitness, and fineness, and generally, they find their application in apparel or upholstery and, to some extent, reinforcements of textile composites [1,2]. Textiles are the second basic need of people after food, which makes their study worthwhile, as humankind's daily life is connected to them [3]. The use of textile material dates to the stone age; it is used for shelter and has been a source of identity, showing one's social status, gender, or culture [4]. Due to the rapid development in computer science in the last decades, there has also been much advancement in manufacturing, testing, and analyzing textiles [5]. Artificial intelligence finds its application from fiber development to fiber assembly in slivers, yarns, fabrics, or garments [6].

There are two ways to summarize research publications, i.e., review papers and bibliometric research analysis [7]. The bibliometric analysis may be used as a predictive measurement tool for experimental study and choosing the research direction for new coming researchers. The statistics derived from the bibliometric analysis quantify the contribution of scientific articles to a particular subject. They reflect current scientific developments and may be used to identify potential developments; thus, the next science pattern may be forecast by bibliometric analysis. Less often, the bibliometric analysis has been published in the field of textiles and garments. Tian and Jun have recently published a bibliometric analysis on protective clothing research [8]; Yan and Xu analyzed the textile patentometrics [9]; Feng and coworkers studied textile and clothing footprint [10]; however, their research tool was CiteSpace. One more interesting bibliometric analysis published in textiles is related to textile schools [11]. In previous literature, diverse disciplines have used VOSviewer to conduct bibliometric analysis, sustainable supply-chain management [12], sustainable design for users [13], international entrepreneurship [14], plant-based dyes [15], health promotion using Twitter [16,17], industrial marketing management [18], applied mathematical modeling [19], circular economy [20], exchange rate and volatility [21], and so on. Thus, there are very few studies related to textiles, and according to our best knowledge, none of the authors have studied artificial intelligence in Textiles using the VOSviewer software tool. This study was aimed to explore the following key questions.

- What is the annual growth of publications in the field of artificial intelligence in textiles? What are their citation trends and usage counts in the database of Web of Science?
- How are the publications related to artificial intelligence in textiles distributed? What are the most influential countries, journals, and institutes?
- Which research group, country, and organization are most productive based on citations and bibliographies?
- What are the emerging topics related to artificial intelligence in textiles?
- How is the existing publication spread? What keywords are related to each other?

Through this bibliometric analysis research, this paper profoundly analyses the topic, i.e., artificial intelligence in textiles and its publication and citation worldwide, usage count and citation time analysis, the choice of authors, cooperation relationship between subjects, co-occurrence analysis of the words used in the abstracts, and cluster analysis of the manuscripts in this field. We believe that this research would help new researchers wisely select the research domain and provide a basic understanding of artificial intelligence's current status in textiles.

2. Data Collection and Research Methodology

2.1. Data Source

Web of Science is considered the most reliable scientific and technical literature indexing platform capable of introducing the most important scientific and technological research fields. The data were retrieved on 31 January 2021, from Science Citation Index (SCI) Core collections, using search query in Appendix A. A total of 996 research papers related to textile image processing were published between 2000 and 2020. The citation counts for top-cited manuscripts were exported based on the SCI citation search method since it guarantees that the citing literature has gone through the scientific evaluation process before publication. The manuscripts with 100 citations were considered as top-cited manuscripts in this research. The journals' impact factor is in accordance with the Journal Citation Reports published in 2019 since it is the most recent available data.

2.2. Bibliometric Methods

Here, we used VOSviewer to develop the mapping of the dataset. Since VOSviewer is a free software tool for constructing and visualizing bibliometric networks, these networks can include journals, researchers, or personal publications, which can be constructed based on citation, bibliographic coupling, co-citation, or co-author relationships. VOSviewer also provides a text mining function, which can be used to construct and visualize the co-occurrence network of essential terms extracted from a large number of scientific pieces of literature.

2.3. Inclusion and Exclusion Criteria

We started as a query string for topics related to artificial intelligence (i.e., image processing, image recognition, pattern recognition, machine learning, deep neural network, object recognition, and computer vision) and textiles (yarn, weave, knitted fabrics, hosiery, woven fabrics, drape, drapability, garments, and nonwovens) at the WoS Core Collection database. A total of 1145 results appeared in the timespan of 2007 to 2020. The publications comprised 11 languages: English (1127), Chinese (4), Spanish (3), Turkish (3), German (2), Japanese (1), Russian (1), Portuguese (1), Slovenian (1), and Croatian (1). As there was a significant difference among languages, we refined data to manuscripts written in English only. The remaining 1129 manuscripts comprised articles (724), early access (26), proceedings papers (398), reviews (21), data papers (2), editorial materials (3), and a correction (1), and thus, data paper, correction, and editorial material were excluded. Finally, 1123 papers were manually checked for their relevance to the topic. It was found that there were some overlapping terms, and thus, these articles were manually excluded from the dataset. For example, drape [22], weave [23], weaves [24], weaving [25,26], woven [27,28], weave ethics [29], weave the advantages [30], systems weave computing and communication [31], piece of music by weaving [32], "Weaving, Swerving, Sideslipping" [33], weaving sensible plots [34], WEAVE and 4MOST spectrographs [35], drape full-motion video, traffic-weaving [36], yarn [37–39], Hadoop yarn [40], yarn cluster [41], social fabric [42], and pressure-sensitive textiles [43]. It should be noted that the contextual meaning of these terms was not the same as the research domain of this manuscript. Thus, in total, 996 manuscripts were refined, and their distribution concerning the Web of Science Index was as follows: Science Citation Index Expanded (637), Social Sciences Citation Index (28), Conference Proceedings Citation Index-Social Sciences and Humanities (13), Conference Proceedings Citation Index-Science (339), Emerging Sources Citation Index (26), and Arts and Humanities Citation Index (5). The dataset of a final chosen manuscript might be requested from the corresponding author of this manuscript; its summary according to document type is presented in Table 1.

Table 1. Document types of final manuscripts.

Document Type	Article	Proceedings Paper	Early Access	Review
Records	649	343	24	21
Rate %	65.161	34.438	2.41	2.108

2.4. Data Analysis

The final dataset was exported from WoS and was analyzed in detail. This bibliometric research analyzed in-depth articles, topics, partnerships, times cited, co-words, and cluster analysis of papers. The VOSviewer was used to examine the co-occurrence mapping.

3. Results and Discussion
3.1. Global Publications and Citation Output

The publication output has been summarized in Figure 1a. It can be seen that the global publication of manuscripts in the field of artificial intelligence has risen. This rise in the number of publications has been divided into three time periods, i.e., 2007 to 2011 (old papers), 2012 to 2015 (recent papers), and 2016 to 2020 (current papers). It was observed that from 2007 to 2011, there was a sudden rise in publications, which became stable from 2012 to 2015, and then, currently, from 2016 to 2020, it has been increased dramatically. However, there was a continuous exponential increase in the number of citations, from 1 citation in 2007 to 1815 citations in 2020, as shown in Figure 1b (only citations from WoS Core collections were considered here). This research domain has shown to be promising, with an h-index of 36 and 7.37 citations per item. The total sum of citations is 7239 up to December 2020.

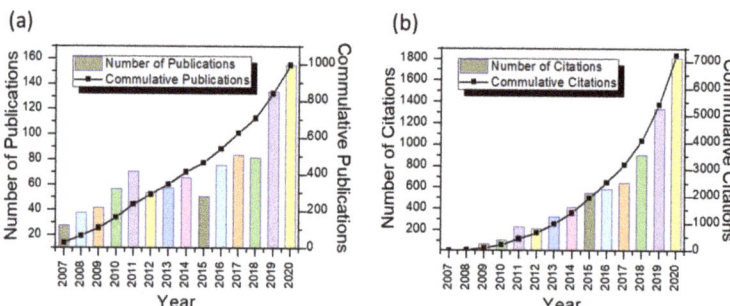

Figure 1. Annual publications and citation output: (**a**) number of publications and their cumulative from 2007 to 2020 and (**b**) number of citations and their cumulative from 2007 to 2020.

The usage count of scientific papers is directly related to the preference of readers. In general, readers from the scientific community prefer reading the latest article. However, highly cited manuscripts are often being used for a long time after their publication. This is what can be observed in Figure 2. The old manuscripts (2007–2011) possessed a higher number of citations, whereas they recently possessed a lesser usage count.

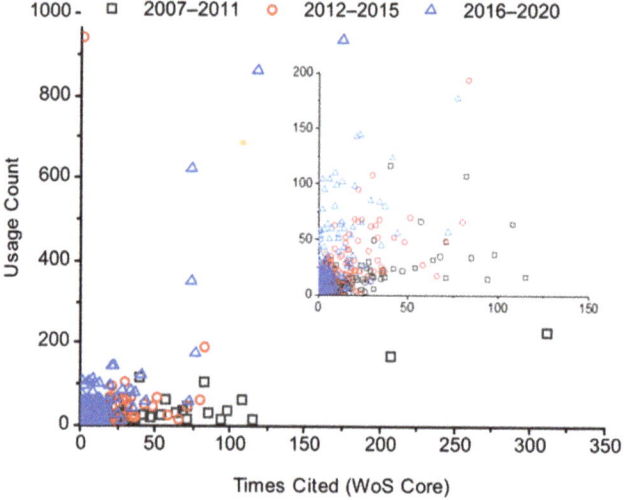

Figure 2. Usage count versus time cited for three different periods, i.e., 2007–2011 (old papers), 2012–2015 (recent papers), and 2016–2020 (current papers).

On the other hand, the current papers (2016–2020) have a higher usage count but lesser citations. The recent papers (2012–2015) had a higher number of citations and a higher usage count; they are at the maturity stage. This analysis agreed with the recent analysis related to the usage count versus citation from the Web of Science [44]. Herewith, citations of more than 100 in the Web of Science of all databases, highly cited manuscripts, are summarized in Table 2. We have listed the details, time cited, and usage count here.

Table 2. Highly cited research papers.

No.	Title	Journal	Year	Time Cited WoS Core	Time Cited WoS	Usage Count Since 2013	Reference
1	Computer-vision-based fabric defect detection: A survey	IEEE Trans. Ind. Electron.	2008	312	356	230	[45]
2	Automated fabric defect detection-A review	Image Vis. Comput.	2011	208	240	168	[46]
3	Stretchable Ti3C2Tx MXene/Carbon Nanotube Composite Based Strain Sensor with Ultrahigh Sensitivity and Tunable Sensing Range	ACS Nano	2018	175	177	934	[47]
4	Fiber/Fabric-Based Piezoelectric and Triboelectric Nanogenerators for Flexible/Stretchable and Wearable Electronics and Artificial Intelligence	Adv. Mater.	2020	118	119	864	[48]
5	Exploiting Data Topology in Visualization and Clustering Self-Organizing Maps	IEEE Trans. Ind. Electron.	2009	115	116	17	[49]
6	Autonomic healing of low-velocity impact damage in fiber-reinforced composites	Compos. Part-A Appl. Sci. Manuf.	2010	108	109	64	[50]
7	Majority Voting: Material Classification by Tactile Sensing Using Surface Texture	IEEE Trans. Robot.	2011	98	100	37	[51]

3.2. Distribution of Publications

In total, 61 countries were contributing to the field of artificial intelligence in textiles. It would not be justifiable to compare publications by country; however, this is highlighted here to show the research domain's distribution. According to their participation, the top countries are illustrated in Figure 3. The number of publications in these countries was as follows: China (321), USA (96), Iran (77), Turkey (68), India (67), France (57), Germany (54), Canada (36), Italy (31), Spain (31), and other (158). It was found that China has been the top participant in this research domain, followed by the USA.

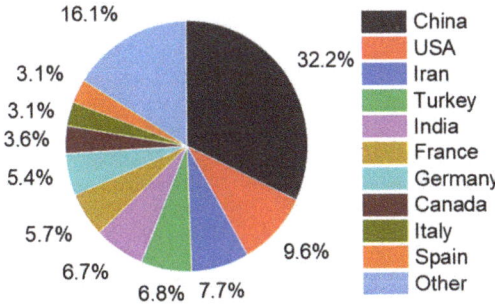

Figure 3. Most influential countries in terms of publications.

In total, 944 organizations or institutes were found to be participating in this research domain. The most influential institutions are summarized in Table 3. Donghua University was considered the top publishing institute in this research domain, followed by Jiangnan University.

Table 3. Top 15 institutes publishing in this research domain.

No.	Name of University	Number of Publications	Rate (%)
1	Donghua University	51	5.115
2	Jiangnan University	41	4.112
3	Isfahan University Technology	32	3.21
4	Soochow University	25	2.508
5	Hong Kong Polytech University	24	2.407
6	Shanghai University Engineering and Science	22	2.207
7	Amirkabir University Technology	21	2.106
8	University Lille Nord France	13	1.304
9	ENSAIT	12	1.204
10	RWTH Aachen	12	1.204
11	University of Minho	12	1.204
12	Indian Institute of Technology	11	1.103
13	Technical University of Liberec	11	1.103
14	Tiangong University	11	1.103
15	Xian Polytech University	11	1.103
16	Other	687	69.007

The most Influential Journals are summarized in Table 4. It can be seen that the *Journal of the Textile Institute* has remained the top choice for the authors for the given field to publish their work, which is followed by the textile research journal.

Table 4. The top 8 journals publishing in this research domain.

No.	Name of Journal	Number of Publications	Impact Factor	Rate (%)
1	Journal of the Textile Institute	73	1.239	7.322
2	Textile Research Journal	59	1.66	5.918
3	Fibres Textiles in Eastern Europe	29	0.76	2.909
4	Fibers and Polymers	23	1.59	2.307
5	Advanced Materials Research	21	–	2.106
6	Proceedings of SPIE	20	0.56	2.006
7	International Journal of Clothing Science and Technology	19	0.92	1.906
8	Indian Journal of Fibre Textile Research	18	0.6	1.805

3.3. Subject Categories of Research Productivity

Based on the classification of subject categories in Web of Science, the publication output data of research related to artificial intelligence in textiles was distributed in 45 subject categories during the last fourteen years. Subject categories containing at least ten articles are shown in Figure 4. Three research fields were prominent for the given research direction, including material science, engineering, and computer science.

The co-occurrence map according to the Web of Science categories was plotted. The circle's size describes the keyword's potential; as shown in Figure 5, there are three clusters: Cluster 1 (Red) is mainly about the subjects related to computer science and artificial intelligence. The related subjects include cybernetic, hardware and architecture, imaging science and photographic technology, information system, information systems, software engineering, telecommunications, theory, and method. The second cluster (Green) is mainly about multidisciplinary fields covering physics, analytical and physical chemistry, nanoscience, and nanotechnology as a multidisciplinary science. The third cluster is related to material science, particularly textiles, polymer science, composite, and mechanics.

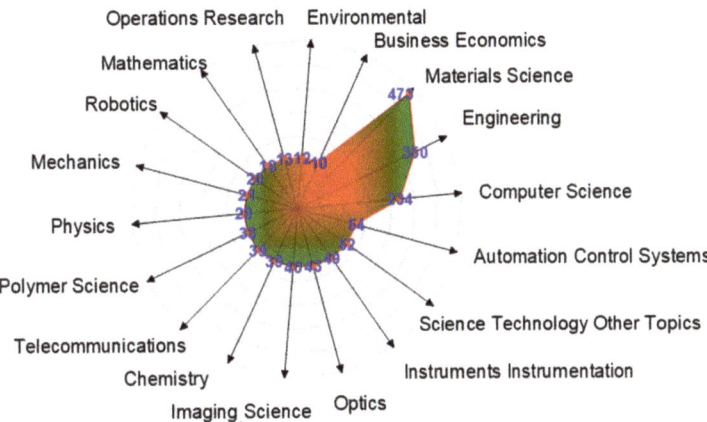

Figure 4. Subject categories according to Research areas in the Web of Science.

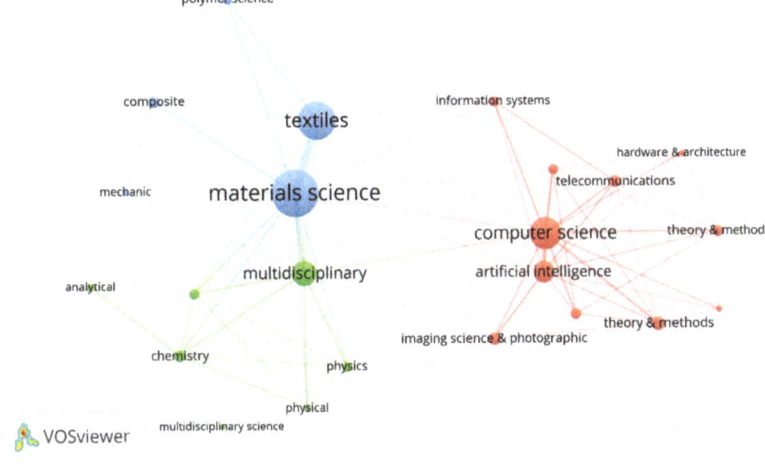

Figure 5. Co-occurrence map based on the Web of Science category.

It should be noted that data mining and machine learning are also considered significant artificial intelligence fields; image processing uses both fields' technology. Image processing's research direction combines image processing and main methods and their application to intelligent detection, recognition, and classification. First, a unique technique is used to capture the required image data, and then, the image data are processed [52]. Finally, computer vision and related digital image tools are used for analysis [53].

3.4. Co-Occurrence of Keywords in the Abstracts

Here, we identified the keywords from the abstracts of all the manuscripts related to artificial intelligence and textiles from the final dataset. The circle's size describes the keyword's potential, as presented in Figure 6, while the line's thickness was kept constant regardless of the link strength. The co-occurrence map based on the text data was plotted using the VOSViewer under the binary counting method with a ten-or-more threshold frequency. The network visualization graph under the association method with weight as the occurrence plotted the 18,883 terms; 403 met the threshold, and the 60% most relevant terms were plotted. It was found that there were three clusters (even with the cluster size = 1). Interestingly, each cluster presented a unique research theme in the field of textile engineering.

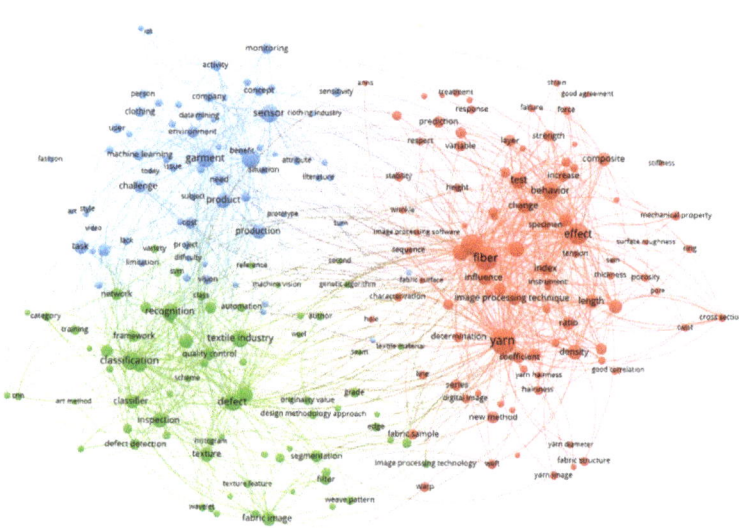

Figure 6. Co-occurrence analysis of research themes from the abstracts of the dataset.

3.4.1. Cluster 1 (Red): Artificial Intelligence for Textile Structures

The textile structure might be understood as the fiber structure; see the main keywords, i.e., ratio, density, diameter, and morphology. It may also be for yarn structure; see the main keywords, i.e., yarn hairiness, yarn count, yarn hairiness, yarn diameter, yarn property, yarn image, and twist. Moreover, the fabric structure may be related to keywords of fabric surface, wrinkle, hole weft, warp, porosity fabric appearance, fabric sample, stiffness, weft direction, surface roughness, and weft yarn. The textile structure cluster also covers the nonwoven and composite mechanical, in which nonwoven is mainly linked to fiber. However, the composite is linked with property, failure, geometry, layer image processing technique, image processing method, image processing software, image processing algorithm, and digital image processing technique are standard terms used to process the data. However, in particular, yarn is linked with the technique of artificial neural networks, i.e., Kalman filtering [54] or yarn color [55]. Property, effect, factor, behavior, distribution, change, test, and influence are common keywords related to every subgroup in the cluster.

3.4.2. Cluster 2 (Green): Artificial Intelligence for Textile Inspection

The fabric defect detection in artificial intelligence in textiles has remained of particular interest for researchers [56,57]. This cluster mainly focuses on the defect, recognition, classification, and inspection in the textile industry that promise quality control. The keywords are primarily related to woven fabric and fabric defect, in which computer vision is used for fabric defect and fabric texture (weave pattern) detection using several approaches such as pattern recognition, design methodology approach, support vector machine, neural network, convolutional neural network, feature extraction, deep learning, image segmentation, and the genetic algorithm of image processing technology. Visual inspection is replaced with automatic detection by training the classifier for color and texture using fabric images [58]. Automation (machine vision) has been proposed for the segmentation of databases into class using the filter to better results without complexity as a novel method to overcome the traditional method.

3.4.3. Cluster 3 (Blue): Artificial Intelligence for Textile and Apparel Production

This cluster focuses on the latest garment technology, particularly sensors for the product and production, to overcome the challenge and tasks. The manuscripts in this cluster mainly discuss the concept of monitoring the environment or activity of patients or persons in real-time. Moreover, it proposes smart textiles for the robot to replace hardware in IoT at low cost as a prototype for the designer in the apparel, clothing, and fashion industry. This cluster discusses data mining, image acquisition, and machine learning algorithms to meet consumer demand for textile products in the future.

4. Conclusions

This study summarizes a vital research domain, artificial intelligence in textiles, in one glance. This study suggests that this research field has remained a good discipline over the last 14 years and shall remain of particular interest in coming years. Thematic analysis revealed that artificial intelligence in textiles is not limited to pattern recognition. The analysis covers all major fields of artificial intelligence, including data mining and machine learning. Some key findings of this research include that the most dominant country was China, with 321 total publications. The USA was the runner-up with 96 total publications. Donghua University and Jiangnan University have the highest number of publications. Kumar [45] from the Indian Institute of Technology, got 312 citations in WoS core collections, while Ngan et al. [46] University of Hong Kong got 138 citations in WoS core collections. Co-occurrence of keywords in the abstracts yielded three major themes, i.e., artificial intelligence for textile structures, textile inspection, textile, and apparel production.

It should be noted that despite the extensive analysis, this research has some limitations. For instance, this analysis is based on the data provided by WoS, which is, of course, one of the authentic and accurate sources of information; however, the trend might be different when adding some other search engines or databases as well as when including manuscripts other than core collections. This study did not cover reviewer's information, such as which institution or reviewers reviewed the manuscripts in this given area. Therefore, the research in this field needs to be further deepened. Moreover, the in-between linkage choice of the journal, editor, and selected reviewer might also be recommended for future research.

Author Contributions: Conceptualization, H.H. and H.M.; methodology, H.H. and S.N.; software, H.H. and S.N.; validation, H.H.; formal analysis, H.H.; investigation, H.H.; resources, H.H.; data curation, H.H.; writing—original draft preparation, H.H.; writing—review and editing, H.M. and T.G.; visualization, H.H. and T.G.; supervision, T.G.; project administration, T.G. and H.M.; funding acquisition, H.M. All authors have read and agreed to the published version of the manuscript.

Funding: This work was supported by the National Natural Science Foundation of China (No. 61673007), Research Fund for International Scientists (RFIS-52150410416), National Natural Science Foundation of China, and the Research Startup grant of ZSTU (20202294-Y).

Institutional Review Board Statement: Not applicable.

Informed Consent Statement: Not applicable.

Data Availability Statement: The corresponding author can provide the data on request.

Conflicts of Interest: The authors declare no potential conflict concerning this article's research, authorship, and publication.

Appendix A

Search Query

(TS=("Image Processing") OR TS=("image recognition") OR TS=("data mining") OR TS=("artificial intelligence") OR TS=("pattern recognition") OR TS=("machine learning") OR TS=("deep neural network") OR TS=("object recognition") OR TS=("computer vision")) AND (TS=(Textiles) OR TS=(Yarn) OR TS=(Weave) OR TS=("Knitted Fabrics") OR

TS=("Hoisery") OR TS=("Woven Fabrics") OR TS=("nonwovens") OR TS=("nonwoven") OR TS=("non-wovens") OR TS=("non-woven") OR TS=(Drape) OR TS=(Drapability) OR TS=(Garments)) NOT (TS=("weave ethics") OR TS=("Hadoop yarn") OR TS=("social fabric") OR TS=("yarn cluster") OR TS=("Weaving, Swerving, Sideslipping") OR TS=("drape full-motion video") OR TS=("Elliot and Symmetric Elliot Extreme Learning Machines for Gaussian Noisy Industrial Thermal Modelling") OR TS=("Skynet Algorithm for Single-dish Radio Mapping. I. Contaminant-cleaning, Mapping, and Photometering Small-scale Structures") OR TS=("What an Entangled Web We Weave: An Information-centric Approach to Time-evolving Socio-technical Systems") OR TS=("MSCS: MeshStereo with Cross-Scale Cost Filtering for fast stereo matching") OR TS=("NPIY: A novel partitioner for improving mapreduce performance") OR TS=("Strategic Environmental Assessment (SEA) Process for Green Materials and Environmental Engineering Systems towards Sustainable Development-Business Excellence Achievements") OR TS=("A Containerized Simulation Platform for Robot Learning Peg-in-Hole Task") OR TS=("Image Processing Strategies and Multiple Paths Toward Solutions") OR TS=("Dynamic Model Evaluation to Accelerate Distributed Machine Learning") OR TS=("A Unified Coded Deep Neural Network Training Strategy based on Generalized PolyDot codes") OR TS=("Chat Sonification Starting from the Polyphonic Model of Natural Language Discourse") OR TS=("BioHIPI: Biomedical Hadoop Image Processing Interface") OR TS=("Metal contamination of bed sediments in the Irwell and Upper Mersey catchments, northwest England: exploring the legacy of industry and urban growth") OR TS=("Multi-scale structural analysis of gas diffusion layers") OR TS=("Characterization of 2D Hybrid Cellular Automata with Periodic Boundary") OR TS=("Granularity based Image processing Eco system in Hadoop to Predict the detailed results for different Medical Images") OR TS=("Research on Data Mining Service and Its Application Case in Complex Industrial Process") OR TS=("Detecting corporate tax evasion using a hybrid intelligent system: A case study of Iran") OR TS=("A Predictive Integrated Genetic-Based Model for Supplier Evaluation and Selection") OR TS=("MORE—a multimodal observation and analysis system for social interaction research") OR TS=("Spatio-temporal Route Mining and Visualization for Busy Waterways") OR TS=("Wove Paper Analysis through Texture Similarities") OR TS=("Research on the Internet Financial Mode Innovation under the Big Data and Multi-Agent Background") OR TS=("CPU Frequency Tuning to Improve Energy Efficiency of MapReduce Systems") OR TS=("The Case of the Strangerationist: Re-interpreting Critical Technical Practice") OR TS=("Large-scale automated proactive road safety analysis using video data") OR TS=("Structural analysis of theridiid spider's testicular cyst using 3D reconstruction rendering") OR TS=("Scalable Data Analytics Using R: Single Machines to Hadoop Spark Clusters") OR TS=("A revised and dated phylogeny of cobweb spiders (Araneae, Araneoidea, Theridiidae): A predatory Cretaceous lineage diversifying in the era of the ants") OR TS=("Improving False Alarm Rate in Intrusion Detection Systems Using Hadoop") OR TS=("Placement Chance Prediction: Clustering and Classification Approach") OR TS=("Development of Monitor System for Dry Eye Symptom") OR TS=("Apache YARN") OR TS=("Resource Elasticity for Large-Scale Machine Learning") OR TS=("The Knowledge Web Meets Big Scholars") OR TS=("Self-Replicating Patterns in 2D Linear Cellular Automata") OR TS=("Particle flow modeling of dry induced roll magnetic separator") OR TS=("Rural landscape planning through spatial modelling and image processing of historical maps") OR TS=("Spider's behavior for ant based clustering algorithm") OR TS=("Acquisition of welding skills in industrial robots") OR TS=("Freeway weaving phenomena observed during congested traffic") OR TS=("Image processing-based method for glass tiles colour matching") OR TS=("Association between post-game recovery protocols, physical and perceived recovery, and performance in elite Australian Football League players") OR TS=("Deep Spatial Pyramid Ensemble for Cultural Event Recognition") OR TS=("Embedding Topical Elements of Parallel Programming, Computer Graphics, and Artificial Intelligence across the Undergraduate CS Required Courses") OR TS=("The Study on Large Scale Image Processing Architecture Based on Hadoop2.0 Clusters") OR TS=("Dynamic Reconfigurable Architectures-

A Boon for Desires of Real Time Systems") OR TS=("A Generic Platform to Automate Legal Knowledge Work Process using Machine Learning") OR TS=("Morphological Investigation and Fractal Properties of Realgar Nanoparticles") OR TS=("Line drawing enhancement of historical architectural plan using difference-of-Gaussians filter") OR TS=("An Earth Imaging Camera Simulation Using Wide-Scale Construction of Reflectance Surfaces") OR TS=("Mood recognition in bipolar patients through the PSYCHE platform: Preliminary evaluations and perspectives") OR TS=("A Comparative Study of Different Segmentation Techniques for Detection of Flaws in NDE Weld Images") OR TS=("Scale-Out Beyond Map-Reduce") OR TS=("Source Apportionment of Water Pollution in the Jinjiang River (China) Using Factor Analysis With Nonnegative Constraints and Support Vector Machines") OR TS=("A squeaky wheel optimisation methodology for two-dimensional strip packing") OR TS=("Adaptive Image Processing Technique for Quality Control in Ceramic Tile Production") OR TS=("Classification of Email using BeaKS: Behavior and Keyword Stemming") OR TS=("Intelligent Decision Support Tools for Multicriteria Product Design") OR TS=("Electric contacts inspection using machine vision") OR TS=("Application of LIDAR to resolving bedrock structure in areas of poor exposure: An example from the STEEP study area, southern Alaska") OR TS=("A New ANFIS for Parameter Prediction With Numeric and Categorical Inputs") OR TS=("Identifying performance bottlenecks in work-stealing computations") OR TS=("Neural Network to Develop Sizing Systems for Production and Logistics via Technology Innovation in Taiwan") OR TS=("Topographic independent component analysis based on fractal theory and morphology applied to texture segmentation") OR TS=("of Form-Active Structures") OR TS=("A Methodology of Export Sectors Identification through Data Mining") OR TS=("Visualization of the spatial and spectral signals of orb-weaving spiders, Nephila pilipes, through the eyes of a honeybee") OR TS=("A generic pigment model for digital painting") OR TS=("Study and application of corner detecting and locating on color aberration analysis of tiles—art. no. 66231B") OR TS=("Laser cleaning of 19th century papers and manuscripts assisted by digital image processing") OR TS=("Investigation of the cutting conditions in milling operations using image texture features") OR TS=("Quality Evaluation of Defects with Indefinite or Unlimited Borders") OR TS=("EVAPORATE MAPPING IN BALA REGION (ANKARA) BY REMOTE SENSING TECHNIQUES") OR TS=("Automatic Inspection of Textured Surfaces by Support Vector Machines") OR TS=("PatchMatch Filter: Edge-Aware Filtering Meets Randomized Search for Visual Correspondence") OR TS=("The NOD3 software package: A graphical user interface-supported reduction package for single-dish radio continuum and polarisation observations") OR TS=("The physics of a popsicle stick bomb") OR TS=("Apache REEF: Retainable Evaluator Execution Framework") OR TS=("AdBench: A Complete Benchmark for Modern Data Pipelines") OR TS=("Imaging Spatially Varying Biomechanical Properties with Neural Networks") OR TS=("Words, words. They're all we have to go on: Image finding without the pictures") OR TS=("traffic-weaving") OR TS=("Feature Maps: A Comprehensible Software Representation for Design Pattern Detection") OR TS=("Study of Distributed Framework Hadoop and Overview of Machine Learning using Apache Mahout") OR TS=("piece of music by weaving") OR TS=("systems weave computing and communication") OR TS=("WEAVE and 4MOST spectrographs") OR PY=(2021) OR DO=(10.1109/TMECH.2020.3022983) OR DO=(10.1080/1369118X.2020.1834603) OR DO=(10.1177/0170840609357380) OR DO=(10.1177/1545968311425908) OR DO=(10.1109/CVPR.2013.242) OR DO=(10.1177/1468794120975988) OR DO=(10.1007/s00354-020-00115-x) OR DO=(10.3390/app10165585) OR DO=(10.1109/JPROC.2020.2986362) OR DO=(10.5334/ijc.1029) OR DO=(10.1101/cshperspect.a034595) OR DO=(10.1007/s11265-019-1438-3) OR DO=(10.1080/13658816.2019.1652304) OR DO=(10.1186/s40537-019-0240-1) OR DO=(10.14778/3352063.3352074) OR DO=(10.1177/0391398819860985) OR DO=(10.1007/s10586-018-2117-z) OR DO=(10.1109/TPDS.2018.2866993)).

References

1. Bowman, S.; Jiang, Q.; Memon, H.; Qiu, Y.; Liu, W.; Wei, Y. Effects of Styrene-Acrylic Sizing on the Mechanical Properties of Carbon Fiber Thermoplastic Towpregs and Their Composites. *Molecules* **2018**, *23*, 547. [CrossRef] [PubMed]
2. Wang, H.; Memon, H.; Hassan, E.A.M.; Elagib, T.H.H.; Hassan, F.E.A.A.; Yu, M. Rheological and Dynamic Mechanical Properties of Abutilon Natural Straw and Polylactic Acid Biocomposites. *Int. J. Polym. Sci.* **2019**, *2019*, 8732520. [CrossRef]
3. Yan, X.; Chen, L.; Memon, H. Introduction. In *Textile and Fashion Education Internationalization: A Promising Discipline from South Asia*; Yan, X., Chen, L., Memon, H., Eds.; Springer Singapore: Singapore, 2022; pp. 1–12.
4. Wijayapala, U.G.S.; Alwis, A.A.P.; Ranathunga, G.M.; Karunaratne, P.V.M. Evolution of Sri Lankan Textile Education from Ancient Times to the 21st Century. In *Textile and Fashion Education Internationalization: A Promising Discipline from South Asia*; Yan, X., Chen, L., Memon, H., Eds.; Springer: Singapore, 2022; pp. 119–144.
5. Siddiqui, M.Q.; Wang, H.; Memon, H. Cotton Fiber Testing. In *Cotton Science and Processing Technology: Gene, Ginning, Garment and Green Recycling*; Wang, H., Memon, H., Eds.; Springer: Singapore, 2020; pp. 99–119.
6. Giri, C.; Jain, S.; Zeng, X.Y.; Bruniaux, P. A Detailed Review of Artificial Intelligence Applied in the Fashion and Apparel Industry. *IEEE Access* **2019**, *7*, 95376–95396. [CrossRef]
7. Xizhen, L.; Zhiqin, S. Research on the knowledge map and visualization of fashion design field in China based on CiteSpace. *J. Silk* **2020**, *57*, 25–34. [CrossRef]
8. Tian, M.; Li, J. Knowledge mapping of protective clothing research—A bibliometric analysis based on visualization methodology. *Text. Res. J.* **2018**, *89*, 3203–3220. [CrossRef]
9. Kuilang, Y.; Qian, X. Research on the innovation frontier of global intelligent textile technology based on patentometrics. *J. Silk* **2021**, *58*, 48–55. [CrossRef]
10. Xiang, F.; Xiaopeng, W.; Xiaoxiao, Q.; Laili, W. Bibliometric analysis of literatures on textile and clothing footprint based on CiteSpace. *Adv. Text. Technol.* **2022**, *30*, 9–17. [CrossRef]
11. Li, Z.; Poon, H.; Chen, W.; Fan, J. A comparative analysis of textile schools by journal publications listed in Web of ScienceTM. *J. Text. Inst.* **2020**, *112*, 1472–1481. [CrossRef]
12. Lis, A.; Sudolska, A.; Tomanek, M. Mapping Research on Sustainable Supply-Chain Management. *Sustainability* **2020**, *12*, 3987. [CrossRef]
13. Geng, D.Y.; Feng, Y.T.; Zhu, Q.H. Sustainable design for users: A literature review and bibliometric analysis. *Environ. Sci. Pollut. Res.* **2020**, *27*, 29824–29836. [CrossRef]
14. Baier-Fuentes, H.; Merigo, J.M.; Amoros, J.E.; Gaviria-Marin, M. International entrepreneurship: A bibliometric overview. *Int. Entrep. Manag. J.* **2019**, *15*, 385–429. [CrossRef]
15. Mei, P.; Lizhu, G.; Zilin, K.; Lan, Z. Research review and prospects of VOSviewer-based textile plant dyeing. *J. Silk* **2021**, *58*, 53–59. [CrossRef]
16. Noor, S.; Guo, Y.; Shah, S.H.H.; Halepoto, H. Bibliometric Analysis of Twitter Knowledge Management Publications Related to Health Promotion. In Proceedings of the 13th International Conference, Hangzhou, China, 28–30 August 2020; pp. 341–354.
17. Syed Hamd Hassan, S.; Saleha, N.; Atif Saleem, B.; Habiba, H. Twitter Research Synthesis for Health Promotion: A Bibliometric Analysis. *Iran. J. Public Health* **2021**, *50*, 2283–2291. [CrossRef]
18. Martinez-Lopez, F.J.; Merigo, J.M.; Gazquez-Abad, J.C.; Ruiz-Real, J.L. Industrial marketing management: Bibliometric overview since its foundation. *Ind. Mark. Manag.* **2020**, *84*, 19–38. [CrossRef]
19. Verma, R.; Lobos-Ossandon, V.; Merigo, J.M.; Cancino, C.; Sienz, J. Forty years of applied mathematical modelling: A bibliometric study. *Appl. Math. Model.* **2021**, *89*, 1177–1197. [CrossRef]
20. Mas-Tur, A.; Guijarro, M.; Carrilero, A. The Influence of the Circular Economy: Exploring the Knowledge Base. *Sustainability* **2019**, *11*, 19. [CrossRef]
21. Flores-Sosa, M.; Avilés-Ochoa, E.; Merigó, J.M. Exchange rate and volatility: A bibliometric review. *Int. J. Financ. Econ.* **2022**, *27*, 1419–1442. [CrossRef]
22. Kara, K.; Wang, Z.K.; Zhang, C.; Alonso, G. doppioDB 2.0: Hardware Techniques for Improved Integration of Machine Learning into Databases. *Proc. Vldb Endow.* **2019**, *12*, 1818–1821. [CrossRef]
23. Lu, J.; Yang, H.; Min, D.; Do, M.N. Patch Match Filter: Efficient Edge-Aware Filtering Meets Randomized Search for Fast Correspondence Field Estimation. In Proceedings of the 2013 IEEE Conference on Computer Vision and Pattern Recognition, Portland, OR, USA, 23–28 June 2013; pp. 1854–1861.
24. Gahegan, M. Fourth paradigm GIScience? Prospects for automated discovery and explanation from data. *Int. J. Geogr. Inf. Sci.* **2020**, *34*, 21. [CrossRef]
25. Ding, H.; Gao, R.X.; Isaksson, A.J.; Landers, R.G.; Parisini, T.; Yuan, Y. State of AI-Based Monitoring in Smart Manufacturing and Introduction to Focused Section. *IEEE/ASME Trans. Mechatron.* **2020**, *25*, 2143–2154. [CrossRef]
26. Dutta, S.; Jeong, H.; Yang, Y.Q.; Cadambe, V.; Low, T.M.; Grover, P. Addressing Unreliability in Emerging Devices and Non-von Neumann Architectures Using Coded Computing. *Proc. IEEE* **2020**, *108*, 1219–1234. [CrossRef]
27. Jacobsen, B.N. Algorithms and the narration of past selves. *Inf. Commun. Soc.* **2020**, 1–16. [CrossRef]
28. Mair, M.; Brooker, P.; Dutton, W.; Sormani, P. Just what are we doing when we're describing AI? Harvey Sacks, the commentator machine, and the descriptive politics of the new artificial intelligence. *Qual. Res.* **2021**, *21*, 341–359. [CrossRef]

29. Croeser, S.; Eckersley, P.; Assoc Comp, M. *Theories of Parenting and Their Application to Artificial Intelligence*; Assoc Computing Machinery: New York, NY, USA, 2019; pp. 423–428.
30. Huang, S.Y.; Wang, Q.Y.; Zhang, S.Y.; Yan, S.P.; He, X.M. Dynamic Context Correspondence Network for Semantic Alignment. In *2019 IEEE/CVF International Conference on Computer Vision*; IEEE Computer Soc: Los Alamitos, CA, USA, 2019; pp. 2010–2019.
31. Schmitt, J.; Hollick, M.; Roos, C.; Steinmetz, R. Adapting the User Context in Realtime: Tailoring Online Machine Learning Algorithms to Ambient Computing. *Mob. Netw. Appl.* **2008**, *13*, 583–598. [CrossRef]
32. Phon-Amnuaisuk, S. Composing Using Heterogeneous Cellular Automata. In *Applications of Evolutionary Computing, Proceedings*; Giacobini, M., Brabazon, A., Cagnoni, S., Di Caro, G.A., Ekart, A., Esparcia Alcazar, A.I., Farooq, M., Fink, A., Machado, P., Eds.; Lecture Notes in Computer Science; Springer: Berlin/Heidelberg, Germany, 2009; Volume 5484, pp. 547–556.
33. Chen, Z.; Yu, J.; Zhu, Y.; Chen, Y.; Li, M. D3: Abnormal driving behaviors detection and identification using smartphone sensors. In Proceedings of the 2015 12th Annual IEEE International Conference on Sensing, Communication, and Networking (SECON), Seattle, WA, USA, 22–25 June 2015; pp. 524–532.
34. Abolafia, M.Y. Narrative Construction as Sensemaking: How a Central Bank Thinks. *Organ. Stud.* **2010**, *31*, 349–367. [CrossRef]
35. Lemasle, B.; Hanke, M.; Storm, J.; Bono, G.; Grebel, E.K. Atmospheric parameters of Cepheids from flux ratios with ATHOS: I. The temperature scale. *Astron. Astrophys.* **2020**, *641*, 15. [CrossRef]
36. Leung, K.; Arechiga, N.; Pavone, M. Backpropagation for Parametric STL. In Proceedings of the 2019 30th IEEE Intelligent Vehicles Symposium, Paris, France, 9–12 June 2019; IEEE: New York, NY, USA, 2019; pp. 185–192.
37. Morfino, V.; Rampone, S.; Weitschek, E. SP-BRAIN: Scalable and reliable implementations of a supervised relevance-based machine learning algorithm. *Soft Comput.* **2020**, *24*, 7417–7434. [CrossRef]
38. Zheng, W.J.; Tynes, M.; Gorelick, H.; Mao, Y.; Cheng, L.; Hou, Y.T.; Assoc Comp, M. FlowCon: Elastic Flow Configuration for Containerized Deep Learning Applications. In Proceedings of the 48th International Conference on Parallel Processing, Kyoto, Japan, 5–8 August 2019; pp. 1–10.
39. Huang, Z. Research on Spark Big Data Recommendation Algorithm under Hadoop Platform. *IOP Conf. Ser. Earth Environ. Sci.* **2019**, *252*, 1–6. [CrossRef]
40. Niu, Z.J.; Tang, S.J.; He, B.S. An Adaptive Efficiency-Fairness Meta-Scheduler for Data-Intensive Computing. *IEEE Trans. Serv. Comput.* **2019**, *12*, 865–879. [CrossRef]
41. Shyamasundar, L.B.; Prathuri, J.R. Processing and Analyzing Big Data Generated from Data Communication and Social Networks: In-terms of Performance Speed and Accuracy. In Proceedings of the 2019 PhD Colloquium on Ethically Driven Innovation and Technology for Society (PhD EDITS), Bangalore, India, 18 August 2019; pp. 1–2.
42. Poblet, M.; Sierra, C. Understanding Help as a Commons. *Int. J. Commons* **2020**, *14*, 481–493. [CrossRef]
43. Dobkin, B.H.; Dorsch, A. The Promise of mHealth: Daily Activity Monitoring and Outcome Assessments by Wearable Sensors. *Neurorehabil. Neural Repair* **2011**, *25*, 788–798. [CrossRef] [PubMed]
44. Wang, X.; Fang, Z.; Sun, X. Usage patterns of scholarly articles on Web of Science: A study on Web of Science usage count. *Scientometrics* **2016**, *109*, 917–926. [CrossRef]
45. Kumar, A. Computer-Vision-Based Fabric Defect Detection: A Survey. *IEEE Trans. Ind. Electron.* **2008**, *55*, 348–363. [CrossRef]
46. Ngan, H.Y.T.; Pang, G.K.H.; Yung, N.H.C. Automated fabric defect detection—A review. *Image Vis. Comput.* **2011**, *29*, 442–458. [CrossRef]
47. Cai, Y.C.; Shen, J.; Ge, G.; Zhang, Y.Z.; Jin, W.Q.; Huang, W.; Shao, J.J.; Yang, J.; Dong, X.C. Stretchable Ti3C2Tx MXene/Carbon Nanotube Composite Based Strain Sensor with Ultrahigh Sensitivity and Tunable Sensing Range. *ACS Nano* **2018**, *12*, 56–62. [CrossRef] [PubMed]
48. Dong, K.; Peng, X.; Wang, Z.L. Fiber/Fabric-Based Piezoelectric and Triboelectric Nanogenerators for Flexible/Stretchable and Wearable Electronics and Artificial Intelligence. *Adv. Mater.* **2020**, *32*, 43. [CrossRef] [PubMed]
49. Tasdemir, K.; Merenyi, E. Exploiting Data Topology in Visualization and Clustering Self-Organizing Maps. *IEEE Trans. Neural Netw.* **2009**, *20*, 549–562. [CrossRef]
50. Patel, A.J.; Sottos, N.R.; Wetzel, E.D.; White, S.R. Autonomic healing of low-velocity impact damage in fiber-reinforced composites. *Compos. Part A Appl. Sci. Manuf.* **2010**, *41*, 360–368. [CrossRef]
51. Jamali, N.; Sammut, C. Majority Voting: Material Classification by Tactile Sensing Using Surface Texture. *IEEE Trans. Robot.* **2011**, *27*, 508–521. [CrossRef]
52. Ting, G.; Meiling, Z.; Lili, S.; Jing, L.; Jingxue, W. An algorithm for extracting inner and outer contours of Korean dress images. *Adv. Text. Technol.* **2022**, *30*, 197–207. [CrossRef]
53. Jianxin, Z.; Gang, H.; Xiaojin, L. Research on measuring method of fabric luster based on computer vision. *J. Silk* **2021**, *58*, 62–68. [CrossRef]
54. Yanmeng, W.; Peng, Q.; Wenguo, Z. Yarn diameter time series prediction based on piecewise polymerization and Kalman filter. *Adv. Text. Technol.* **2022**, *30*, 41–47. [CrossRef]
55. Wang, H.; Halepoto, H.; Hussain, M.A.I.; Noor, S. Cotton Melange Yarn and Image Processing. In *Cotton Science and Processing Technology: Gene, Ginning, Garment and Green Recycling*; Wang, H., Memon, H., Eds.; Springer: Singapore, 2020; pp. 547–565.
56. Shengbao, X.; Liaomo, Z.; Decheng, Y. A method for fabric defect detection based on improved cascade R-CNN. *Adv. Text. Technol.* **2022**, *30*, 48–56. [CrossRef]

57. Min, L.; Shan, Y.; Ruhan, H.; Xun, Y.; Shuqin, C. Jacquard fabric defect detection technology combining context-awareness and convolutional neural network. *Adv. Text. Technol.* **2021**, *29*, 62–66. [CrossRef]
58. Danshu, W.; Yali, Y.; Li, Y. Colored spun fabric pattern recognition based on multi resolution mixed characteristics. *J. Silk* **2021**, *58*, 27–32. [CrossRef]

MDPI
St. Alban-Anlage 66
4052 Basel
Switzerland
Tel. +41 61 683 77 34
Fax +41 61 302 89 18
www.mdpi.com

Materials Editorial Office
E-mail: materials@mdpi.com
www.mdpi.com/journal/materials

www.ingramcontent.com/pod-product-compliance
Lightning Source LLC
LaVergne TN
LVHW070228100526
838202LV00015B/2105